VOLUME SIXTY FIVE

ADVANCES IN MICROBIAL PHYSIOLOGY

Advances in Bacterial Pathogen Biology

VOLUME SIXTY FIVE

ADVANCES IN MICROBIAL PHYSIOLOGY

Advances in Bacterial Pathogen Biology

Edited by

ROBERT K. POOLE

West Riding Professor of Microbiology
Department of Molecular Biology and Biotechnology
The University of Sheffield
Firth Court, Western Bank
Sheffield, UK

ELSEVIER

AMSTERDAM • BOSTON • HEIDELBERG • LONDON
NEW YORK • OXFORD • PARIS • SAN DIEGO
SAN FRANCISCO • SINGAPORE • SYDNEY • TOKYO
Academic Press is an imprint of Elsevier

Academic Press is an imprint of Elsevier
525 B Street, Suite 1800, San Diego, CA 92101-4495, USA
225 Wyman Street, Waltham, MA 02451, USA
32 Jamestown Road, London NW1 7BY, UK
The Boulevard, Langford Lane, Kidlington, Oxford, OX5 1GB, UK

First edition 2014

Notices

Knowledge and best practice in this field are constantly changing. As new research and experience broaden our understanding, changes in research methods, professional practices, or medical treatment may become necessary.

Practitioners and researchers must always rely on their own experience and knowledge in evaluating and using any information, methods, compounds, or experiments described herein. In using such information or methods they should be mindful of their own safety and the safety of others, including parties for whom they have a professional responsibility.

To the fullest extent of the law, neither the Publisher nor the authors, contributors, or editors, assume any liability for any injury and/or damage to persons or property as a matter of products liability, negligence or otherwise, or from any use or operation of any methods, products, instructions, or ideas contained in the material herein.

ISBN: 978-0-12-800142-4
ISSN: 0065-2911

For information on all Academic Press publications
visit our website at store.elsevier.com

CONTENTS

CONTRIBUTORS

Jody Aberdein
Department of Infection and Immunity, University of Sheffield Medical School and Sheffield Teaching Hospitals, Sheffield, United Kingdom

Elaine Allan
Department of Microbial Diseases, UCL Eastman Dental Institute, University College London, London, United Kingdom

Sabela Balboa
Department of Infection and Immunity, University of Sheffield, Sheffield, United Kingdom, and Departamento de Microbiología y Parasitología, Universidad de Santiago de Compostela, Santiago de Compostela, Spain

Michael Berney
Department of Microbiology and Immunology, Albert Einstein College of Medicine, Bronx, New York, USA

Annelie Brauner
Department of Microbiology, Tumor and Cell Biology, Division of Clinical Microbiology, Karolinska Institutet and Karolinska University Hospital, Stockholm, Sweden

Jennifer S. Cavet
Faculty of Life Sciences, University of Manchester, Manchester, United Kingdom

Joby Cole
Department of Infection and Immunity, University of Sheffield Medical School and Sheffield Teaching Hospitals, Sheffield, United Kingdom

Gregory M. Cook
Department of Microbiology and Immunology, Otago School of Medical Sciences, University of Otago, Dunedin, and Maurice Wilkins Centre for Molecular Biodiscovery, The University of Auckland, Auckland, New Zealand

David H. Dockrell
Department of Infection and Immunity, University of Sheffield Medical School and Sheffield Teaching Hospitals, Sheffield, United Kingdom

C.W. Ian Douglas
Integrated BioSciences Group, School of Clinical Dentistry, University of Sheffield, Sheffield, United Kingdom

Thomas Farmilo
Integrated BioSciences Group, School of Clinical Dentistry, University of Sheffield, Sheffield, United Kingdom

Andrew M. Frey
Integrated BioSciences Group, School of Clinical Dentistry, University of Sheffield, Sheffield, United Kingdom

Chris Greening
Department of Microbiology and Immunology, Otago School of Medical Sciences, University of Otago, Dunedin, New Zealand

Kiel Hards
Department of Microbiology and Immunology, Otago School of Medical Sciences, University of Otago, Dunedin, New Zealand

Helen E. Jesse
Faculty of Life Sciences, University of Manchester, Manchester, United Kingdom

Jamil Jubrail
Department of Infection and Immunity, University of Sheffield Medical School and Sheffield Teaching Hospitals, Sheffield, United Kingdom

Rebecca Lowry
Department of Infection and Immunity, University of Sheffield, Sheffield, United Kingdom

Petra Lüthje
Department of Microbiology, Tumor and Cell Biology, Division of Clinical Microbiology, Karolinska Institutet and Karolinska University Hospital, Stockholm, Sweden

Peter Mullany
Department of Microbial Diseases, UCL Eastman Dental Institute, University College London, London, United Kingdom

Kathryn Naylor
Integrated BioSciences Group, School of Clinical Dentistry, University of Sheffield, Sheffield, United Kingdom

Jennifer L. Parker
Department of Infection and Immunity, University of Sheffield, Sheffield, United Kingdom

Chatchawal Phansopa
Integrated BioSciences Group, School of Clinical Dentistry, University of Sheffield, Sheffield, United Kingdom

Adam P. Roberts
Department of Microbial Diseases, UCL Eastman Dental Institute, University College London, London, United Kingdom

Ian S. Roberts
Faculty of Life Sciences, University of Manchester, Manchester, United Kingdom

Jonathan G. Shaw
Department of Infection and Immunity, University of Sheffield, Sheffield, United Kingdom

Graham P. Stafford
Integrated BioSciences Group, School of Clinical Dentistry, University of Sheffield, Sheffield, United Kingdom

CHAPTER ONE

Energetics of Pathogenic Bacteria and Opportunities for Drug Development

Gregory M. Cook[*,†,1,2], **Chris Greening**[*,1], **Kiel Hards**[*,1], **Michael Berney**[‡,1]

*Department of Microbiology and Immunology, Otago School of Medical Sciences, University of Otago, Dunedin, New Zealand
†Maurice Wilkins Centre for Molecular Biodiscovery, The University of Auckland, Auckland, New Zealand
‡Department of Microbiology and Immunology, Albert Einstein College of Medicine, Bronx, New York, USA
[1]These authors contributed equally to this work.
[2]Corresponding author: e-mail address: gregory.cook@otago.ac.nz

Contents

Abstract

The emergence and spread of drug-resistant pathogens and our inability to develop new antimicrobials to overcome resistance has inspired scientists to consider new targets for drug development. Cellular bioenergetics is an area showing promise for the development of new antimicrobials, particularly in the discovery of new anti-tuberculosis drugs where several new compounds have entered clinical trials. In this review, we have examined the bioenergetics of various bacterial pathogens, highlighting the versatility of electron donor and acceptor utilisation and the modularity of electron transport chain components in bacteria. In addition to re-examining classical concepts, we explore new literature that reveals the intricacies of pathogen energetics, for example, how *Salmonella enterica* and *Campylobacter jejuni* exploit host and microbiota to derive powerful electron donors and sinks; the strategies *Mycobacterium tuberculosis* and *Pseudomonas*

Advances in Microbial Physiology
ISSN 0065-2911
http://dx.doi.org/10.1016/bs.ampbs.2014.08.001

aeruginosa use to persist in lung tissues; and the importance of sodium energetics and electron bifurcation in the chemiosmotic anaerobe *Fusobacterium nucleatum*. A combination of physiological, biochemical, and pharmacological data suggests that, in addition to the clinically-approved target F_1F_O-ATP synthase, NADH dehydrogenase type II, succinate dehydrogenase, hydrogenase, cytochrome *bd* oxidase, and menaquinone biosynthesis pathways are particularly promising next-generation drug targets. The realisation of cellular energetics as a rich target space for the development of new antimicrobials will be dependent upon gaining increased understanding of the energetic processes utilised by pathogens in host environments and the ability to design bacterial-specific inhibitors of these processes.

1. INTRODUCTION

The majority of current antimicrobials were developed during the golden era of antimicrobial discovery. These compounds target a number of essential processes for the growth of microbial cells, including peptidoglycan biosynthesis, RNA and protein synthesis, DNA replication, and folic acid metabolism. During this period, antimicrobial use became widespread, not only in hospitals but also in agricultural environments. As quickly as new antimicrobials were developed, however, resistance followed increasing the demand for new derivatives through optimisation of existing molecular scaffolds. The burden of antimicrobial resistance was further compounded by the lack of new drugs with unique targets to overcome resistance and by the increasing cost of antimicrobial discovery and development. The number of new antibiotic approvals by the FDA continues to decline contributing to the withdrawal of pharmaceutical companies in this area (Boucher et al., 2013).

To address the emergence and spread of drug-resistant bacterial pathogens, new drug targets and drugs with a novel mode of action are urgently required to expand our antimicrobial armoury. The development of narrow spectrum agents to prevent widespread resistance developing remains a priority. A key to the development of the next generation of antimicrobials will be increased understanding of how new targets function in the physiological context of the pathogen. Deciphering the essential and non-essential roles of these targets in response to the host environment will be an important question to address.

2. BACTERIAL ENERGETICS AS A TARGET SPACE FOR DRUG DEVELOPMENT

A major structural component of bacterial cells is the cytoplasmic membrane made up of a lipid bilayer that forms a continuous barrier around the cell. The cytoplasmic membrane imparts structure to the cell and allows for the selective (filter) passage of nutrients and wastes into and out of the cell. The membrane also plays an essential role in cellular homeostasis and energy transduction. Several new antimicrobials have been developed that target the bacterial membrane (e.g. daptomycin and lipoglycopeptides), leading to disruption of bacterial membrane integrity, but this target space has remained largely unexplored (Hurdle, O'Neill, Chopra, & Lee, 2011).

2.1. Generation of the proton motive force: An essential property of all bacterial cells

All bacteria require a proton motive force (*pmf*) to grow and remain viable under replicating and non-replicating conditions. During respiration, energy is conserved by the generation of a *pmf* across a proton-impermeable membrane. The electron transport chain components are membrane-bound and asymmetrically arranged across the membrane to achieve net consumption of protons from the cytoplasm and net release of protons on the outside the cell. The *pmf* (electrochemical potential) consists of two gradients: an electrical potential ($\Delta\psi$), due to the charge separation across the membrane ($positive_{outside}/negative_{inside}$) and a chemical transmembrane gradient of protons (ΔpH, $acidic_{outside}/alkaline_{inside}$). At neutral pH, the *pmf* is predominantly in the form of a $\Delta\psi$, but as the external pH drops, the ΔpH increases, and the $\Delta\psi$ decreases to maintain a constant *pmf*. Dissipation of the *pmf* leads to a rapid loss of cell viability and cell death.

A variety of mechanisms are used to generate the *pmf* in bacteria (Fig. 1). In obligately aerobic bacteria, the generation of a *pmf* is mediated primarily by the proton-pumping components of the electron transport chain (Fig. 1, mechanism 4). In facultative anaerobes, when alternative electron acceptors are available (e.g. nitrate and fumarate), proton release is coupled to a terminal reductase (e.g. nitrate reductase) via a *pmf* redox-loop mechanism (Jormakka, Byrne, & Iwata, 2003b; Fig. 1, mechanism 2). Under strictly fermentative conditions, the F_1F_o-ATP synthase can operate as a reversible ATP-driven proton pump to generate the *pmf* (Dimroth & Cook, 2004; Fig. 1, mechanism 3). Furthermore, in some bacteria, end-products (e.g. lactate) efflux can generate a *pmf* (Otto, Sonnenberg, Veldkamp, & Konings, 1980; Fig. 1, mechanism 1). The flexibility of respiration in bacteria under anaerobic conditions is further highlighted by the discovery that endogenous phenazine production by *Pseudomonas aeruginosa* enhances anaerobic

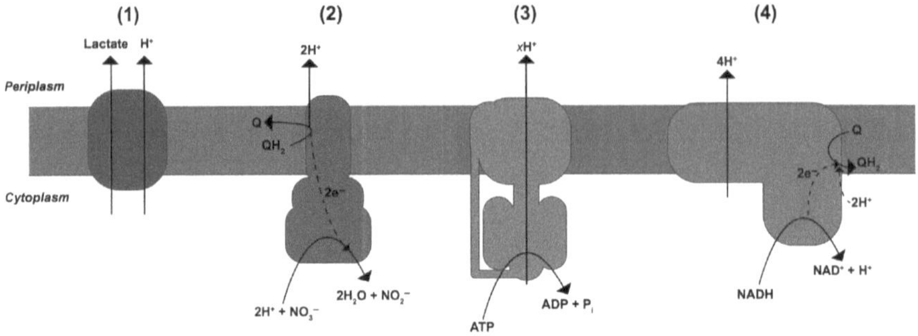

Figure 1 *Mechanisms (1–4) by which a proton motive force can be generated in bacteria.* (1) Co-transport of protons driven by solute (lactate) symport into the periplasm. (2) Redox-loop separation of charge; quinol oxidation results in proton release into the periplasm by virtue of quinol site proximity to the periplasm, while electrons are transferred to reduce a terminal electron acceptor in the cytoplasm that results in neutralisation of charge. (3) Proton export driven by ATP hydrolysis, i.e., ATP synthase working in the reverse direction. (4) Proton translocation mediated by primary proton-pumping complexes. (See the color plate.)

survival through maintenance of the *pmf* (and ATP production) via a redox homeostasis mechanism (Glasser, Kern, & Newman, 2014).

There are a wide range of compounds that target the *pmf* in bacteria (Fig. 2), including agents that inhibit the major proton pumps (e.g. rotenone) (Fig. 2, mechanism 3) and those that facilitate proton transport through the cytoplasmic membrane (protonophores, e.g. carbonyl cyanide *m*-chlorophenyl hydrazine—CCCP) (Fig. 2, mechanism 5). The majority of protonophores are non-specific and functional in both prokaryotic and eukaryotic cell membranes. Individual components of the *pmf* can be collapsed using specific inhibitors. For example, the $\Delta\psi$ can be collapsed by compounds that catalyse electrogenic cation transport across the cell membrane (e.g. valinomycin) Valinomycin is a dodecadepsipeptide that forms a macrocyclic molecule allowing for rapid K^+ movement down its electrochemical gradient (Fig. 2, mechanism 1). The chemical transmembrane gradient of protons (ΔpH) can be collapsed by nigericin through its K^+/H^+ antiporter (electroneutral) activity (Fig. 2, mechanism 2). Nigericin has similar properties to monensin, a Na^+/H^+ exchanger widely used in livestock as a feed additive. Gramicidin is a channel-forming ionophore, making the membrane more permeable to ions (Fig. 2, mechanism 3).

Some bacterial pathogens generate a considerable ΔpH in response to acidification of host tissues (e.g. *Helicobacter pylori*, *Salmonella enterica*, and *Streptococcus pneumoniae*), and collapsing the pH gradient would be an

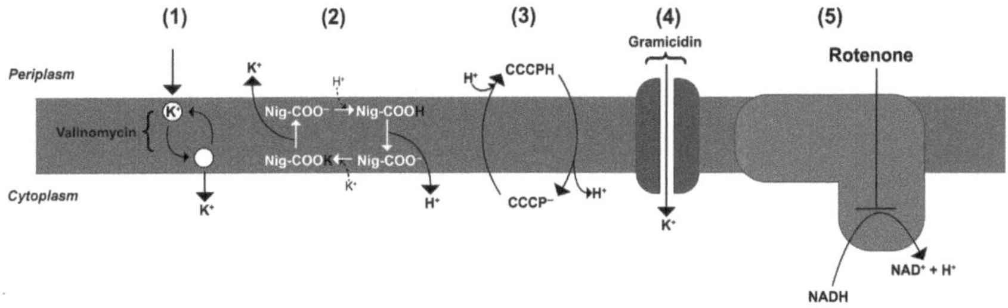

Figure 2 *Traditional inhibitors of proton motive force generation.* (1) Valinomycin is an ionophore, selective for potassium ions, which equilibrates the potassium gradient—dissipating the $\Delta\psi$ (electrogenic). (2) Nigericin is a hydrophobic weak carboxylic acid, which can traverse the membrane as its either protonated acid or neutral salt. It dissipates chemical gradients (i.e. ΔpH) but maintains the charge (one positive charge exchanged for one positive charge—electroneutral). (3) Carbonyl cyanide *m*-chlorophenyl hydrazine (CCCP) is an electrogenic protonophore. $CCCP^-$ is driven to the periplasm by the $\Delta\psi$, while CCCPH is driven to the cytoplasm by the ΔpH. It can equilibrate both $\Delta\psi$ and ΔpH. (4) Gramicidin is a channel-forming ionophore, making the membrane more permeable to ions. (5) Rotenone inhibits primary proton pumping—preventing the initial generation of a proton motive force. (See the color plate.)

effective strategy in acidic tissues to eradicate these bacteria (Hall, Karem, & Foster, 1995; Matin, Zychlinsky, Keyhan, & Sachs, 1996). The *pmf* has recently been screened as a target for methicillin-resistant *Staphylococcus aureus* using high-throughput screening to identify compounds that dissipate individual components of the *pmf*, i.e., the $\Delta\psi$ or ΔpH and synergistic combinations thereof (Farha, Verschoor, Bowdish, & Brown, 2013).

2.2. Diversity and flexibility of electron transport chains in bacteria

The main pathogens discussed in this review are summarised in Table 1. The electron transport chains both within and between these bacteria show a remarkable diversity with regard to both electron donor and electron acceptor utilisation, enabling growth and persistence in a wide variety of environmental niches (Fig. 3). Bacteria are able to use a range of primary dehydrogenases to deliver electrons from central metabolism into the respiratory chain to generate energy. These electrons pass through various redox carriers to the quinone/quinol pool. In bacteria, the electron transport chain is often branched with multiple routes to terminal respiratory oxidases or reductases (Fig. 3). For example, *Escherichia coli* uses a low-affinity (μM for oxygen) proton-pumping cytochrome bo_3 (haem–copper) oxidase

Table 1 Classification and characteristics of the pathogens discussed in this review

Organism	Classification	Metabolism	Major diseases	Primary tissues
Escherichia coli (pathogenic strains)	γ-Proteobacteria Enterobacteriales	Heterotroph Facultative anaerobe	Gastroenteritis Urinary tract infections	Gastrointestinal tract Urinary tract
Salmonella enterica	γ-Proteobacteria Enterobacteriales	Heterotroph Facultative anaerobe	Gastroenteritis Typhoid fever	Gastrointestinal tract
Pseudomonas aeruginosa	γ-Proteobacteria Pseudomonadales	Heterotroph Facultative anaerobe	Opportunistic infections (e.g. pneumonia)	Cystic fibrosis lungs
Neisseria gonorrhoeae *Neisseria meningtidis*	β-Proteobacteria Neisserales	Heterotroph Facultative aerobe	Gonorrhoea Meningitis	Urinary tract Meninges
Campylobacter jejuni	ε-Proteobacteria Campylobacterales	Heterotroph Microaerobe	Gastroenteritis	Gastrointestinal tract
Helicobacter pylori	ε-Proteobacteria Campylobacterales	Heterotroph Microaerobe	Stomach ulcers Stomach cancer	Stomach
Staphylococcus aureus	Firmicutes Bacilliales	Heterotroph Facultative anaerobe	Opportunistic infections (e.g. skin infections)	Skin Respiratory tract
Mycobacterium tuberculosis	Actinobacteria Actinomycetales	Heterotroph Obligate aerobe	Tuberculosis	Lungs
Fusobacterium nucleatum	Fusobacteria Fusobacteriales	Heterotroph Obligate anaerobe	Periodontitis Lemierre's syndrome	Oral cavity
Treponema pallidum	Spirochaetes Spirochaetales	Heterotroph Obligate anaerobe	Syphilis Yaws	Urinary tract Skin

It only lists those pathogens where a relatively complete overview of their energetics is provided.

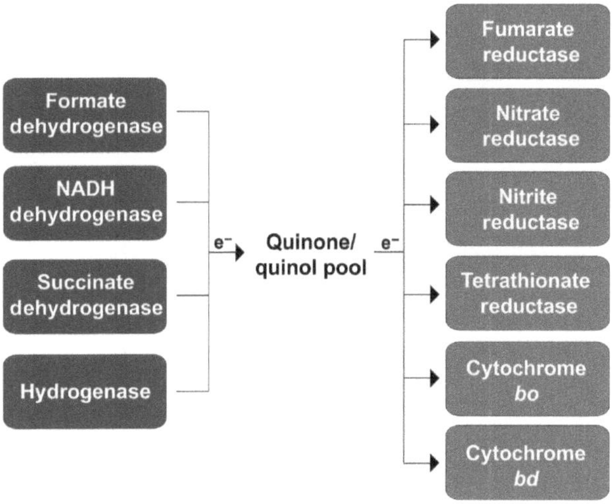

Figure 3 *Generalised schematic overview of relevant electron transfer components.* Complexes indicated in blue oxidise various substrates to reduce quinones. The resulting quinol molecules can be oxidised to result in reduction of various terminal electron acceptors, mediated by the complexes indicated in green. For some electron transfer pathways intermediate complexes and molecules exist, for example, complex III will generally reduce cytochrome *c*, which will serve as the electron donor for complex IV. The complexes used, types of quinones, and intermediates thereof are highly variable between genera. Only complexes relevant to this review are indicated. (See the color plate.)

growing at high oxygen tensions, but switches to a high-affinity (nM for oxygen) non-proton-translocating cytochrome *bd* oxidase when growing at low oxygen tensions (Cotter, Chepuri, Gennis, & Gunsalus, 1990; D'Mello, Hill, & Poole, 1995, 1996; Fig. 3). In *S. enterica*, the electron transport chain shows considerable diversity in response to oxygen tension and will be highlighted throughout this review.

A number of compounds have been shown to inhibit the major components of mitochondrial and bacterial electron transport chains (Fig. 4). However, few if any studies have assessed how specific these compounds are across different bacterial genera.

2.3. Primary respiratory dehydrogenases

2.3.1 NADH dehydrogenases: The roles of bacterial NDH-1 and NDH-2

In many bacterial pathogens, the major entry point to the electron transport chain is the transfer of electrons from reduced nicotinamide dinucleotide (NADH) (reduced by the oxidation of organic carbon) to quinones (e.g.

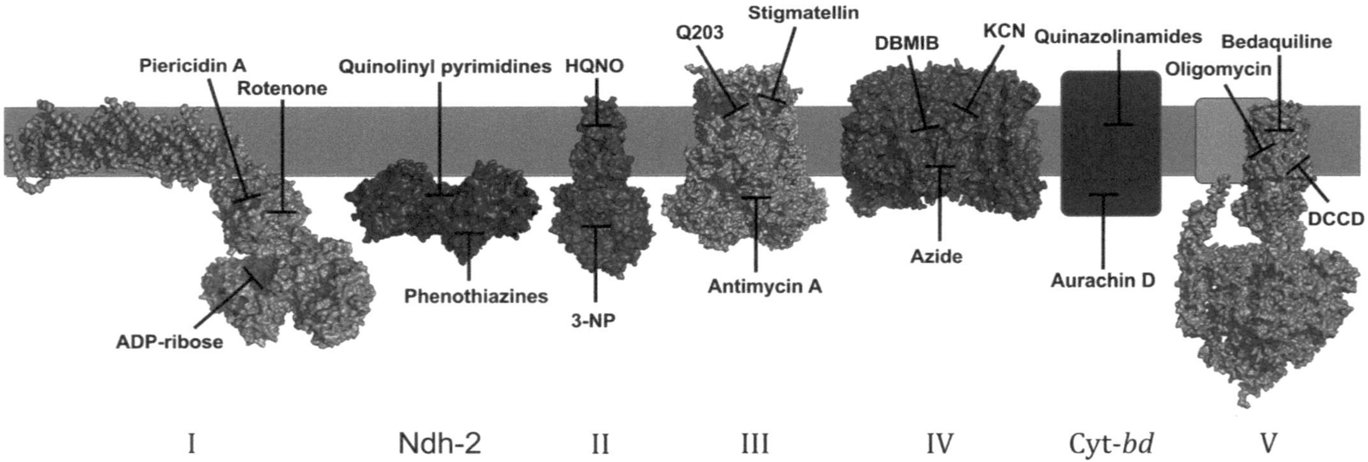

Figure 4 *Diversity of electron transport chain inhibitors*. Structural surface representations of electron transport chain components are indicated where possible. Selected inhibitors of these complexes are indicated with flathead arrows and do not reflect the binding site of the inhibitors. Crystal structures were obtained from RCSB protein data bank from the following accession numbers: complex I, 3M9S; Ndh-2, 4NWZ; complex II, 2WDV; complex III, 3H1J; complex IV, 3ASN; and complex V, 4B2Q. Models were generated using the PyMOL molecular graphics system. (See the color plate.)

ubiquinone or menaquinone; Fig. 3). Three different types of respiratory NADH dehydrogenases have been identified and characterised on the basis of reaction mechanism, subunit composition, and protein architecture (Kerscher, Drose, Zickermann, & Brandt, 2008): the proton-pumping type I NADH dehydrogenase (NDH-1, complex I), the non-proton-pumping type II NADH dehydrogenase (NDH-2; Fig. 4), and the sodium-pumping NADH dehydrogenase (NQR, discussed in Section 2.5). Homologous to mitochondrial complex I, bacterial NDH-1 is encoded by the *nuo* operon and transfers electrons to quinone, conserving energy by translocating protons across the membrane to generate a *pmf*. This multimeric enzyme uses flavin adenine dinucleotide (FAD) and nine iron–sulphur clusters to transport electrons from NADH to the quinone pool. The release of the two electrons during the NADH oxidation produces enough energy to pump four protons across the membrane to generate a *pmf* (Baradaran, Berrisford, Minhas, & Sazanov, 2013).

NDH-2 is more relevant to drug discovery. This small cytoplasmically oriented monotopic membrane protein (40–60 kDa; Fig. 4) catalyses electron transfer from NADH via the flavin cofactor to quinone (Heikal et al., 2014). NDH-2 enzymes are widespread in bacteria and, while also encoded in some eukaryotes (Melo, Bandeiras, & Teixeira, 2004), have not been reported in mammalian mitochondria. This has resulted in the proposal that they may represent a potential drug target for the treatment of pathogenic bacteria (Rao, Alonso, Rand, Dick, & Pethe, 2008; Teh, Yano, & Rubin, 2007; Warman et al., 2013; Weinstein et al., 2005; Yano, Li, Weinstein, Teh, & Rubin, 2006), as well as protozoa (Biagini, Viriyavejakul, O'Neill, Bray, & Ward, 2006; Warman et al., 2013).

In many pathogens, there are copies of both types NDH-1 and NDH-2 in the genome (Melo et al., 2004). In the enteric pathogens *E. coli* and *S. enterica*, these enzymes are differentially expressed, with NDH-2 primarily being synthesised aerobically and NDH-1 being active during anaerobic respiration (Calhoun, Oden, Gennis, de Mattos, & Neijssel, 1993; Unden & Bongaerts, 1997). One potential explanation for the dominant role of NDH-2, even in the presence of NDH-1, is that lack of proton translocation may be desirable during some conditions. NADH oxidation by NDH-2 would not be impeded by a high *pmf*, as would be the case with NDH-1, which could ultimately slow metabolic flux due to back-pressure on the system; NDH-2-mediated NADH oxidation would therefore allow for a higher metabolic flux and increased carbon flow into biosynthetic pathways and ultimately higher rates of ATP synthesis, at the expense of low energetic efficiency of the respiratory chain. The use of NDH-1 could also be deleterious during aerobic conditions due to the increased production of reactive oxygen species (ROS) generated via this complex (Knuuti, Belevich, Sharma, Bloch, & Verkhovskaya, 2013). By contrast, when the oxidant for terminal oxidases is unavailable under anoxic conditions, alternative mechanisms to generate *pmf* are required and hence enterobacteria select for the energetic efficiency of NDH-1 (Unden & Bongaerts, 1997).

There is significant variety in the physiological roles of the NADH dehydrogenases in different pathogens. *Mycobacterium tuberculosis* encodes an NDH-1 complex and two copies of NDH-2 (Ndh and NdhA; Cook, Hards, Vilchèze, Hartman, & Berney, 2014). Transcriptome and biochemical studies indicate that NDH-2 activity is responsible for the majority of NADH oxidation in mycobacteria (Shirude et al., 2012; Weinstein et al., 2005; Yano et al., 2006). Mutagenesis screens suggest that, whereas NdhA is dispensable, Ndh is essential for growth *in vitro* and cannot be compensated for (Griffin et al., 2011; Sassetti, Boyd, & Rubin, 2003). The role of NDH-1 is more enigmatic. In contrast to *S. enterica*, the complex is downregulated during infection, hypoxia, and starvation in *M. tuberculosis* (Betts, Lukey, Robb, McAdam, & Duncan, 2002; Schnappinger et al., 2003; Shi et al., 2005). The enzyme is dispensable for growth and survival *in vitro* (Griffin et al., 2011), though *nuoG* has nevertheless been shown to be critical for virulence in murine and macrophage models (Velmurugan et al., 2007).

S. aureus also encodes two copies of NDH-2, but lacks NDH-1; the isozymes are kinetically distinct, though the differential roles of the individual enzymes are currently unclear (Schurig-Briccio, Yano, Rubin, & Gennis, 2014). Among other pathogens, the minimal genome of *Mycobacterium leprae* encodes a single NDH-2, which is predicted to be the major input into the respiratory chain of this organism (Cole et al., 2001). A contrasting scenario is presented by *Campylobacter jejuni* and *H. pylori*, which lack NDH-2 and instead encode NDH-1 variants (Chen, Andersen, Zhai, & Kharazmi, 1999; Smith, Finel, Korolik, & Mendz, 2000). These organisms have low rates of NADH oxidation (Hoffman & Goodman, 1982) and recent studies suggest that flavodoxin is in fact the reductant for their NDH-1 variants (Weerakoon & Olson, 2008). Consistently, the electron-input subunits of these complexes are novel in both bacteria and have been shown to be essential for growth of *C. jejuni* (Smith et al., 2000; Weerakoon & Olson, 2008).

Several classes of compounds are proposed to target NDH-2 at micromolar concentrations (e.g. phenothiazine analogues, platanetin, and quinolinyl pyrimidines; Fig. 4; Shirude et al., 2012; Weinstein et al., 2005; Yano et al., 2006). Despite poor activity against NDH-2, drugs of the phenothiazine family (trifluoroperazine and chlorpromazine) have potent activity *in vitro* against drug-susceptible and drug-resistant *M. tuberculosis* strains (Amaral, Kristiansen, Abebe, & Millett, 1996; Amaral, Viveiros, & Kristiansen, 2006; Ordway et al., 2003). A phenothiazine analogue was also tested in a mouse model of acute *M. tuberculosis* infection and was found to reduce by 90% the *M. tuberculosis* bacterial load in the lungs after 11 days of treatment compared to a 3- to 4-log reduction in colony forming units with the isoniazid (INH) or rifampin control (Weinstein et al., 2005). From a library of microbial products, two compounds, scopafungin and gramicidin S, were identified as inhibitors of *Mycobacterium smegmatis* NDH-2, with IC_{50} values better than trifluoperazine (Mogi, Matsushita, et al., 2009). The mechanism of phenothiazine action remains unclear, but does not appear to involve competitive inhibition of NADH or quinone binding (Weinstein et al., 2005; Yano et al., 2006). Phenothiazines have been shown to inhibit the NDH-2 isozymes of *S. aureus* (Amaral et al., 2006; Ordway, Viveiros, Leandro, Arroz, & Amaral, 2002; Ordway et al., 2002; Schurig-Briccio et al., 2014). The authors propose that, in addition to inhibition of NDH-2 activity at low micromolar concentrations, phenothiazines are uncouplers of oxidative phosphorylation (Schurig-Briccio et al., 2014). The mechanism of uncoupling remains unclear.

Despite the potential of NDH-2 as a drug target, no potent nanomolar inhibitors of NDH-2 have been reported. The first high-resolution structure of a bacterial NDH-2 was recently solved, revealing unique binding sites for quinone and NADH, allowing concomitant oxidation of NADH from the aqueous cytoplasm and reduction of hydrophobic quinone in the membrane with both substrates accessing the FAD cofactor sequentially (Heikal et al., 2014). The structure shows a homodimeric organisation with a unique dimer interface, but whether the enzyme forms a higher molecular weight species physiologically remains to be ascertained. The enzyme is localised to the cytoplasmic membrane by two separated C-terminal membrane-anchoring regions that are essential for membrane localisation and FAD binding, but not dimerisation. This structure of bacterial NDH-2 provides for the first time a molecular framework for the development of inhibitors of the bacterial enzyme (Heikal et al., 2014). That said, while NDH-2 appears a very attractive target for drug development, this may not be the case in all instances. For example, reduction in NDH-2 activity has been linked to isoniazid (INH) and ethionamide resistance in both slow- and fast-growing mycobacteria (Vilcheze et al., 2005), and phenothiazines have been shown to be antagonistic with INH (Warman et al., 2013). Therefore, the development of NDH-2 inhibitors will have to ascertain that interference with current drug therapy does not occur.

An alternative approach is to target the quinones, the membrane-diffusible redox carriers that accept the electrons from NADH dehydrogenases and other primary dehydrogenases; this would theoretically prevent compensation between respiratory complexes when a specific complex is targeted. There is particular potential for the development of antimycobacterials that target the maturation factors encoding menaquinone; as the sole quinone in *M. tuberculosis*, synthesis of this cofactor is essential for growth and recovery from persistence (Dhiman et al., 2009). There is promise in developing inhibitors against MenA, which catalyses the prenylation step in this mycobacterial menaquinone biogenesis pathway (Kurosu & Crick, 2009); several derivatives of the natural product aurachin RE have recently been shown to be bactericidal against drug-resistant *M. tuberculosis in vitro* (Debnath et al., 2012). There may also be merit in targeting the novel menaquinone biosynthesis pathways of *H. pylori* and *C. jejuni* (Hiratsuka et al., 2008; Seto et al., 2008).

2.3.2 Succinate dehydrogenase: Enzyme variation and essentiality in bacterial pathogens

Succinate dehydrogenases (succinate:quinone reductases, also known as complex II or SQR) catalyse the reaction succinate + Q $\rightleftharpoons$ fumarate + QH_2, which serves as a vital link between the tricarboxylic acid cycle and oxidative phosphorylation. The reverse reaction can be performed by fumarate reductases (quinol:fumarate reductases, QFR), in which fumarate would serve as a terminal electron acceptor. QFR and SQR are closely related enzymes and the reaction catalysed cannot be predicted based on the primary amino acid sequence alone. In *E. coli*, which encodes both a SQR and a QFR, one enzyme can in fact partially compensate for the loss of the other (Guest, 1981; Maklashina, Berthold, & Cecchini, 1998), implying functional redundancy between the two enzymes. All complexes contain the A subunit, where a flavoprotein that acts as the site of dicarboxylate catalysis, and the B subunit, containing iron–sulphur clusters that facilitate electron movement. Additionally, either one or two transmembrane subunits are present, which bind varying amounts of haems depending on the class of enzyme. Due to the existence of eukaryotic complex II, these enzymes are often overlooked for their role in human pathogens, as the potential for useful drug discovery would seem unlikely. However, human complex II is a phylogenetically distinct class of enzyme to the majority of the succinate dehydrogenases and fumarate reductases essential for bacterial pathogenesis (Hägerhäll & Hederstedt, 1996; Hederstedt, 1999; Lancaster, 2001; Lancaster & Kröger, 2000; Lemos, Fernandes, Pereira, Gomes, & Teixeira, 2002).

There are several examples of succinate dehydrogenases that are essential for virulence and can be discriminated for drug targeting. *S. enterica*, for instance, encodes an annotated succinate dehydrogenase and a fumarate reductase (McClelland et al., 2001; Spector et al., 1999). The succinate dehydrogenase is expressed under aerobic conditions, where it is proposed to input electrons metabolised through the tricarboxylic acid cycle into the ubiquinone pool. Consistent with this central role, the deletion of the succinate dehydrogenase in *S. enterica* resulted in a defect in its ability to colonise mice (Mercado-Lubo, Gauger, Leatham, Conway, & Cohen, 2008; Yimga et al., 2006). However, the defect was not nearly as extreme in a deletion in succinyl-CoA synthetase, suggesting functional complementation by the fumarate reductase also encoded in this organism. A single deletion in fumarate reductase gave no substantial difference in mouse survival, but a double mutant was completely attenuated (Mercado-Lubo et al., 2008). As such, the succinate oxidation activity of this organism is important for its viability in host tissues. It is energetically favourable to transfer electrons from succinate ($E^{0\prime} = +33$ mV) to ubiquinone ($E^{0\prime} = +113$ mV), as in *S. enterica*, but not menaquinone ($E^{0\prime} = -74$ mV). Hence, organisms that use menaquinone as the sole electron acceptor, for example, *C. jejuni*, *M. tuberculosis*, and *S. aureus* (Collins & Jones, 1981; Unden & Schirawski, 1997), must energise this reaction. Studies in environmental organisms suggest that the *pmf* drives this reaction, as reviewed in Lancaster (2013) and Lancaster, Haas, Madej, and Mileni (2006). It seems therefore that driving succinate oxidation to menaquinone reduction is an 'Achilles heel'; perhaps targeting the menaquinone reduction site in these succinate dehydrogenases, in concert with inhibitors of menaquinone synthesis and energy generation, would be lethal.

Siccarin serves as an example that selectivity can be achieved in succinate dehydrogenase inhibitors (Mogi, Kawakami, et al., 2009). This compound inhibits the succinate dehydrogenases from *P. aeruginosa* and murine mitochondria, but not those from *E. coli* and porcine mitochondria. Siccarin is believed to be a ubiquinone analogue specific to succinate dehydrogenases (Mogi, Kawakami, et al., 2009); while this compound is unsuitable as a specific antimicrobial, it raises the possibility that menaquinone analogues can be developed against succinate dehydrogenase specific for the bacterial menaquinone-reducing enzymes. Caution should be exercised in an organism-dependent manner, however, as deletion of succinate dehydrogenase in *S. aureus* resulted in a small colony variant phenotype (Gaupp, Schlag, Liebeke, Lalk, & Götz, 2010) and such cells are known to become more drug-resistant because of their slowed growth (McNamara & Proctor, 2000; Proctor et al., 1998; von Eiff, 2008). As such succinate dehydrogenase inhibitors against *S. aureus* may in fact be deleterious.

2.3.3 Formate dehydrogenase: A major electron donor for anaerobic respiratory chains

Formate serves as a major electron donor during the anaerobic respiration of certain pathogens (e.g. *Salmonella*). Formate dehydrogenases are metalloenzymes capable of oxidising formate to carbon dioxide; with a mid-point potential of −410 mV, formate yields low-energy electrons that drive reduction of anaerobic acceptors such as nitrate, concomitant with *pmf* generation through a redox-loop mechanism (Jormakka, Byrne, & Iwata, 2003a, Jormakka et al., 2003b; Jormakka, Tornroth, Byrne, & Iwata, 2002). *E. coli* and *S. enterica* encode three molybdopterin-containing formate dehydrogenases: Fdh-N donates electrons for nitrate respiration; Fdh-O is aerobically expressed and may serve as a redox valve; and Fdh-H is part of a formate hydrogenlyase complex that fermentatively evolves hydrogen. Their activities are particularly important following onset of anaerobiosis, during which glycolytic pyruvate is converted to formate by pyruvate formate lyase; depending on electron acceptor availability, the formate can be respired by Fdh-N, decomposed by Fdh-H, or excreted from the cell (Sawers, 1994). While these enzymes have not been extensively studied in the context of enterobacterial pathogenesis, it is probable that formate serves as a major electron donor in the anaerobic environment of the gastrointestinal tract. Phenotypic studies on *S. enterica* indicate that formate has a protective role during stress (Barker, Kinsella, Jaspe, Friedrich, & O'Connor, 2000) and is an important substrate for host invasion (Huang, Suyemoto, Garner, Cicconi, & Altier, 2008). However, the role of formate dehydrogenases in these processes is still ambiguous and the individual enzymes can be deleted without causing significant phenotypes *in vivo* (Becker et al., 2006; Huang et al., 2008).

In vivo studies have shown that formate is also one of the main electron donors of the food-borne pathogen *C. jejuni*. The organism displays chemotactic behaviour towards formate (Tareen, Dasti, Zautner, Gross, & Lugert,

2010; Vegge, Brondsted, Li, Bang, & Ingmer, 2009) and rapidly consumes it (Hoffman & Goodman, 1982) using the membrane-bound tungsten-containing formate dehydrogenase FdhABCD (Shaw et al., 2012; Smart, Cliff, & Kelly, 2009). Mutants of the enzyme had abnormal immunopathology in a mouse model (Bereswill et al., 2011) and, in combination with a hydrogenase mutant, were unable to colonise its preferred host (Weerakoon, Borden, Goodson, Grimes, & Olson, 2009). Formate alongside hydrogen and sulphite is therefore predicted to be a major electron input in its intestinal niche (Hoffman & Goodman, 1982; Myers & Kelly, 2005; Weerakoon et al., 2009). However, redundancy in respiratory chains has rendered individual formate-oxidising enzymes dispensable in *C. jejuni*, as with *S. enterica* (Becker et al., 2006; Weerakoon et al., 2009). Formate oxidation is also crucial for the pathogenesis of *Aggregatibacter actinomycetemcomitans*, an opportunistic pathogen associated with periodontitis. RNA-seq analysis showed that formate dehydrogenase H was upregulated *in vivo*, together with other anaerobic metabolic enzymes (Jorth, Trivedi, Rumbaugh, & Whiteley, 2013). The mutant strain was attenuated in a murine model (Jorth et al., 2013). More fundamental research is clearly needed into the *in vivo* roles of formate and formate dehydrogenases to elucidate whether there is any therapeutic potential in targeting these enzymes.

2.3.4 Hydrogenase: Consuming a dependable reduced gas

While respiratory pathogens primarily rely on organic electron donors, reduced gases also represent significant energy reservoirs. For example, there is evidence that the *M. tuberculosis* complex liberates electrons for aerobic respiratory chains using a carboxydobacteria-type, molybdenum-containing carbon monoxide dehydrogenase (King, 2003; Park et al., 2007; Zacharia et al., 2013). More generally, molecular hydrogen (H_2) has proven a particularly dependable fuel source for pathogens, given its electronegativity ($E^{0'} = -420$ mV), diffusibility, and abundance in animal systems. As a result of activity of fermentative gut bacteria, this gas is supersaturated by five logs in animal systems compared to the atmosphere; in murine models, microelectrode measurements show that H_2 levels vary from 170 (small intestines) to 40 μM (stomach, liver, and spleen; Maier, 2005; Olson & Maier, 2002). It is known that some respiratory pathogens can use the electrons derived from H_2 to generate a *pmf*. This primarily depends on the activity of the membrane-bound [NiFe]-hydrogenases, which liberate electrons from H_2 at a specialised [NiFe] centre and relay them into the quinone pool via a *b*-type cytochrome. These electrons can then be relayed to terminal oxidases during aerobic respiration or fumarate reductase during anaerobic respiration, resulting in vectorial or scalar proton translocation.

The Maier group first illuminated the importance of hydrogen metabolism in pathogenesis by showing that the membrane-bound hydrogenase of *H. pylori* is essential for virulence (Olson & Maier, 2002). A strain lacking the enzyme was viable *in vitro*, but colonised mice at efficiencies up to 85% lower than wild-type strains (Olson & Maier, 2002). With a high whole-cell activity and an apparent K_m of 1.8 μM, the hydrogenase is adapted to rapidly consume the micromolar concentrations of H_2 available in the stomach. Though physiological and biochemical data on this enzyme remain scarce, it is known that the hydrogenase is optimally expressed during preferred microaerobic conditions. Maier proposes that *H. pylori* grows mixotrophically in the human host through co-metabolising organic carbon sources with molecular hydrogen; in light of both the nutritional deprivation of the gastric mucosa and the limited capacity for the pathogen to catabolise sugars, H_2 oxidation may be responsible for a particularly large proportion of electron flux through the respiratory chain (Maier et al., 1996; Olson & Maier, 2002). It is thought that the electrons derived from H_2 are transferred to O_2, though fumarate may substitute as an electron acceptor (Benoit & Maier, 2008; Maier et al., 1996). Also in line with a central role in the virulence, the hydrogenase appears to be under the control of the two central response regulators in *H. pylori*, the nickel-sensing NikR, and the iron-sensing Fur (Contreras, Thiberge, Mandrand-Berthelot, & Labigne, 2003; Merrell et al., 2003).

Hydrogen oxidation is similarly important but more complex in *S. enterica*. Three homologous membrane-bound [NiFe]-hydrogenases, Hya, Hyb, and Hyd, are collectively required for the organism's virulence in the murine model (Maier, Olczak, Maier, Soni, & Gunn, 2004). The aerobically expressed, oxygen-tolerant Hyd sustains aerobic respiration by coupling H_2 oxidation to oxygen reduction (Bowman et al., 2014; Parkin et al., 2012). Hya and Hyb, in contrast, are anaerobically expressed, oxygen-sensitive enzymes; it is proposed that Hyb sustains anaerobic respiration by coupling H_2 oxidation to fumarate reduction (Lamichhane-Khadka, Kwiatkowski, & Maier, 2010; Sawers, Jamieson, Higgins, & Boxer, 1986), whereas Hya recycles fermentative H_2 produced by the hydrogen formate lyase complex (Sawers et al., 1986; Zbell & Maier, 2009). In line with these roles, *in vitro* and *in vivo* studies show that the enzymes are differentially controlled by the global regulatory systems FNR, ArcBA, and IscR in response to oxygen tension and redox state (Greening & Cook, 2014; Jamieson, Sawers, Rugman, Boxer, & Higgins, 1986; Zbell, Benoit, & Maier, 2007; Zbell, Maier, & Maier, 2008). The ability to sustain three modes of energy conservation using hydrogen metabolism may contribute the metabolic flexibility required for *S. enterica* to invade and disseminate at a range of oxygen tensions (Maier et al., 2004). Consistently, researchers have observed significant reductions in the growth of a Hyb mutant during initial murine infection (Maier et al., 2013), as well as the survival of a Hya mutant in macrophages (Zbell et al., 2008). However, Maier's observation that triple, but not double, hydrogenase mutants were avirulent in a mouse model suggests that there is some overlap or compensation between the hydrogenases (Maier et al., 2004).

The role of hydrogen metabolism in the energetics of other pathogens remains unclear. For example, it has also been shown that the opportunistic pathogen *Bilophila wadsworthia* can rapidly consume hydrogen, but the clinical significance of this observation has not been investigated (da Silva, Venceslau, Fernandes, Valente, & Pereira, 2008). Likewise, studies are resolving the physiological role of the formate hydrogen lyase-related Ehr complex of *M. tuberculosis* (Berney, Greening, Hards, Collins, & Cook, 2014; Berney, Greening, Conrad, Jacobs, & Cook, 2014); this putative complex is regulated by the hypoxic response regulator DosR (Galagan et al., 2013; He, Bretl, Penoske, Anderson, & Zahrt, 2011) and is predicted to be essential for growth (Sassetti et al., 2003). BLAST searches reveal that putative hydrogenases are encoded in numerous other human pathogens, among them obligate parasites *Neisseria meningitidis* and *Yersinia pestis*. However, no comprehensive study has related hydrogen metabolism to their pathogenesis to date.

Hydrogenases have also been proposed and patented as drug targets to treat certain pathogens (Nie et al., 2012). These enzymes have many desirable traits as next-generation targets: they can be essential for virulence, are absent in humans, and are often accessibly membrane-bound. Theoretically, it would also be challenging for organisms to develop point mutations resisting inhibitors specifically targeting the [NiFe]-active site. If used in conjunction with traditional antibiotics, there is a possibility that hydrogenase inhibitors may impede the growth or persistence of certain pathogens. In addition to the hydrogenase structural studies, studies on *H. pylori* suggest that there could be merit in targeting the maturation factors (Olson, Mehta, & Maier, 2001), nickel transporter (Nolan et al., 2002), and the tat translocase (Benoit & Maier, 2014) required for the assembly of a functional hydrogenase; targeting such components might also generate the desired pleiotropic effect on the three hydrogenases of *S. enterica*. Due to the strong understanding of the structure and function of membrane-bound [NiFe]-hydrogenases, there are opportunities to discover hydrogenase inhibitors through activity-based screening and structure-based design. However, no significant research or commercial development in this area has been reported to date.

2.4. Terminal respiratory reductases

2.4.1 Haem–copper terminal oxidases: Proton translocation at a range of oxygen partial pressures

Terminal oxidases are an essential component of aerobic respiratory chains (Figs. 3 and 5A). These enzymes catalyse the four-electron reduction of oxygen to water using electrons derived from organic and inorganic donors. Given the electropositivity of the O_2/H_2O redox couple ($E^{0'} = +825$ mV), O_2 is the most energetically favourable electron acceptor available to pathogens and has the potential to energise vectorial proton translocation. However, terminal oxidases must also employ mechanisms to minimise the release of partially reduced intermediates that cause oxidative stress. There are two major types of terminal oxidases relevant to bacterial pathogens (Fig. 5A). The cytochrome *bd* oxidases catalyse oxygen reduction at an active site comprising haem b_{595} and haem *d*; these bacterial-specific, non-proton translocating oxidases are phylogenetically distinct to the mitochondrial enzymes (Borisov, Gennis, Hemp, & Verkhovsky, 2011; Junemann, 1997; Poole & Cook, 2000). In contrast, the haem–copper oxidases are a diverse group of terminal oxidases that are related to mitochondrial complex IV; these complexes couple the reduction of O_2, at a catalytic site comprising a high-spin haem and a copper ion, to the translocation of protons across the cytoplasmic membrane. There is much variation in the primary sequences and haem content of the haem–copper enzymes, which can be further-divided into the cytochrome aa_3 oxidases, cytochrome bo_3 oxidases, and cytochrome bcc_3 oxidases (Buschmann et al., 2010; Pereira, Santana, & Teixeira, 2001). The *bd* and bo_3 oxidases use quinones as their electron donors, whereas the principal electron donors for the bcc_3 and aa_3 oxidases are *c*-type cytochromes reduced by the cytochrome bc_1 complex (Borisov, Gennis, et al., 2011; Junemann, 1997; Poole & Cook, 2000).

The major respiratory pathogens each encode at least one haem–copper oxidase. The facultative aerobes *Neisseria meningitidis* and *Neisseria gonorrhoeae*

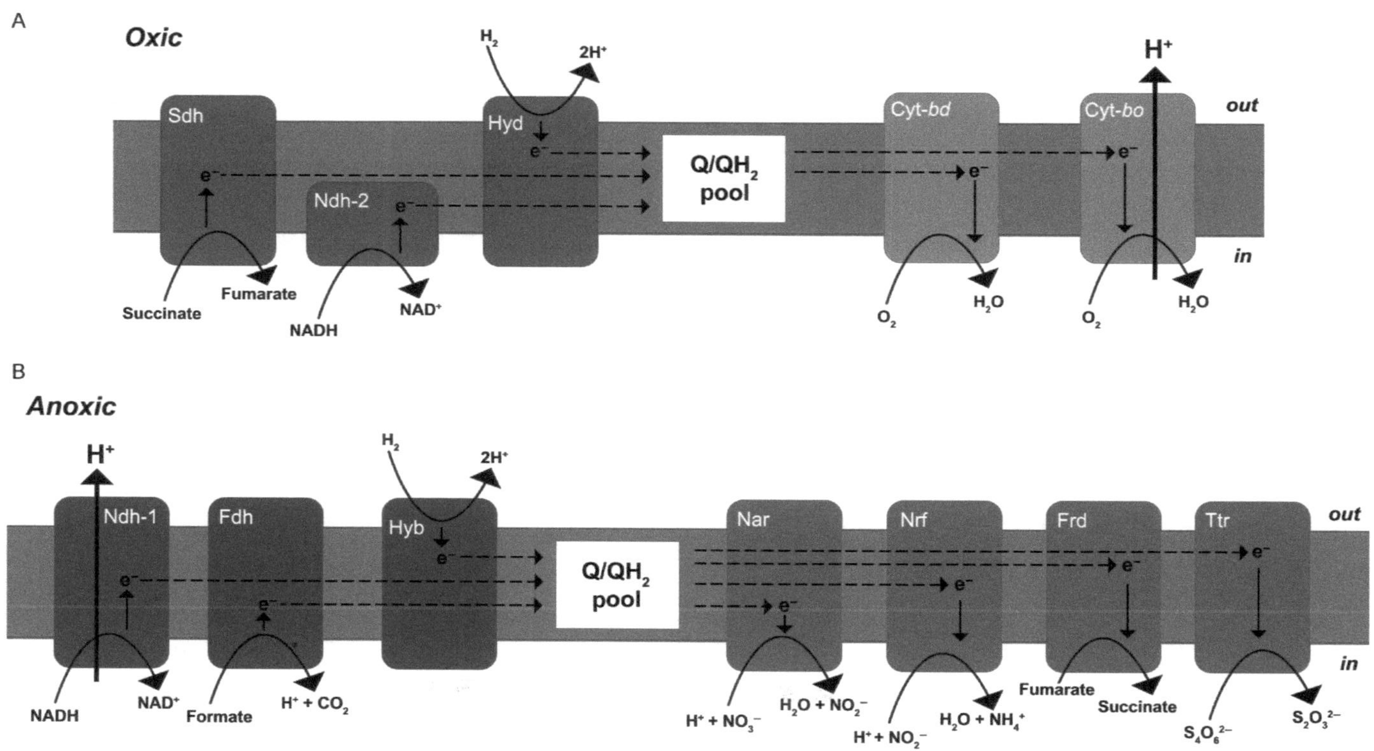

Figure 5 *Diversity of electron transport chain composition under different oxygen tensions*. The major components of the *Salmonella enterica* electron transport chain under (A) oxic and (B) anoxic conditions. As with Fig. 3, not all intermediate steps in electron transport are indicated. While represented as oxic, cytochrome *bd* is generally expressed under microoxic conditions. Additionally, nitrate and nitrite reductases are used preferentially to the other anoxic electron acceptor complexes. This diversity serves as a general model for that which is observed in other bacteria, culturable under both oxic and anoxic conditions. (See the color plate.)

each encode single cytochrome *bcc*$_3$ oxidases that are essential for growth (Aspholm et al., 2010). Enzymes of this class typically have nanomolar affinities for O_2 and hence may confer *Neisseria* the ability to colonise microoxic tissues (Pitcher & Watmough, 2004). Likewise, the microaerophile *H. pylori* is predicted to solely rely on a cytochrome *bcc*$_3$ oxidase (Nagata, Tsukita, Tamura, & Sone, 1996; Smith et al., 2000; Tomb et al., 1997). Its close relative *C. jejuni* combines a cyanide-insensitive cytochrome *bd* oxidase (*cioAB*) with a cytochrome *bcc*$_3$ oxidase (*ccoNOQP*; Jackson et al., 2007); whereas *cioA* was shown to be dispensable, Olson showed that a *ccoN* mutant was unable to colonise chicken caeca (Guccione et al., 2010; Weingarten, Taveirne, & Olson, 2009). Given *bcc*$_3$-type oxidases are essential for several pathogens, yet distantly related to human complex IV, there may be pharmaceutical potential in specifically targeting them. Among other pathogens, enterobacteria switch between the micromolar-affinity, proton-translocating cytochrome *bo*$_3$ oxidase, and the nanomolar-affinity cytochrome *bd* oxidase during shifts in oxygen partial pressure (Cotter et al., 1990; D'Mello et al., 1995, 1996); *in vivo* studies have shown that mutations in either oxidase attenuates virulence of *S. enterica* in galline and murine models (Turner et al., 2003; Zhang-Barber et al., 1997). A recent study has also shown that *S. aureus* strains containing deletions in its cytochrome *aa*$_3$ oxidase or cytochrome *bd* oxidase were unable to disseminate between tissues in mice (Hammer et al., 2013).

The ability of *P. aeruginosa* to use some five terminal oxidases appears to be particularly important for its adaptation in the cystic fibrosis lung (Arai, 2011; Williams, Zlosnik, & Ryall, 2007). Three of the terminal oxidases accept electrons from cytochrome *c*, namely, a cytochrome *aa*$_3$ oxidase (*coxABC*) and two cytochrome *cbb*$_3$ oxidases (*ccoNOQP1* and *ccoNOQP2*). Whereas Cox has the highest efficiency of the oxidases (H^+/O ratio$=6$), Cco1 and Cco2 have the highest affinities for O_2 ($K_m < 40$ nM) of the five enzymes (Arai, 2011). Consistent with this biochemistry, these enzymes are known to be differentially expressed *in vitro*; whereas Cco1 is constitutively active, the ANR-controlled Cco2 (Ray & Williams, 1997) scavenges oxygen during hypoxia and the RpoS-activated Cox (Kawakami, Kuroki, Ishii, Igarashi, & Arai, 2010) maximises proton translocation during nutrient starvation (Comolli & Donohue, 2004). The organism also encodes a stress-induced cytochrome *bo*$_3$ oxidase (*cyoABCD*; Matsushita, Shinagawa, Adachi, & Ameyama, 1982) and a cyanide-insensitive cytochrome *bd* oxidase variant (*cioAB*; Cunningham, Pitt, & Williams, 1997), which directly accept electrons from quinones. Despite having lower affinities and efficiencies than the cytochrome *c* oxidases, these enzymes are proposed to be important during iron starvation and cyanogenesis, respectively (Arai, 2011). *In vivo*, Arai and Williams postulate that this multitude of oxidases facilitates the adaptation of *P. aeruginosa* to decreasing oxygen availability during the invasion of mucosal tissues (Arai, 2011; Jewes & Spencer, 1990; Williams et al., 2007; Worlitzsch et al., 2002). Consistently, despite some overlap and redundancy between the oxidases, a triple oxidase mutant was defective in microaerophilic growth and biofilm formation (Alvarez-Ortega & Harwood, 2007).

Terminal oxidases are also central to the virulence of the obligate aerobe *M. tuberculosis*. Mycobacteria combine a cytochrome *bd* oxidase with a fused cytochrome *bcc*–aa_3 supercomplex (Cook et al., 2014; Megehee, Hosler, & Lundrigan, 2006; Shi et al., 2005). The constitutively expressed *bcc*–aa_3 pathway serves as the major respiratory route in mycobacteria under standard culturing conditions; consistently, groups have not been able to disrupt either component of this supercomplex in *M. tuberculosis* (Matsoso et al., 2005). A recent high-throughput screen (600,000 compounds) targeting ATP homeostasis in *M. tuberculosis* has uncovered inhibitors belonging to imidazopyridines (Mak et al., 2012). A representative of this class of inhibitors (i.e. imidazo[1,2-*a*]pyridine) has been shown to target the *bcc* complex encoded by the *qcrCAB* operon of *M. tuberculosis* (Abrahams et al., 2012; Kang et al., 2014; Pethe et al., 2013). The lead compound from these studies Q203 is antimycobacterial at low micromolar concentrations against pure cultures, in macrophages, and in the murine model of TB infection (Kang et al., 2014; Pethe et al., 2013).

2.4.2 *Cytochrome* bd *oxidase: A bacterial-specific next-generation drug target*

Cytochrome *bd* oxidases are a unique component of prokaryotic electron transport chains. In general, these enzymes have a higher affinity for oxygen but a lower energetic efficiency than the cytochrome aa_3 oxidases and cytochrome bo_3 oxidases (Borisov, Gennis, et al., 2011; Junemann, 1997; Poole & Cook, 2000). The structure of cytochrome *bd* oxidase has yet to be determined, but it is known that the enzyme is a heterodimeric complex composed of a large (CydA) and small subunit (CydB) that hold together the haems b_{558}, b_{595}, and *d*. A third protein CydX, which is important for enzyme activity of the *bd* oxidase, was proposed to be a third subunit of the complex in *E. coli* (Sun et al., 2012; VanOrsdel et al., 2013). It is generally thought that haems b_{595} and *d* form a di-haem site for the reduction of

O_2 (Borisov, Gennis, et al., 2011). Cytochrome *bd* oxidase does not pump protons like cytochrome *c* oxidases, but generates a *pmf* by transmembrane charge separation at an H^+/e^- ratio of 1 (Borisov, Forte, Sarti, & Giuffre, 2011). Ubiquinone and menaquinone are accepted as electron donors depending on the species. For example, CydAB of *E. coli* accepts both quinols (Unden & Bongaerts, 1997), whereas in other species like mycobacteria, only menaquinone is present.

While the physiological role of cytochrome *bd* oxidase has not yet been fully elucidated, the major function is thought to be the reduction of oxygen under low oxygen conditions. This contributes to the maintenance of *pmf* and allows the scavenging of oxygen to colonise oxygen poor niches or to protect oxygen labile enzymes. It is therefore of no surprise that many human pathogens harbour this enzyme because bacteria residing in human tissues are faced with low oxygen conditions either due to the immune response or competition with other bacteria (e.g. in the gut). Many enteric pathogens like *S. enterica* (Fig. 5A), *Shigella flexneri*, *E. coli*, and *Enterococcus faecalis* express the enzyme, as do other infectious agents of the genera *Brucella*, *Bacteroides*, *Streptococcus*, *Listeria*, and *Mycobacterium* (Borisov, Gennis, et al., 2011). Consistent with these physiological roles, *E. coli* and *S. enterica* differentially express the *cydAB* and *cyoABCD* operons according to the oxygen tension and redox state of the cell. To facilitate oxygen scavenging under hypoxic conditions, the two-component system ArcBA activates *cydAB* but represses *cyoABCD* transcription (Cotter, Melville, Albrecht, & Gunsalus, 1997; Iuchi, Chepuri, Fu, Gennis, & Lin, 1990). By contrast, both enzymes are repressed by FNR at the onset of anaerobiosis in favour of nitrate reductase and fumarate reductase (Cotter et al., 1990, 1997; Fink et al., 2007; Sellars, Hall, & Kelly, 2002). The *cydAB* genes in mycobacteria likewise responds to oxygen depletion (Berney & Cook, 2010; Kana et al., 2001); though no redox or oxygen-sensing regulator has been found to control *cydAB* transcription thus far, SenX3–RegX3 has been shown to be a regulator (Roberts, Vadrevu, Madiraju, & Parish, 2011). Nanaerobes also express cytochrome *bd* oxidase when O_2 is available at nanomolar partial pressures. The term nanaerobes was created when the opportunistic pathogen *Bacteroides fragilis*, a so-called strict anaerobe, was found to grow in the presence of nanomolar O_2 concentrations facilitated by cytochrome *bd* oxidase (Baughn & Malamy, 2004). Subsequently, a plethora of so-called strict anaerobes were found to harbour this enzyme, suggesting that many of these are capable of growth under nano-aerobic conditions (Baughn & Malamy, 2004).

Several additional functions have been proposed for this enzyme. Increasing evidence suggests the involvement of CydAB in protecting against the host's oxidative stress and nitrosative stress defences (Giuffre, Borisov, Arese, Sarti, & Forte, 2014; Poole & Cook, 2000). For example, bacterial pathogens face NO stress in the phagosome of macrophages triggered by the innate immune system. In *E. coli*, the enzyme shows lower sensitivity to NO inhibition than cytochrome *bo* and this could be of particular advantage under low oxygen conditions where many NO detoxification mechanisms are less efficient (Giuffre, Borisov, Mastronicola, Sarti, & Forte, 2012; Mason et al., 2009). Several reports also indicate that CydAB is involved in the detoxification of hydrogen peroxide (Borisov, Gennis, & Konstantinov, 1995; Giuffre et al., 2014; Poole & Williams, 1988). *E. coli* cytochrome *bd* oxidase expression increases in response to H_2O_2 and a *cydAB* mutant displayed enhanced expression of catalase, indicating an increase in intracellular H_2O_2 concentration (Lindqvist, Membrillo-Hernandez, Poole, & Cook, 2000). The pleiotropic effects of cytochrome *bd* oxidase deletion in *E. coli* and *Brucella abortus* could be rescued by the addition of catalase and superoxide dismutase (SOD; Endley, McMurray, & Ficht, 2001; Goldman, Gabbert, & Kranz, 1996). Superoxide is a by-product of several electron transport chain components such as NDH-1 and the cytochrome bc_1 complex and is converted to H_2O_2 by SOD (Turrens, 1997). Hence, it is tempting to speculate that cytochrome *bd* oxidase also scavenges ROS produced from the respiratory chain when imbalances in the electron flux occur. Consistently, transcription of the *cydAB* operon responds to other environmental triggers like high pH, high pressure, high temperature, and oxidative stress, as well as respiratory poisons and uncouplers (Borisov, Gennis, et al., 2011; Poole & Cook, 2000). It is also of note that the cytochrome *bd* oxidase variants of *P. aeruginosa* and *C. jejuni* appear to be unusual in that they are low-affinity enzymes, but are exceptionally cyanide-insensitive (Cunningham et al., 1997; Jackson et al., 2007); though their physiological roles are still incomplete-understood, they may be particularly important during stress-inducing conditions.

An additional role that has been proposed for CydAB is supplying oxidising power for biosynthetic processes. For example, haem biosynthesis and disulphide-bond formation have been shown to depend on a functional electron transport chain under oxic, microoxic, and anoxic conditions (Bader, Muse, Ballou, Gassner, & Bardwell, 1999; Mobius et al., 2010). Quinols are good electron acceptors and are available as long as terminal oxidases are functional. Cytochrome *bd* oxidase was shown to be required under microoxic conditions to sustain the activity of protoporphyrinogen IX oxidase (HemG) of *E. coli* (Mobius et al., 2010), as well as for the reoxidation of DsbB, a protein needed for disulphide-bond formation that supports protein folding (Bader et al., 1999). The ABC transporter CydCD was shown to transport glutathione and cysteine in *E. coli* (Pittman, Robinson, & Poole, 2005), contributing to the maintenance of a reduced periplasm. The CydDC proteins are also essential for the assembly of CydAB (Bebbington & Williams, 1993; Siegele, Imlay, & Imlay, 1996).

Some recent literature points towards an important role of this oxygen reductase in pathogenesis. *Brucella suis* and *B. abortus*, the causative agents of brucellosis in swine and cattle, were shown to depend on cytochrome *bd* oxidase for replication and survival in macrophages and mice (Endley et al., 2001; Loisel-Meyer, Jimenez de Bagues, Kohler, Liautard, & Jubier-Maurin, 2005). Other studies have shown that cytochrome *bd* mutants of *S. typhimurium* (Zhang-Barber et al., 1997), *S. flexneri* (Way, Sallustio, Magliozzo, & Goldberg, 1999), and *M. tuberculosis* (Zhang et al., 2012) are at a competitive disadvantage in murine models when co-infected with a wild-type strain. *M. tuberculosis* resides in so-called granuloma in the lungs of its host and this tissue is hypoxic. CydAB was expressed transiently in a mouse infection model (Shi et al., 2005), but given that mice do not form hypoxic granulomas, the full importance of cytochrome *bd* oxidase might be underestimated. Experiments in animal models with hypoxic granuloma could shed more light on the *in vivo* importance of this enzyme. In *M. tuberculosis*, the *bd* oxidase cannot sustain growth as the sole terminal oxidase because the aforementioned cytochrome *bcc*–aa_3 supercomplex is essential (Matsoso et al., 2005). However, a defect in the cytochrome *c* biosynthesis pathway causes increased expression of *cydAB* (Small et al., 2013); this suggests that, while CydAB activity does not yield enough energy to support growth, it might be sufficient to maintain a membrane potential when the haem–copper oxidase is downregulated or inhibited.

Cytochrome *bd* oxidase has not yet been explored as a therapeutic target. Nevertheless, several compounds are known to inhibit the enzyme and have been used for studying structural and mechanistic aspects of respiratory chains in bacteria. Comprehensive lists of known inhibitors have been published (Borisov, Gennis, et al., 2011; Meunier, Madgwick, Reil, Oettmeier, & Rich, 1995). However, most inhibitors are unspecific and inhibit haem–copper oxidases as well. This could be problematic as cytochrome aa_3 oxidases are also present in mitochondria. Aurachin D was the only inhibitor that showed specificity for cytochrome *bd* oxidase, albeit only at lower concentrations (Meunier et al., 1995). On the contrary, many inhibitors are known that have higher affinity for aa_3 oxidases rather than *bd* oxidases. For example, the *bd* oxidase is more resistant to the action of cyanide, azide, nitric oxide, and Zn ion stress. Based on the *in vitro* data available to date, it is tempting to speculate that cytochrome *bd* oxidase plays a role in the adaptation, colonisation, and survival of human pathogens in the host environment. However, evidence for an essential function of this enzyme during pathogenesis is still scarce and proof of such would undoubtedly spark more interest for developing a specific drug. Cytochrome *bd* oxidase might be an imperfect single drug target (due to the presence of other terminal oxidases), but an inhibitor could potentially increase killing rates in combination with other respiratory chain inhibitors due to the enzymes proposed role in oxidative stress defence. Studies aimed at elucidating the physiological role of cytochrome *bd* oxidase during bacteria–host interactions, as well as structural information of the enzyme, will guide potential initiatives for targeted drug development in the future.

2.4.3 Respiratory nitrate and nitrite reductases: Exploiting host immune defences

Oxygen partial pressures vary by orders of magnitude between tissues, as a result of the high rates of aerobic respiration by host, commensal, and pathogen cells. Most pathogens therefore employ anaerobic respiratory chains to grow or survive in hypoxic or anoxic tissues (Fig. 3). Nitrate is the most energetically favourable anaerobic electron acceptor given its relatively positive mid-point potential ($E^{0'} = +430$ mV) compared to alternatives such as fumarate ($E^{0'} = +30$ mV), trimethylamine *N*-oxide (TMAO; $E^{0'} = +130$ mV), and tetrathionate ($E^{0'} = +170$ mV). As a result, many pathogens upregulate preferentially nitrate reductases and denitrification pathways when oxygen is limited and substrate is available (Fig. 5B). During infection, many pathogens can derive nitrate from the nitric oxide produced by inducible nitric oxide synthase (iNOS) during the innate immune response. For example, *S. enterica* uses the flavohaemoglobin/nitric oxide dioxygenase Hmp to convert nitric oxide to nitrate in the macrophage; exploiting the host's defences, the bacterium both detoxifies a poison and provides a strong oxidant for anaerobic respiration (Mills, Rowley, Spiro, Hinton, & Richardson, 2008).

There is growing evidence that denitrification confers a competitive advantage to *S. enterica* and pathogenic *E. coli*. Under anoxic conditions, these enterobacteria employ the regulatory systems FNR, NarXL, and NarQP to ensure that nitrate is used above energetically inferior acceptors such as fumarate (Gilberthorpe & Poole, 2008; Stewart, 2003). The reduction of nitrate to nitrite is catalysed by three nitrate reductases (i.e. respiratory membrane-bound *narGHJI*, assimilatory periplasmic *napFDAGHBC*, and starvation-induced membrane-bound *narZYWV*; McClelland et al., 2001; Spector et al., 1999), and the six-electron reduction of nitrite to ammonium is mediated by two nitrite reductases (i.e. periplasmic cytochrome *c*-dependent *nrfABCDEFG* and cytosolic NADH-dependent *nirBDC*; Wang & Gunsalus, 2000). *S. enterica* can also detoxify macrophage-derived nitric oxide to either nitrate using flavohaemoglobin Hmp, nitrous oxide using flavorubredoxin NorV, or ammonium using NrfA (Mills et al., 2008; Poole, 2005). The nitrate reductases appear to be individually dispensable, but collectively confer a growth advantage (Knuth, Niesalla, Hueck, & Fuchs, 2004; Winter et al., 2013). Using a murine colitis model, the Bäumler group recently demonstrated that wild-type *E. coli* could outcompete a nitrate reductase triple mutant using iNOS-derived nitrate (Winter et al., 2013). Equivalent studies in *S. enterica* showed that nitrate enhances growth at least 10-fold and is also a stimulus for Tsr-mediated energy taxis (Bliska & van der Velden, 2012; Lopez et al., 2012; Rivera-Chavez et al., 2013).

M. tuberculosis is unable to grow using electron acceptors such as nitrate and nitrite. However, Sohaskey has demonstrated that the organism enhances its long-term viability by reducing these compounds (Sohaskey, 2008). *M. tuberculosis* expresses a respiratory nitrate reductase (*narGHJI*) and nitrite reductase (*nirBD*) homologous to those of enterobacteria (Malm et al., 2009). In addition, it encodes the nitrate importer NarK2 (Sohaskey, 2005; Sohaskey & Wayne, 2003), an unresolved nitrite exporter (Griffin et al., 2012), and the potentially non-functional fusion protein NarX (Sohaskey & Wayne, 2003). While *narGHJI* and *nirBD* expression is constitutive, *nark2x* is upregulated during hypoxic and nitrosative stress by DosR (Shi et al., 2005). In line with increased nitrate import and accumulation of reduced cofactors, the rate of nitrate reduction increases in hypoxic cells *in vitro* (Sohaskey & Wayne, 2003). Though the physiological role of the nitrate reductase has not been fully resolved, it may generate *pmf* and reoxidise cofactors in non-growing hypoxic cells; consistently, the enzyme enhanced adaptation to sudden anaerobiosis, but not gradual oxygen depletion *in vitro* (Sohaskey, 2008). While the nitrite produced is partly exported (Griffin et al., 2012), it can also be reduced to ammonium by NirBD; this process is essential for survival in both the *in vitro* Wayne model and in human macrophages (Akhtar, Khan, Sohaskey, Jagannath, & Sarkar, 2013). Like *S. enterica*, there is evidence to suggest that denitrification in *M. tuberculosis* depends on nitric oxide produced by iNOS in the human macrophage (Jung et al., 2013); this molecule serves to both provide the substrate nitrate (Arya et al., 2013) and induce its import through activating the haem-based histidine kinases for DosR (Voskuil et al., 2003).

There is significant evidence that *P. aeruginosa* also uses denitrification during human infection. This opportunistic pathogen chronically infects the cystic fibrosis lung, forming biofilms (Singh et al., 2000) and harnessing the amino acids available in the viscous sputum (Ohman & Chakrabarty, 1982; Palmer, Mashburn, Singh, & Whiteley, 2005). Likely as a result of iNOS activity, the sputum is supersaturated with nitrate and nitrite in cystic fibrosis versus normal lungs (Linnane et al., 1998). Encoding a complete denitrification pathway, *P. aeruginosa* sequentially reduces nitrate to dinitrogen using a membrane-bound nitrate reductase (NarGJ), a cytochrome cd_1-type nitrite reductase (NirS), a cytochrome *bc*-type nitric oxide reductase (NorBC), and a two-copper centre nitrous oxide reductase (NosZ; Williams et al., 2007). Due to the elevated epithelial oxygen uptake (Stutts, Knowles, Gatzy, & Boucher, 1986) and high bacterial cell densities (Wessel et al., 2014) in cystic fibrosis lung tissues, the sputum contains hypoxic or anoxic microenvironments that may promote nitrate respiration (Jewes & Spencer, 1990; Worlitzsch et al., 2002). The Whiteley group have also showed that the nitrate reductase is required for anaerobic growth in an *in vitro* cystic fibrosis sputum medium (Palmer, Brown, & Whiteley, 2007). Consistently, transcription of the denitrifying enzymes is activated by nitrate, the O_2 sensor ANR, and the NO sensor DNR (Arai, Kodama, & Igarashi, 1997; Ye et al., 1995). However, it has become apparent that *P. aeruginosa* is also capable of aerobic denitrification. This process may be particularly important in the context of the host (Alvarez-Ortega & Harwood, 2007; Chen, Xia, & Ju, 2003; Filiatrault et al., 2005); while energetically less efficient, co-respiration of nitrate and oxygen may provide the flexibility needed to withstand temporal and spatial shifts in the concentrations of electron acceptors.

Contrary to its original classification as an obligate aerobe, the sexually transmitted pathogen *N. gonorrhoeae* employs a partial denitrification pathway to grow anaerobically and microaerobically (Knapp & Clark, 1984). It harbours the capacity to reduce nitrite to nitrous oxide, but nitrate reductase is absent and the nitrous oxide reductase genes contain nonsense mutations (Cardinale & Clark, 2005). The copper-containing outer membrane nitrite reductase AniA reduces nitrite to nitric oxide using electrons donated from *c*-type cytochromes in the respiratory chain (Hopper, Tovell, & Cole, 2009; Mellies, Jose, & Meyer, 1997). NorB, a cytoplasmic quinol-dependent nitrous oxide reductase, in turn maintains nitric oxide at anti-inflammatory nanomolar concentrations (Cardinale & Clark, 2005; Householder, Fozo, Cardinale, & Clark, 2000). Both *aniA* and *norB* are induced by nitric oxide as a result of inactivation of the repressor NsrR and hence activation of NarP; as the most strongly upregulated gene under anaerobiosis, *aniA* is also regulated by the oxygen sensor FNR (Householder et al., 2000; Overton et al., 2006; Whitehead et al., 2007). In the human host, denitrification may confer the metabolic flexibility for *N. gonorrhoeae* to generate *pmf* and maintain redox balance in anoxic microenvironments (Householder et al., 2000). Moreover, there is compelling *in vivo* evidence that detoxification of nitric oxide at a range of oxygen tensions enables the bacterium to survive nitrosative stress and suppress inflammatory responses effected by the host's iNOS (Cardinale & Clark, 2005; Stevanin, Moir, & Read, 2005; Tunbridge et al., 2006). Equivalent pathways have been characterised in causative agent of meningococcal disease, *N. meningitidis* (Anjum, Stevanin, Read, & Moir, 2002; Rock et al., 2005), though clinical isolates suggest that AniA is dispensable for pathogenesis (Stefanelli et al., 2008).

Denitrification appears to be a widespread mechanism for the anaerobic adaptation of human pathogens. While *H. pylori* lacks the capacity for denitrification, the periplasm of the related ε-proteobacterium *C. jejuni* encodes a nitrate reductase and a nitrite reductase; nitrate is incapable of sustaining anaerobic growth, though co-respiration with oxygen confers a small advantage *in vitro* (Sellars et al., 2002). *S. aureus* can grow using nitrate and nitrite as alternative electron acceptors (Burke & Lascelles, 1975); microarray analysis has shown that NarGH and NirBD in this organism is induced under hypoxia (Fuchs, Pane-Farre, Kohler, Hecker, & Engelmann, 2007) by the oxygen-sensitive two-component system NreBC (Schlag et al., 2008). Looking beyond the respiratory chain, it is also important to note that nitric oxide enhances the oxidative stress response and antibiotic resistance of both *S. aureus* and *Bacillus anthracis in vivo* (Gusarov, Shatalin, Starodubtseva, & Nudler, 2009; Shatalin et al., 2008); this depends on NO derived from the bacterial nitric oxide synthase (bNOS) reactivating catalase and suppressing the Fenton reaction (Gusarov & Nudler, 2005).

Given the central role of denitrification in many pathogens, it is conceivable that inhibiting these pathways may have therapeutic outcomes. However, much research would be required to develop a specific, penetrable, and potent antimicrobial. Inhibition of nitrate reductase or nitrite reductase may promote membrane depolarisation, redox imbalance, and, in the latter case, accumulation of toxic intermediates; this could impair both fast growers such as *S. enterica* and persisters such as *M. tuberculosis* and *P. aeruginosa*. In addition to the structural subunits, there may also be potential in targeting the proteins involved in the unique pathways for the biosynthesis of their cofactors, for example, sirohaem, haem d_1, or *c*-type cytochromes. There may also be potential in modulating nitric oxide metabolism through inhibiting iNOS, bNOS, or flavohaemoglobin. A new study identified two lead compounds that specifically inhibited bNOS and reduced the survival of *Bacillus subtilis*; there are opportunities for this proof-of-concept study to be extended to pathogens (Holden et al., 2013). However, given the multifaceted role of nitric oxide in both the cytotoxicity and cytoprotection of pathogens, inhibiting nitric oxide production may also yield deleterious effects. For example, administration of the iNOS-2 inhibitor N^6-(1-iminoethyl)-L-lysine caused accelerated progression of *M. tuberculosis* infection to chronic tuberculosis in murine lungs (MacMicking et al., 1997).

2.4.4 Fumarate reductase: Anaerobic respiration using an endogenous electron acceptor

While fumarate reductases cannot be readily distinguished from succinate dehydrogenases on a sequence level, these enzymes have generally been selected for distinct purposes in bacterial cells (Fig. 3). For example, whereas the aforementioned succinate dehydrogenase of *S. enterica* primarily inputs electrons from the tricarboxylic acid into the aerobic respiratory chain, its fumarate reductase has a role in anaerobic respiration (Encheva, Shah, & Gharbia, 2009; Fig. 4B). Consistently, these enzymes are differentially regulated by ArcBA and FNR in *S. enterica* and *E. coli*, with succinate dehydrogenase being expressed under oxic conditions and fumarate reductase being expressed under anoxic conditions in the absence of nitrate (Fink et al., 2007; Sellars et al., 2002)

The microaerophile *H. pylori* lacks a succinate dehydrogenase but encodes a fumarate reductase (Ge, Jiang, Kalisiak, & Taylor, 1997; Pitson, Mendz, Srinivasan, & Hazell, 1999). In this case, fumarate serves as a terminal electron acceptor, with succinate more akin to a fermentative end-product. The central role of fumarate reductase for the growth of this organism is demonstrated by the fact that it is essential for the colonisation of the mouse stomach (Ge et al., 2000) and that nizatidine enhances *Helicobacter* killing in concert with metronidazole (Chen et al., 2002), in a manner that is correlated with the ability of both compounds to independently inhibit fumarate reductase activity. A high percentage of infected patients generate antibodies against this protein, validating that targeting this protein is unlikely to be toxic to humans (Birkholz & Knipp, 1994). Its more flexible relative *C. jejuni* encodes a distinct type of fumarate reductase, as well as an enzyme that has the ability to function physiologically as both a fumarate reductase and a succinate dehydrogenase. Whereas the unidirectional enzyme seems to have a marginal role in growth (Guccione et al., 2010), deletion of the bidirectional enzyme results in major growth defects in microaerobic growth, lethality during oxygen limitation, and the inability to colonise chicken caeca (Guccione et al., 2010; Weingarten et al., 2009).

M. tuberculosis encodes two annotated succinate dehydrogenases (*sdh1* and *sdh2*) and one fumarate reductase (*frd*; Cook et al., 2014). These enzymes must therefore fill roles more complicated than being one dedicated succinate dehydrogenase and one dedicated fumarate reductase, although it is highly likely that considerable functional redundancy exists between the three enzymes. Consistently, the deletion of the *frd* genes resulted in no phenotype (Watanabe et al., 2011), likely a response of compensation. Homologues of *sdh1* and *sdh2* in *M. smegmatis* have been found to be differentially expressed, being upregulated under energy-limiting and oxygen-limiting conditions, respectively (Berney & Cook, 2010). In agreement with this, transposon site hybridisation screens find that *M. tuberculosis sdh1* is essential under aerobic conditions (Griffin et al., 2011), whereas *sdh2* is essential under hypoxia (Baek, Li, & Sassetti, 2011). It may be that *sdh1* is the preferred succinate dehydrogenase for growth and *sdh2* is a bidirectional enzyme that behaves according to the availability of reduced/oxidised substrates (Pecsi et al., 2014). During hypoxia, studies have found that succinate is produced and excreted as an end-product and that this may effect membrane potential generation (Eoh & Rhee, 2013; Watanabe et al., 2011). Rhee proposed that succinate is produced in large quantities as a multifunctional compound that can maintain succinate dehydrogenase activity, be used for ATP synthesis, or be excreted to generate membrane potential when terminal electron acceptors are lacking (Eoh & Rhee, 2013). These results would be consistent with *sdh2* being a bidirectional enzyme and the compensation observed when *frd* was deleted. The functional redundancy and structural differences within the three enzymes may make it challenging to develop inhibitors against succinate dehydrogenase or fumarate dehydrogenase activity in *M. tuberculosis*; however, the development of menaquinone biosynthesis inhibitors or menaquinone analogue inhibitors may be able to overcome this.

2.4.5 *Tetrathionate reductase and other alternative reductases: Emerging roles in host colonisation*

Enterobacteria exploit inflammation to outcompete intestinal microbiota (Stecher et al., 2007). In particular, *S. enterica* drives its host to produce both reactive nitrogen species and ROS using the type III secretion systems T3SS-1 and T3SS-2 (Tsolis, Adams, Ficht, & Baumler, 1999). In addition to converting nitric oxide to nitrate, the organism can detoxify ROS to yield the anaerobic electron acceptor, tetrathionate (Fig. 5B); specifically, tetrathionate is produced when inflammatory oxygen radicals react with thiosulphate derived from the hydrogen sulphide produced by fermentative gut commensals (Winter & Baumler, 2011). Tetrathionate is reduced by the molybdopterin-containing tetrathionate reductase TtrABC; this enzyme is induced under hypoxia by FNR (Hensel, Hinsley, Nikolaus, Sawers, & Berks, 1999) and uses electrons derived from organic electron donors, including alternative donors such as ethanolamine (Thiennimitr et al., 2011). Winter et al. (2010) showed that deletion of the tetrathionate reductase resulted in a 100-fold reduction in intestinal lumen bacterial cell counts in a murine model. Adding to cumulative evidence that tetrathionate is a central anaerobic electron acceptor *in vivo*, it was recently shown that *S. enterica* can mediate energy taxis towards the compound using the chemoreceptor Aer (Rivera-Chavez et al., 2013). There is some *in silico* and *in vitro* evidence that tetrathionate respiration may also have a role in other pathogens, for example, *C. jejuni* and *Citrobacter freundii* (Liu, Denkmann, Kosciow, Dahl, & Kelly, 2013; Lupp et al., 2007; Winter & Baumler, 2011). Collectively, these findings emphasise the overlap between pathogenesis, inflammation, and diet in the infection of *S. enterica* and likely other gastrointestinal pathogens; this raises possibilities for using pharmaceuticals, immunomodulation, and dietary control to control the energetics of such pathogens for preventative or therapeutic outcomes.

It is known that pathogens can reduce a spectrum of other sulphur compounds *in vitro*. For example, thiosulphate is a potential electron acceptor for enteric bacteria and is produced by the oxidation of hydrogen sulphide by mitochondria of the colonic mucosa (Hildebrandt & Grieshaber, 2008). During growth on formate as an electron donor, pathogens such as *S. enterica* and *C. freundii* can drive a menaquinol–thiosulphate oxidoreductase to reduce this electronegative molecule ($E^{0'} = -402$ mV) using the *pmf* (Kapralek, 1972; Stoffels, Krehenbrink, Berks, & Unden, 2012). *E. coli*, *S. enterica*, and *C. jejuni* also express specific reductases that reduce the more potent electrophiles dimethyl sulphoxide (DMSO) and TMAO when available (Fink et al., 2007; Sellars et al., 2002). It is of note that anaerobic sulphate-reducing bacteria of the genus *Desulfovibrio* occasionally present as opportunistic pathogens, for example, *D. desulfuricans* in bacteremia (Goldstein, Citron, Peraino, & Cross, 2003). Looking beyond respiration, it has been shown that SirA of *M. tuberculosis* (previously misannotated as NirA) is a ferredoxin-sulphite oxidoreductase; in addition to an assimilatory role, it may serve to dissipate excess reductant as hydrogen sulphide (Berney, Greening, Hards, et al., 2014; Pinto, Harrison, Hsu, Jacobs, & Leyh, 2007; Schelle & Bertozzi, 2006).

2.5. Generators of sodium motive force in bacterial pathogens

Bacteria can also switch to sodium energetics under conditions where the generation of a *pmf* is impeded by environmental conditions (Fig. 6). Some pathogens generate an electrochemical gradient of sodium ions across their cell membranes, i.e., a sodium motive force (*smf*). The *smf* can be generated by two major mechanisms (Fig. 6). Many pathogens use primary sodium pumps driven by oxidative phosphorylation (e.g. the Na^+-translocating

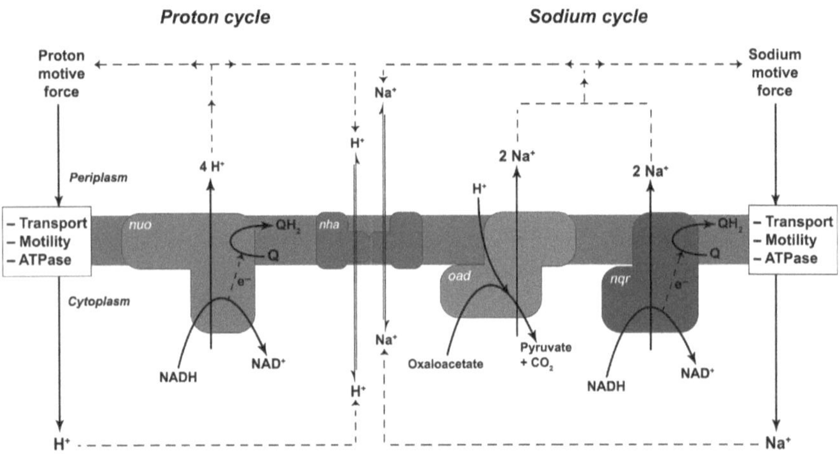

Figure 6 *Generalised role of sodium and proton cycles at the level of the membrane.* Only representative proton- (*nuo*) and sodium- (*oad, nqr*) pumping complexes are shown for simplicity. The sodium and proton motive forces can be used to drive cellular processes such as flagellar motion and ATP synthesis. Antiporters provide an overlap between the two cycles as they can theoretically operate to generate one electrochemical gradient at the expense of the other, depending on cellular needs. Nha is used to refer to antiporters in general, but in reality, antiporters can operate on variable stoichiometry (not pictured) which results in either electrogenic or electroneutral exchange of H^+ and Na^+ depending on the protein. (See the color plate.)

NADH–quinone oxidoreductase NQR; Verkhovsky & Bogachev, 2010) or decarboxylative phosphorylation (e.g. oxaloacetate decarboxylase; Buckel, 2001; Dimroth, 1994). In addition, almost all pathogens can interconvert a *pmf* to a *smf* using sodium/proton antiporters (Mulkidjanian, Dibrov, & Galperin, 2008; Padan, Venturi, Gerchman, & Dover, 2001; Fig. 6).

Phylogenetic evidence posits that sodium may have preceded protons in the evolution of chemiosmosis (Lane, Allen, & Martin, 2010; Mulkidjanian et al., 2008). Consistently, sodium has been retained as a coupling ion in pathogens across diverse phyla, though the relative importance of *pmf* and *smf* greatly varies between organisms (Häse, Fedorova, Galperin, & Dibrov, 2001). The enterobacteria *Klebsiella pneumoniae*, *E. coli*, and *S. enterica* rely on *pmf* for ATP synthesis and other cellular processes, but power specific processes such as the symport of amino acids and citrate using *smf* (Deguchi, Yamato, & Anraku, 1990; van der Rest et al., 1992). Energy generation in *Vibrio cholerae* also depends on a H^+-specific F_1F_o-ATP synthase, though its sodium gradients have a central role in pathogenesis; the ion both regulates the expression of virulence factors (Häse & Mekalanos, 1999) and drives multidrug resistance pumps (Begum et al., 2005; Brown, Paulsen, & Skurray, 1999; Huda, Morita, Kuroda, Mizushima, & Tsuchiya, 2001), toxin exporters (Häse & Mekalanos, 1999), and flagellar motors (Kojima, Yamamoto, Kawagishi, & Homma, 1999; Yorimitsu & Homma, 2001). At the other end of the spectrum, *smf* has the dominant role in the energetics in some strictly anaerobic pathogens, including the periodontal pathogen *Fusobacterium nucleatum* (Kapatral et al., 2002; Schulz et al., 2013) and the syphilis spirochete *Treponema pallidum* (Fraser et al., 1998).

Dimroth established a new paradigm in bioenergetics with his discovery that oxaloacetate decarboxylase of *K. pneumoniae* is in fact a primary sodium pump (Dimroth, 1980). During the catalytic cycle of this trimeric decarboxylase, the carboxylate group from oxaloacetate is transferred to and activated by the biotin cofactor of the β-subunit; the subsequent elimination of the group as CO_2 drives Na^+ binding and translocation via a conserved aspartate residue (Buckel, 2001; Di Berardino & Dimroth, 1996; Dimroth & Thomer, 1993). Thus, whereas oxidative phosphorylation couples redox reactions to ion translocation, decarboxylative phosphorylation uses the free-energy change from decarboxylation for sodium translocation. In *K. pneumoniae* and other enterobacteria, the oxaloacetate decarboxylase enhances the uptake and catabolism of citrate. However, the enzyme is otherwise dispensable for the viability of this metabolically flexible organism

(Bott, Meyer, & Dimroth, 1995). By contrast, the genome of the unculturable *T. pallidum* suggests that its only primary ionic pump is oxaloacetate decarboxylase. Thus, decarboxylative phosphorylation is predicted to solely provide the potential energy for both ATP synthesis and substrate import in this organism and hence may represent a novel target for the development of antimicrobials against syphilis (Häse et al., 2001).

F. nucleatum, another *smf*-based pathogen, conserves free energy by fermenting amino acids such as glutamate via the glutaconyl-CoA pathway (Bakken, Hogh, & Jensen, 1989; Beatrix, Bendrat, Rospert, & Buckel, 1990). Through landmark studies of this pathway in fusobacteria and clostridia, Buckel has concluded that energy generation depends on the concurrent processes of decarboxylative phosphorylation and electron bifurcation. The membrane-bound, biotin-dependent complex glutaconyl-CoA decarboxylase generates *smf* from the decarboxylation of glutaconyl-CoA to crotonyl-CoA (Beatrix et al., 1990). It is proposed that the high-potential crotonoyl-CoA ($E^{0'} = -10$ mV) and low-potential ferredoxin ($E^{0'} = -400$ mV) are thereafter simultaneously reduced by electrons derived from the mid-potential NADH ($E^{0'} = -280$ mV); this electron bifurcation is mediated by the multimeric complex Bcd/EcfBC and depends on flavin serving as a two-electron gate. The low-energy electrons of ferredoxin can either drive reductive cellular processes or be transduced to *smf* via a further primary sodium pump, the ferredoxin-NAD^+ oxidoreductase Rnf (Herrmann, Jayamani, Mai, & Buckel, 2008; Li et al., 2008). These processes are reviewed in more detail by Buckel and Thauer (2013). This efficient coupling of these anaerobic chemiosmotic processes generates a large *smf* in this organism and maximises the output of the Na^+-translocating F_1F_o-ATP synthase (Schulz et al., 2013).

Primary sodium pumps can also be driven by oxidative phosphorylation, notably the Na^+-translocating NADH–quinone oxidoreductase (Verkhovsky & Bogachev, 2010). Phylogenetically and functionally distinct from the previously described H^+-translocating NDH-1 and non-translocating NDH-2, NQR appears to be the major primary sodium pump in aerobic pathogens (Häse et al., 2001). Though an atomic resolution structure of this six-subunit enzyme is unavailable, the chemiosmotic coupling mechanism depends on redox reactions via multiple cofactors (Juarez & Barquera, 2012). First studied in the opportunistic pathogen *Vibrio alginolyticus* (Tokuda & Unemoto, 1981, 1984), NQR generates the *smf* required for transport and motility. Raising opportunities for antimicrobial development, the natural product korormicin has a cytotoxic effect and potently inhibits NQR in this organism (Hayashi, Shibata, Nakayama, Yoshikawa, & Unemoto, 2002; Yoshikawa, Nakayama, Hayashi, Unemoto, & Mochida, 1999). In *K. pneumoniae*, it remains controversial whether *smf* is entirely generated by NQR; studies have not resolved whether NDH-1 translocates H^{+}, Na^{+}, or both ions (Batista, Marreiros, & Pereira, 2012; Bertsova & Bogachev, 2004; Krebs, Steuber, Gemperli, & Dimroth, 1999; Steuber, 2001). Nevertheless, the *smf* is particularly important for citrate fermentation. The utilisation of *smf* both energises the citrate symporter CitS (van der Rest et al., 1992) and enables cofactor regeneration though mediating reversed electron transfer from formate to NAD^{+} (Pfenninger-Li & Dimroth, 1992).

The majority of pathogens also encode multiple Na^{+}/H^{+} antiporters that allow the interconversion of *pmf* and *smf*. For example, *V. cholerae* expresses a wide array of antiporters (Häse et al., 2001) that have a high degree of functional redundancy (Herz, Vimont, Padan, & Berche, 2003; Vimont & Berche, 2000). Likewise, *P. aeruginosa* operates a robust sodium cycle (Häse et al., 2001) using cation:proton antiporters that have been implicated for virulence in mice (Kosono et al., 2005; Ueda & Wood, 2008). It is generally appreciated that Na^{+}/H^{+} antiporters operate homeostatically to modulate chemiosmotic forces to cellular needs (Häse & Barquera, 2001; Häse et al., 2001). As such, it seems likely that any therapeutic strategies targeting the sodium cycle will need to consider potential compensation by antiporter complexes. In this regard, inhibition of Na^{+}/H^{+} antiporters may be best achieved using derivatives of amiloride compounds. They effectively inhibit mammalian CPA1-type antiporters (Noël & Pouysségur, 1995), but concentrations 10 times higher are required to inhibit the antiport activity of *V. parahaemolyticus* (Häse & Barquera, 2001; Kuroda, Shimamoto, Mizushima, & Tsuchiya, 1997). Many improvements in drug design against these targets remain to be made, but finding pathogen-specific inhibitors may prove valuable in combination therapy with other sodium or proton cycle inhibitors.

2.6. ATP homeostasis and the F_1F_o-ATP synthase: A clinically approved drug target

ATP synthases are membrane-bound enzymes found in all kingdoms of life. The ATP synthase has remained remarkably conserved throughout evolution and is believed to be an example of modular evolution during which two functionally independent subunits, a DNA helicase with ATPase activity and a proton-powered motor, became associated and gained new functionality (Doering, Ermentrout, & Oster, 1995). From a structural and functional perspective, three distinct classes of the enzyme are recognised, F-type, A(archaeal)-type, and V(vacuolar)-type. The F-type enzymes, found in prokaryotes and eukaryotes, can synthesise and hydrolyse ATP and this process is generally coupled to protons. In some bacterial species, the enzyme is able to translocate sodium ions (von Ballmoos, Cook, & Dimroth, 2008). The A-type enzyme, found in Archaeal genomes, also shares similar properties. The V-type enzymes are primarily known for their ATP hydrolysis activity in eukaryotic cells, but have been reported in some prokaryotes (Yokoyama, Akabane, Ishii, & Yoshida, 1994), including the pathogens *Enterococcus hirae* (Kakinuma, Yamato, & Murata, 1999) and *T. pallidum* (Fraser et al., 1998).

The F-type ATP synthase consists of two subcomplexes: the water-soluble F_1 sector, which harbours the catalytic centres, and the membrane-embedded F_o complex, which mediates ion translocation across the membrane (Fig. 4). Detailed knowledge on structure and function is available for the water-soluble F_1 headpiece with the subunit composition $\alpha_3\beta_3\gamma\delta\varepsilon$ (Abrahams, Leslie, Lutter, & Walker, 1994; Cingolani & Duncan, 2011; Stock, Leslie, & Walker, 1999). Alternating α and β subunits form a cylinder around subunit γ. Part of the γ subunit protrudes from the bottom of the cylinder and forms the central stalk together with the ε subunit. The central stalk is connected with the oligomeric c-ring of the membrane-intrinsic F_o moiety. The other F_o subunits of bacterial ATP synthases are *a* and b_2, which abut the c-ring laterally. The oligomeric c-ring harbours a series of identical ion-binding sites and varies in size from 8 to 15 subunits depending on the species. The reasons for this variability and the factors governing c-ring size are incompletely understood. The molecular chaperones for ATP synthase in bacteria remain unknown, but an important area of research, given the essentiality of the enzyme in some bacterial pathogens.

The F_1F_o-ATP synthase catalyses ATP synthesis by utilising the electrochemical gradient of protons or sodium ions (*smf*) to generate ATP from ADP and inorganic phosphate (P_i) and operates under conditions of a high *pmf*/*smf* and low intracellular ATP. The enzyme is also capable of working as an ATPase under conditions of high intracellular ATP and an overall low *pmf* (von Ballmoos et al., 2008). Bacterial pathogens are often faced with the challenge of adapting to environmental conditions during transitions from normoxia to hypoxia where the *pmf* drops rapidly (von Ballmoos et al., 2008). Under these conditions, the ATP synthase shifts gears to ATP hydrolysis activity to combat the decreasing *pmf*. As an ATPase, the enzyme hydrolyses ATP while pumping protons or sodium ions from the cytoplasm to the outside the cell. Regulation of ATPase activity under these conditions is essential to prevent wasteful (futile) ATP hydrolysis. The regulatory mechanisms bacteria use to block ATP hydrolysis in response to a suboptimal *pmf* remain incompletely understood, but represent a potential avenue for disrupting ATP homeostasis. Some bacterial F_1F_o-ATPase preparations exhibit high rates of ATP hydrolysis activity, whereas other preparations show extreme latency in ATP hydrolysis activity and can only be activated by detergents, solvents, or proteases. The molecular mechanisms responsible for this inhibition remain to be identified.

F_1F_o-ATP synthase has been shown to be essential for the optimal growth of mycobacteria, including *M. tuberculosis*, on fermentable and non-fermentable carbon sources (Griffin et al., 2011; Tran & Cook, 2005; Zhang et al., 2012). The reasons for this remain unclear. One explanation is that the extraordinary amount of ATP required to synthesise a mycobacterial cell might exceed what can be supplied by substrate-level phosphorylation alone (Cox & Cook, 2007). Alternatively, the organism may not be able to support uncoupled respiration; either they lack a conduit for proton re-entry in the absence of the F_1F_o-ATP synthase or they are unable to adjust the proton permeability of the cytoplasmic membrane to allow a futile cycle of protons to operate. Consistently, the cytoplasmic membrane of *M. smegmatis* is known to be extremely impermeable to protons (Tran et al., 2005). The essentiality of F_1F_o-ATP synthase has also been demonstrated in several other Gram-positive pathogens, namely, *S. pneumoniae* (Ferrandiz & de la Campa, 2002), *Listeria monocytogenes* (Cotter, Gahan, & Hill, 2000), and *S. aureus* (Balemans et al., 2012). There is also evidence that the complex is essential in the Gram-negative *H. pylori* (McGowan, Cover, & Blaser, 1997). This may reflect the cellular ATP demands of this strictly respiratory pathogen. Alternatively, ATP-driven proton extrusion may contribute to the maintenance of intracellular pH homeostasis and a high ΔpH in its acidic stomach niche (Matin et al., 1996).

The F_1F_o-ATP synthase has thus only been shown experimentally to be essential for growth in a relatively small number of bacteria (Cotter et al., 2000; Koebmann, Nilsson, Kuipers, & Jensen, 2000; McGowan et al., 1997; Tran & Cook, 2005). This is an intriguing observation given the universal distribution of this enzyme in bacterial cells and its central role in cellular ATP homeostasis. This may reflect the ability of ATP generation by substrate-level phosphorylation to sustain the growth of most bacterial cells on fermentable carbon sources, where increased glycolytic flux can compensate for the loss of oxidative phosphorylation (Friedl et al., 1983). For organisms that can generate energy through fermentation, for example, *S. enterica* (mixed-acid and hydrogen fermentation; Sawers et al., 1986; Zbell & Maier, 2009) and *P. aeruginosa* (arginine and pyruvate fermentation; Williams et al., 2007), F_1F_o-ATP synthase is generally dispensable when fermentable carbon sources are available in the culture medium; hence, substrate-level phosphorylation is sufficient to maintain the basic needs of such cells. However, oxidative phosphorylation still has a major role in the pathogenesis of some of these facultative fermenters (Terry, Pina, & Mattingly, 1992; Turner et al., 2003); for example, strains of *S. enterica* containing deletions in *atpB* and *atpH* were highly attenuated in murine and galline models (Turner et al., 2003).

Whereas proton-coupled ATP synthases are widespread in prokaryotes, sodium-coupled enzymes are restricted to a few anaerobic bacterial genera (Dimroth & Cook, 2004). A sodium-coupled F_1F_o-ATP synthase has recently been reported in *F. nucleatum*, a Gram-negative bacterium implicated in the etiology of periodontal diseases (Schulz et al., 2013). We have shown that inhibiting this enzyme has a bactericidal effect, suggesting that it is essential for viability (M. Iglesias-Cans & G.M. Cook, unpublished data). Sodium-translocating V_1V_o-ATP synthases have also been characterised in *E. hirae* (Kakinuma et al., 1999) and predicted in *T. pallidum* (Fraser et al., 1998). While these pumps seem to primarily be involved in sodium extrusion, studies have not ruled out roles in ATP synthesis. *T. pallidum* encodes two such pumps that are speculated to differ in their directionality or ion selectivity; as previously discussed, Na^+-translocating oxaloacetate decarboxylase is the only predicted primary pump in this organism and hence *smf* is likely to drive ATP synthesis (Häse et al., 2001). Further work is clearly needed to determine if this bacterium and other pathogens harbour Na^+-coupled ATP synthases.

Taking these observations into account would argue against the F_1F_o-ATP synthase being a good broad-spectrum drug target for bacterial pathogens, but could be exploited in the small number of bacteria where the enzyme is essential for growth even on fermentable carbon sources. A large number of natural and synthetic inhibitors have been identified that target the F_1 and F_o motors of F_1F_o-ATP synthase (reviewed in Hong & Pedersen, 2008), but the majority of these inhibitors also target mitochondrial ATP synthase. A key to the druggability of bacterial ATP synthase will be the identification of bacterial-specific inhibitors and determining the essentiality/non-essentiality of this enzyme in bacterial growth and persistence. Diarylquinolines have been shown to target the F_1F_o-ATP synthase of mycobacterial species and inhibit

ATP synthesis (Andries et al., 2005; Koul et al., 2007, 2008; Lounis et al., 2010). Genome sequencing of both *M. tuberculosis* and *M. smegmatis* mutants that are resistant to diarylquinolines, i.e., bedaquiline (TMC207, BDQ), revealed that the target of these compounds is the oligomeric c-ring (encoded by *atpE*) of the enzyme (Andries et al., 2005; Huitric, Verhasselt, Andries, & Hoffner, 2007; Huitric et al., 2010). The purified c-ring from *M. smegmatis* binds BDQ with a K_D of 500 nM, and modeling/docking and kinetic studies suggest that BDQ blocks rotary movement of the c-ring during catalysis by mimicking key residues in the proton transfer chain (de Jonge, Koymans, Guillemont, Koul, & Andries, 2007; Haagsma et al., 2011). Further investigations with inverted membrane vesicles of *M. smegmatis* and BDQ have revealed that the drug acts independently of the *pmf* and that electrostatic forces play an important role in its interaction with the ATP synthase (Haagsma et al., 2011).

In 2012, the FDA approved the drug bedaquiline for the treatment of multidrug-resistant *M. tuberculosis*, the first drug licensed in 40 years for the treatment of tuberculosis disease. Bedaquiline exhibits *in vitro* bactericidal activity against mycobacterial strains that are either susceptible or resistant to all first-line and many second-line drugs and has extraordinary *in vivo* activity against several mycobacterial species in different animal models and in TB patients (Andries et al., 2005; Lounis et al., 2006, 2010). When mycobacterial cells (growing or non-growing) are treated with BDQ, time-dependent (not dose-dependent) killing is observed (Andries et al., 2005). The mechanism of killing is not clear, but it does not involve the dissipation of the membrane potential, which is lethal to all living cells. A dose-dependent decrease in intracellular ATP has been observed when *M. tuberculosis* cells are treated with BDQ (Koul et al., 2007, 2008), but these data do not explain cell death, because mycobacterial cells can be depleted of ATP and yet remain viable (Frampton, Aggio, Villas-Boas, Arcus, & Cook, 2012). BDQ is bactericidal towards most species of mycobacteria but is only bacteriostatic against *M. avium* (Lounis, Gevers, Van den Berg, Vranckx, & Andries, 2009). The identification of the mechanisms underlying this sensitivity will be important in understanding how BDQ exerts its anti-tuberculosis activity. Tran and Cook (2005) originally hypothesised that BDQ may act as a respiratory plug blocking ATP synthase, leading to hyperpolarisation of the *pmf* and backpressure on NADH turnover and proton pumping. Consistent with this hypothesis is recent data reporting that non-proton-translocating cytochrome *bd* oxidase (*cydAB*) is upregulated in response to BDQ challenge (Koul et al., 2014). The upregulation of *cydAB* expression is required for continued electron transport chain activity (and NADH turnover), as electron transport to this terminal oxidase is not impeded by a high *pmf*. In support of these data, we have shown that *cydAB* mutants of *M. smegmatis* show greater sensitivity to BDQ compared to wild-type cells (Robson & Cook, unpublished data). Further work is required to understand the precise mechanism of cell death caused by bedaquiline.

The notion that decreasing intracellular ATP levels in *M. tuberculosis* leads to cell death has directed scientists to perform high-throughout screens (600,000 compounds) targeting ATP homeostasis. These screens have uncovered 140 new compounds that reduce ATP levels in *M. tuberculosis*, among them the aforementioned Q203 that targets the cytochrome *bcc*–aa_3 supercomplex (Mak et al., 2012). There is also potential in administering bedaquiline and other inhibitors of ATP homeostasis for the treatment of other pathogens, notably drug-resistant *S. aureus* (Balemans et al., 2012) and *F. nucleatum* (M. Iglesias-Cans & G.M. Cook, unpublished data).

3. CONCLUSIONS AND FUTURE PERSPECTIVES

Table 2 summarises the major components of the respiratory chains of the key pathogens in this review. The discovery that a compound targeting a central enzyme in cellular energy generation found in both prokaryotes and eukaryotes would lead to the first drug (bedaquiline) licensed for tuberculosis in 40 years has inspired scientists to consider bacterial energy generation and energetics as a target space for antimicrobial development despite the risk of human toxicity. The future and development of bacterial energetics for antimicrobial discovery will be dependent on increased understanding of the energetic processes operating in bacterial pathogens in host tissues during both replication and persistence. For example, the ability of *S. enterica* to stimulate the host to produce an electron acceptor (tetrathionate) for anaerobic respiration to outgrow the fermenting microbiota (Winter et al., 2010) provides fascinating insight into the novel energetic strategies employed by bacterial pathogens. Important energetic machineries where no mammalian homologues exist like NDH-2, cytochrome *bd* oxidase, and hydrogenases are clearly attractive targets for drug development. Advancing these targets will be dependent on both identifying high-affinity specific inhibitors of these complexes and elucidating the role of these components in the target pathogen. The ability of the ATP synthase inhibitor bedaquiline to kill both replicating and non-replicating cells, resulting in shortened TB chemotherapy, highlights the importance of this area for drug discovery and expansion

Table 2 Summary of the organisation and flexibility of the respiratory chains of the key pathogens discussed in this review

Organism	Primary dehydrogenases and decarboxylases	Terminal oxidases and reductases	Redox carriers	ATP synthase
Escherichia coli	NADH dehydrogenase I NADH dehydrogenase II Na^+-NADH dehydrogenase Succinate dehydrogenase Formate dehydrogenases (2) Hydrogenase (2)	Cytochrome bo_3 oxidase Cytochrome *bd* oxidase Nitrate reductase (3) Nitrite reductase (2) Fumarate reductase DMSO reductase TMAO reductase	Ubiquinone Menaquinone Cytochrome *c*	H^+-F_1F_o
Salmonella enterica	NADH dehydrogenase I NADH dehydrogenase II Na^+-NADH dehydrogenase Succinate dehydrogenase Formate dehydrogenases (2) Hydrogenase (3)	Cytochrome bo_3 oxidase Cytochrome *bd* oxidase Nitrate reductase (3) Nitrite reductase (2) Fumarate reductase DMSO reductase TMAO reductase Tetrathionate reductase Thiosulphate reductase	Menaquinone Ubiquinone Cytochrome *c*	H^+-F_1F_o
Pseudomonas aeruginosa	NADH dehydrogenase I NADH dehydrogenase II Na^+-NADH dehydrogenase	Cytochrome cbb_3 oxidase (2) Cytochrome aa_3 oxidase Cytochrome bo_3 oxidase	Ubiquinone Cytochrome *c*	H^+-F_1F_o

Continued

Table 2 Summary of the organisation and flexibility of the respiratory chains of the key pathogens discussed in this review—cont'd

Organism	Primary dehydrogenases and decarboxylases	Terminal oxidases and reductases	Redox carriers	ATP synthase
	Succinate dehydrogenase Formate dehydrogenase	Cyanide-insensitive oxidase Nitrate reductase Nitrite reductase		
Neisseria gonorrhoeae *Neisseria meningtidis*	NADH dehydrogenase I Na^+-NADH dehydrogenase Succinate dehydrogenase Hydrogenase?	Cytochrome *cbb*$_3$ oxidase Nitrite reductase	Ubiquinone Cytochrome *c*	H^+-F_1F_o
Campylobacter jejuni	Flavodoxin dehydrogenase? Succinate dehydrogenase Formate dehydrogenase Hydrogenase Sulphite oxidase	Cytochrome *cbb*$_3$ oxidase Cyanide-insensitive oxidase Fumarate reductase Nitrate reductase Nitrite reductase Thiosulphate reductase DMSO reductase TMAO reductase	Menaquinone Cytochrome *c*	H^+-F_1F_o
Helicobacter pylori	Flavodoxin dehydrogenase? Hydrogenase	Cytochrome *cbb*$_3$ oxidase Fumarate reductase	Menaquinone Cytochrome *c*	H^+-F_1F_o
Staphylococcus aureus	NADH dehydrogenase II (2)	Cytochrome *aa*$_3$ oxidase	Menaquinone Cytochrome *c*	H^+-F_1F_o

Table 2 Summary of the organisation and flexibility of the respiratory chains of the key pathogens discussed in this review—cont'd

Organism	Primary dehydrogenases and decarboxylases	Terminal oxidases and reductases	Redox carriers	ATP synthase
	Succinate dehydrogenase Formate dehydrogenase	Cytochrome *bd* oxidase Nitrate reductase Nitrite reductase		
Mycobacterium tuberculosis	NADH dehydrogenase I NADH dehydrogenase II (2) Succinate dehydrogenase (2) Formate dehydrogenase? Carbon monoxide dehydrogenase	Cytochrome *bcc*–*aa*$_3$ complex Cytochrome *bd* oxidase Fumarate reductase Nitrate reductase	Menaquinone	H^+-F_1F_o
Fusobacterium nucleatum	Glutaconyl-CoA decarboxylase Ferredoxin-NAD^+ oxidoreductase	N/A	N/A	Na^+-F_1F_o
Treponema pallidum	Oxaloacetate decarboxylase	N/A	N/A	Na^+-V_1V_o (2)

Only chemiosmotic complexes are listed; hence, fermentation pathways are omitted. The number of isozymes present is shown in brackets when more than one. Some alternative electron transport chain components (e.g. proline dehydrogenase, lactate dehydrogenase, transhydrogenase, and cytochrome *c* peroxidise) were omitted from this review for the sake of conciseness and are not listed in this table.

of our antimicrobial armoury. There are now clinical opportunities to extend this approach to other human pathogens and electron transport chain components.

ACKNOWLEDGEMENTS

Research in the Cook Laboratory is supported by the Health Research Council and the Marsden Fund from the Royal Society of New Zealand. K. H. and C. G. were supported by an Otago Postgraduate Scholarship from the University of Otago. G. M. C. is supported by a James Cook Fellowship from the Royal Society of New Zealand.

REFERENCES

Abrahams, K. A., Cox, J. A., Spivey, V. L., Loman, N. J., Pallen, M. J., Constantinidou, C., et al. (2012). Identification of novel imidazo[1,2-*a*]pyridine inhibitors targeting *M. tuberculosis* QcrB. *PLoS One*, 7, e52951.

Abrahams, J. P., Leslie, A. G., Lutter, R., & Walker, J. E. (1994). Structure at 2.8 Å resolution of F_1-ATPase from bovine heart mitochondria. *Nature*, *370*, 621–628.

Akhtar, S., Khan, A., Sohaskey, C. D., Jagannath, C., & Sarkar, D. (2013). Nitrite reductase NirBD is induced and plays an important role during *in vitro* dormancy of *Mycobacterium tuberculosis*. *Journal of Bacteriology*, *195*, 4592–4599.

Alvarez-Ortega, C., & Harwood, C. S. (2007). Responses of *Pseudomonas aeruginosa* to low oxygen indicate that growth in the cystic fibrosis lung is by aerobic respiration. *Molecular Microbiology*, *65*, 153–165.

Amaral, L., Kristiansen, J. E., Abebe, L. S., & Millett, W. (1996). Inhibition of the respiration of multi-drug resistant clinical isolates of *Mycobacterium tuberculosis* by thioridazine: Potential use for initial therapy of freshly diagnosed tuberculosis. *Journal of Antimicrobial Chemotherapy*, *38*, 1049–1053.

Amaral, L., Viveiros, M., & Kristiansen, J. E. (2006). "Non-Antibiotics": Alternative therapy for the management of MDRTB and MRSA in economically disadvantaged countries. *Current Drug Targets*, 7, 887–891.

Andries, K., Verhasselt, P., Guillemont, J., Göhlmann, H. W. H., Neefs, J.-M., Winkler, H., et al. (2005). A diarylquinoline drug active on the ATP synthase of *Mycobacterium tuberculosis*. *Science*, *307*, 223–227.

Anjum, M. F., Stevanin, T. M., Read, R. C., & Moir, J. W. (2002). Nitric oxide metabolism in *Neisseria meningitidis*. *Journal of Bacteriology*, *184*, 2987–2993.

Arai, H. (2011). Regulation and function of versatile aerobic and anaerobic respiratory metabolism in *Pseudomonas aeruginosa*. *Frontiers in Microbiology*, *2*, 103.

Arai, H., Kodama, T., & Igarashi, Y. (1997). Cascade regulation of the two CRP/FNR-related transcriptional regulators (ANR and DNR) and the denitrification enzymes in *Pseudomonas aeruginosa*. *Molecular Microbiology*, *25*, 1141–1148.

Arya, S., Sethi, D., Singh, S., Hade, M. D., Singh, V., Raju, P., et al. (2013). Truncated hemoglobin, HbN, is post-translationally modified in *Mycobacterium tuberculosis* and modulates host–pathogen interactions during intracellular infection. *Journal of Biological Chemistry*, *288*, 29987–29999.

Aspholm, M., Aas, F. E., Harrison, O. B., Quinn, D., Vik, A., Viburiene, R., et al. (2010). Structural alterations in a component of cytochrome c oxidase and molecular evolution of pathogenic *Neisseria* in humans. *PLoS Pathogens*, *6*, e1001055.

Bader, M., Muse, W., Ballou, D. P., Gassner, C., & Bardwell, J. C. (1999). Oxidative protein folding is driven by the electron transport system. *Cell*, *98*, 217–227.

Baek, S.-H., Li, A. H., & Sassetti, C. M. (2011). Metabolic regulation of mycobacterial growth and antibiotic sensitivity. *PLoS Biology*, *9*, e1001065.

Bakken, V., Hogh, B. T., & Jensen, H. B. (1989). Utilization of amino acids and peptides by *Fusobacterium nucleatum*. *Scandinavian Journal of Dental Research*, *97*, 43–53.

Balemans, W., Vranckx, L., Lounis, N., Pop, O., Guillemont, J., Vergauwen, K., et al. (2012). Novel antibiotics targeting respiratory ATP synthesis in Gram-positive pathogenic bacteria. *Antimicrobial Agents and Chemotherapy*, *56*, 4131–4139.

Baradaran, R., Berrisford, J. M., Minhas, G. S., & Sazanov, L. A. (2013). Crystal structure of the entire respiratory complex I. *Nature*, *494*, 443–448.

Barker, H. C., Kinsella, N., Jaspe, A., Friedrich, T., & O'Connor, C. D. (2000). Formate protects stationary-phase Escherichia coli and Salmonella cells from killing by a cationic antimicrobial peptide. *Molecular Microbiology*, *35*, 1518–1529.

Batista, A. P., Marreiros, B. C., & Pereira, M. M. (2012). The role of proton and sodium ions in energy transduction by respiratory complex I. *IUBMB Life*, *64*, 492–498.
Baughn, A. D., & Malamy, M. H. (2004). The strict anaerobe *Bacteroides fragilis* grows in and benefits from nanomolar concentrations of oxygen. *Nature*, *427*, 441–444.
Beatrix, B., Bendrat, K., Rospert, S., & Buckel, W. (1990). The biotin-dependent sodium ion pump glutaconyl-CoA decarboxylase from *Fusobacterium nucleatum* (subsp. *nucleatum*). Comparison with the glutaconyl-CoA decarboxylases from gram-positive bacteria. *Archives of Microbiology*, *154*, 362–369.
Bebbington, K. J., & Williams, H. D. (1993). Investigation of the role of the *cydD* gene product in production of a functional cytochrome *d* oxidase in *Escherichia coli*. *FEMS Microbiology Letters*, *112*, 19–24.
Becker, D., Selbach, M., Rollenhagen, C., Ballmaier, M., Meyer, T. F., Mann, M., et al. (2006). Robust *Salmonella* metabolism limits possibilities for new antimicrobials. *Nature*, *440*, 303–307.
Begum, A., Rahman, M. M., Ogawa, W., Mizushima, T., Kuroda, T., & Tsuchiya, T. (2005). Gene cloning and characterization of four MATE family multidrug efflux pumps from *Vibrio cholerae* Non-O1. *Microbiology and Immunology*, *49*, 949–957.
Benoit, S. L., & Maier, R. J. (2008). Hydrogen and nickel metabolism in *Helicobacter* species. *Annals of the New York Academy of Sciences*, *1125*, 242–251.
Benoit, S. L., & Maier, R. J. (2014). Twin-arginine translocation system in *Helicobacter pylori*: TatC, but not TatB, is essential for viability. *mBio*, *5*, e01016-13.
Bereswill, S., Fischer, A., Plickert, R., Haag, L. M., Otto, B., Kuhl, A. A., et al. (2011). Novel murine infection models provide deep insights into the "menage a trois" of *Campylobacter jejuni*, microbiota and host innate immunity. *PLoS One*, *6*, e20953.
Berney, M., & Cook, G. M. (2010). Unique flexibility in energy metabolism allows mycobacteria to combat starvation and hypoxia. *PLoS One*, *5*, e8614.
Berney, M., Greening, C., Hards, K., Collins, D., & Cook, G. M. (2014). Three different [NiFe] hydrogenases confer metabolic flexibility in the obligate aerobe *Mycobacterium smegmatis*. *Environmental Microbiology*, *16*, 318–330.
Berney, M., Greening, C., Conrad, R., Jacobs W. R., Jr., Cook, G. M. (2014). An obligately aerobic soil bacterium activates fermentative hydrogen production to survive reductive stress during hypoxia. *Proceedings of the National Academy of Sciences of the United States of America*, *111*, 11479–11484.
Bertsova, Y. V., & Bogachev, A. V. (2004). The origin of the sodium-dependent NADH oxidation by the respiratory chain of *Klebsiella pneumoniae*. *FEBS Letters*, *563*, 207–212.
Betts, J. C., Lukey, P. T., Robb, L. C., McAdam, R. A., & Duncan, K. (2002). Evaluation of a nutrient starvation model of *Mycobacterium tuberculosis* persistence by gene and protein expression profiling. *Molecular Microbiology*, *43*, 717–731.
Biagini, G. A., Viriyavejakul, P., O'Neill, P. M., Bray, P. G., & Ward, S. A. (2006). Functional characterization and target validation of alternative complex I of *Plasmodium falciparum* mitochondria. *Antimicrobial Agents and Chemotherapy*, *50*, 1841–1851.
Birkholz, S., & Knipp, U. (1994). Fumarate reductase of *Helicobacter pylori*—An immunogenic protein. *Journal of Medical Microbiology*, *41*, 56–62.
Bliska, J. B., & van der Velden, A. W. (2012). Salmonella "sops" up a preferred electron receptor in the inflamed intestine. *mBio*, *3*, e00226-12.
Borisov, V. B., Forte, E., Sarti, P., & Giuffre, A. (2011). Catalytic intermediates of cytochrome *bd* terminal oxidase at steady-state: Ferryl and oxy-ferrous species dominate. *Biochimica et Biophysica Acta*, *1807*, 503–509.
Borisov, V. B., Gennis, R. B., Hemp, J., & Verkhovsky, M. I. (2011). The cytochrome *bd* respiratory oxygen reductases. *Biochimica et Biophysica Acta*, *1807*, 1398–1413.

Borisov, V., Gennis, R., & Konstantinov, A. A. (1995). Peroxide complex of cytochrome *bd*: Kinetics of generation and stability. *International Journal of Biochemistry and Molecular Biology*, *37*, 975–982.

Bott, M., Meyer, M., & Dimroth, P. (1995). Regulation of anaerobic citrate metabolism in *Klebsiella pneumoniae*. *Molecular Microbiology*, *18*, 533–546.

Boucher, H. W., Talbot, G. H., Benjamin, D. K., Jr., Bradley, J., Guidos, R. J., Jones, R. N., et al. (2013). 10 × 20 Progress—Development of new drugs active against gram-negative bacilli: An update from the Infectious Diseases Society of America. *Clinical Infectious Diseases*, *56*, 1685–1694.

Bowman, L., Flanagan, L., Fyfe, P. K., Parkin, A., Hunter, W. N., & Sargent, F. (2014). How the structure of the large subunit controls function in an oxygen-tolerant [NiFe]-hydrogenase. *Biochemical Journal*, *458*, 449–458.

Brown, M. H., Paulsen, I. T., & Skurray, R. A. (1999). The multidrug efflux protein NorM is a prototype of a new family of transporters. *Molecular Microbiology*, *31*, 394–395.

Buckel, W. (2001). Sodium ion-translocating decarboxylases. *Biochimica et Biophysica Acta*, *1505*, 15–27.

Buckel, W., & Thauer, R. K. (2013). Energy conservation via electron bifurcating ferredoxin reduction and proton/Na^+ translocating ferredoxin oxidation. *Biochimica et Biophysica Acta*, *1827*, 94–113.

Burke, K. A., & Lascelles, J. (1975). Nitrate reductase system in *Staphylococcus aureus* wild type and mutants. *Journal of Bacteriology*, *123*, 308–316.

Buschmann, S., Warkentin, E., Xie, H., Langer, J. D., Ermler, U., & Michel, H. (2010). The structure of cbb_3 cytochrome oxidase provides insights into proton pumping. *Science*, *329*, 327–330.

Calhoun, M. W., Oden, K. L., Gennis, R. B., de Mattos, M. J., & Neijssel, O. M. (1993). Energetic efficiency of *Escherichia coli*: Effects of mutations in components of the aerobic respiratory chain. *Journal of Bacteriology*, *175*, 3020–3025.

Cardinale, J. A., & Clark, V. L. (2005). Determinants of nitric oxide steady-state levels during anaerobic respiration by *Neisseria gonorrhoeae*. *Molecular Microbiology*, *58*, 177–188.

Chen, M., Andersen, L. P., Zhai, L., & Kharazmi, A. (1999). Characterization of the respiratory chain of *Helicobacter pylori*. *FEMS Immunology and Medical Microbiology*, *24*, 169–174.

Chen, M., Jensen, B., Zhai, L., Colding, H., Kharazmi, A., Kristiansen, J. E., et al. (2002). Nizatidine and omeprazole enhance the effect of metronidazole on *Helicobacter pylori in vitro*. *International Journal of Antimicrobial Agents*, *19*, 195–200.

Chen, F., Xia, Q., & Ju, L. K. (2003). Aerobic denitrification of *Pseudomonas aeruginosa* monitored by online NAD(P)H fluorescence. *Applied and Environmental Microbiology*, *69*, 6715–6722.

Cingolani, G., & Duncan, T. M. (2011). Structure of the ATP synthase catalytic complex (F_1) from *Escherichia coli* in an autoinhibited conformation. *Nature Structural & Molecular Biology*, *18*, 701–707.

Cole, S. T., Eiglmeier, K., Parkhill, J., James, K. D., Thomson, N. R., Wheeler, P. R., et al. (2001). Massive gene decay in the leprosy bacillus. *Nature*, *409*, 1007–1011.

Collins, M. D., & Jones, D. (1981). Distribution of isoprenoid quinone structural types in distribution of isoprenoid quinone structural types in bacteria and their taxonomic implications. *Microbiological Reviews*, *45*, 316–354.

Comolli, J. C., & Donohue, T. J. (2004). Differences in two *Pseudomonas aeruginosa* cbb_3 cytochrome oxidases. *Molecular Microbiology*, *51*, 1193–1203.

Contreras, M., Thiberge, J. M., Mandrand-Berthelot, M. A., & Labigne, A. (2003). Characterization of the roles of NikR, a nickel-responsive pleiotropic autoregulator of *Helicobacter pylori*. *Molecular Microbiology*, *49*, 947–963.

Cook, G. M., Hards, K., Vilchèze, C., Hartman, T., & Berney, M. (2014). Energetics of respiration and oxidative phosphorylation in mycobacteria. In G. Hatfull, & W. R. Jacobs, Jr. (Eds.), *Molecular genetics of mycobacteria*. Washington, DC: ASM Press.

Cotter, P. A., Chepuri, V., Gennis, R. B., & Gunsalus, R. P. (1990). Cytochrome *o* (*cyoABCDE*) and *d* (*cydAB*) oxidase gene expression in *Escherichia coli* is regulated by oxygen, pH, and the *fnr* gene product. *Journal of Bacteriology*, *172*, 6333–6338.

Cotter, P. D., Gahan, C. G., & Hill, C. (2000). Analysis of the role of the *Listeria monocytogenes* F_0F_1-ATPase operon in the acid tolerance response. *International Journal of Food Microbiology*, *60*, 137–146.

Cotter, P. A., Melville, S. B., Albrecht, J. A., & Gunsalus, R. P. (1997). Aerobic regulation of cytochrome *d* oxidase (*cydAB*) operon expression in Escherichia coli: Roles of Fnr and ArcA in repression and activation. *Molecular Microbiology*, *25*, 605–615.

Cox, R. A., & Cook, G. M. (2007). Growth regulation in the mycobacterial cell. *Current Molecular Medicine*, *7*, 231–245.

Cunningham, L., Pitt, M., & Williams, H. D. (1997). The *cioAB* genes from *Pseudomonas aeruginosa* code for a novel cyanide-insensitive terminal oxidase related to the cytochrome *bd* quinol oxidases. *Molecular Microbiology*, *24*, 579–591.

da Silva, S. M., Venceslau, S. S., Fernandes, C. L., Valente, F. M., & Pereira, I. A. (2008). Hydrogen as an energy source for the human pathogen *Bilophila wadsworthia*. *Antonie Van Leeuwenhoek*, *93*, 381–390.

de Jonge, M. R., Koymans, L. H., Guillemont, J. E., Koul, A., & Andries, K. (2007). A computational model of the inhibition of *Mycobacterium tuberculosis* ATPase by a new drug candidate R207910. *Proteins*, *67*, 971–980.

Debnath, J., Siricilla, S., Wan, B., Crick, D. C., Lenaerts, A. J., Franzblau, S. G., et al. (2012). Discovery of selective menaquinone biosynthesis inhibitors against *Mycobacterium tuberculosis*. *Journal of Medicinal Chemistry*, *55*, 3739–3755.

Deguchi, Y., Yamato, I., & Anraku, Y. (1990). Nucleotide sequence of *gltS*, the Na^+/glutamate symport carrier gene of *Escherichia coli* B. *Journal of Biological Chemistry*, *265*, 21704–21708.

Dhiman, R. K., Mahapatra, S., Slayden, R. A., Boyne, M. E., Lenaerts, A., Hinshaw, J. C., et al. (2009). Menaquinone synthesis is critical for maintaining mycobacterial viability during exponential growth and recovery from non-replicating persistence. *Molecular Microbiology*, *72*, 85–97.

Di Berardino, M., & Dimroth, P. (1996). Aspartate 203 of the oxaloacetate decarboxylase beta-subunit catalyses both the chemical and vectorial reaction of the Na^+ pump. *EMBO Journal*, *15*, 1842–1849.

Dimroth, P. (1980). A new sodium-transport system energized by the decarboxylation of oxaloacetate. *FEBS Letters*, *122*, 234–236.

Dimroth, P. (1994). Bacterial sodium ion-coupled energetics. *Antonie Van Leeuwenhoek*, *65*, 381–395.

Dimroth, P., & Cook, G. M. (2004). Bacterial Na^+- or H^+-coupled ATP synthases operating at low electrochemical potential. *Advances in Microbial Physiology*, *49*, 175–218.

Dimroth, P., & Thomer, A. (1993). On the mechanism of sodium ion translocation by oxaloacetate decarboxylase of *Klebsiella pneumoniae*. *Biochemistry*, *32*, 1734–1739.

D'Mello, R., Hill, S., & Poole, R. K. (1995). The oxygen affinity of cytochrome *bo*' in *Escherichia coli* determined by the deoxygenation of oxyleghemoglobin and oxymyoglobin: K_m values for oxygen are in the submicromolar range. *Journal of Bacteriology*, *177*, 867–870.

D'Mello, R., Hill, S., & Poole, R. K. (1996). The cytochrome *bd* quinol oxidase in *Escherichia coli* has an extremely high oxygen affinity and two oxygen-binding haems: Implications for regulation of activity *in vivo* by oxygen inhibition. *Microbiology*, *142*, 755–763.

Doering, C., Ermentrout, B., & Oster, G. (1995). Rotary DNA motors. *Biophysical Journal, 69*, 2256–2267.
Encheva, V., Shah, H. N., & Gharbia, S. E. (2009). Proteomic analysis of the adaptive response of *Salmonella enterica* serovar Typhimurium to growth under anaerobic conditions. *Microbiology, 155*, 2429–2441.
Endley, S., McMurray, D., & Ficht, T. A. (2001). Interruption of the *cydB* locus in *Brucella abortus* attenuates intracellular survival and virulence in the mouse model of infection. *Journal of Bacteriology, 183*, 2454–2462.
Eoh, H., & Rhee, K. Y. (2013). Multifunctional essentiality of succinate metabolism in adaptation to hypoxia in *Mycobacterium tuberculosis*. *Proceedings of the National Academy of Sciences of the United States of America, 110*, 6554–6559.
Farha, M. A., Verschoor, C. P., Bowdish, D., & Brown, E. D. (2013). Collapsing the proton motive force to identify synergistic combinations against *Staphylococcus aureus*. *Chemistry & Biology, 20*, 1168–1178.
Ferrandiz, M. J., & de la Campa, A. G. (2002). The membrane-associated F_0F_1 ATPase is essential for the viability of *Streptococcus pneumoniae*. *FEMS Microbiology Letters, 212*, 133–138.
Filiatrault, M. J., Wagner, V. E., Bushnell, D., Haidaris, C. G., Iglewski, B. H., & Passador, L. (2005). Effect of anaerobiosis and nitrate on gene expression in *Pseudomonas aeruginosa*. *Infection and Immunity, 73*, 3764–3772.
Fink, R. C., Evans, M. R., Porwollik, S., Vazquez-Torres, A., Jones-Carson, J., Troxell, B., et al. (2007). FNR is a global regulator of virulence and anaerobic metabolism in *Salmonella enterica* serovar *Typhimurium* (ATCC 14028s). *Journal of Bacteriology, 189*, 2262–2273.
Frampton, R., Aggio, R. B., Villas-Boas, S. G., Arcus, V. L., & Cook, G. M. (2012). Toxin-antitoxin systems of *Mycobacterium smegmatis* are essential for cell survival. *Journal of Biological Chemistry, 287*, 5340–5356.
Fraser, C. M., Norris, S. J., Weinstock, G. M., White, O., Sutton, G. G., Dodson, R., et al. (1998). Complete genome sequence of *Treponema pallidum*, the syphilis spirochete. *Science, 281*, 375–388.
Friedl, P., Hoppe, J., Gunsalus, R. P., Michelsen, O., von Meyenburg, K., & Schairer, H. U. (1983). Membrane integration and function of the three F_0 subunits of the ATP synthase of *Escherichia coli* K12. *The EMBO Journal, 2*, 99–103.
Fuchs, S., Pane-Farre, J., Kohler, C., Hecker, M., & Engelmann, S. (2007). Anaerobic gene expression in *Staphylococcus aureus*. *Journal of Bacteriology, 189*, 4275–4289.
Galagan, J. E., Minch, K., Peterson, M., Lyubetskaya, A., Azizi, E., Sweet, L., et al. (2013). The *Mycobacterium tuberculosis* regulatory network and hypoxia. *Nature, 499*, 178–183.
Gaupp, R., Schlag, S., Liebeke, M., Lalk, M., & Götz, F. (2010). Advantage of upregulation of succinate dehydrogenase in *Staphylococcus aureus* biofilms. *Journal of Bacteriology, 192*, 2385–2394.
Ge, Z., Feng, Y., Dangler, C. A., Xu, S., Taylor, N. S., & Fox, J. G. (2000). Fumarate reductase is essential for *Helicobacter pylori* colonization of the mouse stomach. *Microbial Pathogenesis, 29*, 279–287.
Ge, Z., Jiang, Q., Kalisiak, M. S., & Taylor, D. E. (1997). Cloning and functional characterization of *Helicobacter pylori* fumarate reductase operon comprising three structural genes coding for subunits C, A and B. *Gene, 204*, 227–234.
Gilberthorpe, N. J., & Poole, R. K. (2008). Nitric oxide homeostasis in *Salmonella typhimurium*: Roles of respiratory nitrate reductase and flavohemoglobin. *Journal of Biological Chemistry, 283*, 11146–11154.
Giuffre, A., Borisov, V. B., Arese, M., Sarti, P., & Forte, E. (2014). Cytochrome *bd* oxidase and bacterial tolerance to oxidative and nitrosative stress. *Biochimica et Biophysica Acta, 1837*, 1178–1187.

Giuffre, A., Borisov, V. B., Mastronicola, D., Sarti, P., & Forte, E. (2012). Cytochrome *bd* oxidase and nitric oxide: From reaction mechanisms to bacterial physiology. *FEBS Letters*, *586*, 622–629.

Glasser, N. R., Kern, S. E., & Newman, D. K. (2014). Phenazine redox cycling enhances anaerobic survival in *Pseudomonas aeruginosa* by facilitating generation of ATP and a proton-motive force. *Molecular Microbiology*, *92*, 399–412.

Goldman, B. S., Gabbert, K. K., & Kranz, R. G. (1996). The temperature-sensitive growth and survival phenotypes of Escherichia coli *cydDC* and *cydAB* strains are due to deficiencies in cytochrome *bd* and are corrected by exogenous catalase and reducing agents. *Journal of Bacteriology*, *178*, 6348–6351.

Goldstein, E. J., Citron, D. M., Peraino, V. A., & Cross, S. A. (2003). *Desulfovibrio desulfuricans* bacteremia and review of human *Desulfovibrio* infections. *Journal of Clinical Microbiology*, *41*, 2752–2754.

Greening, C., & Cook, G. M. (2014). Integration of hydrogenase expression and hydrogen sensing in bacterial cell physiology. *Current Opinion in Microbiology*, *18*, 30–38.

Griffin, J. E., Gawronski, J. D., Dejesus, M. A., Ioerger, T. R., Akerley, B. J., & Sassetti, C. M. (2011). High-resolution phenotypic profiling defines genes essential for mycobacterial growth and cholesterol catabolism. *PLoS Pathogens*, *7*, e1002251.

Griffin, J. E., Pandey, A. K., Gilmore, S. A., Mizrahi, V., McKinney, J. D., Bertozzi, C. R., et al. (2012). Cholesterol catabolism by *Mycobacterium tuberculosis* requires transcriptional and metabolic adaptations. *Chemistry & Biology*, *19*, 218–227.

Guccione, E., Hitchcock, A., Hall, S. J., Mulholland, F., Shearer, N., van Vliet, A. H. M., et al. (2010). Reduction of fumarate, mesaconate and crotonate by Mfr, a novel oxygen-regulated periplasmic reductase in *Campylobacter jejuni*. *Environmental Microbiology*, *12*, 576–591.

Guest, J. R. (1981). Partial replacement of succinate dehydrogenase function by phage- and plasmid-specified fumarate reductase in *Escherichia coli*. *Microbiology*, *122*, 171–179.

Gusarov, I., & Nudler, E. (2005). NO-mediated cytoprotection: Instant adaptation to oxidative stress in bacteria. *Proceedings of the National Academy of Sciences of the United States of America*, *102*, 13855–13860.

Gusarov, I., Shatalin, K., Starodubtseva, M., & Nudler, E. (2009). Endogenous nitric oxide protects bacteria against a wide spectrum of antibiotics. *Science*, *325*, 1380–1384.

Haagsma, A. C., Podasca, I., Koul, A., Andries, K., Guillemont, J., Lill, H., et al. (2011). Probing the interaction of the diarylquinoline TMC207 with its target mycobacterial ATP synthase. *PLoS One*, *6*, e23575.

Hägerhäll, C., & Hederstedt, L. (1996). A structural model for the membrane-integral domain of succinate: Quinone oxidoreductases. *FEBS Letters*, *389*, 25–31.

Hall, H. K., Karem, K. L., & Foster, J. W. (1995). Molecular responses of microbes to environmental pH stress. *Advances in Microbial Physiology*, *37*, 229–272.

Hammer, N. D., Reniere, M. L., Cassat, J. E., Zhang, Y., Hirsch, A. O., Indriati Hood, M., et al. (2013). Two heme-dependent terminal oxidases power *Staphylococcus aureus* organ-specific colonization of the vertebrate host. *mBio*, *4*, e00241-13.

Häse, C. C., & Barquera, B. (2001). Role of sodium bioenergetics in *Vibrio cholerae*. *Biochimica et Biophysica Acta*, *1505*, 169–178.

Häse, C. C., Fedorova, N. D., Galperin, M. Y., & Dibrov, P. A. (2001). Sodium ion cycle in bacterial pathogens: Evidence from cross-genome comparisons. *Microbiology and Molecular Biology Reviews*, *65*, 353–370.

Häse, C. C., & Mekalanos, J. J. (1999). Effects of changes in membrane sodium flux on virulence gene expression in *Vibrio cholerae*. *Proceedings of the National Academy of Sciences of the United States of America*, *96*, 3183–3187.

Hayashi, M., Shibata, N., Nakayama, Y., Yoshikawa, K., & Unemoto, T. (2002). Korormicin insensitivity in *Vibrio alginolyticus* is correlated with a single point mutation of Gly-140 in the NqrB subunit of the Na^+-translocating NADH-quinone reductase. *Archives of Biochemistry and Biophysics, 401*, 173–177.

He, H., Bretl, D. J., Penoske, R. M., Anderson, D. M., & Zahrt, T. C. (2011). Components of the Rv0081–Rv0088 locus, which encodes a predicted formate hydrogenlyase complex, are coregulated by Rv0081, MprA, and DosR in *Mycobacterium tuberculosis*. *Journal of Bacteriology, 193*, 5105–5118.

Hederstedt, L. (1999). Respiration without O_2. *Science, 284*, 1941–1942.

Heikal, A., Nakatani, Y., Dunn, E., Weimar, M. R., Day, C. L., Baker, E. N., et al. (2014). Structure of the bacterial type II NADH dehydrogenase: A monotopic membrane protein with an essential role in energy generation. *Molecular Microbiology, 91*, 950–964.

Hensel, M., Hinsley, A. P., Nikolaus, T., Sawers, G., & Berks, B. C. (1999). The genetic basis of tetrathionate respiration in *Salmonella typhimurium*. *Molecular Microbiology, 32*, 275–287.

Herrmann, G., Jayamani, E., Mai, G., & Buckel, W. (2008). Energy conservation via electron-transferring flavoprotein in anaerobic bacteria. *Journal of Bacteriology, 190*, 784–791.

Herz, K., Vimont, S., Padan, E., & Berche, P. (2003). Roles of NhaA, NhaB, and NhaD Na^+/H^+ antiporters in survival of *Vibrio cholerae* in a saline environment. *Journal of Bacteriology, 185*, 1236–1244.

Hildebrandt, T. M., & Grieshaber, M. K. (2008). Three enzymatic activities catalyze the oxidation of sulfide to thiosulfate in mammalian and invertebrate mitochondria. *FEBS Journal, 275*, 3352–3361.

Hiratsuka, T., Furihata, K., Ishikawa, J., Yamashita, H., Itoh, N., Seto, H., et al. (2008). An alternative menaquinone biosynthetic pathway operating in microorganisms. *Science, 321*, 1670–1673.

Hoffman, P. S., & Goodman, T. G. (1982). Respiratory physiology and energy conservation efficiency of *Campylobacter jejuni*. *Journal of Bacteriology, 150*, 319–326.

Holden, J. K., Li, H., Jing, Q., Kang, S., Richo, J., Silverman, R. B., et al. (2013). Structural and biological studies on bacterial nitric oxide synthase inhibitors. *Proceedings of the National Academy of Sciences of the United States of America, 110*, 18127–18131.

Hong, S., & Pedersen, P. L. (2008). ATP synthase and the actions of inhibitors utilized to study its roles in human health, disease, and other scientific areas. *Microbiology and Molecular Biology Reviews, 72*, 590–641.

Hopper, A., Tovell, N., & Cole, J. (2009). A physiologically significant role in nitrite reduction of the CcoP subunit of the cytochrome oxidase cbb_3 from *Neisseria gonorrhoeae*. *FEMS Microbiology Letters, 301*, 232–240.

Householder, T. C., Fozo, E. M., Cardinale, J. A., & Clark, V. L. (2000). Gonococcal nitric oxide reductase is encoded by a single gene, *norB*, which is required for anaerobic growth and is induced by nitric oxide. *Infection and Immunity, 68*, 5241–5246.

Huang, Y., Suyemoto, M., Garner, C. D., Cicconi, K. M., & Altier, C. (2008). Formate acts as a diffusible signal to induce *Salmonella* invasion. *Journal of Bacteriology, 190*, 4233–4241.

Huda, M. N., Morita, Y., Kuroda, T., Mizushima, T., & Tsuchiya, T. (2001). Na^+-driven multidrug efflux pump VcmA from *Vibrio cholerae* Non-O1, a non-halophilic bacterium. *FEMS Microbiology Letters, 203*, 235–239.

Huitric, E., Verhasselt, P., Andries, K., & Hoffner, S. E. (2007). In vitro antimycobacterial spectrum of a diarylquinoline ATP synthase inhibitor. *Antimicrobial Agents and Chemotherapy, 51*, 4202–4204.

Huitric, E., Verhasselt, P., Koul, A., Andries, K., Hoffner, S., & Andersson, D. I. (2010). Rates and mechanisms of resistance development in *Mycobacterium tuberculosis* to a novel diarylquinoline ATP synthase inhibitor. *Antimicrobial Agents and Chemotherapy, 54*, 1022–1028.

Hurdle, J. G., O'Neill, A. J., Chopra, I., & Lee, R. E. (2011). Targeting bacterial membrane function: An underexploited mechanism for treating persistent infections. *Nature Reviews. Microbiology, 9*, 62–75.

Iuchi, S., Chepuri, V., Fu, H. A., Gennis, R. B., & Lin, E. C. (1990). Requirement for terminal cytochromes in generation of the aerobic signal for the *arc* regulatory system in *Escherichia coli*: Study utilizing deletions and *lac* fusions of *cyo* and *cyd*. *Journal of Bacteriology, 172*, 6020–6025.

Jackson, R. J., Elvers, K. T., Lee, L. J., Gidley, M. D., Wainwright, L. M., Lightfoot, J., et al. (2007). Oxygen reactivity of both respiratory oxidases in *Campylobacter jejuni*: The *cydAB* genes encode a cyanide-resistant, low-affinity oxidase that is not of the cytochrome *bd* type. *Journal of Bacteriology, 189*, 1604–1615.

Jamieson, D. J., Sawers, R. G., Rugman, P. A., Boxer, D. H., & Higgins, C. F. (1986). Effects of anaerobic regulatory mutations and catabolite repression on regulation of hydrogen metabolism and hydrogenase isoenzyme composition in *Salmonella typhimurium*. *Journal of Bacteriology, 168*, 405–411.

Jewes, L. A., & Spencer, R. C. (1990). The incidence of anaerobes in the sputum of patients with cystic fibrosis. *Journal of Medical Microbiology, 31*, 271–274.

Jormakka, M., Byrne, B., & Iwata, S. (2003a). Formate dehydrogenase—A versatile enzyme in changing environments. *Current Opinion in Structural Biology, 13*, 418–423.

Jormakka, M., Byrne, B., & Iwata, S. (2003b). Protonmotive force generation by a redox loop mechanism. *FEBS Letters, 545*, 25–30.

Jormakka, M., Tornroth, S., Byrne, B., & Iwata, S. (2002). Molecular basis of proton motive force generation: Structure of formate dehydrogenase-N. *Science, 295*, 1863–1868.

Jorth, P., Trivedi, U., Rumbaugh, K., & Whiteley, M. (2013). Probing bacterial metabolism during infection using high-resolution transcriptomics. *Journal of Bacteriology, 195*, 4991–4998.

Juarez, O., & Barquera, B. (2012). Insights into the mechanism of electron transfer and sodium translocation of the Na^+-pumping NADH:quinone oxidoreductase. *Biochimica et Biophysica Acta, 1817*, 1823–1832.

Junemann, S. (1997). Cytochrome *bd* terminal oxidase. *Biochimica et Biophysica Acta, 1321*, 107–127.

Jung, J. Y., Madan-Lala, R., Georgieva, M., Rengarajan, J., Sohaskey, C. D., Bange, F. C., et al. (2013). The intracellular environment of human macrophages that produce nitric oxide promotes growth of mycobacteria. *Infection and Immunity, 81*, 3198–3209.

Kakinuma, Y., Yamato, I., & Murata, T. (1999). Structure and function of vacuolar Na^+-translocating ATPase in *Enterococcus hirae*. *Journal of Bioenergetics and Biomembranes, 31*, 7–14.

Kana, B. D., Weinstein, E. A., Avarbock, D., Dawes, S. S., Rubin, H., & Mizrahi, V. (2001). Characterization of the *cydAB*-encoded cytochrome *bd* oxidase from *Mycobacterium smegmatis*. *Journal of Bacteriology, 183*, 7076–7086.

Kang, S., Kim, R. Y., Seo, M. J., Lee, S., Kim, Y. M., Seo, M., et al. (2014). Lead optimization of a novel series of imidazo[1,2-a]pyridine amides leading to a clinical candidate (Q203) as a multi- and extensively-drug resistant antituberculosis agent. *Journal of Medicinal Chemistry, 57*(12), 5293–5305.

Kapatral, V., Anderson, I., Ivanova, N., Reznik, G., Los, T., Lykidis, A., et al. (2002). Genome sequence and analysis of the oral bacterium *Fusobacterium nucleatum* strain ATCC 25586. *Journal of Bacteriology, 184*, 2005–2018.

Kapralek, F. (1972). The physiological role of tetrathionate respiration in growing *Citrobacter*. *Journal of General Microbiology, 71*, 133–139.

Kawakami, T., Kuroki, M., Ishii, M., Igarashi, Y., & Arai, H. (2010). Differential expression of multiple terminal oxidases for aerobic respiration in *Pseudomonas aeruginosa*. *Environmental Microbiology, 12*, 1399–1412.

Kerscher, S., Drose, S., Zickermann, V., & Brandt, U. (2008). The three families of respiratory NADH dehydrogenases. *Results and Problems in Cell Differentiation, 45*, 185–222.

King, G. M. (2003). Uptake of carbon monoxide and hydrogen at environmentally relevant concentrations by mycobacteria. *Applied and Environmental Microbiology, 69*, 7266–7272.

Knapp, J. S., & Clark, V. L. (1984). Anaerobic growth of *Neisseria gonorrhoeae* coupled to nitrite reduction. *Infection and Immunity, 46*, 176–181.

Knuth, K., Niesalla, H., Hueck, C. J., & Fuchs, T. M. (2004). Large-scale identification of essential *Salmonella* genes by trapping lethal insertions. *Molecular Microbiology, 51*, 1729–1744.

Knuuti, J., Belevich, G., Sharma, V., Bloch, D. A., & Verkhovskaya, M. (2013). A single amino acid residue controls ROS production in the respiratory complex I from *Escherichia coli*. *Molecular Microbiology, 90*, 1190–1200.

Koebmann, B. J., Nilsson, D., Kuipers, O. P., & Jensen, P. R. (2000). The membrane-bound H^+-ATPase complex is essential for growth of *Lactococcus lactis*. *Journal of Bacteriology, 182*, 4738–4743.

Kojima, S., Yamamoto, K., Kawagishi, I., & Homma, M. (1999). The polar flagellar motor of *Vibrio cholerae* is driven by an Na^+ motive force. *Journal of Bacteriology, 181*, 1927–1930.

Kosono, S., Haga, K., Tomizawa, R., Kajiyama, Y., Hatano, K., Takeda, S., et al. (2005). Characterization of a multigene-encoded sodium/hydrogen antiporter (sha) from *Pseudomonas aeruginosa*: Its involvement in pathogenesis. *Journal of Bacteriology, 187*, 5242–5248.

Koul, A., Dendouga, N., Vergauwen, K., Molenberghs, B., Vranckx, L., Willebrords, R., et al. (2007). Diarylquinolines target subunit c of mycobacterial ATP synthase. *Nature Chemical Biology, 3*, 323–324.

Koul, A., Vranckx, L., Dendouga, N., Balemans, W., Van den Wyngaert, I., Vergauwen, K., et al. (2008). Diarylquinolines are bactericidal for dormant mycobacteria as a result of disturbed ATP homeostasis. *Journal of Biological Chemistry, 283*, 25273–25280.

Koul, A., Vranckx, L., Dhar, N., Gohlmann, H. W., Ozdemir, E., Neefs, J. M., et al. (2014). Delayed bactericidal response of *Mycobacterium tuberculosis* to bedaquiline involves remodelling of bacterial metabolism. *Nature Communications, 5*, 3369.

Krebs, W., Steuber, J., Gemperli, A. C., & Dimroth, P. (1999). Na^+ translocation by the NADH:ubiquinone oxidoreductase (complex I) from *Klebsiella pneumoniae*. *Molecular Microbiology, 33*, 590–598.

Kuroda, T., Shimamoto, T., Mizushima, T., & Tsuchiya, T. (1997). Mutational analysis of amiloride sensitivity of the NhaA Na^+/H^+ antiporter from *Vibrio parahaemolyticus*. *Journal of Bacteriology, 179*, 7600–7602.

Kurosu, M., & Crick, D. C. (2009). MenA is a promising drug target for developing novel lead molecules to combat *Mycobacterium tuberculosis*. *Medicinal Chemistry, 5*, 197–207.

Lamichhane-Khadka, R., Kwiatkowski, A., & Maier, R. J. (2010). The Hyb hydrogenase permits hydrogen-dependent respiratory growth of *Salmonella enterica* serovar *Typhimurium*. *mBio, 1*, e00284-10.

Lancaster, C. R. (2001). Succinate:quinone oxidoreductases—What can we learn from Wolinella succinogenes quinol:fumarate reductase? *FEBS Letters, 504*, 133–141.

Lancaster, C. R. D. (2013). The di-heme family of respiratory complex II enzymes. *Biochimica et Biophysica Acta, 1827*, 679–687.

Lancaster, C. R. D., Haas, A. H., Madej, M. G., & Mileni, M. (2006). Recent progress on obtaining theoretical and experimental support for the "E-pathway hypothesis" of coupled transmembrane electron and proton transfer in dihaem-containing quinol:fumarate reductase. *Biochimica et Biophysica Acta, 1757*, 988–995.

Lancaster, C. R., & Kröger, A. (2000). Succinate:quinone oxidoreductases: New insights from X-ray crystal structures. *Biochimica et Biophysica Acta, 1459*, 422–431.

Lane, N., Allen, J. F., & Martin, W. (2010). How did LUCA make a living? Chemiosmosis in the origin of life. *Bioessays*, *32*, 271–280.

Lemos, R. S., Fernandes, A. S., Pereira, M. M., Gomes, C. M., & Teixeira, M. (2002). Quinol:fumarate oxidoreductases and succinate:quinone oxidoreductases: Phylogenetic relationships, metal centres and membrane attachment. *Biochimica et Biophysica Acta*, *1553*, 158–170.

Li, F., Hinderberger, J., Seedorf, H., Zhang, J., Buckel, W., & Thauer, R. K. (2008). Coupled ferredoxin and crotonyl coenzyme A (CoA) reduction with NADH catalyzed by the butyryl-CoA dehydrogenase/Etf complex from *Clostridium kluyveri*. *Journal of Bacteriology*, *190*, 843–850.

Lindqvist, A., Membrillo-Hernandez, J., Poole, R. K., & Cook, G. M. (2000). Roles of respiratory oxidases in protecting *Escherichia coli* K12 from oxidative stress. *Antonie Van Leeuwenhoek*, *78*, 23–31.

Linnane, S. J., Keatings, V. M., Costello, C. M., Moynihan, J. B., O'Connor, C. M., Fitzgerald, M. X., et al. (1998). Total sputum nitrate plus nitrite is raised during acute pulmonary infection in cystic fibrosis. *American Journal of Respiratory and Critical Care Medicine*, *158*, 207–212.

Liu, Y. W., Denkmann, K., Kosciow, K., Dahl, C., & Kelly, D. J. (2013). Tetrathionate stimulated growth of *Campylobacter jejuni* identifies a new type of bi-functional tetrathionate reductase (TsdA) that is widely distributed in bacteria. *Molecular Microbiology*, *88*, 173–188.

Loisel-Meyer, S., Jimenez de Bagues, M. P., Kohler, S., Liautard, J. P., & Jubier-Maurin, V. (2005). Differential use of the two high-oxygen-affinity terminal oxidases of *Brucella suis* for *in vitro* and intramacrophagic multiplication. *Infection and Immunity*, *73*, 7768–7771.

Lopez, C. A., Winter, S. E., Rivera-Chavez, F., Xavier, M. N., Poon, V., Nuccio, S. P., et al. (2012). Phage-mediated acquisition of a type III secreted effector protein boosts growth of *Salmonella* by nitrate respiration. *mBio*, *3*, e00143-12.

Lounis, N., Gevers, T., Van den Berg, J., Vranckx, L., & Andries, K. (2009). ATP synthase inhibition of *Mycobacterium avium* is not bactericidal. *Antimicrobial Agents and Chemotherapy*, *53*, 4927–4929.

Lounis, N., Guillemont, J., Veziris, N., Koul, A., Jarlier, V., & Andries, K. (2010). R207910 (TMC207): A new antibiotic for the treatment of tuberculosis. *Médecine et Maladies Infectieuses*, *40*, 383–390.

Lounis, N., Veziris, N., Chauffour, A., Truffot-Pernot, C., Andries, K., & Jarlier, V. (2006). Combinations of R207910 with drugs used to treat multidrug-resistant tuberculosis have the potential to shorten treatment duration. *Antimicrobial Agents and Chemotherapy*, *50*, 3543–3547.

Lupp, C., Robertson, M. L., Wickham, M. E., Sekirov, I., Champion, O. L., Gaynor, E. C., et al. (2007). Host-mediated inflammation disrupts the intestinal microbiota and promotes the overgrowth of *Enterobacteriaceae*. *Cell Host & Microbe*, *2*, 119–129.

MacMicking, J. D., North, R. J., LaCourse, R., Mudgett, J. S., Shah, S. K., & Nathan, C. F. (1997). Identification of nitric oxide synthase as a protective locus against tuberculosis. *Proceedings of the National Academy of Sciences of the United States of America*, *94*, 5243–5248.

Maier, R. J. (2005). Use of molecular hydrogen as an energy substrate by human pathogenic bacteria. *Biochemical Society Transactions*, *33*, 83–85.

Maier, R. J., Fu, C., Gilbert, J., Moshiri, F., Olson, J., & Plaut, A. G. (1996). Hydrogen uptake hydrogenase in *Helicobacter pylori*. *FEMS Microbiology Letters*, *141*, 71–76.

Maier, R. J., Olczak, A., Maier, S., Soni, S., & Gunn, J. (2004). Respiratory hydrogen use by *Salmonella enterica* serovar *Typhimurium* is essential for virulence. *Infection and Immunity*, *72*, 6294–6299.

Maier, L., Vyas, R., Cordova, C. D., Lindsay, H., Schmidt, T. S., Brugiroux, S., et al. (2013). Microbiota-derived hydrogen fuels *Salmonella typhimurium* invasion of the gut ecosystem. *Cell Host & Microbe, 14*, 641–651.

Mak, P. A., Rao, S. P., Ping Tan, M., Lin, X., Chyba, J., Tay, J., et al. (2012). A high-throughput screen to identify inhibitors of ATP homeostasis in non-replicating *Mycobacterium tuberculosis*. *ACS Chemical Biology, 7*, 1190–1197.

Maklashina, E., Berthold, D. A., & Cecchini, G. (1998). Anaerobic expression of *Escherichia coli* succinate dehydrogenase: Functional replacement of fumarate reductase in the respiratory chain during anaerobic growth. *Journal of Bacteriology, 180*, 5989–5996.

Malm, S., Tiffert, Y., Micklinghoff, J., Schultze, S., Joost, I., Weber, I., et al. (2009). The roles of the nitrate reductase NarGHJI, the nitrite reductase NirBD and the response regulator GlnR in nitrate assimilation of *Mycobacterium tuberculosis*. *Microbiology, 155*, 1332–1339.

Mason, M. G., Shepherd, M., Nicholls, P., Dobbin, P. S., Dodsworth, K. S., Poole, R. K., et al. (2009). Cytochrome *bd* confers nitric oxide resistance to *Escherichia coli*. *Nature Chemical Biology, 5*, 94–96.

Matin, A., Zychlinsky, E., Keyhan, M., & Sachs, G. (1996). Capacity of *Helicobacter pylori* to generate ionic gradients at low pH is similar to that of bacteria which grow under strongly acidic conditions. *Infection and Immunity, 64*, 1434–1436.

Matsoso, L. G., Kana, B. D., Crellin, P. K., Lea-Smith, D. J., Pelosi, A., Powell, D., et al. (2005). Function of the cytochrome bc_1-aa_3 branch of the respiratory network in mycobacteria and network adaptation occurring in response to its disruption. *Journal of Bacteriology, 187*, 6300–6308.

Matsushita, K., Shinagawa, E., Adachi, O., & Ameyama, M. (1982). *o*-Type cytochrome oxidase in the membrane of aerobically grown *Pseudomonas aeruginosa*. *FEBS Letters, 139*, 255–258.

McClelland, M., Sanderson, K. E., Spieth, J., Clifton, S. W., Latreille, P., Courtney, L., et al. (2001). Complete genome sequence of *Salmonella enterica* serovar *Typhimurium* LT2. *Nature, 413*, 852–856.

McGowan, C. C., Cover, T. L., & Blaser, M. J. (1997). Analysis of F_1F_0-ATPase from *Helicobacter pylori*. *Infection and Immunity, 65*, 2640–2647.

McNamara, P. J., & Proctor, R. A. (2000). *Staphylococcus aureus* small colony variants, electron transport and persistent infections. *International Journal of Antimicrobial Agents, 14*, 117–122.

Megehee, J. A., Hosler, J. P., & Lundrigan, M. D. (2006). Evidence for a cytochrome bcc-aa_3 interaction in the respiratory chain of *Mycobacterium smegmatis*. *Microbiology, 152*, 823–829.

Mellies, J., Jose, J., & Meyer, T. F. (1997). The *Neisseria gonorrhoeae* gene aniA encodes an inducible nitrite reductase. *Molecular and General Genetics, 256*, 525–532.

Melo, A. M. P., Bandeiras, T. M., & Teixeira, M. (2004). New insights into type II NAD(P)H:quinone oxidoreductases. *Microbiology and Molecular Biology Reviews, 68*, 603–616.

Mercado-Lubo, R., Gauger, E. J., Leatham, M. P., Conway, T., & Cohen, P. S. (2008). A *Salmonella enterica* serovar *typhimurium* succinate dehydrogenase/fumarate reductase double mutant is avirulent and immunogenic in BALB/c mice. *Infection and Immunity, 76*, 1128–1134.

Merrell, D. S., Thompson, L. J., Kim, C. C., Mitchell, H., Tompkins, L. S., Lee, A., et al. (2003). Growth phase-dependent response of *Helicobacter pylori* to iron starvation. *Infection and Immunity, 71*, 6510–6525.

Meunier, B., Madgwick, S. A., Reil, E., Oettmeier, W., & Rich, P. R. (1995). New inhibitors of the quinol oxidation sites of bacterial cytochromes *bo* and *bd*. *Biochemistry, 34*, 1076–1083.

Mills, P. C., Rowley, G., Spiro, S., Hinton, J. C., & Richardson, D. J. (2008). A combination of cytochrome *c* nitrite reductase (NrfA) and flavorubredoxin (NorV) protects *Salmonella enterica* serovar *Typhimurium* against killing by NO in anoxic environments. *Microbiology, 154*, 1218–1228.
Mobius, K., Arias-Cartin, R., Breckau, D., Hannig, A. L., Riedmann, K., Biedendieck, R., et al. (2010). Heme biosynthesis is coupled to electron transport chains for energy generation. *Proceedings of the National Academy of Sciences of the United States of America, 107*, 10436–10441.
Mogi, T., Kawakami, T., Arai, H., Igarashi, Y., Matsushita, K., Mori, M., et al. (2009). Siccanin rediscovered as a species-selective succinate dehydrogenase inhibitor. *Journal of Biochemistry, 146*, 383–387.
Mogi, T., Matsushita, K., Murase, Y., Kawahara, K., Miyoshi, H., Ui, H., et al. (2009). Identification of new inhibitors for alternative NADH dehydrogenase (NDH-II). *FEMS Microbiology Letters, 291*, 157–161.
Mulkidjanian, A. Y., Dibrov, P., & Galperin, M. Y. (2008). The past and present of sodium energetics: May the sodium-motive force be with you. *Biochimica et Biophysica Acta, 1777*, 985–992.
Myers, J. D., & Kelly, D. J. (2005). A sulphite respiration system in the chemoheterotrophic human pathogen *Campylobacter jejuni*. *Microbiology, 151*, 233–242.
Nagata, K., Tsukita, S., Tamura, T., & Sone, N. (1996). A *cb*-type cytochrome-*c* oxidase terminates the respiratory chain in *Helicobacter pylori*. *Microbiology, 142*, 1757–1763.
Nie, W., Tang, H., Fang, Z., Chen, J., Chen, H., & Xiu, Q. (2012). Hydrogenase: The next antibiotic target? *Clinical Science, 122*, 575–580.
Noël, J., & Pouysségur, J. (1995). Hormonal regulation, pharmacology, and membrane sorting of vertebrate Na^{+}/H^{+} exchanger isoforms. *The American Journal of Physiology, 268*, C283–C296.
Nolan, K. J., McGee, D. J., Mitchell, H. M., Kolesnikow, T., Harro, J. M., O'Rourke, J., et al. (2002). *In vivo* behavior of a *Helicobacter pylori* SS1 nixA mutant with reduced urease activity. *Infection and Immunity, 70*, 685–691.
Ohman, D. E., & Chakrabarty, A. M. (1982). Utilization of human respiratory secretions by mucoid *Pseudomonas aeruginosa* of cystic fibrosis origin. *Infection and Immunity, 37*, 662–669.
Olson, J. W., & Maier, R. J. (2002). Molecular hydrogen as an energy source for *Helicobacter pylori*. *Science, 298*, 1788–1790.
Olson, J. W., Mehta, N. S., & Maier, R. J. (2001). Requirement of nickel metabolism proteins HypA and HypB for full activity of both hydrogenase and urease in *Helicobacter pylori*. *Molecular Microbiology, 39*, 176–182.
Ordway, D., Viveiros, M., Leandro, C., Arroz, M. J., & Amaral, L. (2002). Intracellular activity of clinical concentrations of phenothiazines including thioridiazine against phagocytosed *Staphylococcus aureus*. *International Journal of Antimicrobial Agents, 20*, 34–43.
Ordway, D., Viveiros, M., Leandro, C., Arroz, M. J., Molnar, J., Kristiansen, J. E., et al. (2002). Chloropromazine has intracellular killing activity against phagocytosed *Staphylococcus aureus* at clinical concentrations. *Journal of Infection and Chemotherapy, 8*, 227–231.
Ordway, D., Viveiros, M., Leandro, C., Bettencourt, R., Almeida, J., Martins, M., et al. (2003). Clinical concentrations of thioridazine kill intracellular multidrug-resistant *Mycobacterium tuberculosis*. *Antimicrobial Agents and Chemotherapy, 47*, 917–922.
Otto, R., Sonnenberg, A. S., Veldkamp, H., & Konings, W. N. (1980). Generation of an electrochemical proton gradient in *Streptococcus cremoris* by lactate efflux. *Proceedings of the National Academy of Sciences of the United States of America, 77*, 5502–5506.
Overton, T. W., Whitehead, R., Li, Y., Snyder, L. A., Saunders, N. J., Smith, H., et al. (2006). Coordinated regulation of the *Neisseria gonorrhoeae*-truncated denitrification

pathway by the nitric oxide-sensitive repressor, NsrR, and nitrite-insensitive NarQ-NarP. *Journal of Biological Chemistry*, *281*, 33115–33126.
Padan, E., Venturi, M., Gerchman, Y., & Dover, N. (2001). Na^+/H^+ antiporters. *Biochimica et Biophysica Acta*, *1505*, 144–157.
Palmer, K. L., Brown, S. A., & Whiteley, M. (2007). Membrane-bound nitrate reductase is required for anaerobic growth in cystic fibrosis sputum. *Journal of Bacteriology*, *189*, 4449–4455.
Palmer, K. L., Mashburn, L. M., Singh, P. K., & Whiteley, M. (2005). Cystic fibrosis sputum supports growth and cues key aspects of *Pseudomonas aeruginosa* physiology. *Journal of Bacteriology*, *187*, 5267–5277.
Park, S. W., Song, T., Kim, S. Y., Kim, E., Oh, J. I., Eom, C. Y., et al. (2007). Carbon monoxide dehydrogenase in mycobacteria possesses a nitric oxide dehydrogenase activity. *Biochemical and Biophysical Research Communications*, *362*, 449–453.
Parkin, A., Bowman, L., Roessler, M. M., Davies, R. A., Palmer, T., Armstrong, F. A., et al. (2012). How *Salmonella* oxidises H_2 under aerobic conditions. *FEBS Letters*, *586*, 536–544.
Pecsi, I., Hards, K., Ekayanaka, N., Berney, M., Hartman, T., Jacobs, W. R., Jr., et al. (2014). Essentiality of succinate dehydrogenase in *Mycobacterium smegmatis* and its role in the generation of the membrane potential under hypoxia. *mBio*, *5*, e01093–14.
Pereira, M. M., Santana, M., & Teixeira, M. (2001). A novel scenario for the evolution of haem-copper oxygen reductases. *Biochimica et Biophysica Acta*, *1505*, 185–208.
Pethe, K., Bifani, P., Jang, J., Kang, S., Park, S., Ahn, S., et al. (2013). Discovery of Q203, a potent clinical candidate for the treatment of tuberculosis. *Nature Medicine*, *19*, 1157–1160.
Pfenninger-Li, X. D., & Dimroth, P. (1992). NADH formation by Na(+)-coupled reversed electron transfer in *Klebsiella pneumoniae*. *Molecular Microbiology*, *6*, 1943–1948.
Pinto, R., Harrison, J. S., Hsu, T., Jacobs, W. R., Jr., & Leyh, T. S. (2007). Sulfite reduction in mycobacteria. *Journal of Bacteriology*, *189*, 6714–6722.
Pitcher, R. S., & Watmough, N. J. (2004). The bacterial cytochrome cbb_3 oxidases. *Biochimica et Biophysica Acta*, *1655*, 388–399.
Pitson, S. M., Mendz, G. L., Srinivasan, S., & Hazell, S. L. (1999). The tricarboxylic acid cycle of *Helicobacter pylori*. *European Journal of Biochemistry*, *260*, 258–267.
Pittman, M. S., Robinson, H. C., & Poole, R. K. (2005). A bacterial glutathione transporter (Escherichia coli CydDC) exports reductant to the periplasm. *Journal of Biological Chemistry*, *280*, 32254–32261.
Poole, R. K. (2005). Nitric oxide and nitrosative stress tolerance in bacteria. *Biochemical Society Transactions*, *33*, 176–180.
Poole, R. K., & Cook, G. M. (2000). Redundancy of aerobic respiratory chains in bacteria? Routes, reasons and regulation. *Advances in Microbial Physiology*, *43*, 165–224.
Poole, R. K., & Williams, H. D. (1988). Formation of the 680 nm-absorbing form of the cytochrome *bd* oxidase complex of Escherichia coli by reaction of hydrogen peroxide with the ferric form. *FEBS Letters*, *231*, 243–246.
Proctor, R. A., Kahl, B., von Eiff, C., Vaudaux, P. E., Lew, D. P., & Peters, G. (1998). Staphylococcal small colony variants have novel mechanisms for antibiotic resistance. *Clinical Infectious Diseases*, *27*, S68–S74.
Rao, S. P., Alonso, S., Rand, L., Dick, T., & Pethe, K. (2008). The protonmotive force is required for maintaining ATP homeostasis and viability of hypoxic, nonreplicating *Mycobacterium tuberculosis*. *Proceedings of the National Academy of Sciences of the United States of America*, *105*, 11945–11950.
Ray, A., & Williams, H. D. (1997). The effects of mutation of the *anr* gene on the aerobic respiratory chain of *Pseudomonas aeruginosa*. *FEMS Microbiology Letters*, *156*, 227–232.

Rivera-Chavez, F., Winter, S. E., Lopez, C. A., Xavier, M. N., Winter, M. G., Nuccio, S. P., et al. (2013). *Salmonella* uses energy taxis to benefit from intestinal inflammation. *PLoS Pathogen, 9*, e1003267.

Roberts, G., Vadrevu, I. S., Madiraju, M. V., & Parish, T. (2011). Control of CydB and GltA1 expression by the SenX3 RegX3 two component regulatory system of *Mycobacterium tuberculosis*. *PLoS One, 6*, e21090.

Rock, J. D., Mahnane, M. R., Anjum, M. F., Shaw, J. G., Read, R. C., & Moir, J. W. (2005). The pathogen *Neisseria meningitidis* requires oxygen, but supplements growth by denitrification. Nitrite, nitric oxide and oxygen control respiratory flux at genetic and metabolic levels. *Molecular Microbiology, 58*, 800–809.

Sassetti, C. M., Boyd, D. H., & Rubin, E. J. (2003). Genes required for mycobacterial growth defined by high density mutagenesis. *Molecular Microbiology, 48*, 77–84.

Sawers, G. (1994). The hydrogenases and formate dehydrogenases of *Escherichia coli*. *Antonie Van Leeuwenhoek, 66*, 57–88.

Sawers, R. G., Jamieson, D. J., Higgins, C. F., & Boxer, D. H. (1986). Characterization and physiological roles of membrane-bound hydrogenase isoenzymes from *Salmonella typhimurium*. *Journal of Bacteriology, 168*, 398–404.

Schelle, M. W., & Bertozzi, C. R. (2006). Sulfate metabolism in mycobacteria. *ChemBioChem*, 7, 1516–1524.

Schlag, S., Fuchs, S., Nerz, C., Gaupp, R., Engelmann, S., Liebeke, M., et al. (2008). Characterization of the oxygen-responsive NreABC regulon of *Staphylococcus aureus*. *Journal of Bacteriology, 190*, 7847–7858.

Schnappinger, D., Ehrt, S., Voskuil, M. I., Liu, Y., Mangan, J. A., Monahan, I. M., et al. (2003). Transcriptional adaptation of *Mycobacterium tuberculosis* within macrophages: Insights into the phagosomal environment. *Journal of Experimental Medicine, 198*, 693–704.

Schulz, S., Iglesias-Cans, M., Krah, A., Yildiz, O., Leone, V., Matthies, D., et al. (2013). A new type of Na^+-driven ATP synthase membrane rotor with a two-carboxylate ion-coupling motif. *PLoS Biology, 11*, e1001596.

Schurig-Briccio, L. A., Yano, T., Rubin, H., & Gennis, R. B. (2014). Characterization of the type 2 NADH:menaquinone oxidoreductases from *Staphylococcus aureus* and the bactericidal action of phenothiazines. *Biochimica et Biophysica Acta, 1837*, 954–963.

Sellars, M. J., Hall, S. J., & Kelly, D. J. (2002). Growth of *Campylobacter jejuni* supported by respiration of fumarate, nitrate, nitrite, trimethylamine-N-oxide, or dimethyl sulfoxide requires oxygen. *Journal of Bacteriology, 184*, 4187–4196.

Seto, H., Jinnai, Y., Hiratsuka, T., Fukawa, M., Furihata, K., Itoh, N., et al. (2008). Studies on a new biosynthetic pathway for menaquinone. *Journal of the American Chemical Society, 130*, 5614–5615.

Shatalin, K., Gusarov, I., Avetissova, E., Shatalina, Y., McQuade, L. E., Lippard, S. J., et al. (2008). *Bacillus anthracis*-derived nitric oxide is essential for pathogen virulence and survival in macrophages. *Proceedings of the National Academy of Sciences of the United States of America, 105*, 1009–1013.

Shaw, F. L., Mulholland, F., Le Gall, G., Porcelli, I., Hart, D. J., Pearson, B. M., et al. (2012). Selenium-dependent biogenesis of formate dehydrogenase in *Campylobacter jejuni* is controlled by the fdhTU accessory genes. *Journal of Bacteriology, 194*, 3814–3823.

Shi, L., Sohaskey, C. D., Kana, B. D., Dawes, S., North, R. J., Mizrahi, V., et al. (2005). Changes in energy metabolism of *Mycobacterium tuberculosis* in mouse lung and under *in vitro* conditions affecting aerobic respiration. *Proceedings of the National Academy of Sciences of the United States of America, 102*, 15629–15634.

Shirude, P. S., Paul, P., Choudhury, N. R., Kedari, C., Bandodkar, B., & Ugarkar, B. G. (2012). Quinolinyl pyrimidines: Potent inhibitors of NDH-2 as a novel class of anti-TB agents. *ACS Medicinal Chemistry Letters, 3*, 736–740.

Siegele, D. A., Imlay, K. R., & Imlay, J. A. (1996). The stationary-phase-exit defect of *cydC* (*surB*) mutants is due to the lack of a functional terminal cytochrome oxidase. *Journal of Bacteriology, 178*, 6091–6096.
Singh, P. K., Schaefer, A. L., Parsek, M. R., Moninger, T. O., Welsh, M. J., & Greenberg, E. P. (2000). Quorum-sensing signals indicate that cystic fibrosis lungs are infected with bacterial biofilms. *Nature, 407*, 762–764.
Small, J. L., O'Donoghue, A. J., Boritsch, E. C., Tsodikov, O. V., Knudsen, G. M., Vandal, O., et al. (2013). Substrate specificity of MarP, a periplasmic protease required for resistance to acid and oxidative stress in *Mycobacterium tuberculosis*. *Journal of Biological Chemistry, 288*(18), 12489–12499.
Smart, J. P., Cliff, M. J., & Kelly, D. J. (2009). A role for tungsten in the biology of *Campylobacter jejuni*: Tungstate stimulates formate dehydrogenase activity and is transported via an ultra-high affinity ABC system distinct from the molybdate transporter. *Molecular Microbiology, 74*, 742–757.
Smith, M. A., Finel, M., Korolik, V., & Mendz, G. L. (2000). Characteristics of the aerobic respiratory chains of the microaerophiles *Campylobacter jejuni* and *Helicobacter pylori*. *Archives of Microbiology, 174*, 1–10.
Sohaskey, C. D. (2005). Regulation of nitrate reductase activity in *Mycobacterium tuberculosis* by oxygen and nitric oxide. *Microbiology, 151*, 3803–3810.
Sohaskey, C. D. (2008). Nitrate enhances the survival of *Mycobacterium tuberculosis* during inhibition of respiration. *Journal of Bacteriology, 190*, 2981–2986.
Sohaskey, C. D., & Wayne, L. G. (2003). Role of narK2X and narGHJI in hypoxic upregulation of nitrate reduction by *Mycobacterium tuberculosis*. *Journal of Bacteriology, 185*, 7247–7256.
Spector, M. P., Garcia del Portillo, F., Bearson, S. M., Mahmud, A., Magut, M., Finlay, B. B., et al. (1999). The *rpoS*-dependent starvation-stress response locus *stiA* encodes a nitrate reductase (*narZYWV*) required for carbon-starvation-inducible thermotolerance and acid tolerance in *Salmonella typhimurium*. *Microbiology, 145*, 3035–3045.
Stecher, B., Robbiani, R., Walker, A. W., Westendorf, A. M., Barthel, M., Kremer, M., et al. (2007). Salmonella enterica serovar typhimurium exploits inflammation to compete with the intestinal microbiota. *PLoS Biology, 5*, 2177–2189.
Stefanelli, P., Colotti, G., Neri, A., Salucci, M. L., Miccoli, R., Di Leandro, L., et al. (2008). Molecular characterization of nitrite reductase gene (*aniA*) and gene product in *Neisseria meningitidis* isolates: Is *aniA* essential for meningococcal survival? *IUBMB Life, 60*, 629–636.
Steuber, J. (2001). Na^+ translocation by bacterial NADH:quinone oxidoreductases: An extension to the complex-I family of primary redox pumps. *Biochimica et Biophysica Acta, 1505*, 45–56.
Stevanin, T. M., Moir, J. W., & Read, R. C. (2005). Nitric oxide detoxification systems enhance survival of *Neisseria meningitidis* in human macrophages and in nasopharyngeal mucosa. *Infection and Immunity, 73*, 3322–3329.
Stewart, V. (2003). Biochemical Society Special Lecture. Nitrate- and nitrite-responsive sensors NarX and NarQ of proteobacteria. *Biochemical Society Transactions, 31*, 1–10.
Stock, D., Leslie, A. G. W., & Walker, J. E. (1999). Molecular architecture of the rotary motor in ATP synthase. *Science, 286*, 1700–1705.
Stoffels, L., Krehenbrink, M., Berks, B. C., & Unden, G. (2012). Thiosulfate reduction in *Salmonella enterica* is driven by the proton motive force. *Journal of Bacteriology, 194*, 475–485.
Stutts, M. J., Knowles, M. R., Gatzy, J. T., & Boucher, R. C. (1986). Oxygen consumption and ouabain binding sites in cystic fibrosis nasal epithelium. *Pediatric Research, 20*, 1316–1320.

Sun, Y. H., de Jong, M. F., den Hartigh, A. B., Roux, C. M., Rolan, H. G., & Tsolis, R. M. (2012). The small protein CydX is required for function of cytochrome *bd* oxidase in *Brucella abortus*. *Frontiers in Cellular and Infection Microbiology*, *2*, 47.

Tareen, A. M., Dasti, J. I., Zautner, A. E., Gross, U., & Lugert, R. (2010). *Campylobacter jejuni* proteins Cj0952c and Cj0951c affect chemotactic behaviour towards formic acid and are important for invasion of host cells. *Microbiology*, *156*, 3123–3135.

Teh, J. S., Yano, T., & Rubin, H. (2007). Type II NADH:menaquinone oxidoreductase of *Mycobacterium tuberculosis*. *Infectious Disorders Drug Targets*, *7*, 169–181.

Terry, J. M., Pina, S. E., & Mattingly, S. J. (1992). Role of energy metabolism in conversion of nonmucoid *Pseudomonas aeruginosa* to the mucoid phenotype. *Infection and Immunity*, *60*, 1329–1335.

Thiennimitr, P., Winter, S. E., Winter, M. G., Xavier, M. N., Tolstikov, V., Huseby, D. L., et al. (2011). Intestinal inflammation allows *Salmonella* to use ethanolamine to compete with the microbiota. *Proceedings of the National Academy of Sciences of the United States of America*, *108*, 17480–17485.

Tokuda, H., & Unemoto, T. (1981). A respiration-dependent primary sodium extrusion system functioning at alkaline pH in the marine bacterium *Vibrio alginolyticus*. *Biochemical and Biophysical Research Communications*, *102*, 265–271.

Tokuda, H., & Unemoto, T. (1984). Na^+ is translocated at NADH:quinone oxidoreductase segment in the respiratory chain of *Vibrio alginolyticus*. *Journal of Biological Chemistry*, *259*, 7785–7790.

Tomb, J. F., White, O., Kerlavage, A. R., Clayton, R. A., Sutton, G. G., Fleischmann, R. D., et al. (1997). The complete genome sequence of the gastric pathogen *Helicobacter pylori*. *Nature*, *388*, 539–547.

Tran, S. L., & Cook, G. M. (2005). The F_1F_o-ATP synthase of *Mycobacterium smegmatis* is essential for growth. *Journal of Bacteriology*, *187*, 5023–5028.

Tran, S. L., Rao, M., Simmers, C., Gebhard, S., Olsson, K., & Cook, G. M. (2005). Mutants of *Mycobacterium smegmatis* unable to grow at acidic pH in the presence of the protonophore carbonyl cyanide m-chlorophenylhydrazone. *Microbiology*, *151*, 665–672.

Tsolis, R. M., Adams, L. G., Ficht, T. A., & Baumler, A. J. (1999). Contribution of *Salmonella typhimurium* virulence factors to diarrheal disease in calves. *Infection and Immunity*, *67*, 4879–4885.

Tunbridge, A. J., Stevanin, T. M., Lee, M., Marriott, H. M., Moir, J. W., Read, R. C., et al. (2006). Inhibition of macrophage apoptosis by *Neisseria meningitidis* requires nitric oxide detoxification mechanisms. *Infection and Immunity*, *74*, 729–733.

Turner, A. K., Barber, L. Z., Wigley, P., Muhammad, S., Jones, M. A., Lovell, M. A., et al. (2003). Contribution of proton-translocating proteins to the virulence of *Salmonella enterica* serovars Typhimurium, Gallinarum, and Dublin in chickens and mice. *Infection and Immunity*, *71*, 3392–3401.

Turrens, J. F. (1997). Superoxide production by the mitochondrial respiratory chain. *Bioscience Reports*, *17*, 3–8.

Ueda, A., & Wood, T. K. (2008). Potassium and sodium transporters of *Pseudomonas aeruginosa* regulate virulence to barley. *Applied Microbiology and Biotechnology*, *79*, 843–858.

Unden, G., & Bongaerts, J. (1997). Alternative respiratory pathways of *Escherichia coli*: Energetics and transcriptional regulation in response to electron acceptors. *Biochimica et Biophysica Acta*, *1320*, 217–234.

Unden, G., & Schirawski, J. (1997). The oxygen-responsive transcriptional regulator FNR of *Escherichia coli*: The search for signals and reactions. *Molecular Microbiology*, *25*, 205–210.

van der Rest, M. E., Siewe, R. M., Abee, T., Schwarz, E., Oesterhelt, D., & Konings, W. N. (1992). Nucleotide sequence and functional properties of a sodium-dependent citrate transport system from *Klebsiella pneumoniae*. *Journal of Biological Chemistry*, *267*, 8971–8976.

VanOrsdel, C. E., Bhatt, S., Allen, R. J., Brenner, E. P., Hobson, J. J., Jamil, A., et al. (2013). The *Escherichia coli* CydX protein is a member of the CydAB cytochrome *bd* oxidase complex and is required for cytochrome *bd* oxidase activity. *Journal of Bacteriology, 195*, 3640–3650.

Vegge, C. S., Brondsted, L., Li, Y. P., Bang, D. D., & Ingmer, H. (2009). Energy taxis drives *Campylobacter jejuni* toward the most favorable conditions for growth. *Applied and Environmental Microbiology, 75*, 5308–5314.

Velmurugan, K., Chen, B., Miller, J. L., Azogue, S., Gurses, S., Hsu, T., et al. (2007). *Mycobacterium tuberculosis nuoG* is a virulence gene that inhibits apoptosis of infected host cells. *PLoS Pathogens, 3*, e110.

Verkhovsky, M. I., & Bogachev, A. V. (2010). Sodium-translocating NADH:quinone oxidoreductase as a redox-driven ion pump. *Biochimica et Biophysica Acta, 1797*, 738–746.

Vilcheze, C., Weisbrod, T. R., Chen, B., Kremer, L., Hazbon, M. H., Wang, F., et al. (2005). Altered NADH/NAD^+ ratio mediates coresistance to isoniazid and ethionamide in mycobacteria. *Antimicrobial Agents and Chemotherapy, 49*, 708–720.

Vimont, S., & Berche, P. (2000). NhaA, an Na^+/H^+ antiporter involved in environmental survival of *Vibrio cholerae*. *Journal of Bacteriology, 182*, 2937–2944.

von Ballmoos, C., Cook, G. M., & Dimroth, P. (2008). Unique rotary ATP synthase and its biological diversity. *Annual Review of Biophysics, 37*, 43–64.

von Eiff, C. (2008). *Staphylococcus aureus* small colony variants: A challenge to microbiologists and clinicians. *International Journal of Antimicrobial Agents, 31*, 507–510.

Voskuil, M. I., Schnappinger, D., Visconti, K. C., Harrell, M. I., Dolganov, G. M., Sherman, D. R., et al. (2003). Inhibition of respiration by nitric oxide induces a *Mycobacterium tuberculosis* dormancy program. *Journal of Experimental Medicine, 198*, 705–713.

Wang, H., & Gunsalus, R. P. (2000). The *nrfA* and *nirB* nitrite reductase operons in *Escherichia coli* are expressed differently in response to nitrate than to nitrite. *Journal of Bacteriology, 182*, 5813–5822.

Warman, A. J., Rito, T. S., Fisher, N. E., Moss, D. M., Berry, N. G., O'Neill, P. M., et al. (2013). Antitubercular pharmacodynamics of phenothiazines. *The Journal of Antimicrobial Chemotherapy, 68*, 869–880.

Watanabe, S., Zimmermann, M., Goodwin, M. B., Sauer, U., Barry, C. E., 3rd., & Boshoff, H. I. (2011). Fumarate reductase activity maintains an energized membrane in anaerobic *Mycobacterium tuberculosis*. *PLoS Pathogens*, 7, e1002287.

Way, S. S., Sallustio, S., Magliozzo, R. S., & Goldberg, M. B. (1999). Impact of either elevated or decreased levels of cytochrome bd expression on *Shigella flexneri* virulence. *Journal of Bacteriology, 181*, 1229–1237.

Weerakoon, D. R., Borden, N. J., Goodson, C. M., Grimes, J., & Olson, J. W. (2009). The role of respiratory donor enzymes in *Campylobacter jejuni* host colonization and physiology. *Microbial Pathogenesis, 47*, 8–15.

Weerakoon, D. R., & Olson, J. W. (2008). The *Campylobacter jejuni* NADH:ubiquinone oxidoreductase (complex I) utilizes flavodoxin rather than NADH. *Journal of Bacteriology, 190*, 915–925.

Weingarten, R. A., Taveirne, M. E., & Olson, J. W. (2009). The dual-functioning fumarate reductase is the sole succinate:quinone reductase in *Campylobacter jejuni* and is required for full host colonization. *Journal of Bacteriology, 191*, 5293–5300.

Weinstein, E. A., Yano, T., Li, L. S., Avarbock, D., Avarbock, A., Helm, D., et al. (2005). Inhibitors of type II NADH:menaquinone oxidoreductase represent a class of antitubercular drugs. *Proceedings of the National Academy of Sciences of the United States of America, 102*, 4548–4553.

Wessel, A. K., Arshad, T. A., Fitzpatrick, M., Connell, J. L., Bonnecaze, R. T., Shear, J. B., et al. (2014). Oxygen limitation within a bacterial aggregate. *mBio, 5*, e00992.

Whitehead, R. N., Overton, T. W., Snyder, L. A., McGowan, S. J., Smith, H., Cole, J. A., et al. (2007). The small FNR regulon of *Neisseria gonorrhoeae*: Comparison with the larger *Escherichia coli* FNR regulon and interaction with the NarQ-NarP regulon. *BMC Genomics, 8*, 35.

Williams, H. D., Zlosnik, J. E., & Ryall, B. (2007). Oxygen, cyanide and energy generation in the cystic fibrosis pathogen *Pseudomonas aeruginosa*. *Advances in Microbial Physiology, 52*, 1–71.

Winter, S. E., & Baumler, A. J. (2011). A breathtaking feat: To compete with the gut microbiota, *Salmonella* drives its host to provide a respiratory electron acceptor. *Gut Microbes, 2*, 58–60.

Winter, S. E., Thiennimitr, P., Winter, M. G., Butler, B. P., Huseby, D. L., Crawford, R. W., et al. (2010). Gut inflammation provides a respiratory electron acceptor for *Salmonella*. *Nature, 467*, 426–429.

Winter, S. E., Winter, M. G., Xavier, M. N., Thiennimitr, P., Poon, V., Keestra, A. M., et al. (2013). Host-derived nitrate boosts growth of *E. coli* in the inflamed gut. *Science, 339*, 708–711.

Worlitzsch, D., Tarran, R., Ulrich, M., Schwab, U., Cekici, A., Meyer, K. C., et al. (2002). Effects of reduced mucus oxygen concentration in airway *Pseudomonas* infections of cystic fibrosis patients. *The Journal of Clinical Investigation, 109*, 317–325.

Yano, T., Li, L. S., Weinstein, E., Teh, J. S., & Rubin, H. (2006). Steady-state kinetics and inhibitory action of antitubercular phenothiazines on *Mycobacterium tuberculosis* type-II NADH-menaquinone oxidoreductase (NDH-2). *Journal of Biological Chemistry, 281*, 11456–11463.

Ye, R. W., Haas, D., Ka, J. O., Krishnapillai, V., Zimmermann, A., Baird, C., et al. (1995). Anaerobic activation of the entire denitrification pathway in *Pseudomonas aeruginosa* requires Anr, an analog of Fnr. *Journal of Bacteriology, 177*, 3606–3609.

Yimga, M. T., Leatham, M. P., Allen, J. H., Laux, D. C., Conway, T., & Cohen, P. S. (2006). Role of gluconeogenesis and the tricarboxylic acid cycle in the virulence of *Salmonella enterica* serovar typhimurium in BALB/c mice. *Infection and Immunity, 74*, 1130–1140.

Yokoyama, K., Akabane, Y., Ishii, N., & Yoshida, M. (1994). Isolation of prokaryotic V_0V_1-ATPase from a thermophilic eubacterium *Thermus thermophilus*. *Journal of Biological Chemistry, 269*, 12248–12253.

Yorimitsu, T., & Homma, M. (2001). Na^+-driven flagellar motor of *Vibrio*. *Biochimica et Biophysica Acta, 1505*, 82–93.

Yoshikawa, K., Nakayama, Y., Hayashi, M., Unemoto, T., & Mochida, K. (1999). Korormicin, an antibiotic specific for gram-negative marine bacteria, strongly inhibits the respiratory chain-linked Na^+-translocating NADH:quinone reductase from the marine *Vibrio alginolyticus*. *Journal of Antibiotics, 52*, 182–185.

Zacharia, V. M., Manzanillo, P. S., Nair, V. R., Marciano, D. K., Kinch, L. N., Grishin, N. V., et al. (2013). *cor*, a novel carbon monoxide resistance gene, is essential for *Mycobacterium tuberculosis* pathogenesis. *mBio, 4*, e00721-13.

Zbell, A. L., Benoit, S. L., & Maier, R. J. (2007). Differential expression of NiFe uptake-type hydrogenase genes in *Salmonella enterica* serovar *Typhimurium*. *Microbiology, 153*, 3508–3516.

Zbell, A. L., & Maier, R. J. (2009). Role of the Hya hydrogenase in recycling of anaerobically produced H_2 in *Salmonella enterica* serovar *Typhimurium*. *Applied and Environmental Microbiology, 75*, 1456–1459.

Zbell, A. L., Maier, S. E., & Maier, R. J. (2008). *Salmonella enterica* serovar *Typhimurium* NiFe uptake-type hydrogenases are differentially expressed *in vivo*. *Infection and Immunity*, *76*, 4445–4454.
Zhang, Y. J., Ioerger, T. R., Huttenhower, C., Long, J. E., Sassetti, C. M., Sacchettini, J. C., et al. (2012). Global assessment of genomic regions required for growth in *Mycobacterium tuberculosis*. *PLoS Pathogens*, *8*, e1002946.
Zhang-Barber, L., Turner, A. K., Martin, G., Frankel, G., Dougan, G., & Barrow, P. A. (1997). Influence of genes encoding proton-translocating enzymes on suppression of *Salmonella typhimurium* growth and colonization. *Journal of Bacteriology*, *179*, 7186–7190.

CHAPTER TWO

The Impact of Horizontal Gene Transfer on the Biology of *Clostridium difficile*

Adam P. Roberts[1], Elaine Allan, Peter Mullany
Department of Microbial Diseases, UCL Eastman Dental Institute, University College London, London, United Kingdom
[1]Corresponding author: e-mail address: adam.roberts@ucl.ac.uk

Contents

Abstract

Clostridium difficile infection (CDI) is now recognised as the main cause of healthcare associated diarrhoea. Over the recent years there has been a change in the epidemiology of CDI with certain related strains dominating infection. These strains have been termed hyper-virulent and have successfully spread across the globe. Many *C. difficile* strains have had their genomes completely sequenced allowing researchers to build up a very detailed picture of the contribution of horizontal gene transfer to the adaptive potential, through the acquisition of mobile DNA, of this organism. Here, we review and discuss the contribution of mobile genetic elements to the biology of this clinically important pathogen.

Advances in Microbial Physiology
ISSN 0065-2911
http://dx.doi.org/10.1016/bs.ampbs.2014.08.002

1. INTRODUCTION TO *CLOSTRIDIUM DIFFICILE* AND CDI

Clostridium difficile is an anaerobic, Gram-positive, endospore-forming, motile, rod-shaped bacterium which was originally isolated from the stool of a healthy infant (Hall & O'Toole, 1935). It is responsible for the disease in many different animals including economically important food animals, such as pigs (Songer, 2004), companion animals, such as dogs (Keel, Brazier, Post, Weese, & Songer, 2007), and humans (Arroyo et al., 2005). The past decade has seen the emergence of strains of the hyper-virulent PCR ribotype 027 complex, which is associated with increased incidence and severity of disease. CDI caused by other important ribotypes, including 017, 023, 078, has also increased (Barbut, Jones, & Eckert, 2011).

C. difficile causes toxin-mediated gastrointestinal diseases ranging from mild, self-limiting diarrhoea to life threatening, and sometimes fatal, pseudomembranous colitis, and toxic megacolon (Norén, 2010; Rupnik, Wilcox, & Gerding, 2009). The main virulence factors of *C. difficile* are the toxins A and B, designated TcdA and TcdB, respectively, which glycosylate Rho family proteins within host cells and lead to depolymerisation of the actin cytoskeleton. The toxin genes, *tcdA* and *tcdB*, are both located on the 19-kb pathogenicity locus, PaLoc (Braun, Hundsberger, Leukel, Sauerborn, & von Eichel-Streiber, 1996) which encodes three other proteins; the sigma factor, TcdR, which positively regulates the expression of *tcdA* and *tcdB*, TcdE which is related to bacteriophage holin proteins, and TcdC which was thought to be a negative regulator (anti-sigma factor) of toxin production although more recent studies have cast doubt on this role (Bakker, Smits, Kuijper, & Corver, 2012).

In addition to the PaLoc-encoded toxins, some *C. difficile* strains produce a third toxin called the binary toxin (designated cytolethal distending toxin, CDT), which also acts by actin modification (Gerding, Johnson, Rupnik, & Aktories, 2014). The exact role of each of the toxins in CDI remains the subject of debate as strains are isolated from CDI patients carrying different genetic variations and combinations of the three genes (Dingle et al., 2011). Furthermore, mutational analysis of the toxin genes has been carried out by different groups using different experimental strategies and different descendants of the same strain, potentially contributing to conflicting results about the exact role each toxin plays in pathogenesis (Kuehne et al., 2010; Lyras et al., 2009).

CDI usually follows exposure of the patient to antimicrobial therapy which disrupts the intestinal microbiota sufficiently to destroy its protective colonisation resistance and allows environmentally acquired *C. difficile* spores to germinate and colonise the colon. There are multiple groups of related *C. difficile* strains, known as ribotypes, responsible for CDI which are grouped according to the sequence similarities between the 16*S* and 23*S* rDNA intergenic regions. Certain PCR ribotypes have emerged as hyper-virulent, e.g., 027 and 078. As well as hyper-virulence, another emerging property of *C. difficile* is resistance to antibiotics, many of which have been acquired on MGEs (Rupnik et al., 2009; Smits, 2013).

Integrated mobile genetic elements, as opposed to self-replicating plasmids, can cause heritable changes to the host in multiple ways. They can cause mutations simply by disrupting the genes at their target site upon insertion and they can form gene fusions upon insertion as seen for the conjugative transposon CTn*5* in strain 630 (Sebaihia et al., 2006). Introduction of foreign DNA, from the previous host, can occur upon transfer and removal of host DNA upon excision is also possible. Various mobile elements have been shown to undergo inversion reactions within their target site (e.g., O'Keeffe, Hill, & Ross, 1999) and this can cause both upstream and downstream polar effects resulting from interrupted transcription from the host across the insertion site and differing expression levels resulting from MGE originating transcripts reading out of the MGE into the host genome. Polar effects also occur following insertion into a new site. There are also instances of deletion events leaving part of an MGE isolated, and fixed, in the host's chromosome. Furthermore, there are examples where interactions between different MGEs result in *trans*-activation of elements due to the activity of another, e.g., high frequency recombination of the genome due to activation of an *oriT* located in a chromosomal copy of a conjugative element.

Highly related MGEs are common within *C. difficile* genomes and exhibit variation in structure and properties, which allow them to be grouped into families (discussed below). Because of the variety of MGEs apparent in the hundreds of *C. difficile* genome sequences available, a detailed description of each element will not be provided but representatives from each different group will be discussed. The individual families of MGEs commonly found in *C. difficile* are ordered according to increasing genetic complexity with the "simplest" MGEs discussed first.

2. INTRONS

Introns were first discovered in eukaryotes and are ubiquitous in higher eukaryotes such as humans. In prokaryotes, some introns are capable of transposition and are often associated with other MGEs such as conjugative transposons and conjugative plasmids. Introns are classified into either group one (I) or two (II) according to their highly conserved secondary structure (Belfort & Perlman, 1995). Group II introns have six conserved helices which emanate from a central wheel. This secondary structure is required for splicing. In one of the conserved helix (number IV), there is often an ORF encoding a multifunctional protein which has a reverse transcriptase domain which, as part of a riboprotein complex, can catalyse transposition of the intron to an allelic site via retrohoming or transposition to a different site (Belfort & Perlman, 1995).

In *C. difficile*, there is a group II intron inserted into the conjugative transposon, Tn*5397* in strain 630 (Mullany, Pallen, Wilks, Stephen, & Tabaqchali, 1996). This group II intron is found within the gene within *orf14*, which is believed to be essential for conjugation of the element. Splicing of the group II intron has been demonstrated by PCR with primers reading across the predicted splice site (Roberts, Braun, von Eichel-Streiber, & Mullany, 2001; Roberts, Johanesen, Lyras, Mullany, & Rood, 2001) demonstrating that this group II intron is capable of splicing from the mRNA. However, when the reverse transcriptase encoding domain was disrupted by insertion of a kanamycin resistance gene, the ability of the intron to splice was abolished but Tn*5397* was still capable of transfer. One explanation for this is that the intron has inserted very close to the 3′ end of the gene and splicing is not required to obtain a functional Orf14. Another group II intron has been found associated with a Tn*5397*-like element in strain QCD-63Q42 (Brouwer, Warburton, Roberts, Mullany, & Allan, 2011), which inserted within a different gene compared to the strain 630 group II intron. It is not yet known if splicing is required for conjugation of this element.

The contribution of group II introns to the overall biology of *C. difficile* is likely to be low due to the fact that they can perfectly splice out of the pre-mRNA transcript rather than creating a mutation. However, they do offer the possibility of a regulatory step, if they only splice under certain environmental conditions and/or differentially splice under certain conditions then different proteins can be produced. This ability to generate a variety of proteins from a small genome is one of the hallmarks of bacteria. Furthermore, due to their small size (3 kb), they are likely to have little biological cost to their host cell.

3. IStrons

A family of chimeric elements have been discovered in *C. difficile*. These have been termed IStrons as they combine part of a group I intron (with the maintenance of the characteristic long range tertiary structure) at their 5′ ends together with a partial, and a full length, transposase from an IS sequence at their 3′ ends (Hasselmayer, Nitsche, Braun, & von Eichel-Streiber, 2004). One of these unusual elements was first found inserted into the *tcdA* gene in *C. difficile* strain C34 and is labelled CdISt1 (Hasselmayer et al., 2004). The sequence of the partial and full transposase genes are homologous to those of IS*200* and IS*1341* transposases, respectively. The combination of the partial IS*200* and the full length IS*1341* are reminiscent of the complex IS*605* family of elements (Depardieu, Reynolds, & Courvalin, 2003). Multiple copies of CdISt1 were found within the genome of strain C34 and all were shown to splice, demonstrating that they are all functional ribozymes. In addition to the ability to auto-catalytically splice from the mRNA transcripts, evidence suggests that they are also able to spread by transposition. Neither a group I intron or an IS of this family has been found in any of the *C. difficile* strains sequenced to date which suggests that the element has been formed elsewhere, prior to invasion of the *C. difficile* genomes (Hasselmayer et al., 2004). Unlike the group II intron present in Tn*5397*, analysis of splicing by PCR revealed the existence of alternative PCR products, likely to be the result of alternative splicing at the 3′ splice site. This genetic variation could be important for the evolution and regulation of the *C. difficile* genome as natural selection can act on any of the alternative splicing products and may select for these variations (Hasselmayer et al., 2004).

4. MOBILISABLE TRANSPOSONS

This group of MGEs are more complex as they are capable of both transposition and transfer between cells. They were originally discovered as a result of the antibiotic resistance that they encode in *C. difficile* strains SGC0545 and W1 (Wren, Mullany, Clayton, & Tabaqchali, 1998, 1999). The 6.3 kb elements, Tn*4453*a and Tn*4453*b, both encode chloramphenicol resistance via the chloramphenicol acetyltransferase gene, *catD*, in strain W1 (Lyras et al., 1998). Characterisation of these elements, which differ only by a few base pairs, and the highly related *Clostridium perfringens* element, Tn*4451*, has shown that these elements undergo precise excision from the target site to form a circular molecule. Interestingly, when the joint of the ends of Tn*4453* are ligated together a promoter is formed which drives the expression of the large serine recombinase, TnpX (Lyras & Rood, 2000) which is the only transposon-encoded protein required for transposition of these elements (Lyras, Adams, Lucet, & Rood, 2004). In

C. perfringens, experiments have shown that the transposon-encoded protein, TnpZ, has similarity with various plasmid mobilisation proteins of the Mob/Pre family and Tn*4451* has the cognate *oriT* site upstream of *tnpZ*. Mobilisation experiments were performed in an *Escherichia coli* strain containing the conjugative plasmid, RP4, which showed that Tn*4451* and Tn*4453*, present on pUC18 or pUC19 could be conjugated to a *C. perfringens* host using the RP4 conjugation apparatus. Furthermore, a 2-bp mutation in the conserved *oriT* within Tn*4451* reduced the transfer frequency by almost 400-fold and a *tnpZ* mutant abolished transfer completely. These experiments confirmed Tn*4451* and Tn*4453* as the first mobilisable, but non-conjugative, transposons isolated in Gram-positive bacteria (Crellin & Rood, 1998; Lyras et al., 1998).

A recent study using BLAST software to analyse genome sequences of *C. difficile* strains for orthologues of known recombinase genes from MGEs resulted in the identification of a 15-kb putative mobilisable transposon in the genome of strain QCD-63Q42 from Canada, which was termed Tn*6115* (Brouwer et al., 2011). This element contains genes encoding a serine recombinase, a predicted ABC transporter, a protein containing a virulence associated E domain (PFAM; PF05272), and a predicted two component signal transduction system. The GC content of Tn*6115* is 47.5% compared to <30% for the rest of the chromosome, suggesting that it may be a relatively recent acquisition to the QCD-63Q42 genome.

The final element which will be discussed in this section is the transferable 9.6 kb erythromycin resistance-encoding element, Tn*5398*, discovered in strain 630 (Farrow, Lyras, & Rood, 2001; Mullany, Wilks, & Tabaqchali, 1995). This element contains two copies of the rRNA adenine *N*-6-methyltransferase gene; *erm*(B), which confers the MLS resistance phenotype to the *m*acrolide, *l*incosamide, and *s*treptogramin B antibiotics. Tn*5398* has been shown to transfer between *C. difficile* strains and from *C. difficile* to *Staphylococcus aureus* and *Bacillus subtilis* (Hächler, Berger-Bächi, & Kayser, 1987; Mullany et al., 1995; Wüst & Hardegger, 1983). Unlike all of the other mobilisable transposons investigated to date, Tn*5398* does not have a candidate gene encoding a recombinase so the excision reaction which precedes conjugative transfer is thought to be catalysed *in trans* by another of the many recombinases present in *C. difficile* 630 (Adams, Lyras, Farrow, & Rood, 2002; Sebaihia et al., 2006). Similarly, Tn*5398* does not appear to encode any proteins involved in mobilisation. It does, however, have two potential *oriT* sequences which share identity at the *nic* site with the *oriT* of the conjugative transposon Tn*5397* and, therefore, it is possible that the Tn*5397* proteins could nick the Tn*5398 oriT* and initiate transfer to a recipient. Following transfer, there would have to be another suitable recombinase supplied *in trans* in order to catalyse insertion or insertion could occur by RecA-mediated recombination if suitable homology occurs between the host genome and the transferred DNA (Adams et al., 2002). An alternative explanation is that it can transfer by a high frequency of recombination (Hfr) mechanism, as has recently been shown for the PaLoc (see below; Brouwer et al., 2013).

Recently Brouwer et al. (2011) found a Tn*5398*-like element in the genome sequence of strain ATCC 43255 which has 98% sequence identity to Tn*5398* excluding the *erm*(B) and downstream *ORF*298 region. In addition, the target site for the *erm*(B)–*ORF*298 region is within an ORF which is disrupted upon insertion of this section of DNA, suggesting that the region of Tn*5398* encoding Erm(B) is a recent arrival into an older element. A recent report (Wasels, Spigaglia, Barbanti, & Mastrantonio, 2013) has described the Tn*5398 erm*(B) and *ORF*298 inserted in other MGEs in several clinical strains of *C. difficile* (strains F17, CI17, and 2007855). It is interesting that one of the two predicted *oriT* sites in Tn*5398* is located within the *ORF*298 which is always found associated with *erm*(B) and is also found in the *Enterococcus faecalis* plasmid, pAMβ1 (Adams et al., 2002; Wasels et al., 2013). Bi-directional horizontal gene transfer of the conjugative transposon, Tn*5397*, has been recently demonstrated between *C. difficile* strains and *E. faecalis* JH2-2 (Jasni, Mullany, Hussain, & Roberts, 2010) so it is possible that the *erm*(B)-*ORF*298 region of DNA present in Tn*5398* and other MGEs has been mobilised between *C. difficile* and *E. faecaelis*.

5. CONJUGATIVE TRANSPOSONS

Conjugative transposons are capable of both intracellular transposition and intercellular conjugative transfer. They are also referred to as integrative conjugative elements; however, these terms are interchangeable (Roberts et al., 2006). There are many different classes of conjugative transposons; however, within *C. difficile*, there are two main families known as the Tn*916*- and the Tn*1549*-like elements (Brouwer, Roberts, Mullany, & Allan, 2012; Brouwer et al., 2011). Tn*916* is an 18-kb conjugative transposon which encodes resistance to tetracycline and minocycline via the ribosomal protection protein, Tet(M). It was first discovered in *E. faecalis* (Franke & Clewell, 1981) who observed transferable tetracycline resistance in the absence of transferable plasmids. Tn*916* contains 24 potential ORFs which were arranged into functional modules. This is also the case for the 34-kb vancomycin resistance element, Tn*1549*, which was also originally found in *E. faecalis* (Garnier, Taourit, Glaser, Courvalin, & Galimand, 2000). The modules of genes within these elements encode proteins involved in recombination, conjugation, transcriptional regulation, and accessory functions (i.e., not involved in the previous three roles). The accessory module which, in Tn*916* contains *tet*(M) and in Tn*1549* contains the *vanB* operon (Garnier et al., 2000), is likely to have an important impact on *C. difficile* biology as many different Tn*916*- and Tn*1549*-like elements have been found in *C. difficile* genomes and almost all of them have different accessory modules which have presumably led to selection of the host conjugative transposon (Brouwer, Roberts, et al., 2012; Brouwer et al., 2011; Dingle et al., 2014; He et al., 2010; Sebaihia et al., 2006), some of which will be discussed below.

5.1. Tn*916*-like elements

The first Tn*916*-like element to be extensively characterised in *C. difficile* was Tn*5397* from *C. difficile* 630. This 21 kb element encodes tetracycline resistance via *tet*(M) and is highly related to Tn*916* across its length apart from the ends (Roberts, Braun, et al., 2001; Roberts, Johanesen, et al., 2001; Wang et al., 2000). In Tn*916*, the genes, *xisTn* and *intTn*, encode an excisionase and a tyrosine integrase, respectively, responsible for the excision, circularisation, and integration of the element. In Tn*5397*, these genes are absent and are replaced by a gene encoding a large serine recombinase, *tndX*, the product of which is closely related to the large serine recombinase TnpX encoded by the previously mentioned mobilisable transposons Tn*4451* and Tn*4453* (Wang et al., 2000). The other main difference between Tn*916* and Tn*5397* is the presence of a group II intron within *orf14* of Tn*5397* (see above). Tn*5397* is capable of transfer between *C. difficile* strains and to and from *B. subtilis* and *E. faecalis* (Jasni et al., 2010; Mullany et al., 1990). Moreover, Tn*5397* has been found in *Enterococcus faecium* (Agersø, Pedersen, & Aarestrup, 2006) and has been shown to transfer to an oral streptococcal recipient in a mixed species biofilm (Roberts, Pratten, Wilson, & Mullany, 1999).

Six other putative conjugative transposons which belonged either to the Tn*916*-like family (CTn*1*, CTn*6*, CTn*7* plus the previously known Tn*5397* element) or to the Tn*1549*-like family (CTn*2*, CTn*4*, and CTn*5*) have been identified by genome sequencing of strain 630

(Sebaihia et al., 2006). CTn*1* and CTn*6* encode a tyrosine integrase similar to that found in Tn*916*, whereas Tn*5397* and CTn7 encode a large serine recombinase. The accessory region of each of the elements are different in all four of the Tn*916*-like elements; containing genes predicted to encode Tet(M) (Tn*5397*), an ABC transporter (CTn*1*), a magnesium transporting P-type ATPase (CTn*7*), or hypothetical proteins (CTn*6*). Furthermore, all of these elements with the exception of CTn*6* have been shown to transfer to *B. subtilis* (Brouwer et al., 2011). Interrogation of the genome sequences of multiple *C. difficile* strains showed that these elements were extremely common and were found in all of the genome sequences analysed (Brouwer, Allan, Mullany, & Roberts, 2012; Brouwer, Roberts, et al., 2012; Brouwer et al., 2011).

In *C. difficile* R20291, the ribotype 027 strain responsible for the outbreaks at Stoke Mandeville Hospital which began in 2003, two conjugative transposons were detected. One was very similar (82% identical excluding the accessory module) to the CTn*1* element from strain 630 and one was similar to CTn*5* and is discussed below. Every ribotype 027 strain subsequently analysed (Brouwer et al., 2011) contained an element highly related (70–98% sequence identity) to the CTn*1*-like element in R20291.

In the other strains examined (Brouwer, Allan, et al., 2012; Brouwer, Roberts, et al., 2012; Brouwer et al., 2011), there is a vast array of accessory content present with many predicted ABC transporters potentially involved in resistance to antimicrobial agents produced by both the host and the gut microbiota. Strain QCD-63Q42 contains two CTn*1*-like elements, the first of which contains an insertion of a prophage showing 92% sequence similarity to prophage 1 from strain 630. The second CTn*1*-like element in this strain contains an accessory region of 33 kb (compared to a 9.5 kb accessory region in strain 630) which is predicted to encode ABC transporters, two sigma factors and a transcriptional regulator (Brouwer et al., 2011).

5.2. Tn*1549*-like elements

The original Tn*1549* element described in *E. faecalis* encodes resistance to vancomycin via the *vanB* operon. The acquisition of vancomycin resistance by *C. difficile* would be a clinical disaster as vancomycin is often the drug of choice for treatment. Despite all of the Tn*1549*-like elements discovered in *C. difficile* so far, we have not seen one carrying a functional *vanB* operon. Each of the Tn*1549*-like elements discovered to date has a different accessory region, again often predicted to encode ABC transporters as is the case

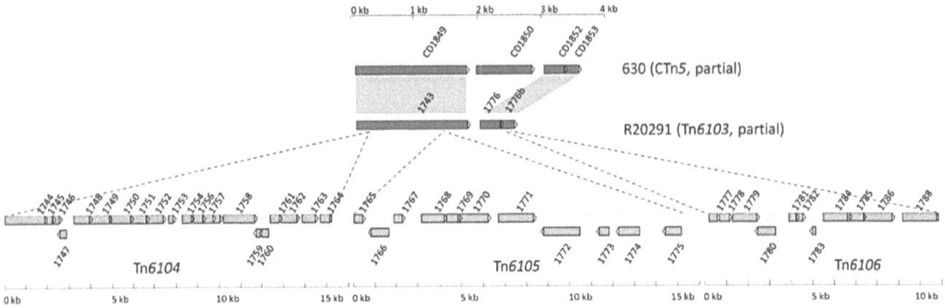

Figure 1 Partial comparison of CTn5 from strain 630 and Tn*6103* from strain R20291. The three insertion sites of the putative mobilisable transposons; Tn*6104*, Tn*6105*, and Tn*6106* can be seen in the R20291, ORFs 1743, and 1776b. *Reproduced from Brouwer et al. (2011).*

with the Tn*916*-like elements. In strain R20291, there is an element which has been designated Tn*6103*. This element is highly related to CTn*5* from *C. difficile* 630; however, it also has three insertions. Within two genes in Tn*6103* (CD1743 and CD1776b) there are three insertions of putative mobilisable transposons which have been designated Tn*6104*, Tn*6105* (both 15 kb and inserted into CD1743), and Tn*6106* (10 Kb inserted into CD1776b; Fig. 1).

Each of these three insertions has the hall marks of independent transposable elements, such as possession of a recombinase gene. In order to see if these were discrete mobile elements, PCRs were carried out with outward facing primers designed to amplify the circular form of the excised mobilisable transposons. This PCR was successful for Tn*6104* which demonstrates it is capable of excision. Tn*6104* contains 21 ORFs including a large serine recombinase related to that found in Tn*4453* and Tn*4451* and a mob gene as discussed above. In contrast to Tn*4453* which carries *catD* as its accessory gene, Tn*6104* carries accessory genes predicted to encode a transcriptional regulator, a two component regulatory system, an ABC transporter, three sigma factor encoding genes and a putative toxin–antitoxin system. The existence of a conjugative transposon carrying multiple transcriptional regulators is interesting since it is possible that, rather than just regulating transcription within the MGE, these might influence the host transcriptome, an intriguing possibility that awaits experimental investigation (Brouwer et al., 2011). In addition to the evidence of excision and circularisation of one of these elements, many other related elements have been found, which do not contain Tn*6104* or Tn*6106* indicating that these are also capable of insertion into the Tn*6103* like elements in different *C. difficile* strains (Brouwer et al., 2011). The details of many of the

Tn*916*- and Tn*1549*-like elements discovered to date can be found in Brouwer, Allan, et al. (2012), Brouwer, Roberts, et al. (2012), Brouwer et al. (2011), and Dingle et al. (2014).

6. OTHER INTEGRATIVE MGEs IN *C. difficile*

There are also some MGEs present within genomes of *C. difficile* which do not fit into the above two groups such as Tn*6164*, a 106-kb insertion found within the genome of the PCR ribotype 078 strain, M120, isolated from an Irish diabetic patient (Corver et al., 2012; He et al., 2010). Bioinformatic analysis of the 90 ORFs within this 106 kb insertion revealed that this element is composed of parts of other MGEs from different bacteria with the likely origin of the majority of genes being *Thermoanaerobacter* sp. and *Streptococcus pneumoniae* (Corver et al., 2012). Tn*6164* contains *tet*(44) and *ant(6)-Ib* predicted to confer resistance to tetracycline and streptomycin, respectively, although no phenotype could be ascribed to these genes. PCR for the circular form was positive and showed that the entire element could excise from the genome of strain M120 and circularise, however, despite extensive experimentation, conjugation to the non-toxigenic recipient strain CD37, could not be detected. Interestingly, the repeated conjugation of a co-resident Tn*916*-like element, designated Tn*6190*, could be detected (Corver et al., 2012). Other sequenced PCR ribotype 078 strains were screened for the presence of Tn*6164* and it was found to be sporadically present among closely related strains. Additionally a small number of strains were found to contain only half of this element, which originates from *Thermoanaerobacter* and is actually a complete prophage (Corver et al., 2012). Analysis of data from the patients from which the different *C. difficile*::Tn*6194* strains originated shows that patients both with and without the element suffered from severe diarrhoea but mortality was more common in patients infected with *C. difficile*::Tn*6194* (29% vs. 3%). It should be stressed that this data is only indicative of a possible link between Tn*6194* and virulence because, as stated by the authors, the numbers are too low for the *C. difficile*::Tn*6194* positive patients ($n=7$) so multivariate analysis could not be carried (Corver et al., 2012).

7. THE *skin*^{Cd} ELEMENT

The 14.6-kb *skin*^{Cd} element is a prophage-like genetic element, which is inserted into the *sigK* gene of *C. difficile* (Haraldsen & Sonenshein, 2003).

For efficient sporulation, *skin*^{Cd} needs to excise from the genome in the mother cell to produce an intact *sigK* gene. Excision of *skin*^{Cd} is mediated by a large serine recombinase. The exact mechanism by which excision mediates efficient sporulation remains to be determined; however; strains which lack the *skin*^{Cd} element have been shown to be defective in sporulation (Haraldsen & Sonenshein, 2003).

8. BACTERIOPHAGES

Prophages are common in the genomes of both clinical (Nale et al., 2012; Shan et al., 2012) and environmental (Hargreaves, Colvin, Patel, Clokie, & Clokie, 2013) *C. difficile* isolates. Viral particles with genomes identical to chromosomally located prophages have also been recovered from filter-sterilised stool samples from patients suffering with CDI (Meessen-Pinard, Sekulovic, & Fortier, 2012) demonstrating that the prophages are induced during infection. Two of these prophages isolated from stool samples, φMMP02 and φMMP04 were found to be induced spontaneously during broth culture of the lysogen as their virions could be detected at levels of 4–5 log plaque forming units (PFU)/ml in the culture supernatant; however; this increased to 8–9 PFU/ml when the lysogens were exposed to the antibiotics ciprofloxacin, moxifloxacin, levofloxacin (all quinolones), or mitomycin C (Meessen-Pinard et al., 2012). These quinolone antibiotics are known risk factors for CDI (Pépin et al., 2005) and may also promote horizontal gene transfer between *C. difficile* strains, mediated by bacteriophages.

While the direct link between bacteriophages and virulence of *C. difficile* is still being explored and the presence of virulence factors on bacteriophage genomes has not yet been proven, there are some recent studies demonstrating a bacteriophage-specific effect on toxin production. Three recent studies have shown that the *C. difficile* phages, φCD119, φCD38-2, and φCD27, can modulate toxin production inferring there must be some interaction with the PaLoc, either directly or via regulatory proteins although this is not yet understood (Govind, Vediyappan, Rolfe, Dupuy, & Fralick, 2009; Sekulovic, Meessen-Pinard, & Fortier, 2011; Williams et al., 2013).

Despite no proven virulence factors (e.g., toxin genes) being found within *C. difficile* bacteriophage genomes, other bacterial genes have been discovered which may themselves modulate virulence factors of *C. difficile*. The bacteriophage φCDHM1 has recently been sequenced revealing the presence of *agr* gene homologues within the 52-kb φCDHM1 genome (Hargreaves, Kropinski, & Clokie, 2014). The *agr* locus, comprising *agrACDB* encodes a quorum-sensing system that is present in many Gram-positive pathogens including *C. difficile*. The chromosomally located *agr* system of strain R20291 was recently characterised. Mutation of this chromosomally located *agrA* showed that this system is responsible for modulating the expression of 75 genes during late exponential growth of *C. difficile* R20291 (Martin et al., 2013) including genes associated with flagella assembly and function and also toxin production. As the *agr*-like genes in φCDHM1 are maintained during the lytic cycle and subsequent lysogeny of a new host, and are expressed in this host, it is likely that the quorum-sensing system is interacting with the host genome but the manner of such an interaction is still currently unknown (Hargreaves et al., 2014).

Another interesting bacteriophage–host interaction is seen with φC2 in *C. difficile* strain CD80. The φC2 bacteriophage is capable of transducing erythromycin resistance to other *C. difficile* strains. The erythromycin resistance gene was localised to a novel 13 Kb transposon designated Tn*6215* (Goh et al., 2013). Tn*6215* has 17 ORFs and is predicted to encode a homologue of the mob protein, TnpZ, from Tn*4453* and Tn*4451* (see above); however, this is truncated and also lacks the *oriT* site so is likely not functional. Tn*6125* contains plasmid-related genes predicted to encode a Rep, related to plasmids from other low GC% Gram-positive bacteria, two toxin/antitoxin systems, one of which is related to that found in Tn*5398* and a CTn*1*-like element found in strains M68 and 2007855 (Brouwer, Roberts, et al., 2012). Tn*6215* was also shown to transfer during filter-mating experiments from a φC2 free donor, albeit at a much lower frequency of transfer (1.2×10^{-6} transconjugants per donor cell for transduction compared to 1.8×10^{-9} during filter mating; Goh et al., 2013), suggesting it is either a mobilisable transposon or it can be transduced at low level during filter-mating experiments by a native phage other than φC2 from this donor.

These studies show that bacteriophages are common in *C. difficile* isolates and have the ability to modulate host gene expression. Additionally, transduction of antibiotic resistance has been demonstrated showing that bacteriophage are responsible for the horizontal gene transfer of DNA other than their own genomes. The overall impact that bacteriophage have on the biology of *C. difficile* is not yet fully understood and continues to be explored.

9. TRANSFER OF THE PaLoc

The PaLoc has long been thought of as either an independent MGE or a remnant of a bacteriophage, based on the similarity of the PaLoc-encoded

TcdE and bacteriophage holins and supported by the observed effect of some bacteriophages on toxin modulation (Govind et al., 2009; Sekulovic et al., 2011; Williams et al., 2013). The PaLoc has other hallmarks of an MGE, not least because the entire 19.6 kb region of DNA is absent in non-toxigenic *C. difficile* strains, and instead there is 115 bp of non-coding DNA at the integration site (Braun et al., 1996; Brouwer, Roberts, et al., 2012). Additionally, the PaLoc is always found at the same chromosomal site in toxigenic strains (Dingle et al., 2011). Large scale genomic studies of *C. difficile* populations suggested that the PaLoc has transferred between *C. difficile* lineages in the past (Dingle et al., 2014; He et al., 2010) and recently it was demonstrated that the PaLoc is transferable from strain 630 (ribotype 012) to the non-toxigenic strains, CD37 (ribotype 009), OX904rif (ribotype 138), and OX 2157rif (ribotype 140) (Brouwer et al., 2013). In one of the CD37 transconjugants, toxin production was found to be comparable to that of the 630 donor by endpoint titration assays using culture supernatant and HFF-1 cells (Brouwer et al., 2013).

Unlike mobilisable and conjugative transposons, which transfer as discrete genetic entities with well-defined ends (Roberts & Mullany, 2011), the PaLoc transfers on variably sized DNA fragments ranging from at least 66 to 272 kbp (Brouwer et al., 2013) (Fig. 2).

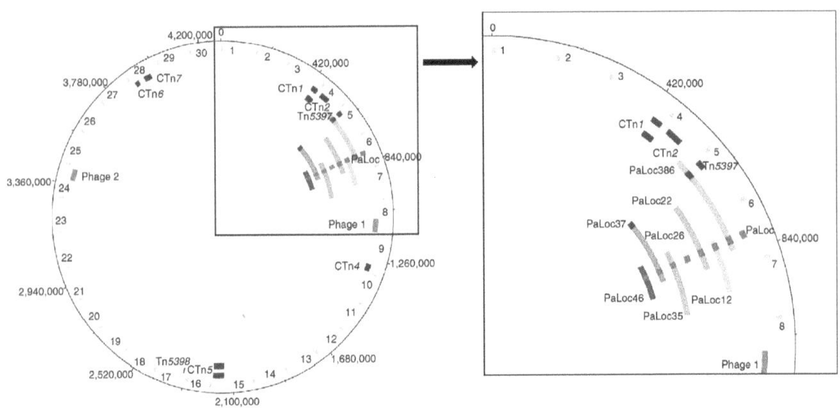

Figure 2 The panel on the left shows a circular schematic representation of the *C. difficile* 630 chromosome. The first inner ring shows the position of the 30 regions of differing DNA, which were used to determine the source of DNA in transconjugants. The second inner ring shows the position of the PaLoc, conjugative transposons and prophage. The next eight inner rings show the extent of the transferred DNA in eight transconjugants, which are enlarged and labelled in the right-hand panel. *Reproduced from Brouwer et al. (2013).* (See the color plate.)

The transfer frequency is relatively low, between 10^{-7} and 10^{-9} transconjugants per donor, similar to that observed with the conjugative transposons, Tn*5397*, CTn*5*, and CTn*1* in strain 630 (Brouwer et al., 2011; Roberts, Braun, et al., 2001; Roberts, Johanesen, et al., 2001). Remarkably, despite this low transfer frequency, co-transfer of the PaLoc with conjugative transposons was observed. In one transconjugant, Tn*5397* was shown to have transferred on the same piece of DNA as the PaLoc where as other PaLoc transconjugants had acquired conjugative transposons as independent transfer events, not linked to the DNA responsible for PaLoc transfer.

This suggested that the transfer of the PaLoc-containing DNA molecules and the separate transfer of the conjugative transposon molecules are linked. If the PaLoc and conjugative transposons were transferring independently of each other, the non-selected element would almost never be seen in transconjugants as the non-selected element is expected to be present in only 1 out of 10^{16} transconjugants (i.e., as PaLoc and conjugative transposons both transfer at a frequency of approximately 1 transconjugant per 10^{8} cells, then the expected frequency of transfer if the two events are not linked is $10^{8} \times 10^{8} = 1$ in 10^{16}).

A hypothesis to explain these observations is that one or more of the conjugative transposons are mediating transfer of the chromosome by a process similar to Hfr in which an *oriT* within the chromosome mediates chromosomal transfer. In the recipient, the incoming chromosomal DNA is integrated into the chromosome by homologous recombination. In theory, the whole chromosome can be transferred but this is not usually the case as conjugation usually stops before full chromosomal transfer due to disruption of the mating pairs or nicking of the incoming DNA strand. Therefore DNA proximal to the integrated *oriT* is transferred at a higher frequency. There are examples of conjugative transposons mediating Hfr type transfer, e.g., in *Vibrio cholerae* (Hochhut, Marrero, & Waldor, 2000) and *Bacteroides* sp. (Whittle, Hamburger, Shoemaker, & Salyers, 2006). In these cases, the conjugative transposons do not excise so the chromosome is mobilised. Three transfer-proficient conjugative transposons in the *C. difficile* genome: CTn*1*, CTn*2*, and Tn*5397* are relatively close to the PaLoc (Brouwer et al., 2013; Fig. 2) and are candidates for mediation of Hfr transfer. Although all can excise (Brouwer et al., 2011) this does not mean that they always do before the *oriT* is nicked. A likely explanation for the frequent co-transfer of genetic elements is that when one or more of the conjugative transposons initiates a transfer event, other elements present in the same cell are activated by *trans*-acting regulatory molecules. This type of interaction has been observed between Tn*916*-like elements, although the regulatory molecule(s) responsible have not been identified (Flannagan & Clewell, 1994; Wang et al., 2000). Another intriguing possibility is that the transfer of a large section of the chromosome confers a selective advantage to *C. difficile* and may result in chromosomal repair by recombination enabling exaggerated genomic plasticity and adaptability similar to that hypothesised for the evolution of competence and DNA uptake by *S. pneumoniae* (Johnston, Campo, Bergé, Polard, & Claverys, 2014). In this case, it would be advantageous to have multiple *oriT*s around the chromosome in order to ensure transfer of all parts of the chromosome which may explain why the conjugative transposons are so prolific and dispersed within the *C. difficile* genomes. Although it is likely that the conjugative transposons are responsible for the transfer of the PaLoc, this is still under investigation.

10. CONCLUSIONS

A common theme throughout the description of the different types of MGE presented in this chapter is that they can associate together. We have described an intron within a conjugative transposon (Tn*5397*), an IStron within the PaLoc, mobilisable transposons (Tn*6104*, Tn*6105*, and Tn*6106*) within a conjugative transposon (Tn*6103*), and a prophage within a transposon (Tn*6164*). *C. difficile* appears to have an inherent ability to acquire MGEs of various different classes and these MGEs can co-localise together within the genome, possibly minimising their fitness costs and/or because insertion into another MGE is less likely to lead to a deleterious effect on the host and is therefore more often selected. Whatever the reason, it gives *C. difficile* the opportunity to build up a complex and often unexpected (e.g., the IStrons) array of MGEs, which have profound effects on its biology. The effects on *C. difficile* range from modulatory effects on gene expression, e.g., the toxin and flagella genes and control of sporulation, to mutational effects of insertion and excision of an element (such as the IStron in *tcdA*) and *trans*-acting effects such as the mobilisation of large chromosomal regions which transfer at the same time as independent conjugative transposons. It is worth highlighting that the production of the main virulence factors, TcdA and TcdB, are both modulated by MGEs and their cognate genes are likely transferred by MGEs demonstrating a fundamental reliance of the pathogen on its mobile DNA. One could wonder whether *C. difficile* is the master of its mobilome or simply its puppet.

REFERENCES

Adams, V., Lyras, D., Farrow, K. A., & Rood, J. I. (2002). The clostridial mobilisable transposons. *Cellular and Molecular Life Sciences*, *59*, 2033–2043.

Agersø, Y., Pedersen, A. G., & Aarestrup, F. M. (2006). Identification of Tn*5397*-like and Tn*916*-like transposons and diversity of the tetracycline resistance gene *tet*(M) in enterococci from humans, pigs and poultry. *Journal of Antimicrobial Chemotherapy*, *57*, 832–839.

Arroyo, L. G., Kruth, S. A., Willey, B. M., Staempfli, H. R., Low, D. E., & Weese, J. S. (2005). PCR ribotyping of *Clostridium difficile* isolates originating from human and animal sources. *Journal of Medical Microbiology*, *54*, 163–166.

Bakker, D., Smits, W. K., Kuijper, E. J., & Corver, J. (2012). TcdC does not significantly repress toxin expression in *Clostridium difficile* 630ΔErm. *PLoS One*, 7, e43247.

Barbut, F., Jones, G., & Eckert, C. (2011). Epidemiology and control of *Clostridium difficile* infections in healthcare settings: An update. *Current Opinion in Infectious Disease*, *24*, 370–376.

Belfort, M., & Perlman, P. S. (1995). Mechanisms of intron mobility. *Journal of Biological Chemistry*, *270*, 30237–30240.

Braun, V., Hundsberger, T., Leukel, P., Sauerborn, M., & von Eichel-Streiber, C. (1996). Definition of the single integration site of the pathogenicity locus in *Clostridium difficile*. *Gene*, *181*, 29–38.

Brouwer, M. S., Allan, E., Mullany, P., & Roberts, A. P. (2012). Draft genome sequence of the nontoxigenic *Clostridium difficile* strain CD37. *Journal of Bacteriology*, *194*, 2125–2126.

Brouwer, M. S., Roberts, A. P., Hussain, H., Williams, R. J., Allan, E., & Mullany, P. (2013). Horizontal gene transfer converts non-toxigenic *Clostridium difficile* strains into toxin producers. *Nature Communications*, *4*, 2601.

Brouwer, M. S., Roberts, A. P., Mullany, P., & Allan, E. (2012). In silico analysis of sequenced strains of *Clostridium difficile* reveals a related set of conjugative transposons carrying a variety of accessory genes. *Mobile Genetic Elements*, *2*, 8–12.

Brouwer, M. S., Warburton, P. J., Roberts, A. P., Mullany, P., & Allan, E. (2011). Genetic organisation, mobility and predicted functions of genes on integrated, mobile genetic elements in sequenced strains of *Clostridium difficile*. *PLoS One*, *6*(8), e23014. http://dx.doi.org/10.1371/journal.pone.0023014.

Corver, J., Bakker, D., Brouwer, M. S., Harmanus, C., Hensgens, M. P., Roberts, A. P., et al. (2012). Analysis of a *Clostridium difficile* PCR ribotype 078 100 kilobase island reveals the presence of a novel transposon, Tn*6164*. *BMC Microbiology*, *12*, 130.

Crellin, P. K., & Rood, J. I. (1998). Tn*4451* from *Clostridium perfringens* is a mobilizable transposon that encodes the functional Mob protein, TnpZ. *Molecular Microbiology*, *27*, 631–642.

Depardieu, F., Reynolds, P. E., & Courvalin, P. (2003). VanD-type vancomycin-resistance *Enterococcus faecium* 10/96A. *Antimicrobial Agents and Chemotherapy*, *47*, 7–18.

Dingle, K. E., Elliott, B., Robinson, E., Griffiths, D., Eyre, D. W., Stoesser, N., et al. (2014). Evolutionary history of the *Clostridium difficile* pathogenicity locus. *Genome Biology and Evolution*, *6*, 36–52.

Dingle, K. E., Griffiths, D., Didelot, X., Evans, J., Vaughan, A., Kachrimanidou, M., et al. (2011). Clinical *Clostridium difficile*: Clonality and pathogenicity locus diversity. *PLoS One*, *6*, e19993.

Farrow, K. A., Lyras, D., & Rood, J. I. (2001). Genomic analysis of the erythromycin resistance element Tn*5398* from *Clostridium difficile*. *Microbiology*, *147*, 2717–2728.

Flannagan, S. E., & Clewell, D. B. (1994). Conjugative transfer of Tn*916* in *Enterococcus faecalis*: Trans activation of homologous transposons. *Journal of Bacteriology*, *173*, 7136–7141.

Franke, A. E., & Clewell, D. B. (1981). Evidence for a chromosome-borne resistance transposon (Tn*916*) in *Streptococcus faecalis* that is capable of "conjugal" transfer in the absence of a conjugative plasmid. *Journal of Bacteriology, 145*, 494–502.

Garnier, F., Taourit, S., Glaser, P., Courvalin, P., & Galimand, M. (2000). Characterization of transposon Tn*1549*, conferring VanB-type resistance in *Enterococcus* spp. *Microbiology, 146*, 1481–1489.

Gerding, D. N., Johnson, S., Rupnik, M., & Aktories, K. (2014). *Clostridium difficile* binary toxin CDT: Mechanism, epidemiology, and potential clinical importance. *Gut Microbes, 5*, 15–27.

Goh, S., Hussain, H., Chang, B. J., Emmett, W., Riley, T. V., & Mullany, P. (2013). Phage φC2 mediates transduction of Tn*6215*, encoding erythromycin resistance, between *Clostridium difficile* strains. *MBio, 4*, e00840-13.

Govind, R., Vediyappan, G., Rolfe, R. D., Dupuy, B., & Fralick, J. A. (2009). Bacteriophage-mediated toxin gene regulation in *Clostridium difficile*. *Journal of Virology, 83*, 12037–12045.

Hächler, H., Berger-Bächi, B., & Kayser, F. H. (1987). Genetic characterization of a Clostridium difficile erythromycin-clindamycin resistance determinant that is transferable to *Staphylococcus* aureus. *Antimicrobial Agents and Chemotherapy, 31*(7), 1039–1045.

Hall, J. C., & O'Toole, E. (1935). Intestinal flora in new-born infants with a description of a new pathogenic anaerobe *Bacillus difficilis*. *American Journal of Diseases of Children, 49*, 390–402.

Haraldsen, J. D., & Sonenshein, A. L. (2003). Efficient sporulation in *Clostridium difficile* requires disruption of the sigmaK gene. *Molecular Microbiology, 48*, 811–821.

Hargreaves, K. R., Colvin, H. V., Patel, K. V., Clokie, J. J., & Clokie, M. R. (2013). Genetically diverse *Clostridium difficile* strains harboring abundant prophages in an estuarine environment. *Applied and Environmental Microbiology, 79*, 6236–6243.

Hargreaves, K. R., Kropinski, A. M., & Clokie, M. R. (2014). What does the talking?: Quorum sensing signalling genes discovered in a bacteriophage genome. *PLoS One, 9*, e85131.

Hasselmayer, O., Nitsche, C., Braun, V., & von Eichel-Streiber, C. (2004). The IStron CdISt1 of *Clostridium difficile*: Molecular symbiosis of a group I intron and an insertion element. *Anaerobe, 10*, 85–92.

He, M., Sebaihia, M., Lawley, T. D., Stabler, R. A., Dawson, L. F., Martin, M. J., et al. (2010). Evolutionary dynamics of *Clostridium difficile* over short and long time scales. *Proceedings of the National Academy of Sciences of the United States of America, 107*, 7527–7532.

Hochhut, B., Marrero, J., & Waldor, M. K. (2000). Mobilization of plasmids and chromosomal DNA mediated by the SXT element, a constin found in *Vibrio cholerae* O139. *Journal of Bacteriology, 183*, 1124–1132.

Jasni, A. S., Mullany, P., Hussain, H., & Roberts, A. P. (2010). Demonstration of conjugative transposon (Tn*5397*)-mediated horizontal gene transfer between *Clostridium difficile* and *Enterococcus faecalis*. *Antimicrobial Agents and Chemotherapy, 54*, 4924–4926.

Johnston, C., Campo, N., Bergé, M. J., Polard, P., & Claverys, J. P. (2014). *Streptococcus pneumoniae*, le transformiste. *Trends on Microbiology, 22*, 113–119.

Keel, K., Brazier, J. S., Post, K. W., Weese, S., & Songer, J. G. (2007). Prevalence of PCR ribotypes among *Clostridium difficile* isolates from pigs, calves, and other species. *Journal of Clinical Microbiology, 45*, 1963–1964.

Kuehne, S. A., Cartman, S. T., Heap, J. T., Kelly, M. L., Cockayne, A., & Minton, N. P. (2010). The role of toxin A and toxin B in *Clostridium difficile* infection. *Nature, 467*, 711–713.

Lyras, D., Adams, V., Lucet, I., & Rood, J. I. (2004). The large resolvase TnpX is the only transposon-encoded protein required for transposition of the Tn*4451/3* family of integrative mobilizable elements. *Molecular Microbiology, 51*, 1787–1800.

Lyras, D., O'Connor, J. R., Howarth, P. M., Sambol, S. P., Carter, G. P., Phumoonna, T., et al. (2009). Toxin B is essential for virulence of *Clostridium difficile*. *Nature, 458*, 1176–1179.

Lyras, D., & Rood, J. I. (2000). Transposition of Tn*4451* and Tn*4453* involves a circular intermediate that forms a promoter for the large resolvase, TnpX. *Molecular Microbiology, 38*, 588–601.

Lyras, D., Storie, C., Huggins, A. S., Crellin, P. K., Bannam, T. L., & Rood, J. I. (1998). Chloramphenicol resistance in *Clostridium difficile* is encoded on Tn*4453* transposons that are closely related to Tn*4451* from *Clostridium perfringens*. *Antimicrobial Agents and Chemotherapy, 42*, 1563–1567.

Martin, M. J., Clare, S., Goulding, D., Faulds-Pain, A., Barquist, L., Browne, H. P., et al. (2013). The *agr* locus regulates virulence and colonization genes in *Clostridium difficile* 027. *Journal of Bacteriology, 195*, 3672–3681.

Meessen-Pinard, M., Sekulovic, O., & Fortier, L. C. (2012). Evidence of *in vivo* prophage induction during *Clostridium difficile* infection. *Applied and Environmental Microbiology, 78*, 7662–7670.

Mullany, P., Pallen, M., Wilks, M., Stephen, J. R., & Tabaqchali, S. (1996). A group II intron in a conjugative transposon from the gram-positive bacterium, *Clostridium difficile*. *Gene, 174*, 145–150.

Mullany, P., Wilks, M., Lamb, I., Clayton, C., Wren, B., & Tabaqchali, S. (1990). Genetic analysis of a tetracycline resistance element from *Clostridium difficile* and its conjugal transfer to and from *Bacillus subtilis*. *Journal of General Microbiology, 136*, 1343–1349.

Mullany, P., Wilks, M., & Tabaqchali, S. (1995). Transfer of macrolide-lincosamide-streptogramin B (MLS) resistance in *Clostridium difficile* is linked to a gene homologous with toxin A and is mediated by a conjugative transposon, Tn*5398*. *Journal of Antimicrobial Chemotherapy, 35*, 305–315.

Nale, J. Y., Shan, J., Hickenbotham, P. T., Fawley, W. N., Wilcox, M. H., & Clokie, M. R. (2012). Diverse temperate bacteriophage carriage in *Clostridium difficile* 027 strains. *PLoS One, 7*, e37263.

Norén, T. (2010). *Clostridium difficile* and the disease it causes. *Methods in Molecular Biology, 646*, 9–35.

O'Keeffe, T., Hill, C., & Ross, R. P. (1999). In situ inversion of the conjugative transposon Tn*916* in *Enterococcus faecium* DPC3675. *FEMS Microbiology Letters, 173*, 265–271.

Pépin, J., Saheb, N., Coulombe, M. A., Alary, M. E., Corriveau, M. P., Authier, S., et al. (2005). Emergence of fluoroquinolones as the predominant risk factor for *Clostridium difficile*-associated diarrhea: A cohort study during an epidemic in Quebec. *Clinical Infectious Disease, 41*, 1254–1260.

Roberts, A. P., Braun, V., von Eichel-Streiber, C., & Mullany, P. (2001). Demonstration that the group II intron from the clostridial conjugative transposon Tn*5397* undergoes splicing *in vivo*. *Journal of Bacteriology, 183*, 1296–1299.

Roberts, A. P., Chandler, M., Courvalin, P., Guédon, G., Mullany, P., Pembroke, T., et al. (2006). Revised nomenclature for transposable genetic elements. *Plasmid, 60*, 167–173.

Roberts, A. P., Johanesen, P. A., Lyras, D., Mullany, P., & Rood, J. I. (2001). Comparison of Tn*5397* from *Clostridium difficile*, Tn*916* from *Enterococcus faecalis* and the CW*459tet*(-M) element from *Clostridium perfringens* shows that they have similar conjugation regions but different insertion and excision modules. *Microbiology, 147*, 1243–1251.

Roberts, A. P., & Mullany, P. (2011). Tn*916*-like genetic elements: A diverse group of modular mobile elements conferring antibiotic resistance. *FEMS Microbiology Reviews, 35*, 856–871.

Roberts, A. P., Pratten, J., Wilson, M., & Mullany, P. (1999). Transfer of a conjugative transposon, Tn*5397* in a model oral biofilm. *FEMS Microbiology Letters, 177*, 63–66.

Rupnik, M., Wilcox, M. H., & Gerding, D. N. (2009). *Clostridium difficile* infection: New developments in epidemiology and pathogenesis. *Nature Reviews in Microbiology*, *7*, 526–536.

Sebaihia, M., Wren, B. W., Mullany, P., Fairweather, N. F., Minton, N., Stabler, R., et al. (2006). The multidrug-resistant human pathogen *Clostridium difficile* has a highly mobile, mosaic genome. *Nature Genetics*, *38*, 779–786.

Sekulovic, O., Meessen-Pinard, M., & Fortier, L. C. (2011). Prophage-stimulated toxin production in *Clostridium difficile* NAP1/027 lysogens. *Journal of Bacteriology*, *193*, 2726–2734.

Shan, J., Patel, K. V., Hickenbotham, P. T., Nale, J. Y., Hargreaves, K. R., & Clokie, M. R. (2012). Prophage carriage and diversity within clinically relevant strains of *Clostridium difficile*. *Applied and Environmental Microbiology*, *78*, 6027–6034.

Smits, W. K. (2013). Hype or Hypervirulence: A reflection on problematic *C. difficile* strains. *Virulence*, *4*, 592–596.

Songer, J. G. (2004). The emergence of *Clostridium difficile* as a pathogen of food animals. *Animal Health Research Reviews*, *5*, 321–326.

Wang, H., Roberts, A. P., Lyras, D., Rood, J. I., Wilks, M., & Mullany, P. (2000). Characterization of the ends and target sites of the novel conjugative transposon Tn*5397* from *Clostridium difficile*: Excision and circularization is mediated by the large resolvase, TndX. *Journal of Bacteriology*, *182*, 3775–3783.

Wasels, F., Spigaglia, P., Barbanti, F., & Mastrantonio, P. (2013). *Clostridium difficile erm*(B)-containing elements and the burden on the in vitro fitness. *Journal of Medical Microbiology*, *62*, 1461–1467.

Whittle, G., Hamburger, N., Shoemaker, N. B., & Salyers, A. A. (2006). A bacteroides conjugative transposon, CTnERL, can transfer a portion of itself by conjugation without excising from the chromosome. *Journal of Bacteriology*, *188*, 1169–1174.

Williams, R., Meader, E., Mayer, M., Narbad, A., Roberts, A. P., & Mullany, P. (2013). Determination of the attP and attB sites of phage CD27 from *Clostridium difficile* NCTC 12727. *Journal of Medical Microbiology*, *62*, 1439–1443.

Wren, B. W., Mullany, P., Clayton, C., & Tabaqchali, S. (1998). Molecular cloning and genetic analysis of a chloramphenicol acetyltransferase determinant from *Clostridium difficile*. *Antimicrobial Agents and Chemotherapy*, *32*, 1213–1217.

Wren, B. W., Mullany, P., Clayton, C., & Tabaqchali, S. (1999). Nucleotide sequence of a chloramphenicol acetyltransferase gene from *Clostridium difficile*. *Nucleic Acids Research*, *17*, 4877.

Wüst, J., & Hardegger, U. (1983). Transferable resistance to clindamycin, erythromycin, and tetracycline in *Clostridium difficile*. *Antimicrobial Agents and Chemotherapy*, *23*, 784–786.

CHAPTER THREE

Metal Ion Homeostasis in *Listeria monocytogenes* and Importance in Host–Pathogen Interactions

Helen E. Jesse, Ian S. Roberts, Jennifer S. Cavet[1]
Faculty of Life Sciences, University of Manchester, Manchester, United Kingdom
[1]Corresponding author: e-mail address: jennifer.s.cavet@manchester.ac.uk

Contents

Abstract

Listeria monocytogenes is responsible for one of the most life-threatening food-borne infections and the leading cause of food-poisoning associated deaths in the UK. Infection may be of the unborn/newly born infant where disease may manifest as listeric abortion, stillbirth or late-onset neonatal listeriosis, while in adults, infection usually affects the central nervous system causing meningitis. Crucial to the survival of *L. monocytogenes*, both inside and outside the host, is its ability to acquire metals which act as cofactors for a broad range of its cellular proteins. However, *L. monocytogenes* must also protect itself against the innate toxicity of metals. The importance of metals in host–pathogen interactions is illustrated by the restriction of metals (including zinc and iron) in vertebrates in response to infection and the use of high levels of metals (copper and zinc) as part of the antimicrobial defences within host phagocytes. As such, *L. monocytogenes* is equipped with various mechanisms to tightly control its cellular metal pools and avoid metal poisoning. These include multiple DNA-binding metal-responsive transcription factors, metal-acquisition, metal-detoxification and metal-storage systems, some of which represent key *L. monocytogenes* virulence determinants.

Advances in Microbial Physiology
ISSN 0065-2911
http://dx.doi.org/10.1016/bs.ampbs.2014.08.003

This review discusses current knowledge of the role of metals in *L. monocytogenes* infections, with a focus on the mechanisms that contribute to zinc and copper homeostasis in this organism. The requirement to precisely control cellular metal levels may impose a vulnerability to *L. monocytogenes* which can be exploited in antimicrobials and therapeutics.

1. INTRODUCTION

Bacterial survival is dependent upon their ability to acquire metals that are needed to cofactor approximately a third of their proteins (Waldron & Robinson, 2009). However, an over-supply of these metals can also be toxic to bacteria by competitively inhibiting correct metal acquisition by metal-requiring proteins or, in the case of redox active metals, by driving Fenton chemistry to produce deadly hydroxyl radicals (Imlay, Chin, & Linn, 1988; Macomber & Imlay, 2009; Varghese, Wu, Park, Imlay, & Imlay, 2007). To be successful within a particular niche, bacteria must therefore effectively compete for sufficient amounts of each metal to supply their metal-requiring proteins while avoiding an excess of any metal. Host innate immune defences are known to exploit this requirement in a process termed nutritional immunity; a term initially used to refer to the host restriction of iron availability (Weinberg, 2009), but that now extends to include the manipulation of other metals (including manganese, zinc and copper) to attack invading microbes by both metal restriction and exposure to metal excess (for a recent review of this paradigm, refer to Hood & Skaar, 2012). However, bacterial pathogens may respond to this assault by employing strategies to maintain metal homeostasis.

Metal restriction by the host is mediated by the expression of metal/metal-complex binding proteins; including lipocalin-2 (alias siderocalin or neutrophil gelatinase-associated lipocalin), lactoferrin, transferrin and intracellular ferritin that can withhold iron, haptoglobin and haemopexin that bind haem and haemoglobin, respectively, metallothioneins that sequester intracellular zinc, and calprotectin that sequesters zinc and manganese (Hood & Skaar, 2012; Subramanian Vignesh, Landero Figueroa, Porollo, Caruso, & Deepe, 2013). Generalised changes in metal homeostasis also occur in response to infection, such as those resulting from the release of pro-inflammatory cytokines and the hormone hepcidin that lead to reduced systemic iron availability (Cassat & Skaar, 2013; Collins, 2008; Wessling-Resnick, 2010). Metal availability for intracellular pathogens can be further manipulated by the altered expression and/or localisation of host metal transporters that lead to metal fluxes within microbe-containing compartments to either limit availability (iron, manganese and zinc) or intoxicate with excess copper and zinc (Achard et al., 2012; Cellier, Courville, & Campion, 2007; Ding et al., 2013; Kehl-Fie & Skaar, 2010; Osman et al., 2010; Shi et al., 2014; Subramanian Vignesh et al., 2013; Ward, Abomoelak, Hoye, Steinberg, & Talaat, 2010; White, Lee, Kambe, Fritsche, & Petris, 2009).

For bacterial pathogens residing within a host, the levels and sources of metals that they encounter and can access will also vary according to a variety of factors such as their route of entry, their particular host niche, their adaptive strategies for survival within the host and their array of metal uptake and storage systems. The host represents a diverse array of ecological niches (both intracellular and extracellular) with respect to the bioavailability of metals, this being affected not only by total metal content and the strategies for nutritional immunity, but also by factors such as metal speciation, pH, oxygen levels, abundance of metal-binding ligands and the presence of other competing microbes. The specific metal demands of a pathogen will reflect their metalloproteome, but can also alter due to changes in their metabolic and enzymatic requirements within a particular niche, such as the expression of virulence factors, which can in turn be affected by the metal supply (Gabriel & Helmann, 2009; McHugh et al., 2003; Rowe, Starnes, & Chivers, 2005). Bacterial pathogens have thus evolved multiple different strategies to protect against an under-supply or over-supply of any particular metal and possess different repertoires of metal-sensing proteins, metal-acquisition strategies, metal-storage and trafficking proteins, and metal-efflux systems that reflect their individual needs (Diaz-Ochoa, Jellbauer, Klaus, & Raffatellu, 2014; Hood & Skaar, 2012). These metal homeostatic systems may thus equip bacterial pathogens with the ability to survive in a diverse array of host tissues and to resist the host's antimicrobial arsenal associated with altering metal supply.

Listeria monocytogenes has become an important model for studying intracellular parasitism as well as being a paradigm for studying the associated immune responses to intracellular parasites. Much is known about the virulence of *L. monocytogenes* and certain aspects of the host response to infection are also well documented (Cossart, 2011; Mostowy & Cossart, 2012; Serbina, Shi, & Pamer, 2012; Witte et al., 2012). However, the events surrounding the adaptation of this pathogen to different environments encountered both inside and outside the host are less well described. This review focuses on the strategies employed by *L. monocytogenes* to control its cellular metal pools to maintain metal homeostasis; this being of utmost importance for survival in both the external environment and during infections where *L. monocytogenes* must overcome components of host nutritional immunity. In particular, this review describes recent progress in understanding the mechanisms for handling zinc and copper and their roles in *Listeria*–host interactions, while iron and haem transport by *L. monocytogenes* has recently been reviewed elsewhere (Jin et al., 2006; Klebba, Charbit, Xiao, Jiang, & Newton, 2012; McLaughlin, Hill, & Gahan, 2011).

2. OVERVIEW OF *LISTERIA MONOCYTOGENES* DISEASE PROGRESSION

L. monocytogenes is a facultatively anaerobic Gram positive bacterium that is the causative agent of the food-borne infection listeriosis. There are 13 serotypes of *L. monocytogenes*, although most cases of human listeriosis are caused by just four of these; 1/2a, 1/2b, 1/2c and 4b (Cossart, 2011).

L. monocytogenes is prevalent in the environment and is commonly present in the soil, decaying plant matter and the faeces of mammals (Gray & Killinger, 1966; Schuchat, Swaminathan, & Broome, 1991). Indeed, this bacterium has been found in approximately 5% of faecal samples from healthy adults (Schlech et al., 1983; Schuchat et al., 1991). However, human listeriosis is rare, with only 147 reported cases in England and Wales in 2011 (Pubic Health England). Infections are caused by ingestion of contaminated food, including raw vegetables, dairy products, meat and fish. The infective dose is thought to be greater than 10^9 organisms (Lorber, 1997), and the incubation period is 11–70 days, with a mean of 31 days (Linnan et al., 1988). In healthy individuals, the effects are often limited to febrile gastroenteritis, with symptoms including fever, diarrhoea, nausea, aching muscles and headache (Ooi & Lorber, 2005). However, in the elderly and those with a compromised immune system, the consequences of an infection can be much more serious, typically leading to meningitis, septicaemia, endocarditis and encephalitis, with a mortality rate of 20–30% (Cossart, 2011; Nieman & Lorber, 1980). *L. monocytogenes* infections are also a particular risk for pregnant women and their unborn infants, with spontaneous abortion, stillbirth and neonatal meningitis common following infection (Lorber, 1997).

L. monocytogenes has adapted to survive in many different environmental niches, which in a host include the various microenvironments of the gastrointestinal tract and within epithelial cells and phagocytes (Schuppler

& Loessner, 2010). This pathogen is able to efficiently invade and replicate within host cells, which allows it to evade many of the immune defences of the host. Moreover, *L. monocytogenes* is particularly efficient at crossing host barriers, including the intestinal epithelium, the placenta and the blood/brain barrier (Lecuit, 2007; Ooi & Lorber, 2005).

2.1. The intracellular life cycle of *L. monocytogenes*

Following ingestion, *L. monocytogenes* enters the epithelial cells of the intestinal tract and is then able to escape the internalisation vacuole and migrate intracellularly (Fig. 1). Once in the cytosol, *L. monocytogenes* is able to cause

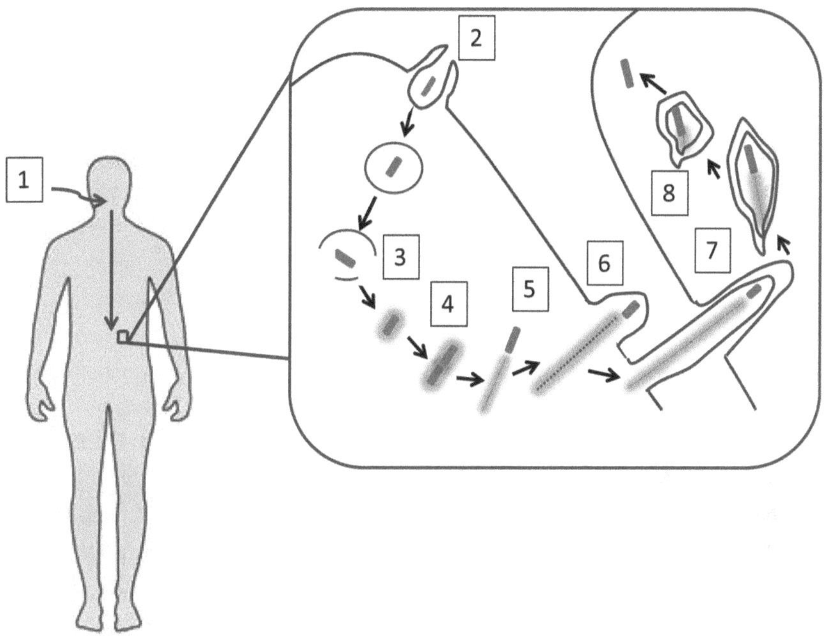

Figure 1 The life cycle of *L. monocytogenes* within the host. *L. monocytogenes* is ingested [1] and enters the epithelial cells of the intestinal tract through interaction of the surface proteins, InlA and InlB, with the host cell-surface receptor E-cadherin and hepatocyte growth factor receptor tyrosine kinase c-Met, respectively [2]. The bacteria are then able to escape the internalisation vacuole in a process that involves the pore-forming toxin listeriolysin O (LLO) and activation of the phospholipases PlcA and PlcB by a zinc-metalloprotease (Mpl) [3]. Once in the cytosol, *L. monocytogenes* is able to replicate [4] and then recruit actin by production of ActA [5], which facilitates motility and cell to cell spread [6]. The actin and the associated bacterium pushes through the cytoplasmic membrane of a neighbouring cell [7], forming a double-membrane phagosome [8], from which *L. monocytogenes* can escape, by again employing LLO, PlcA and PlcB. *Adapted from Tilney and Portnoy (1989).*

nucleation and polymerisation of host cell actin allowing it to spread from cell to cell and evade the immune system of the host (Tilney & Portnoy, 1989). There are several important virulence factors in *L. monocytogenes*, some of which are involved in more than one stage of infection (Lecuit, 2007). The genes encoding many of the key virulence factors are coordinately regulated by the positive transcriptional regulator PrfA, in response to temperature and other environmental cues during the transition from an environmental to intracellular lifestyle, thereby allowing a rapid shift to the 'virulent state' (de las Heras, Cain, Bielecka, & Vazquez-Boland, 2011; Deshayes et al., 2012; Johansson et al., 2002).

L. monocytogenes is able to enter eukaryotic cells by the interaction of one of several surface proteins, including internalins and the actin polymerising protein ActA, with a eukaryotic receptor (Garcia-del Portillo, Calvo, D'Orazio, & Pucciarelli, 2011). The best characterised of these interactions is that of the listerial surface protein internalin A (InlA) with the human transmembrane glycoprotein receptor E-cadherin, which triggers internalisation of *L. monocytogenes* by host cells that would not otherwise be phagocytic (Mengaud et al., 1996; Vazquez-Boland, Dominguez, Fernandez-Garayzabal, & Suarez, 1992). Alternatively, *L. monocytogenes* can enter host cells by interaction of internalin B (InlB) with the hepatocyte growth factor receptor tyrosine kinase c-Met. This interaction instigates phosphorylation of c-Met and subsequent signalling promotes uptake of *L. monocytogenes* by epithelial cells (Shen, Naujokas, Park, & Ireton, 2000). *L. monocytogenes* is also able to enter mammalian cells non-specifically and independently of InlA, InlB and haemolysin via M cells, leading to an accumulation of this organism in Peyer's patches in the small intestine of infected individuals. Established virulence factors may therefore not be essential for *L. monocytogenes* invasion by this route (Corr, Hill, & Gahan, 2006).

The escape of *L. monocytogenes* from the host vacuole involves the activation of the phospholipases PlcA and PlcB by a metalloprotease (Mpl), in conjunction with the haemolysin listeriolysin O, a pore forming toxin which alters vacuolar pH and calcium concentrations and disrupts the membrane of the vacuole (Vazquez-Boland et al., 1992). Upon release from the vacuole, *L. monocytogenes* produces ActA, which promotes actin polymerisation to allow the bacteria to move through the cytoplasm and push through the cytoplasmic membranes of neighbouring cells, facilitating cell to cell spread (Yoshikawa, Ogawa, Hain, Chakraborty, & Sasakawa, 2009).

The metal challenges faced by *L. monocytogenes* are likely to be diverse as it transitions from the external environment to its pathogenic lifestyle.

Within a host, *L. monocytogenes* must initially resist the hostile environment of the gastrointestinal tract, where metal availability will be substantially influenced by dietary intake, the acid pH in the stomach, the presence of microbes and microbial products in the intestines, as well as host metal-binding ligands that include components contributing to nutritional immunity. Following uptake or invasion into host cells, *L. monocytogenes* escapes from the internalisation phagosome into the host cell cytosol where the strict regulation of metal availability is likely to present a further challenge to *L. monocytogenes* in terms of acquiring sufficient of each metal to meet its cellular demands.

3. CONTROL OF METAL LEVELS IN BACTERIA

To cope with the need to both acquire metals and protect against metal toxicity, bacteria have evolved elaborate metal homeostatic mechanisms that act to tightly control cellular metal pools and ensure the correct metallation of their metal-requiring proteins. These include metal-responsive transcriptional regulators (metal sensors), metal-importers, metal-exporters, metal-sequestration proteins and metallochaperones (Waldron & Robinson, 2009).

Bacterial metal-sensor proteins have been described as the arbiters of metal sufficiency (Moore & Helmann, 2005), since they detect sufficiency of each metal to control the expression of genes encoding proteins involved in metal homeostasis or to alter the demand for a specific metal. As such, the metal affinities of the sensors of a specific metal have been hypothesised to determine the buffered intracellular concentration of that metal and hence its accessibility to proteins (Waldron, Rutherford, Ford, & Robinson, 2009). For example, *Escherichia coli* possesses two zinc sensors, ZntR and Zur, which regulate expression of a zinc exporter (ZntA) and zinc importer (ZnuABC), respectively, and which have femtomolar affinities for zinc (Outten & O'Halloran, 2001). When the intracellular level of zinc becomes sufficiently high to form zinc-ZntR and zinc-Zur, this will trigger zinc export and halt zinc import, thereby retaining the buffered zinc levels in a range determined by the femtomolar affinities of the sensors.

Since metals tend to associate with the metal-binding sites of proteins according to the Irving–Williams series (Irving & Williams, 1948), it is anticipated that the sensors of the more competitive metals, such as zinc and copper, will have tighter affinities for their detected metals than those of the less competitive metals, such as manganese and iron, thus restricting the access of the former to proteins requiring other metals (Waldron & Robinson, 2009). For the most part this appears to be the case; the femtomolar and zeptomolar sensitivities of the *E. coli* zinc and copper sensors, respectively, contrasting with the micromolar sensitivity of the iron sensor Fur (Changela et al., 2003; Foster, Pernil, Patterson, & Robinson, 2014; Outten & O'Halloran, 2001; Patterson et al., 2013; Waldron & Robinson, 2009). Thus, the buffered concentrations of the different metals in a cell and their availability to proteins appear, at least partly, to be determined by the relative metal affinities of the set of sensors in that cell.

The speciation of metals in cells is also important in determining metal accessibility to proteins. The zeptomolar and femtomolar sensitivities of the *E. coli* copper and zinc sensors, respectively, are a measure of the dissociation constants in free solution and infer a lack of fully aqueous (or free) copper or zinc inside cells (Changela et al., 2003; Outten & O'Halloran, 2001). Instead, any metal that is not stably incorporated into proteins is thought to be present as a labile pool bound to metabolites (such as free amino acids and glutathione) and macromolecular surfaces, with metal-requiring proteins acquiring their metals from this buffered metal pool via ligand exchange reactions. The metallation status of metal-requiring proteins may thus be determined by the thermodynamic equilibrium with this readily exchangeable metal pool. In addition, metals may be channelled to some proteins using a specific metal shuttle, thereby allowing them to overcome thermodynamic constraints, for example, a few proteins are known to acquire their metals (copper or nickel) directly from dedicated metallochaperones (Kaluarachchi, Chan Chung, & Zamble, 2010; Osman et al., 2013; Robinson & Winge, 2010).

The following sections focus on the mechanisms used by *L. monocytogenes* to control the intracellular availability of zinc and copper, and thereby ensure the acquisition of correct metal cofactors by its metal-requiring proteins.

4. *L. MONOCYTOGENES* AND ZINC

Zinc is essential to all organisms and is required by a wide range of proteins that perform vital roles including DNA replication, transcription, translation, cell division, glycolysis, pH regulation and oxidative stress resistance. The diverse protein requirements for zinc stems from the importance of this metal in both structural roles, such as within zinc fingers, and as a catalytic cofactor, due to its ability to act as a Lewis acid (Patzer & Hantke, 1998). Bacteria contain zinc at a concentration of approximately 0.2 mM

(Outten & O'Halloran, 2001), and zinc-binding proteins have been estimated to comprise 5–6% of the total proteome, of which 80% are enzymes, including RNA and DNA polymerases, alcohol dehydrogenases, lyases and hydrolases (Andreini, Bertini, & Rosato, 2009; Waldron & Robinson, 2009). Despite the abundance of zinc within cells, the cytosolic availability of zinc is thought to be buffered to very low levels, reflecting its position at the top of the Irving–Williams stability series for protein–metal binding and propensity to bind to adventitious protein sites (Outten & O'Halloran, 2001; Waldron & Robinson, 2009). This is achieved by the strict regulation of zinc import and export, as well as the sequestration of intracellular zinc by zinc-binding proteins and other ligands (Waldron et al., 2009). For *L. monocytogenes*, the challenge of maintaining zinc homeostasis is likely to be exacerbated during infections as it is thought to encounter environments in which zinc availability is severely limited as well as potentially being exposed to high levels of zinc, for example, from pancreatic fluids in the intestine (Carrigan et al., 2007; Kehl-Fie & Skaar, 2010).

4.1. Zinc availability within the host

After iron, zinc is the most abundant transition metal in the human body and zinc-binding proteins have been predicted to comprise 8.8% of eukaryotic proteins (Andreini, Banci, Bertini, & Rosato, 2006). Nonetheless, zinc availability is tightly controlled within mammalian systems and total zinc content varies between tissues. The labile zinc level (not stably incorporated into proteins) within plasma is estimated as 8 mM (Zalewski et al., 2006), indicating its potential availability for invading microbes. In contrast, zinc availability within the cytosol of eukaryotic cells is thought to be buffered in the picomolar range (Colvin, Holmes, Fontaine, & Maret, 2010), due to the strict regulation of zinc uptake, zinc export and intracellular sequestration in specialised vesicles (zincosomes) or zinc sequestering metallothioneins (Sekler, Sensi, Hershfinkel, & Silverman, 2007).

Zinc plays an important role in host immunity (Prasad, 2013). Zinc acts as a second messenger of immune cells and affects their activity, with massive fluxes in zinc levels occurring within infected cells (Haase & Rink, 2007; Hirano et al., 2008; Kitamura et al., 2006; Marvin et al., 2012). Zinc has also been shown to play a vital role in controlling the innate immune response to polymicrobial sepsis, with zinc deficiency promoting susceptibility to bacterial infection by enhancing NF-κB activity which leads to enhanced inflammation and lung injury (Bao et al., 2010). In response to bacterial molecules and cytokines, the host is known to induce a rapid reduction in the level of zinc in blood plasma, or hypozincaemia (Gaetke, McClain, Talwalkar, & Shedlofsky, 1997), and can decrease serum zinc levels by up to 80% (Liuzzi et al., 2005), thus reducing its availability for microbes. Furthermore, the release of zinc (and manganese)-chelating proteins at epithelial surfaces in response to infection and within tissue abscesses inhibits invasion by some bacteria, but can also allow the growth of some pathogens that express high affinity zinc-import systems enabling them to out compete commensals (Corbin et al., 2008; Davis, Kakuda, & DiRita, 2009; Liu et al., 2012; Rosadini, Gawronski, Raimunda, Arguello, & Akerley, 2011). With regard to the intracellular environment, the availability of both cytosolic and lysosomal zinc has also been shown to be reduced in activated macrophages, with depletion of zinc in macrophage phagosomes associated with limiting the survival of intracellular pathogens (Ammendola et al., 2007; Subramanian Vignesh et al., 2013). In contrast, intoxication with high zinc in macrophage phagosomes has been reported as an antimicrobial strategy against *Mycobacterium tuberculosis* (Botella, Stadthagen, Lugo-Villarino, de Chastellier, & Neyrolles, 2012) and may also act as a defence strategy within neutrophils (Ong, Gillen, Barnett, Walker, & McEwan, 2014). It is therefore not surprising that zinc-sensing and both zinc-import and zinc-efflux systems have been linked to the host colonisation or virulence of several bacterial pathogens including *L. monocytogenes*, *E. coli*, *Salmonella enterica* sv. *Typhimurium*, *Campylobacter jejuni*, *Haemophilus influenza*, *Haemophilus ducreyi*, *Streptococcus pneumonia*, *Streptococcus pyogenes*, *Brucella abortus*, *Pasteurella multocida* and the plant pathogen *Xanthomonas* sp. (Aarestrup & Hasman, 2004; Ammendola et al., 2007; Bayle et al., 2011; Botella et al., 2012; Campoy et al., 2002; Corbett et al., 2012; Corbin et al., 2008; Dahiya & Stevenson, 2010; Davis et al., 2009; Dowd, Casey, Begley, Hill, & Gahan, 2012; Gabbianelli et al., 2011; Garrido et al., 2003; Hood et al., 2012; Kim, Watanabe, Shirahata, & Watarai, 2004; Lewis et al., 1999; Murphy et al., 2013; Ong et al., 2014; Sabri, Houle, & Dozois, 2009; Tang et al., 2005; Weston, Brenot, & Caparon, 2009; Yang et al., 2007). There are some exceptions, however, zinc-sensing Zur mutants of *S. enterica* and *Staphylococcus aureus* are not attenuated (Campoy et al., 2002; Lindsay & Foster, 2001). Moreover, mutants of *Yersinia pestis* lacking the primary high affinity zinc-uptake system ZnuABC display similar virulence as the wild-type strain in mouse models of bubonic and pneumonic plague (Desrosiers et al., 2010). It is possible that *Y. pestis* possesses a second high-affinity zinc-uptake system that can compensate for the loss of ZnuABC *in vivo*. Alternatively, *Y. pestis* may have access to different zinc sources within the host environment.

4.1.1 Extracellular zinc chelation

The host produces metal-chelating proteins at mucosal and epithelial surfaces in response to bacterial infection. These include the S100 family of calcium-binding proteins that have been shown to sequester a range of transition metals with high affinity, as well as being involved in the generation of superoxide species (Hood & Skaar, 2012). They have been implicated in the defence against invading pathogens, including *S. aureus* (Corbin et al., 2008), *E. coli* (Glaser et al., 2005), *Candida albicans* (Clohessy & Golden, 1995; Urban et al., 2009) and *Aspergillus fumigatus* (Bianchi, Niemiec, Siler, Urban, & Reichenbach, 2011; McCormick et al., 2010). A pro-inflammatory role has also been suggested for some members of the S100 family (Hood & Skaar, 2012). The structure of these proteins and their interaction with metal ions has been thoroughly explored by others (Moroz et al., 2003a; Moroz, Blagova, Wilkinson, Wilson, & Bronstein, 2009a; Moroz et al., 2009b; Moroz, Dodson, Wilson, Lukanidin, & Bronstein, 2003b) and are not discussed in detail herein.

One such S100 protein, psoriacin (also known as S100A7), is induced in keratinocytes by *E. coli* infection (Glaser et al., 2005). It has been hypothesised that psoriacin (Michalek et al., 2009) prevents the survival of *E. coli* in a zinc-dependent manner as this property is attenuated when the protein is saturated with zinc. There is also some evidence that this protein may prevent bacterial growth by the permeabilisation of bacterial membranes (Michalek et al., 2009). Another S100 protein, calprotectin (also known as calgranulin A/B), is a heterodimer consisting of S100A8 and S100A9 and accumulates at sites of inflammation (Gebhardt, Nemeth, Angel, & Hess, 2006). The antimicrobial activity of calprotectin against *C. albicans* was initially suggested to be due to zinc chelation (Clohessy & Golden, 1995), with calprotectin also shown to limit the growth of *A. fumigatus* by sequestering zinc (McCormick et al., 2010). However, the antimicrobial activity of calprotectin towards *S. aureus* has since been shown to be due to both manganese and zinc chelation, with abscesses caused by *S. aureus* infection containing large accumulations of this protein but being depleted of both manganese and zinc (Corbin et al., 2008). It is notable, however, that mice devoid of calprotectin were not found to have significantly different zinc contents compared with wild-type mice, while abscesses from such mice had high levels of manganese which coincided with increased bacterial load at these sites (Corbin et al., 2008). Metal chelation by calprotectin is mediated by two high-affinity metal-binding sites which can both bind zinc with nanomolar affinity, whereas only one can bind manganese with similar affinity (Damo et al., 2013; Kehl-Fie et al., 2011). The possession of these two sites is suggested to account for the broad spectrum antimicrobial activity of calprotectin compared to other S100 proteins. Remarkably, *S. enterica* sv. *Typhimurium* appears able to avoid the deleterious effects of zinc sequestration by calprotectin within the intestine due to its high affinity ZnuABC zinc-uptake system which provides it with a selective advantage over commensal intestinal bacteria (Liu et al., 2012). These findings thus emphasise the importance of zinc uptake in bacterial pathogenesis

In addition to calprotectin accumulating in extracellular fluids in response to inflammation, calprotectin is also abundant in the cytosol of neutrophils, making up approximately 40% of the total protein composition (Hessian, Edgeworth, & Hogg, 1993). However, calprotectin appears to only have a high affinity for zinc when it encounters extracellular calcium, implying that it may lack the ability to sequester zinc intracellularly (Brophy, Hayden, & Nolan, 2012). Nonetheless, calprotectin has been reported to effectively inhibit the growth of *L. monocytogenes in vitro* and within keratinocytes (Zaia et al., 2009). However, at a high multiplicity of infection (100 bacteria to 1 epithelial cell), *L. monocytogenes* can cause the mobilisation of calprotectin in invaded epithelial cells from a diffuse cytoplasmic distribution to a filamentous pattern that colocalises with the microtubule network (Zaia et al., 2009). This occurs in a listeriolysin O and calcium-dependent manner and can abrogate the antilisterial activity of calprotectin (Zaia et al., 2009). The relevance of these findings to *in vivo* infections is currently unknown. Nonetheless, it is predicted that either one or both of the two high-affinity zinc-uptake systems of *L. monocytogenes* (see below) may contribute to overcoming zinc chelation by S100 proteins during infections.

4.1.2 Control of intracellular zinc availability

Mammalian transporters of the ZIP (alias SLC39, Zrt or Irt-like) and ZnT (alias SLC30, CDF) families have a role in altering intracellular zinc levels in response to infection (Begum et al., 2002; Besecker et al., 2008; Kitamura et al., 2006; Liu et al., 2013); the ZIP transporters have a role in zinc influx into the cytosol, while the ZnT transporters efflux zinc from the cytosol (Eide, 2006). Increased expression of the zinc transporter ZIP8 is observed in cells (monocytes, macrophages and epithelial cells) treated with bacterial components or inflammatory cytokines and it associates with the lysosomal marker Lamp1, indicating a role for ZIP8 in transporting zinc from the phagolysosome to the cytosol to limit zinc availability to the vacuole dwelling bacteria (Begum et al., 2002; Besecker et al., 2008; Liu et al., 2013). ZIP8 is also highly expressed in activated T cells and is associated with increased INF-γ secretion (Aydemir, Liuzzi, McClellan, & Cousins, 2009). However, there is also evidence of reduced expression of other ZIP importers (Zip6 and Zip10), and increased expression of ZnT zinc exporters (Znt1, Znt4 and Znt10), causing a reduction in cytosolic zinc following stimulation of dendritic cells with lipopolysaccharide via Toll-like receptor-4 (Kitamura et al., 2006). In addition, activation of macrophages with the cytokine granulocyte macrophage-colony stimulating factor has been reported to enhance the antimicrobial defences against the intracellular fungal pathogen *Histoplasma capsulatum* by cytosolic zinc limitation (Subramanian Vignesh et al., 2013). This was associated with enhanced cytosolic zinc sequestration by metallothioneins and increased expression of ZnT exporters to shuttle zinc into the Golgi apparatus, and which also coincided with increased reactive oxygen species generation (Subramanian Vignesh et al., 2013).

In addition to zinc limitation, it has also been suggested that elevated zinc levels may be used as an antimicrobial defence within phagocytes (Botella et al., 2012; Ong et al., 2014). Recently, infection of macrophages with *M. tuberculosis* or *E. coli* was shown to cause zinc accumulation within bacterial-containing phagosomes (Botella et al., 2011). This coincided with increased expression of the *M. tuberculosis* metal-exporting P_1-type ATPase CtpC (Botella et al., 2011). CtpC is known to be expressed in response to high zinc levels and *M. tuberculosis* mutants lacking *ctpC* over-accumulate zinc and have impaired intracellular growth (Botella et al., 2011). Similarly *E. coli* mutants lacking the zinc exporter ZntA were killed more rapidly than wild-type bacteria (Botella et al., 2011). These findings therefore imply that exposure to toxic zinc levels may contribute to bacterial killing within macrophage phagosomes. This however is in contrast to studies with other pathogens, such as *S. enterica* sv. *Typhimurium,* where possession of the high-affinity zinc-import system ZnuABC is required for intraphagosomal survival in both epithelial cells and macrophages and phagosomal killing is associated with zinc limitation (Ammendola et al., 2007; Campoy et al., 2002). It is notable, however, that *M. tuberculosis* CtpC has also been associated with manganese export and manganese incorporation into secreted proteins, with *ctpC* mutants having reduced Fe/Mn-superoxide dismutase activity and increased sensitivity to oxidative stress (Padilla-Benavides,

Long, Raimunda, Sassetti, & Arguello, 2013). Thus, decreased survival of an *M. tuberculosis ctpC* mutant in macrophage phagosomes may, at least in part, relate to a loss of essential enzyme activity and sensitivity to oxidative stress, rather than a direct effect of zinc toxicity in this compartment. Nonetheless, zinc toxicity has also been associated with the killing of *S. pyogenes* in human neutrophils, with zinc-sensitive mutants of *S. pyogenes* mutants, lacking the CzcD zinc-efflux pump, having increased susceptibility to neutrophil killing and reduced virulence in a mouse infection model (Ong et al., 2014). Furthermore, *S. pyogenes* infection of neutrophils was associated with the mobilisation of zinc in these cells (Ong et al., 2014).

Thus, it appears that both zinc import and zinc efflux may be required by bacterial pathogens to survive within host cells and the dependency on these systems may vary for different pathogens and within different cell types. Following uptake or invasion into host cells, *L. monocytogenes* must initially overcome the zinc stresses encountered within the internalisation phagosome (low and/or high zinc), prior to its escape into the host cell cytosol. The strict regulation of cytosolic free zinc within eukaryotic cells is likely to present a further challenge for *L. monocytogenes* in terms of acquiring sufficient zinc to supply its zinc-requiring proteins, which include virulence factors associated with intracellular survival and cell to cell spread.

4.2. Zinc sensing and homeostasis in *L. monocytogenes*

It is vital that the zinc import, export and storage mechanisms are tightly regulated in bacteria in order to achieve zinc homeostasis across a range of environmental conditions (Gabriel & Helmann, 2009; Hantke, 2001; Natori et al., 2007). A number of zinc homeostatic proteins have now been identified in *L. monocytogenes* (Fig. 2). These include the zinc sensor Zur, which in response to changes in zinc availability regulates the expression of two high-affinity ABC-type zinc-uptake systems (ZurLAM and ZinABC) and a S14 ribosomal protein (Lmo1882) as part of the zinc starvation response of this pathogen. In addition, *L. monocytogenes* is predicted to possess a second zinc sensor, herein designated CzrA, which is thought to trigger P_1-type ATPase-mediated zinc efflux in response to elevated zinc levels and hence protect against zinc toxicity.

4.2.1 The Zur regulon

The first low zinc-induced transport system to be identified in *L. monocytogenes* was that of the *zurA-zurM-zur* zinc-uptake operon, which encodes an ATP-binding component protein (ZurA), a hydrophobic

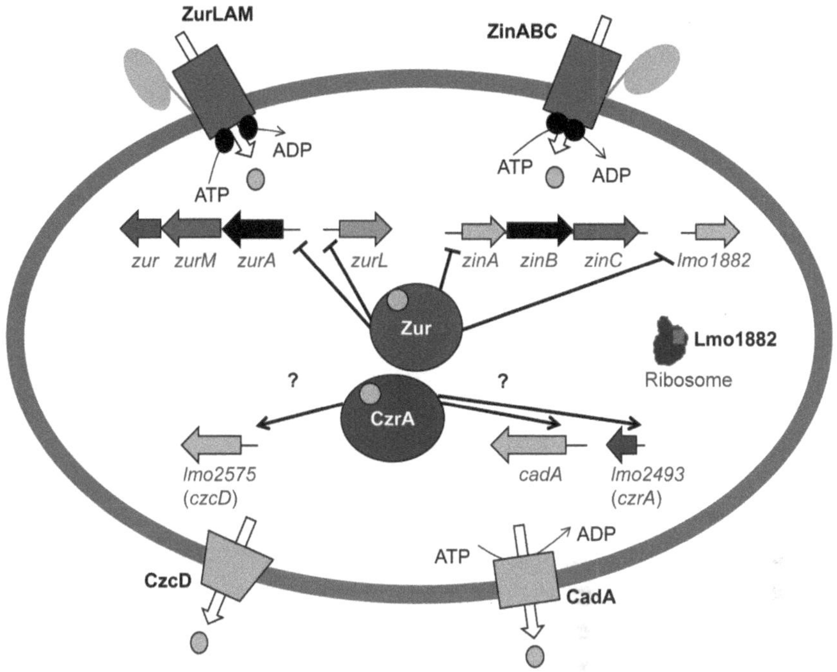

Figure 2 Zinc homeostasis in *L. monocytogenes*. Zinc depletion causes loss of Zur-mediated repression triggering transcription of genes involved in the zinc starvation response, including genes encoding two ABC-type zinc-uptake systems, ZurLAM and ZinABC, and a S14 ribosomal protein Lmo1882. *L. monocytogenes* has a second deduced zinc-responsive transcriptional regulator, CzrA (encoded by *lmo2493*), which belongs to the ArsR/SmtB family. CzrA is predicted (shown as ?) to act as a zinc-sensing derepressor such that in the presence of elevated levels of zinc it triggers expression of genes (*lmo2574* and *cadA*) encoding two deduced metal exporters, designated CzcD (CDF-family) and CadA (P_{1B}-type ATPase), as well as its own expression. Grey circles represent zinc.

membrane permease protein (ZurM) and the transcriptional regulatory protein, Zur (*Z*inc *u*ptake *r*egulator, alias ZurR) (Dalet, Gouin, Cenatiempo, Cossart, & Hechard, 1999). Zur confers zinc-responsive expression of the operon by acting as a co-repressor and binding to DNA in elevated zinc levels to prevent transcription (Corbett et al., 2012; Dalet et al., 1999). Zur from *L. monocytogenes* comprises 144 amino acids and belongs to the Fur family of metal-sensing transcriptional regulators, sharing 60% sequence identity with Zur from *B. subtilis* (Gaballa & Helmann, 1998) and 23% sequence identity with *E. coli* Zur (Outten & O'Halloran, 2001). Like *B. subtilis* Zur (Ma, Gabriel, & Helmann, 2011), Zur from

L. monocytogenes is thought to contain two distinct metal-binding sites per monomer (four per dimer) with different affinities and coordination environments, with one zinc tightly bound in a Cys_4 structural site and the second zinc bound in a $Cys_1Glu_1His_2$ metal-sensing site. There is negative cooperativity between the two metal-sensing sites in the *B. subtilis* Zur dimer which allows sequential activation from an inactive dimer and may allow for graduated changes in gene expression in response to zinc depletion (Ma et al., 2011).

In addition to the *zurA-zurM-zur* operon, *L. monocytogenes* Zur is now known to regulate the expression of a second zinc-uptake operon, designated *zinA-zinB-zinC*, and open reading frame *lmo1671* (herein designated *zurL*) that likely encodes the substrate-binding protein component of the ZurAM ABC-type transporter (Corbett et al., 2012; Dowd et al., 2012). The *zinA-zincB-zinC* operon encodes the substrate-binding lipoprotein, ATPase and membrane permease components, respectively, of a second zinc-uptake ABC-type transporter (Corbett et al., 2012). There is constitutive elevated expression of these genes in *zur* mutants of *L. monocytogenes* (Corbett et al., 2012; Dowd et al., 2012). Consistent with their regulation by Zur, sequences corresponding to the consensus Zur-binding site (Gabriel & Helmann, 2009) are present in the promoter regions of *zurA*, *zurL* and *zinA* (Corbett et al., 2012; Dowd et al., 2012).

A Zur binding site is also present upstream of *lmo1882* (Corbett et al., 2012; Dowd et al., 2012), which is one of two *L. monocytogenes* genes encoding S14 ribosomal subunits, the other being *rpsN*. Zur from *B. subtilis* is known to not only regulate genes involved in zinc import in response to changing zinc levels, but is also responsible for regulating alternative ribosomal proteins, YhzA and YtiA, which are paralogous to ribosomal L31 and S14 proteins, respectively (Gabriel & Helmann, 2009). The Zur-controlled ribosomal proteins lack the two CXXC metal-binding motifs that are present in their cognate zinc-requiring proteins and can functionally replace their cognate proteins during zinc limitation. As such, the Zur-regulated S14 paralogue YhzA allows continued ribosome synthesis when there is insufficient zinc to support S14 function, whereas the displacement of the L31 protein RpmE (containing a zinc-binding motif) with YtiA from assembled ribosomes results in mobilisation of zinc to support growth (Akanuma, Nanamiya, Natori, Nomura, & Kawamura, 2006; Gabriel & Helmann, 2009). The *L. monocytogenes* S14 protein Lmo1882 lacks the two CXXC metal-binding motifs present in its cognate zinc-requiring protein RpsN and hence, with similarity to *B. subtilis* YhzA, Lmo1882 is predicted to replace RpsN to allow continued ribosome synthesis during zinc limitation. However, in contrast to *B. subtilis*, *L. monocytogenes* encodes only one ribosomal L31 protein, Lmo2548, which is not predicted to bind zinc and, consistent with this, there is no identifiable Zur-binding site upstream of its gene (our unpublished observations). This implies that Lmo2548 is required under both zinc-replete and zinc-limiting conditions. Thus, rapid mobilisation of zinc from ribosomes by exchanging L31 proteins does not appear to be a zinc-starvation strategy of *L. monocytogenes* and hence this pathogen may be particularly dependent upon high-affinity zinc uptake for survival in zinc-deplete conditions.

A bioinformatics search was carried out to identify other members of the *zur* regulon in *L. monocytogenes* based on similarity to genes known to be regulated by Zur in *B. subtilis* and with a putative *zur* box within 500 base pairs of their start codon (Dowd et al., 2012). This identified *arpJ* (based on similarity to *B. subtilis yckA*) which has 1.2-fold increased expression in a *zur* mutant and encodes a protein with similarity to amino acid ABC-transporter permease proteins as part of the *arpJ-lmo2251-lmo2252* operon, also encoding a putative ATP-binding protein and an aspartate aminotransferase. In addition, *lmo1849* was identified (based on similarity to *B. subtilis znuC*, alias *ycdI*) which is upregulated 1.3-fold in a *zur* mutant and encodes the ATP-binding protein of a further predicted ABC-type metal transporter encoded by the *lmo1849-lmo1848-lmo1847* operon (Dowd et al., 2012). However, due to the greater similarity of the products of this operon to the *B. subtilis* MntABCD manganese-uptake system (which appears absent from *L. monocytogenes*), rather than the zinc-uptake systems, it is possible that the products of *lmo1849-lmo1848-lmo1847* have specificity towards manganese import as opposed to encoding a third ABC-type zinc-import system in *L. monocytogenes*. The contribution of these additional putative Zur-regulated genes in *L. monocytogenes* zinc homeostasis however remains unknown.

Mutants of *L. monocytogenes* that lack a functional Zur have been found to have a smaller cell size than wild-type cells and form smaller colonies (Dowd et al., 2012). Furthermore, while the *zur* mutant exhibited no growth defect in zinc-depleted conditions, it was hypersensitive to high concentrations of zinc, consistent with Zur acting as a repressor of zinc import (Dowd et al., 2012). This work also highlighted the important role of zinc homeostasis in infection by *L. monocytogenes*, as the *zur* mutant, with uncontrolled expression of zinc import, had reduced virulence in a mouse model of infection. It was hypothesised that this attenuation may be due to the reduced motility also exhibited by this strain (Dowd et al., 2012), with zinc having been shown to affect the expression of genes involved in motility in *E. coli* and a zinc-uptake (*znuB*) mutant of *E. coli* shown to have reduced motility (Lee, Barrett, & Poole, 2005; Sabri et al., 2009; Sigdel et al., 2006). In addition, Zur from *L. monocytogenes* has also been implicated in tolerance to bile (Begley, Gahan, & Hill, 2002), which may also contribute to a role in virulence, although reduced tolerance to bile was not detected for a *zur* mutant in a more recent study (Dowd et al., 2012).

4.2.2 Zinc import by the ZurLAM and ZinABC systems

High-affinity zinc import across the cytoplasmic membrane of both Gram negative and Gram positive bacteria primarily involves the ATP-binding cassette (ABC)-type transporters, which are typified by the well-characterised ZnuABC system in *E. coli* (Patzer & Hantke, 1998). Import systems of the ABC-type comprises a substrate-binding protein (often periplasmic in Gram negative bacteria or a membrane-associated lipoprotein in Gram positive bacteria), a membrane permease and a cytoplasmic ATP-binding component (Berducci, Mazzetti, Rotilio, & Battistoni, 2004; Chandra, Yogavel, & Sharma, 2007). It is not known whether these transporters carry zinc as a free ion, or in a complex with another substance as was recently reported for nickel import [as a nickel-(L-His)$_2$ complex] by *E. coli* (Chivers, Benanti, Heil-Chapdelaine, Iwig, & Rowe, 2012). In some Gram negative bacteria, an additional periplasmic protein (e.g. ZinT in *E. coli*) performs an auxiliary role in capturing zinc for zinc uptake by the ABC-type transporter under conditions of severe zinc shortage (Graham et al., 2009; Petrarca, Ammendola, Pasquali, & Battistoni, 2010).

So far, *L. monocytogenes* has been shown to possess two ABC-type zinc-uptake systems; ZurAM and ZinABC, which are regulated by the repressor Zur in response to zinc (Corbett et al., 2012; Dalet et al., 1999; Dowd et al., 2012). The genes *zurA* and *zinB* each encode the ATPase component of the transporters, *zurM* and *zinC* each encode a membrane permease, while *zinA* encodes a zinc-binding lipoprotein (Corbett et al., 2012; Dalet et al., 1999). Recently, a Zur-regulated gene, *lmo1671* (herein *zurL*), encoding a second deduced zinc-binding lipoprotein was identified in *L. monocytogenes* (Dowd et al., 2012). The *zurL* gene is upregulated 4.6-fold in a *zur* mutant (Dowd et al., 2012) and is hypothesised to encode the substrate binding component that functions as part of a ZurLAM system (Fig. 2). Deletion of either *zurAM* or *zinA* has no detectable effect on the growth of *L. monocytogenes* in defined medium (containing approximately 0.3 μM zinc), yet a double *zurAM zinA* deletion mutant was unable to grow in the absence of zinc supplementation, thus suggesting at least some functional redundancy between the two systems under these conditions (Corbett et al., 2012).

To assess the contributions of the two zinc-uptake systems to the intracellular growth of *L. monocytogenes*, the survival of *zurAM* and *zinA* single mutants or a *zurAM zinA* double mutant was examined during infection of HeLa epithelial cells (Corbett et al., 2012). Deletion of *zinA* had no detectable effect on growth in these cells whereas growth of the *zurAM* mutant was significantly impaired, indicating the particular importance of ZurAM during intracellular growth. Notably, deletion of both *zinA* and *zurAM* severely attenuated intracellular growth, demonstrating a contribution of both zinc-uptake systems to *L. monocytogenes* growth in the cytoplasm of host cells (Corbett et al., 2012). The differences in the survival of the *zurAM* and *zinA* mutants within host cells imply that each system makes a distinct contribution to zinc acquisition. Actin accumulation and polymerisation was also monitored in HeLa cells to determine whether or not the failure of the double mutant to grow in HeLa cells was a consequence of it being trapped in a vacuole within the cell (Corbett et al., 2012). Crucially, while wild-type cells were found either coated in actin, or at the head of actin comet tails, indicating vacuolar escape and actin-based intracellular motility, the number of *zurAM zinA* double mutant cells associated with comet tails was significantly lower (eightfold). Hence, while the double zinc-uptake mutant is not defective in escaping the primary vacuole or in nucleating actin, it is defective in transitioning from actin nucleation to actin polymerisation. A likely consequence of defective actin tail formation is a failure of *L. monocytogenes* to spread from cell to cell, and this was also confirmed in a plaque assay (Corbett et al., 2012). It is possible that the low level of intracellular replication of the *zurAM zinA* mutant contributes to the lack of actin polymerisation and/or is related to impaired cell division, with filamentous forms of the mutant being apparent within the infected cells (Corbett et al., 2012). Alternatively, the failure in cell to cell spread and reduced comet tail formation of the *zurAM zinA* mutant could be a result of reduced phosphatidylcholine phospholipase C, PlcB, and/or ActA activity. PlcB is required for escape of *L. monocytogenes* from the double-membrane vacuole that results from intercellular spread (Vazquez-Boland et al., 1992), whereas ActA mediates polymerisation of host actin filaments on the bacterial surface to generate bacterial movement in the cytosol (Cossart, 2000). Both, PlcB and ActA activation is dependent upon their cleavage by the zinc metalloprotease Mpl (Poyart, Abachin, Razafimanantsoa, & Berche, 1993). It is therefore conceivable that an inability to acquire sufficient zinc might limit the activity of Mpl and thereby the amount of mature PlcB and ActA produced by *L. monocytogenes*.

In contrast to *in vitro* infections, the deletion of either *zurAM* or *zinA* was found to cause a significant reduction in *L. monocytogenes* virulence within an oral mouse model of infection, with substantially increased survival of mice infected with either mutant as compared to the wild type (Corbett et al., 2012), and thus demonstrating that the possession of both zinc-uptake systems is required for *L. monocytogenes* virulence during *in vivo* infections. The double mutant was also attenuated and significantly less virulent than the wild type (Corbett et al., 2012). Clearly, the ZurAM and ZinABC systems play distinct but somewhat reciprocal roles in zinc acquisition by *L. monocytogenes* under conditions of zinc limitation. While both systems are needed for the full virulence of *L. monocytogenes* in HeLa cells and in the mouse model of infection, intriguingly there is some functional redundancy between the two systems that varies depending upon the environment in which the bacterium is growing.

The existence of two ABC-type zinc-uptake systems within an organism is rare, but not unprecedented in bacteria. For example, *H. influenzae* also has two ABC-type zinc-uptake systems, ZnuABC and ZevAB, with the latter being important for growth in lungs (Rosadini et al., 2011). It is tempting to speculate that the possession of the two zinc-uptake systems in *L. monocytogenes* provides a mechanism by which this pathogen can respond to zinc deficiency within a variety of environments and during different stages of infection, whereby each system makes a distinct contribution under different stress conditions. It has recently been shown that zinc competitively inhibits manganese binding to the Psa manganese-uptake permease of *S. pneumonia*, which is an essential virulence determinant, to competitively inhibit zinc uptake (Counago et al., 2014; Eijkelkamp et al., 2014). Similarly, zinc is known to interfere with metal uptake by other ABC transporters for iron and manganese (Desrosiers et al., 2010; Kolenbrander, Andersen, Baker, & Jenkinson, 1998; Sabri et al., 2009). Hence, it is possible that differences in the metal-binding properties of ZinA and ZurL confer differences in metal specificity and hence the antagonistic effects of other metals on zinc import by these permeases. Alternatively, the preferred zinc substrate (zinc complex or free zinc ion) imported by each system may differ.

4.2.3 Other proteins with a role in zinc import

In addition to the ABC-type transporters, some bacteria possess other systems that can contribute to zinc uptake; such as transporters with similarity to eukaryotic ZIP transporters (e.g. *E. coli* and *S. enterica* sv. *Typhimurium* ZupT), the NRAMP-family transporters, albeit being primarily associated with manganese import (e.g. *E. coli, S. enterica* sv. *Typhimurium* and *B. subtilis* MntH), and ZosA from *B. subtilis* that belongs to the P_{1B}-type ATPase family (Gaballa, Wang, Ye, & Helmann, 2002; Grass et al., 2005; Guedon et al., 2003; Hantke, 2005; Karlinsey, Maguire, Becker, Crouch, & Fang, 2010; Kehres, Zaharik, Finlay, & Maguire, 2000; Sabri et al., 2009; Taudte & Grass, 2010; Zaharik et al., 2004). With some similarity *to B. subtilis, L. monocytogenes* possesses proteins with similarity to MntH (Lmo1424) and ZosA (Lm0641, alias FrvA), although the latter has been associated with haem export (McLaughlin et al., 2012). A further protein, YciC, has also been implicated to play a secondary role in *B. subtilis* zinc import, although an alternative role for YciC as an extracytoplasmic zinc metallochaperone has also been suggested (Gabriel, Miyagi, Gaballa, & Helmann, 2008). No *yciC*-related genes are present within the *L. monocytogenes* genome.

4.2.4 The response to elevated zinc and zinc export

There are three types of zinc-export systems that have been described in bacteria: (i) the cadmium/lead/zinc-transporting P_{1B}-type ATPases that utilise ATP-derived energy to flux metals across the cytoplasmic membrane (e.g. ZntA and CadA from *E. coli* and *B. subtilis*, respectively); (ii) the cation-diffusion facilitator (CDF) family (e.g. ZitB and CzcD from *E. coli* and *B. subtilis*, respectively) where metal transport is driven by a concentration, chemiosmotic or potassium gradient; and (iii) the resistance, nodulation and cell division (RND)-driven chemiosmotic-efflux systems (e.g. CzcCBA from *Cupriavidus metallidurans*) that in Gram negative bacteria span both membranes and the periplasmic space to export metals to the outside of the cell (Blencowe & Morby, 2003; Hantke, 2001; Moore, Gaballa, Hui, Ye, & Helmann, 2005; Nies, 2003, 2007). Little is known about the systems used by *L. monocytogenes* to export zinc, although the genome encodes a deduced P_{1B}-type ATPase (Lmo01100, alias CadA) and a CzcD family transporter (Lmo2575, herein designated CzcD) which are likely candidates for a role in zinc export (our unpublished observations). In addition, some *L. monocytogenes* strains possess a plasmid encoded CadA P_{1B}-type ATPase, within the *cadAC* cadmium-resistance determinant (Lebrun, Audurier, & Cossart, 1994), which by analogy to related proteins may also be anticipated to export zinc in these strains. However, while the plasmid-borne *cadAC* genes can confer cadmium-resistance to *L. monocytogenes*, they were found not to be associated with zinc resistance (Lebrun et al., 1994). It is however, likely that this is due to the possession of chromosomally encoded zinc exporters (i.e. CzcD and CadA) in this organism (Fig. 2). In other bacteria, genes encoding zinc-exporting P_{1B}-type ATPases and CzcD family transporters are regulated by zinc-responsive transcriptional regulators belonging to the MerR-family (e.g. ZntR from *E. coli*) or ArsR/SmtB-family (e.g. CzrA from *B. subtilis*) of metal sensors (Binet & Poole, 2000; Brocklehurst et al., 1999; Moore et al., 2005). A gene (*lmo2493*) encoding a deduced ArsR/SmtB-family repressor (herein designated CzrA) is present within the *L. monocytogenes* chromosome, and possesses a characteristic zinc-sensing motif (Osman & Cavet, 2010). With analogy to *B. subtilis* CzrA, this protein is therefore predicted to cause induced expression of CadA and CzcD, as well as its own expression, in response to elevated zinc levels. With the recent discovery that zinc-efflux systems have been associated with the virulence of some bacterial pathogens (Botella et al., 2012; Ong et al., 2014), it will be important to examine the contributions of these systems *to L. monocytogenes* pathogenicity.

5. *L. MONOCYTOGENES* AND COPPER

The redox cycling ability of copper has led to its use as a cofactor in important enzymes with roles including electron transfer, oxidase and oxygenase activities, and detoxification of reactive oxygen species (Fraústo da Silva & Williams, 2001), and includes the respiratory menaquinol oxidase in *L. monocytogenes*. Nonetheless, copper-requiring proteins are far less prevalent than zinc-requiring proteins, with bacterial copper-binding proteins predicted to comprise only 0.3% of bacterial proteomes and all of which appear to be extracytoplasmic (Andreini, Banci, Bertini, & Rosato, 2008). This may reflect its low bioavailability in the ancient world (prior to an oxidising atmosphere) and/or problems associated with the complexity of copper chemistry and copper toxicity (Dupont, Grass, & Rensing, 2011). Indeed, copper can be extremely toxic to cells due to outcompeting most other divalent metal ions at metal-binding sites in proteins, thus displacing their native metals, and mediating the disassembly of solvent exposed iron–sulphur clusters in enzymes (Macomber & Imlay, 2009; Tottey et al., 2012; Waldron & Robinson, 2009). In addition, as a redox active metal, copper may participate in Fenton chemistry to produce highly toxic hydroxyl radicals that kill cells by oxidative damage (Imlay et al., 1988). As is the case for zinc, organisms have therefore evolved elaborate copper homeostatic systems that act to tightly control cellular copper pools and maintain the buffered cytosolic available copper concentrations at very low levels. These systems include transcriptional regulators, membrane transporters and copper metallochaperones, with the latter playing a role in supplying copper to copper-requiring proteins while preventing the adverse side reactions of copper *en route* (Osman et al., 2013; Robinson & Winge, 2010; Tottey et al., 2012; Waldron & Robinson, 2009).

The problem of maintaining copper homeostasis may be particularly acute for *L. monocytogenes*. It inhabits environments including soil, effluents and food where it will be exposed to fluctuating copper levels. The use of copper as an antimicrobial in animal feeds and as a disinfectant in factory-based farming (Aarestrup & Hasman, 2004; Borkow & Gabbay, 2005; Noyce, Michels, & Keevil, 2006b) means *L. monocytogenes* will also have to combat potentially toxic levels of copper during food production and processing. Adapting to fluctuations in the level of copper is also likely to be important during infections as *L. monocytogenes* replicates in a number of host tissues that will offer different challenges with regard to copper availability. Early in infection, *L. monocytogenes* grows in the gall bladder, where there are thought to be high levels of copper due, at least in part, to the action of the Wilson P_{1B}-type ATPase (ATP7B) which transports copper from the liver into bile for excretion (Lutsenko, LeShane, & Shinde, 2007). Furthermore, the copper toxicity in the gastrointestinal tract may be increased due to the acidic anaerobic conditions there causing the reduction of Cu(II) to Cu(I), which may diffuse through the cytoplasmic membrane to cause increased copper accumulation (Outten, Huffman, Hale, & O'Halloran, 2001). Following invasion of host cells, *L. monocytogenes* may encounter high phagosomal copper levels, prior to escaping into the host cell cytoplasm where access to copper will be extremely limited. Adapting to these different niches within the host offers a challenge in terms of maintaining a supply of copper for copper requiring enzymes while avoiding copper toxicity.

5.1. Exploiting the microbicidal activity of copper in host immune defences

Copper has been used to control microbes in the environment for centuries, and copper continues to be used today in numerous settings, including in healthcare, the supply of drinking water, as a biocide in fruit production, as well as in other agricultural practices and in the food industry (Aarestrup & Hasman, 2004; Borkow & Gabbay, 2005; Grass, Rensing, & Solioz, 2011;

Lin, Stout, & Yu, 2011; Noyce, Michels, & Keevil, 2006a, 2006b). It is also apparent that it is used in both innate and adaptive immune responses, with copper deficiency having been linked to increased vulnerability to bacterial infections (Festa et al., 2011; Percival, 1998; Wolschendorf et al., 2011). Changes in copper levels have been detected in the livers and spleens of animals infected with a range of pathogens, including bacteria, trypanosomes and viruses (Crocker, Lee, Aboko-Cole, & Durham, 1992; de Abel et al., 1993; Ilback, Benyamin, Lindh, & Friman, 2003; Tufft, Nockels, & Fettman, 1988). A reduction in interleukin-2 production and T cell proliferation has been reported during copper deficiency (Percival, 1998), and the number of antibody producing cells in mice raised on a copper-deficient diet is lower than those fed on a normal diet and is associated with increased morbidity upon infection (Newberne, Hunt, & Young, 1968; Prohaska & Lukasewycz, 1981; Suttle & Jones, 1989). Copper deficiency has also been associated with a reduction in neutrophil numbers in human peripheral blood and causes a reduction in the respiratory burst oxidase activity and reduced bacterial killing ability of neutrophils and macrophages (Huang & Failla, 2000; Percival, 1998).

More recently, several different lines of research have pointed towards pathogens being exposed to high copper levels within macrophage phagosomes, with copper contributing to the microbicidal activity there. Indeed, pro-inflammatory factors stimulate copper accumulation by macrophages, via increased expression of the copper-importer CTR1 and increased expression and trafficking of the Menkes copper-transporting P_{1B}-type ATPase ATP7A from the Golgi apparatus to phagosome-associated vesicles which contributes to their bactericidal activity (Achard et al., 2012; White et al., 2009). Bacterial infection of macrophages also coincides with increased expression of these and various other host proteins involved in handling copper, and copper accumulates within intracellular vesicles, 'copper hot spots' (Achard et al., 2012). Furthermore, elevated copper levels are sensed by bacterial pathogens within macrophage phagosomes and copper detoxification systems are required for their survival in these compartments, and within animal infection models, with copper-chelation increasing pathogen survival (Achard et al., 2012; Ding et al., 2013; Osman et al., 2010; Shi et al., 2014; Ward et al., 2010; White et al., 2009; Wolschendorf et al., 2011). The precise mechanism of copper toxicity within macrophages remains unknown, but suggestions include copper contributing to respiratory burst oxidase-mediated bacterial killing by participating in Fenton chemistry (White et al., 2009), and/or by potentiating nitrosative stress (Djoko et al., 2012; Schaible & Kaufmann, 2005). Copper may also function directly as a bactericidal agent by the mispopulation of metalloproteins that require other metals and the disruption of protein iron–sulphur clusters.

5.2. Copper sensing and homeostasis in *L. monocytogenes*

The mechanisms of copper uptake are largely unknown in bacteria. For some bacteria, P_{1B}-type ATPases have been suggested to act as specific copper importers (Francis & Thomas, 1997b; Hassani, Astier, Nitschke, & Ouchane, 2010; Lewinson, Lee, & Rees, 2009; Odermatt, Suter, Krapf, & Solioz, 1993; Tottey, Rich, Rondet, & Robinson, 2001; Wunderli-Ye & Solioz, 2001), although it has been suggested that all P_{1B}-type ATPases must mediate metal efflux and hence these proteins more likely have a role in extracytoplasmic cupro-protein assembly by fluxing metals out of the cytosol (Raimunda, Gonzalez-Guerrero, Leeber, & Arguello, 2011). Other reported mechanisms of bacterial copper import include the import of a siderophore-like copper-chelating molecule, methanobactin, in a TonB-dependent manner by methanotrophs and the involvement of a major facilitator superfamily-type transporter, CcoA, in copper import by the phototroph *Rhodobacter capsulatus* (Arguello, Raimunda, & Padilla-Benavides, 2013). Bacterial copper export primarily involves copper-exporting P_{1B}-type ATPases that couple Cu(I)-efflux to the hydrolysis of ATP, and which may acquire their copper for transport from cytosolic copper metallochaperones (Arguello et al., 2013). Other copper-export systems include the RND family transporter CusCFBA, first identified in *E. coli*, and components of plasmid-borne copper-resistance associated Pco systems which have so far been characterised in some Gram negative bacteria (for a recent review, refer to Arguello et al., 2013).

The first copper transporter to be identified in *L. monocytogenes* was plasmid encoded CtpA from *L. monocytogenes* strain DRDC8, and which appears to be restricted to this strain due to its absence from other sequenced listerial genomes (Corbett et al., 2011). Mutants of strain DRDC8 lacking CtpA were found to be cleared much more rapidly from the livers of infected animals than the wild type in mouse infection experiments (Francis & Thomas, 1997b), indicating that copper homeostasis is important for virulence of *L. monocytogenes*. These mutants were also found to be no more sensitive to exogenous copper in the growth media than wild type, but were more sensitive to the effects of copper chelation, thus leading to the suggestion that CtpA has a role in copper uptake (Francis & Thomas, 1997a). However, such a role appears unlikely (Raimunda et al., 2011), and CtpA may alternatively have a role in supplying copper for copper-dependent enzyme activity important for *L. monocytogenes* virulence. A chromosomal copper-resistance operon, *csoR-copA-copZ*, has since been described in *L. monocytogenes* (Corbett et al., 2011), which encodes a copper-sensing transcriptional regulator belonging to the CsoR-RncR family, a copper-exporting P_{1B}-type ATPase and a copper metallochaperone, respectively (Fig. 3). In addition, *L. monocytogenes* possesses a gene, *lmo01018*, with similarity to CutC copper-binding proteins in other bacteria that have been implicated in copper homeostasis (Gupta, Lee, Camakaris, & Wu, 1995;

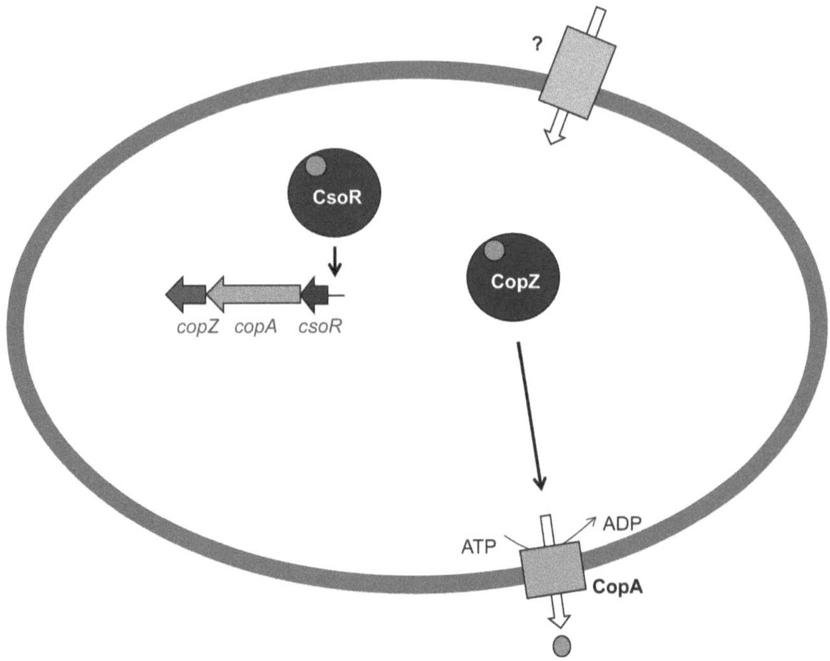

Figure 3 Copper homeostasis in *L. monocytogenes*. In response to elevated copper levels, the cytosolic copper-sensor CsoR dissociates from DNA to trigger expression of the *csoR-copA-copZ* operon. CopA functions as a P_{1B}-type ATPase and contributes to copper-resistance by exporting copper from the cytosol. CopZ functions as a copper-metallochaperone which is proposed to interact with and supply copper to CopA, and thereby contribute to copper-export. In addition, by sequestering cytosolic copper, CopZ is thought to protect the cytosol from the deleterious effects of copper and at the same time restrict the access of CsoR to copper to moderate *csoR-copA-copZ* expression. The mechanism of copper entry into *L. monocytogenes* is unknown (marked ?).

Latorre, Olivares, Reyes-Jara, Lopez, & Gonzalez, 2011; Rodrigues et al., 2008; Zhu et al., 2005), although the precise role of CutC proteins in controlling intracellular copper levels remains unclear.

5.2.1 Copper efflux mediated by the csoR-copA-copZ operon

A bioinformatics analysis of the *L. monocytogenes* (strain EGDe) genome identified *lmo1853* (subsequently designated *copA*) as encoding a probable copper-exporting P_{1B}-type ATPase, possessing features typical of Cu(I)/Ag(I) transporters (Mandal, Yang, Kertesz, & Arguello, 2004), with 43% and 37% identity to CopA from *B. subtilis* and *E. coli*, respectively (Corbett et al., 2011). The *copA* gene forms part of a three gene operon along with *csoR* (previously *lmo1852*) and *copZ* (previously *lmo1854*), which is expressed in response to elevated levels of copper, but not in response to other metals (Corbett et al., 2011). A *copA* mutant of *L. monocytogenes* is hypersensitive to exogenous copper and over-accumulates copper relative to wild-type cells, consistent with a role for CopA in copper efflux (Corbett et al., 2011).

Copper-responsive expression of the *csoR-copA-copZ* operon is mediated by CsoR, which acts as a copper-sensing derepressor (Corbett et al., 2011). Indeed, mutants lacking *csoR* show elevated constitutive expression from the operon promoter and CsoR has been shown to bind to sequences within a single 32-bp site containing an imperfect inverted repeat that spans the predicted -10 and -35 promoter elements (Corbett et al., 2011). Copper alters the assembly of the CsoR–DNA complex, although CsoR cannot be completely displaced from the promoter by copper (Corbett et al., 2011). Similar to the related CsoR proteins from *M. tuberculosis* and *B. subtilis* (Liu et al., 2007; Ma, Cowart, Scott, & Giedroc, 2009a; Ma et al., 2009b), *L. monocytogenes* CsoR senses copper via a conserved Cys_2His_1 site (Corbett et al., 2011). However, in contrast to the other CsoR proteins, *L. monocytogenes* CsoR does not possess the conserved glutamate residue in the α3-helix that is implicated in copper-induced conformational changes in CsoR. Furthermore, mutation of equivalent residues within *L. monocytogenes* CsoR does not appear to affect the copper-mediated switch that leads to derepression of the *csoR-copA-copZ* operon (Corbett et al., 2011). Thus, if *L. monocytogenes* CsoR does use a similar hydrogen bonding to the related proteins in order to stabilise the copper-bound form, this must use alternative residues that remain to be identified.

L. monocytogenes CopZ shares sequence similarity to Atx1/CopZ-like copper metallochaperones (Corbett et al., 2011). In other bacteria, these proteins contribute to copper homeostasis by supplying copper to cuproproteins (including copper-exporting P_{1B}-type ATPases) through copper-mediated protein–protein interactions and, by so doing, can prevent copper from causing damage *en route* (Robinson & Winge, 2010; Tottey et al., 2012). For example *B. subtilis* CopZ is known to interact with and donate copper to the CopA copper exporter (Banci Bertini, Ciofi-Baffoni, Del Conte, & Gonnelli, 2003; Radford et al., 2003), while *Enterococcus hirae* CopZ additionally delivers copper to the copper-responsive repressor CopY to cause derepression of the *cop* operon in this organism (Magnani & Solioz, 2005). Based on similarity to these proteins *L. monocytogenes* CopZ is expected to route cytosolic copper to CopA for efflux. In addition, with some analogy to *E. hirae* CopZ, a role in supplying copper to CsoR to mediate derepression of the *csoR-copA-copZ* operon is also possible. However, *L. monocytogenes* mutants lacking *copZ* display substantially increased expression of the *csoR-copA-copZ* in response to copper (Corbett et al., 2011), thus implying that, in contrast to supplying copper to CsoR, *L. monocytogenes* CopZ actually denies CsoR access to copper and as such can temper the transcriptional response to elevated copper (Corbett et al., 2011). CopZ was also found to confer increased copper resistance in *L. monocytogenes* mutants lacking copper-inducible *csoR-copA-copZ*, consistent with a role in protecting the cytosol from deleterious effects of copper (Corbett et al., 2011). In the current model (Fig. 3), CopZ is suggested to play a role in buffering cytosolic copper, such that CsoR can only sense copper once CopZ has become saturated with this metal (Corbett et al., 2011). It is possible that this provides an advantage to *L. monocytogenes* by moderating expression of the *csoR-copA-copZ* operon in high copper environments.

While bacterial copper resistance determinants have been associated with the virulence of other bacterial pathogens, *copA* mutants of *L. monocytogenes* display similar virulence to wild-type bacteria in an oral mouse infection model; there being no detectable difference in the survival or listerial load of mice infected with the *copA* mutant or wild-type bacteria (Corbett et al., 2011). Similarly, there is no detectable difference in the survival of wild type and the *copA* mutant within macrophages (Corbett et al., 2011). It is possible therefore that the rapid escape of *L. monocytogenes* from the internalisation phagosome to the host cell cytosol (refer to Section 2) circumvents a requirement for copper-resistance mechanisms for intracellular survival, thus contrasting with other pathogens (e.g. *S. enterica* sv. *Typhimurium*) that remain within the phagosome and are dependent on their copper-resistance mechanisms. The copper-resistance system in *L. monocytogenes* may therefore have evolved to combat elevated copper levels encountered outside host cells and within the wider environment.

6. CONCLUDING REMARKS

The complement of metal sensors and metal-homeostatic proteins within different bacterial pathogens likely reflect the different metal challenges faced within their particular niche for proliferation. For example, while elevated copper levels are a feature of macrophage phagosomes, thereby selecting for multiple specialised copper-resistance determinants in bacteria that can reside in this compartment, the buffered levels of copper are extremely low within the cytosol of host cells and hence copper resistance does not appear to be required for the intracellular survival of *L. monocytogenes* which escapes into the host cell cytosol and appears to have a single copper-resistance system. In contrast, *L. monocytogenes* is atypical in possessing two high-affinity zinc-uptake systems, both of which contribute to virulence. While there is some functional redundancy between the two zinc-uptake systems, this varies depending upon the environment in which the bacteria are growing and it is possible that this reflects differences in the substrates transported by each system and the available zinc sources in different environments, and/or their ability to function in the presence of other competing ligands and metal ions. Hence, further knowledge of the different metal challenges faced by different pathogens and of their various assortments of metal-handling systems may offer opportunities for developing more selective antimicrobial treatments in the future, with bacterial metal-homeostatic proteins representing attractive targets for drug development. However, while differences may exist in the available sources of metals to pathogens within different host niches, these are largely unknown, and hence this should be a focus of future research.

The change in population demographics coupled with the increased susceptibility of the elderly to food-borne listerial infections means that there is an increasing need to control *L. monocytogenes* in the food chain. Metals (including zinc and copper) are commonly used as antimicrobial agents within the food industry, as well as being included in animal feed for growth promotion. There is also recent interest in exploiting the bactericidal effects of metal nanoparticles in a range of applications, including food processing (Martirosyan & Schneider, 2014). Clearly, a concern is that the widespread use of metals as biocides could select for metal-resistant strains and support the growth of *L. monocytogenes*. Metal resistance has been associated with the selection of antibiotic-resistance (Baker-Austin, Wright, Stepanauskas, & McArthur, 2006; Bednorz et al., 2013; Seiler & Berendonk, 2012), further raising concerns for public health. As such, an understanding of metal-based mechanisms of host immune defences and of the bacterial systems that sense and adapt to metals, presents opportunities to inform future control strategies and for improving current metal-based treatments used in food production, thus leading to effective regimes to reduce *L. monocytogenes* contamination.

REFERENCES

Aarestrup, F. M., & Hasman, H. (2004). Susceptibility of different bacterial species isolated from food animals to copper sulphate, zinc chloride and antimicrobial substances used for disinfection. *Veterinary Microbiology*, *100*, 83–89.

Achard, M. E., Stafford, S. L., Bokil, N. J., Chartres, J., Bernhardt, P. V., Schembri, M. A., et al. (2012). Copper redistribution in murine macrophages in response to *Salmonella* infection. *The Biochemical Journal*, *444*, 51–57.

Akanuma, G., Nanamiya, H., Natori, Y., Nomura, N., & Kawamura, F. (2006). Liberation of zinc-containing L31 (RpmE) from ribosomes by its paralogous gene product, YtiA, in *Bacillus subtilis*. *Journal of Bacteriology*, *188*, 2715–2720.

Ammendola, S., Pasquali, P., Pistoia, C., Petrucci, P., Petrarca, P., Rotilio, G., et al. (2007). High-affinity Zn^{2+} uptake system ZnuABC is required for bacterial zinc homeostasis in intracellular environments and contributes to the virulence of *Salmonella enterica*. *Infection and Immunity*, *75*, 5867–5876.

Andreini, C., Banci, L., Bertini, I., & Rosato, A. (2006). Zinc through the three domains of life. *Journal of Proteome Research*, *5*, 3173–3178.

Andreini, C., Banci, L., Bertini, I., & Rosato, A. (2008). Occurrence of copper proteins through the three domains of life: A bioinformatic approach. *Journal of Proteome Research*, *7*, 209–216.

Andreini, C., Bertini, I., & Rosato, A. (2009). Metalloproteomes: A bioinformatic approach. *Accounts of Chemical Research*, *42*, 1471–1479.

Arguello, J. M., Raimunda, D., & Padilla-Benavides, T. (2013). Mechanisms of copper homeostasis in bacteria. *Frontiers in Cellular and Infection Microbiology*, *3*, 73.

Aydemir, T. B., Liuzzi, J. P., McClellan, S., & Cousins, R. J. (2009). Zinc transporter ZIP8 (SLC39A8) and zinc influence IFN-γ expression in activated human T cells. *Journal of Leukocyte Biology*, *86*, 337–348.

Baker-Austin, C., Wright, M. S., Stepanauskas, R., & McArthur, J. V. (2006). Co-selection of antibiotic and metal resistance. *Trends in Microbiology*, *14*, 176–182.

Banci, L., Bertini, I., Ciofi-Baffoni, S., Del Conte, R., & Gonnelli, L. (2003). Understanding copper trafficking in bacteria: Interaction between the copper transport protein CopZ and the N-terminal domain of the copper ATPase CopA from *Bacillus subtilis*. *Biochemistry*, *42*, 1939–1949.

Bao, S., Liu, M. J., Lee, B., Besecker, B., Lai, J. P., Guttridge, D. C., et al. (2010). Zinc modulates the innate immune response in vivo to polymicrobial sepsis through regulation of NF-κB. *American Journal of Physiology Lung Cellular and Molecular Physiology*, *298*, L744–L754.

Bayle, L., Chimalapati, S., Schoehn, G., Brown, J., Vernet, T., & Durmort, C. (2011). Zinc uptake by *Streptococcus pneumoniae* depends on both AdcA and AdcAII and is essential for normal bacterial morphology and virulence. *Molecular Microbiology, 82*, 904–916.

Bednorz, C., Oelgeschlager, K., Kinnemann, B., Hartmann, S., Neumann, K., Pieper, R., et al. (2013). The broader context of antibiotic resistance: Zinc feed supplementation of piglets increases the proportion of multi-resistant *Escherichia coli in vivo*. *International Journal of Medical Microbiology, 303*, 396–403.

Begley, M., Gahan, C. G., & Hill, C. (2002). Bile stress response in *Listeria monocytogenes* LO28: Adaptation, cross-protection, and identification of genetic loci involved in bile resistance. *Applied and Environmental Microbiology, 68*, 6005–6012.

Begum, N. A., Kobayashi, M., Moriwaki, Y., Matsumoto, M., Toyoshima, K., & Seya, T. (2002). *Mycobacterium bovis* BCG cell wall and lipopolysaccharide induce a novel gene, *BIGM103*, encoding a 7-TM protein: Identification of a new protein family having Zn-transporter and Zn-metalloprotease signatures. *Genomics, 80*, 630–645.

Berducci, G., Mazzetti, A. P., Rotilio, G., & Battistoni, A. (2004). Periplasmic competition for zinc uptake between the metallochaperone ZnuA and Cu, Zn superoxide dismutase. *FEBS Letters, 569*, 289–292.

Besecker, B., Bao, S., Bohacova, B., Papp, A., Sadee, W., & Knoell, D. L. (2008). The human zinc transporter SLC39A8 (Zip8) is critical in zinc-mediated cytoprotection in lung epithelia. *American Journal of Physiology Lung Cellular and Molecular Physiology, 294*, L1127–L1136.

Bianchi, M., Niemiec, M. J., Siler, U., Urban, C. F., & Reichenbach, J. (2011). Restoration of anti-Aspergillus defense by neutrophil extracellular traps in human chronic granulomatous disease after gene therapy is calprotectin-dependent. *The Journal of Allergy and Clinical Immunology, 127*(1243–1252), e1247.

Binet, M. R., & Poole, R. K. (2000). Cd(II), Pb(II) and Zn(II) ions regulate expression of the metal-transporting P-type ATPase ZntA in *Escherichia coli*. *FEBS Letters, 473*, 67–70.

Blencowe, D. K., & Morby, A. P. (2003). Zn(II) metabolism in prokaryotes. *FEMS Microbiology Reviews, 27*, 291–311.

Borkow, G., & Gabbay, J. (2005). Copper as a biocidal tool. *Current Medicinal Chemistry, 12*, 2163–2175.

Botella, H., Peyron, P., Levillain, F., Poincloux, R., Poquet, Y., Brandli, I., et al. (2011). Mycobacterial P_1-type ATPases mediate resistance to zinc poisoning in human macrophages. *Cell Host & Microbe, 10*, 248–259.

Botella, H., Stadthagen, G., Lugo-Villarino, G., de Chastellier, C., & Neyrolles, O. (2012). Metallobiology of host-pathogen interactions: An intoxicating new insight. *Trends in Microbiology, 20*, 106–112.

Brocklehurst, K. R., Hobman, J. L., Lawley, B., Blank, L., Marshall, S. J., Brown, N. L., et al. (1999). ZntR is a Zn(II)-responsive MerR-like transcriptional regulator of *zntA* in *Escherichia coli*. *Molecular Microbiology, 31*, 893–902.

Brophy, M. B., Hayden, J. A., & Nolan, E. M. (2012). Calcium ion gradients modulate the zinc affinity and antibacterial activity of human calprotectin. *Journal of the American Chemical Society, 134*, 18089–18100.

Campoy, S., Jara, M., Busquets, N., Perez De Rozas, A. M., Badiola, I., & Barbe, J. (2002). Role of the high-affinity zinc uptake znuABC system in *Salmonella enterica* serovar typhimurium virulence. *Infection and Immunity, 70*, 4721–4725.

Carrigan, P. E., Hentz, J. G., Gordon, G., Morgan, J. L., Raimondo, M., Anbar, A. D., et al. (2007). Distinctive heavy metal composition of pancreatic juice in patients with pancreatic carcinoma. *Cancer Epidemiology, Biomarkers and Prevention, 16*, 2656–2663.

Cassat, J. E., & Skaar, E. P. (2013). Iron in infection and immunity. *Cell Host & Microbe, 13*, 509–519.

Cellier, M. F., Courville, P., & Campion, C. (2007). Nramp1 phagocyte intracellular metal withdrawal defense. *Microbes and Infection, 9*, 1662–1670.

Chandra, B. R., Yogavel, M., & Sharma, A. (2007). Structural analysis of ABC-family periplasmic zinc binding protein provides new insights into mechanism of ligand uptake and release. *Journal of Molecular Biology, 367*, 970–982.

Changela, A., Chen, K., Xue, Y., Holschen, J., Outten, C. E., O'Halloran, T. V., et al. (2003). Molecular basis of metal-ion selectivity and zeptomolar sensitivity by CueR. *Science, 301*, 1383–1387.

Chivers, P. T., Benanti, E. L., Heil-Chapdelaine, V., Iwig, J. S., & Rowe, J. L. (2012). Identification of Ni-(L-His)$_2$ as a substrate for NikABCDE-dependent nickel uptake in *Escherichia coli. Metallomics, 4*, 1043–1050.

Clohessy, P. A., & Golden, B. E. (1995). Calprotectin-mediated zinc chelation as a biostatic mechanism in host defence. *Scandinavian Journal of Immunology, 42*, 551–556.

Collins, H. L. (2008). Withholding iron as a cellular defence mechanism—Friend or foe? *European Journal of Immunology, 38*, 1803–1806.

Colvin, R. A., Holmes, W. R., Fontaine, C. P., & Maret, W. (2010). Cytosolic zinc buffering and muffling: Their role in intracellular zinc homeostasis. *Metallomics, 2*, 306–317.

Corbett, D., Schuler, S., Glenn, S., Andrew, P. W., Cavet, J. S., & Roberts, I. S. (2011). The combined actions of the copper-responsive repressor CsoR and copper-metallochaperone CopZ modulate CopA-mediated copper efflux in the intracellular pathogen *Listeria monocytogenes. Molecular Microbiology, 81*, 457–472.

Corbett, D., Wang, J., Schuler, S., Lopez-Castejon, G., Glenn, S., Brough, D., et al. (2012). Two zinc uptake systems contribute to the full virulence of *Listeria monocytogenes* during growth *in vitro* and *in vivo. Infection and Immunity, 80*, 14–21.

Corbin, B. D., Seeley, E. H., Raab, A., Feldmann, J., Miller, M. R., Torres, V. J., et al. (2008). Metal chelation and inhibition of bacterial growth in tissue abscesses. *Science, 319*, 962–965.

Corr, S., Hill, C., & Gahan, C. G. (2006). An *in vitro* cell-culture model demonstrates internalin- and hemolysin-independent translocation of *Listeria monocytogenes* across M cells. *Microbial Pathogenesis, 41*, 241–250.

Cossart, P. (2000). Actin-based motility of pathogens: The Arp2/3 complex is a central player. *Cellular Microbiology, 2*, 195–205.

Cossart, P. (2011). Illuminating the landscape of host-pathogen interactions with the bacterium *Listeria monocytogenes. Proceedings of the National Academy of Sciences of the United States of America, 108*, 19484–19491.

Counago, R. M., Ween, M. P., Begg, S. L., Bajaj, M., Zuegg, J., O'Mara, M. L., et al. (2014). Imperfect coordination chemistry facilitates metal ion release in the Psa permease. *Nature Chemical Biology, 10*, 35–41.

Crocker, A., Lee, C., Aboko-Cole, G., & Durham, C. (1992). Interaction of nutrition and infection: Effect of copper deficiency on resistance to Trypanosoma lewisi. *Journal of the National Medical Association, 84*, 697–706.

Dahiya, I., & Stevenson, R. M. (2010). The ZnuABC operon is important for *Yersinia ruckeri* infections of rainbow trout, Oncorhynchus mykiss (Walbaum). *Journal of Fish Diseases, 33*, 331–340.

Dalet, K., Gouin, E., Cenatiempo, Y., Cossart, P., & Hechard, Y. (1999). Characterisation of a new operon encoding a Zur-like protein and an associated ABC zinc permease in *Listeria monocytogenes. FEMS Microbiology Letters, 174*, 111–116.

Damo, S. M., Kehl-Fie, T. E., Sugitani, N., Holt, M. E., Rathi, S., Murphy, W. J., et al. (2013). Molecular basis for manganese sequestration by calprotectin and roles in the innate immune response to invading bacterial pathogens. *Proceedings of the National Academy of Sciences of the United States of America, 110*, 3841–3846.

Davis, L. M., Kakuda, T., & DiRita, V. J. (2009). A *Campylobacter jejuni znuA* orthologue is essential for growth in low-zinc environments and chick colonization. *Journal of Bacteriology, 191*, 1631–1640.
de Abel, Matousek, de la Cruz, A. J., Burguera, J. L., Burguera, M., & Anez, N. (1993). Changes in the total content of iron, copper, and zinc in serum, heart, liver, spleen, and skeletal muscle tissues of rats infected with Trypanosoma cruzi. *Biological Trace Element Research, 37*, 51–70.
de las Heras, A., Cain, R. J., Bielecka, M. K., & Vazquez-Boland, J. A. (2011). Regulation of *Listeria* virulence: PrfA master and commander. *Current Opinion in Microbiology, 14*, 118–127.
Deshayes, C., Bielecka, M. K., Cain, R. J., Scortti, M., de las Heras, A., Pietras, Z., et al. (2012). Allosteric mutants show that PrfA activation is dispensable for vacuole escape but required for efficient spread and *Listeria* survival *in vivo*. *Molecular Microbiology, 85*, 461–477.
Desrosiers, D. C., Bearden, S. W., Mier, I., Jr., Abney, J., Paulley, J. T., Fetherston, J. D., et al. (2010). Znu is the predominant zinc importer in *Yersinia pestis* during in vitro growth but is not essential for virulence. *Infection and Immunity, 78*, 5163–5177.
Diaz-Ochoa, V. E., Jellbauer, S., Klaus, S., & Raffatellu, M. (2014). Transition metal ions at the crossroads of mucosal immunity and microbial pathogenesis. *Frontiers in Cellular and Infection Microbiology, 4*, 2.
Ding, C., Festa, R. A., Chen, Y. L., Espart, A., Palacios, O., Espin, J., et al. (2013). *Cryptococcus neoformans* copper detoxification machinery is critical for fungal virulence. *Cell Host & Microbe, 13*, 265–276.
Djoko, K. Y., Franiek, J. A., Edwards, J. L., Falsetta, M. L., Kidd, S. P., Potter, A. J., et al. (2012). Phenotypic characterization of a *copA* mutant of *Neisseria gonorrhoeae* identifies a link between copper and nitrosative stress. *Infection and Immunity, 80*, 1065–1071.
Dowd, G. C., Casey, P. G., Begley, M., Hill, C., & Gahan, C. G. (2012). Investigation of the role of ZurR in the physiology and pathogenesis of *Listeria monocytogenes*. *FEMS Microbiology Letters, 327*, 118–125.
Dupont, C. L., Grass, G., & Rensing, C. (2011). Copper toxicity and the origin of bacterial resistance—New insights and applications. *Metallomics, 3*, 1109–1118.
Eide, D. J. (2006). Zinc transporters and the cellular trafficking of zinc. *Biochimica et Biophysica Acta, 1763*, 711–722.
Eijkelkamp, B. A., Morey, J. R., Ween, M. P., Ong, C. L., McEwan, A. G., Paton, J. C., et al. (2014). Extracellular zinc competitively inhibits manganese uptake and compromises oxidative stress management in *Streptococcus pneumoniae*. *PLoS One, 9*, e89427.
Festa, R. A., Jones, M. B., Butler-Wu, S., Sinsimer, D., Gerads, R., Bishai, W. R., et al. (2011). A novel copper-responsive regulon in *Mycobacterium tuberculosis*. *Molecular Microbiology, 79*, 133–148.
Foster, A. W., Pernil, R., Patterson, C. J., & Robinson, N. J. (2014). Metal specificity of cyanobacterial nickel-responsive repressor InrS: Cells maintain zinc and copper below the detection threshold for InrS. *Molecular Microbiology, 92*, 797–812.
Francis, M. S., & Thomas, C. J. (1997a). The *Listeria monocytogenes* gene *ctpA* encodes a putative P-type ATPase involved in copper transport. *Molecular and General Genetics, 253*, 484–491.
Francis, M. S., & Thomas, C. J. (1997b). Mutants in the CtpA copper transporting P-type ATPase reduce virulence of Listeria monocytogenes. *Microbial Pathogenesis, 22*, 67–78.
Fraústo da Silva, J. J. R., & Williams, R. J. P. (2001). *The biological chemistry of the elements: The inorganic chemistry of life* (2nd ed.). Oxford: Oxford University Press.
Gaballa, A., & Helmann, J. D. (1998). Identification of a zinc-specific metalloregulatory protein, Zur, controlling zinc transport operons in *Bacillus subtilis*. *Journal of Bacteriology, 180*, 5815–5821.

Gaballa, A., Wang, T., Ye, R. W., & Helmann, J. D. (2002). Functional analysis of the *Bacillus subtilis* Zur regulon. *Journal of Bacteriology, 184*, 6508–6514.
Gabbianelli, R., Scotti, R., Ammendola, S., Petrarca, P., Nicolini, L., & Battistoni, A. (2011). Role of ZnuABC and ZinT in *Escherichia coli* O157:H7 zinc acquisition and interaction with epithelial cells. *BMC Microbiology, 11*, 36.
Gabriel, S. E., & Helmann, J. D. (2009). Contributions of Zur-controlled ribosomal proteins to growth under zinc starvation conditions. *Journal of Bacteriology, 191*, 6116–6122.
Gabriel, S. E., Miyagi, F., Gaballa, A., & Helmann, J. D. (2008). Regulation of the *Bacillus subtilis yciC* gene and insights into the DNA-binding specificity of the zinc-sensing metalloregulator Zur. *Journal of Bacteriology, 190*, 3482–3488.
Gaetke, L. M., McClain, C. J., Talwalkar, R. T., & Shedlofsky, S. I. (1997). Effects of endotoxin on zinc metabolism in human volunteers. *The American Journal of Physiology, 272*, E952–E956.
Garcia-del Portillo, F., Calvo, E., D'Orazio, V., & Pucciarelli, M. G. (2011). Association of ActA to peptidoglycan revealed by cell wall proteomics of intracellular *Listeria monocytogenes*. *The Journal of Biological Chemistry, 286*, 34675–34689.
Garrido, M. E., Bosch, M., Medina, R., Llagostera, M., Perez de Rozas, A. M., Badiola, I., et al. (2003). The high-affinity zinc-uptake system ZnuACB is under control of the iron-uptake regulator (*fur*) gene in the animal pathogen *Pasteurella multocida*. *FEMS Microbiology Letters, 221*, 31–37.
Gebhardt, C., Nemeth, J., Angel, P., & Hess, J. (2006). S100A8 and S100A9 in inflammation and cancer. *Biochemical Pharmacology, 72*, 1622–1631.
Glaser, R., Harder, J., Lange, H., Bartels, J., Christophers, E., & Schroder, J. M. (2005). Antimicrobial psoriasin (S100A7) protects human skin from *Escherichia coli* infection. *Nature Immunology, 6*, 57–64.
Graham, A. I., Hunt, S., Stokes, S. L., Bramall, N., Bunch, J., Cox, A. G., et al. (2009). Severe zinc depletion of *Escherichia coli*: Roles for high affinity zinc binding by ZinT, zinc transport and zinc-independent proteins. *The Journal of Biological Chemistry, 284*, 18377–18389.
Grass, G., Franke, S., Taudte, N., Nies, D. H., Kucharski, L. M., Maguire, M. E., et al. (2005). The metal permease ZupT from *Escherichia coli* is a transporter with a broad substrate spectrum. *Journal of Bacteriology, 187*, 1604–1611.
Grass, G., Rensing, C., & Solioz, M. (2011). Metallic copper as an antimicrobial surface. *Applied and Environmental Microbiology, 77*, 1541–1547.
Gray, M. L., & Killinger, A. H. (1966). *Listeria monocytogenes* and listeric infections. *Bacteriological Reviews, 30*, 309–382.
Guedon, E., Moore, C. M., Que, Q., Wang, T., Ye, R. W., & Helmann, J. D. (2003). The global transcriptional response of *Bacillus subtilis* to manganese involves the MntR, Fur, TnrA and σ^B regulons. *Molecular Microbiology, 49*, 1477–1491.
Gupta, S. D., Lee, B. T., Camakaris, J., & Wu, H. C. (1995). Identification of *cutC* and *cutF* (*nlpE*) genes involved in copper tolerance in *Escherichia coli*. *Journal of Bacteriology, 177*, 4207–4215.
Haase, H., & Rink, L. (2007). Signal transduction in monocytes: The role of zinc ions. *Biometals, 20*, 579–585.
Hantke, K. (2001). Bacterial zinc transporters and regulators. *Biometals, 14*, 239–249.
Hantke, K. (2005). Bacterial zinc uptake and regulators. *Current Opinion in Microbiology, 8*, 196–202.
Hassani, B. K., Astier, C., Nitschke, W., & Ouchane, S. (2010). CtpA, a copper-translocating P-type ATPase involved in the biogenesis of multiple copper-requiring enzymes. *The Journal of Biological Chemistry, 285*, 19330–19337.
Hessian, P. A., Edgeworth, J., & Hogg, N. (1993). MRP-8 and MRP-14, two abundant Ca^{2+}-binding proteins of neutrophils and monocytes. *Journal of Leukocyte Biology, 53*, 197–204.

Hirano, T., Murakami, M., Fukada, T., Nishida, K., Yamasaki, S., & Suzuki, T. (2008). Roles of zinc and zinc signaling in immunity: Zinc as an intracellular signaling molecule. *Advances in Immunology*, *97*, 149–176.

Hood, M. I., Mortensen, B. L., Moore, J. L., Zhang, Y., Kehl-Fie, T. E., Sugitani, N., et al. (2012). Identification of an *Acinetobacter baumannii* zinc acquisition system that facilitates resistance to calprotectin-mediated zinc sequestration. *PLoS Pathogens*, *8*, e1003068.

Hood, M. I., & Skaar, E. P. (2012). Nutritional immunity: Transition metals at the pathogen-host interface. *Nature Reviews Microbiology*, *10*, 525–537.

Huang, Z. L., & Failla, M. L. (2000). Copper deficiency suppresses effector activities of differentiated U937 cells. *Journal of Nutrition*, *130*, 1536–1542.

Ilback, N. G., Benyamin, G., Lindh, U., & Friman, G. (2003). Sequential changes in Fe, Cu, and Zn in target organs during early Coxsackievirus B3 infection in mice. *Biological Trace Element Research*, *91*, 111–124.

Imlay, J. A., Chin, S. M., & Linn, S. (1988). Toxic DNA damage by hydrogen peroxide through the Fenton reaction *in vivo* and *in vitro*. *Science*, *240*, 640–642.

Irving, H., & Williams, R. J. P. (1948). Order of stability of metal complexes. *Nature*, *162*, 2.

Jin, B., Newton, S. M., Shao, Y., Jiang, X., Charbit, A., & Klebba, P. E. (2006). Iron acquisition systems for ferric hydroxamates, haemin and haemoglobin in *Listeria monocytogenes*. *Molecular Microbiology*, *59*, 1185–1198.

Johansson, J., Mandin, P., Renzoni, A., Chiaruttini, C., Springer, M., & Cossart, P. (2002). An RNA thermosensor controls expression of virulence genes in *Listeria monocytogenes*. *Cell*, *110*, 551–561.

Kaluarachchi, H., Chan Chung, K. C., & Zamble, D. B. (2010). Microbial nickel proteins. *Natural Product Reports*, *27*, 681–694.

Karlinsey, J. E., Maguire, M. E., Becker, L. A., Crouch, M. L., & Fang, F. C. (2010). The phage shock protein PspA facilitates divalent metal transport and is required for virulence of *Salmonella enterica* sv. Typhimurium. *Molecular Microbiology*, *78*, 669–685.

Kehl-Fie, T. E., Chitayat, S., Hood, M. I., Damo, S., Restrepo, N., Garcia, C., et al. (2011). Nutrient metal sequestration by calprotectin inhibits bacterial superoxide defense, enhancing neutrophil killing of *Staphylococcus aureus*. *Cell Host & Microbe*, *10*, 158–164.

Kehl-Fie, T. E., & Skaar, E. P. (2010). Nutritional immunity beyond iron: A role for manganese and zinc. *Current Opinion in Chemical Biology*, *14*, 218–224.

Kehres, D. G., Zaharik, M. L., Finlay, B. B., & Maguire, M. E. (2000). The NRAMP proteins of *Salmonella typhimurium* and *Escherichia coli* are selective manganese transporters involved in the response to reactive oxygen. *Molecular Microbiology*, *36*, 1085–1100.

Kim, S., Watanabe, K., Shirahata, T., & Watarai, M. (2004). Zinc uptake system (*znuA* locus) of *Brucella abortus* is essential for intracellular survival and virulence in mice. *The Journal of Veterinary Medical Science*, *66*, 1059–1063.

Kitamura, H., Morikawa, H., Kamon, H., Iguchi, M., Hojyo, S., Fukada, T., et al. (2006). Toll-like receptor-mediated regulation of zinc homeostasis influences dendritic cell function. *Nature Immunology*, *7*, 971–977.

Klebba, P. E., Charbit, A., Xiao, Q., Jiang, X., & Newton, S. M. (2012). Mechanisms of iron and haem transport by *Listeria monocytogenes*. *Molecular and Membrane Biology*, *29*, 69–86.

Kolenbrander, P. E., Andersen, R. N., Baker, R. A., & Jenkinson, H. F. (1998). The adhesion-associated *sca* operon in *Streptococcus gordonii* encodes an inducible high-affinity ABC transporter for Mn^{2+} uptake. *Journal of Bacteriology*, *180*, 290–295.

Latorre, M., Olivares, F., Reyes-Jara, A., Lopez, G., & Gonzalez, M. (2011). CutC is induced late during copper exposure and can modify intracellular copper content in *Enterococcus faecalis*. *Biochemical and Biophysical Research Communications*, *406*, 633–637.

Lebrun, M., Audurier, A., & Cossart, P. (1994). Plasmid-borne cadmium resistance genes in *Listeria monocytogenes* are similar to *cadA* and *cadC* of *Staphylococcus aureus* and are induced by cadmium. *Journal of Bacteriology*, *176*, 3040–3048.

Lecuit, M. (2007). Human listeriosis and animal models. *Microbes and Infection, 9*, 1216–1225.
Lee, L. J., Barrett, J. A., & Poole, R. K. (2005). Genome-wide transcriptional response of chemostat-cultured *Escherichia coli* to zinc. *Journal of Bacteriology, 187*, 1124–1134.
Lewinson, O., Lee, A. T., & Rees, D. C. (2009). A P-type ATPase importer that discriminates between essential and toxic transition metals. *Proceedings of the National Academy of Sciences of the United States of America, 106*, 4677–4682.
Lewis, D. A., Klesney-Tait, J., Lumbley, S. R., Ward, C. K., Latimer, J. L., Ison, C. A., et al. (1999). Identification of the *znuA*-encoded periplasmic zinc transport protein of *Haemophilus ducreyi*. *Infection and Immunity, 67*, 5060–5068.
Lin, Y. E., Stout, J. E., & Yu, V. L. (2011). Controlling *Legionella* in hospital drinking water: An evidence-based review of disinfection methods. *Infection Control and Hospital Epidemiology, 32*, 166–173.
Lindsay, J. A., & Foster, S. J. (2001). *zur*: A Zn^{2+}-responsive regulatory element of *Staphylococcus aureus*. *Microbiology, 147*, 1259–1266.
Linnan, M. J., Mascola, L., Lou, X. D., Goulet, V., May, S., Salminen, C., et al. (1988). Epidemic listeriosis associated with Mexican-style cheese. *The New England Journal of Medicine, 319*, 823–828.
Liu, M. J., Bao, S., Galvez-Peralta, M., Pyle, C. J., Rudawsky, A. C., Pavlovicz, R. E., et al. (2013). ZIP8 regulates host defense through zinc-mediated inhibition of NF-κB. *Cell Reports, 3*, 386–400.
Liu, J. Z., Jellbauer, S., Poe, A. J., Ton, V., Pesciaroli, M., Kehl-Fie, T. E., et al. (2012). Zinc sequestration by the neutrophil protein calprotectin enhances *Salmonella* growth in the inflamed gut. *Cell Host & Microbe, 11*, 227–239.
Liu, T., Ramesh, A., Ma, Z., Ward, S. K., Zhang, L., George, G. N., et al. (2007). CsoR is a novel *Mycobacterium tuberculosis* copper-sensing transcriptional regulator. *Nature Chemical Biology, 3*, 60–68.
Liuzzi, J. P., Lichten, L. A., Rivera, S., Blanchard, R. K., Aydemir, T. B., Knutson, M. D., et al. (2005). Interleukin-6 regulates the zinc transporter Zip14 in liver and contributes to the hypozincemia of the acute-phase response. *Proceedings of the National Academy of Sciences of the United States of America, 102*, 6843–6848.
Lorber, B. (1997). Listeriosis. *Clinical Infectious Disease, 24*, 1–9, quiz 10–11.
Lutsenko, S., LeShane, E. S., & Shinde, U. (2007). Biochemical basis of regulation of human copper-transporting ATPases. *Archives of Biochemistry and Biophysics, 463*, 134–148.
Ma, Z., Cowart, D. M., Scott, R. A., & Giedroc, D. P. (2009a). Molecular insights into the metal selectivity of the copper(I)-sensing repressor CsoR from *Bacillus subtilis*. *Biochemistry, 48*, 3325–3334.
Ma, Z., Cowart, D. M., Ward, B. P., Arnold, R. J., DiMarchi, R. D., Zhang, L., et al. (2009b). Unnatural amino acid substitution as a probe of the allosteric coupling pathway in a mycobacterial Cu(I) sensor. *Journal of the American Chemical Society, 131*, 18044–18045.
Ma, Z., Gabriel, S. E., & Helmann, J. D. (2011). Sequential binding and sensing of Zn(II) by *Bacillus subtilis* Zur. *Nucleic Acids Research, 39*, 9130–9138.
Macomber, L., & Imlay, J. A. (2009). The iron-sulfur clusters of dehydratases are primary intracellular targets of copper toxicity. *Proceedings of the National Academy of Sciences of the United States of America, 106*, 8344–8349.
Magnani, D., & Solioz, M. (2005). Copper chaperone cycling and degradation in the regulation of the *cop* operon of *Enterococcus hirae*. *Biometals, 18*, 407–412.
Mandal, A. K., Yang, Y., Kertesz, T. M., & Arguello, J. M. (2004). Identification of the transmembrane metal binding site in Cu^{+}-transporting P_{IB}-type ATPases. *The Journal of Biological Chemistry, 279*, 54802–54807.
Martirosyan, A., & Schneider, Y. J. (2014). Engineered nanomaterials in food: Implications for food safety and consumer health. *International Journal of Environmental Research and Public Health, 11*, 5720–5750.

Marvin, R. G., Wolford, J. L., Kidd, M. J., Murphy, S., Ward, J., Que, E. L., et al. (2012). Fluxes in "free" and total zinc are essential for progression of intraerythrocytic stages of *Plasmodium falciparum*. *Chemistry and Biology, 19*, 731–741.

McCormick, A., Heesemann, L., Wagener, J., Marcos, V., Hartl, D., Loeffler, J., et al. (2010). NETs formed by human neutrophils inhibit growth of the pathogenic mold *Aspergillus fumigatus*. *Microbes and Infection, 12*, 928–936.

McHugh, J. P., Rodriguez-Quinones, F., Abdul-Tehrani, H., Svistunenko, D. A., Poole, R. K., Cooper, C. E., et al. (2003). Global iron-dependent gene regulation in *Escherichia coli*. A new mechanism for iron homeostasis. *The Journal of Biological Chemistry, 278*, 29478–29486.

McLaughlin, H. P., Hill, C., & Gahan, C. G. (2011). The impact of iron on *Listeria monocytogenes*; inside and outside the host. *Current Opinion in Biotechnology, 22*, 194–199.

McLaughlin, H. P., Xiao, Q., Rea, R. B., Pi, H., Casey, P. G., Darby, T., et al. (2012). A putative P-type ATPase required for virulence and resistance to haem toxicity in *Listeria monocytogenes*. *PLoS One*, 7, e30928.

Mengaud, J., Lecuit, M., Lebrun, M., Nato, F., Mazie, J. C., & Cossart, P. (1996). Antibodies to the leucine-rich repeat region of internalin block entry of *Listeria monocytogenes* into cells expressing E-cadherin. *Infection and Immunity, 64*, 5430–5433.

Michalek, M., Gelhaus, C., Hecht, O., Podschun, R., Schroder, J. M., Leippe, M., et al. (2009). The human antimicrobial protein psoriasin acts by permeabilization of bacterial membranes. *Developmental and Comparative Immunology, 33*, 740–746.

Moore, C. M., Gaballa, A., Hui, M., Ye, R. W., & Helmann, J. D. (2005). Genetic and physiological responses of *Bacillus subtilis* to metal ion stress. *Molecular Microbiology, 57*, 27–40.

Moore, C. M., & Helmann, J. D. (2005). Metal ion homeostasis in *Bacillus subtilis*. *Current Opinion in Microbiology, 8*, 188–195.

Moroz, O. V., Antson, A. A., Grist, S. J., Maitland, N. J., Dodson, G. G., Wilson, K. S., et al. (2003a). Structure of the human S100A12-copper complex: Implications for host-parasite defence. *Acta Crystallographica Section D: Biological Crystallography, 59*, 859–867.

Moroz, O. V., Blagova, E. V., Wilkinson, A. J., Wilson, K. S., & Bronstein, I. B. (2009a). The crystal structures of human S100A12 in apo form and in complex with zinc: New insights into S100A12 oligomerisation. *Journal of Molecular Biology, 391*, 536–551.

Moroz, O. V., Burkitt, W., Wittkowski, H., He, W., Ianoul, A., Novitskaya, V., et al. (2009b). Both Ca^{2+} and Zn^{2+} are essential for S100A12 protein oligomerization and function. *BMC Biochemistry, 10*, 11.

Moroz, O. V., Dodson, G. G., Wilson, K. S., Lukanidin, E., & Bronstein, I. B. (2003b). Multiple structural states of S100A12: A key to its functional diversity. *Microscopy Research and Technique, 60*, 581–592.

Mostowy, S., & Cossart, P. (2012). Virulence factors that modulate the cell biology of *Listeria* infection and the host response. *Advances in Immunology, 113*, 19–32.

Murphy, T. F., Brauer, A. L., Kirkham, C., Johnson, A., Koszelak-Rosenblum, M., & Malkowski, M. G. (2013). Role of the zinc uptake ABC transporter of *Moraxella catarrhalis* in persistence in the respiratory tract. *Infection and Immunity, 81*, 3406–3413.

Natori, Y., Nanamiya, H., Akanuma, G., Kosono, S., Kudo, T., Ochi, K., et al. (2007). A fail-safe system for the ribosome under zinc-limiting conditions in *Bacillus subtilis*. *Molecular Microbiology, 63*, 294–307.

Newberne, P. M., Hunt, C. E., & Young, V. R. (1968). The role of diet and the reticuloendothelial system in the response of rats to *Salmonella typhimurium* infection. *British Journal of Experimental Pathology, 49*, 448–457.

Nieman, R. E., & Lorber, B. (1980). Listeriosis in adults: A changing pattern. Report of eight cases and review of the literature, 1968-1978. *Reviews of Infectious Diseases, 2*, 207–227.

Nies, D. H. (2003). Efflux-mediated heavy metal resistance in prokaryotes. *FEMS Microbiology Reviews*, *27*, 313–339.

Nies, D. H. (2007). Biochemistry. How cells control zinc homeostasis. *Science*, *317*, 1695–1696.

Noyce, J. O., Michels, H., & Keevil, C. W. (2006a). Potential use of copper surfaces to reduce survival of epidemic meticillin-resistant *Staphylococcus aureus* in the healthcare environment. *Journal of Hospital Infection*, *63*, 289–297.

Noyce, J. O., Michels, H., & Keevil, C. W. (2006b). Use of copper cast alloys to control *Escherichia coli* O157 cross-contamination during food processing. *Applied and Environmental Microbiology*, *72*, 4239–4244.

Odermatt, A., Suter, H., Krapf, R., & Solioz, M. (1993). Primary structure of two P-type ATPases involved in copper homeostasis in *Enterococcus hirae*. *The Journal of Biological Chemistry*, *268*, 12775–12779.

Ong, C. L., Gillen, C. M., Barnett, T. C., Walker, M. J., & McEwan, A. G. (2014). An antimicrobial role for zinc in innate immune defense against group A *Streptococcus*. *The Journal of Infectious Diseases*, *209*, 1500–1508.

Ooi, S. T., & Lorber, B. (2005). Gastroenteritis due to *Listeria monocytogenes*. *Clinical Infectious Diseases*, *40*, 1327–1332.

Osman, D., & Cavet, J. S. (2010). Bacterial metal-sensing proteins exemplified by ArsR-SmtB family repressors. *Natural Product Reports*, *27*, 668–680.

Osman, D., Patterson, C. J., Bailey, K., Fisher, K., Robinson, N. J., Rigby, S. E., et al. (2013). The copper supply pathway to a *Salmonella* Cu, Zn-superoxide dismutase (SodCII) involves P1B-type ATPase copper efflux and periplasmic CueP. *Molecular Microbiology*, *87*, 466–477.

Osman, D., Waldron, K. J., Denton, H., Taylor, C. M., Grant, A. J., Mastroeni, P., et al. (2010). Copper homeostasis in *Salmonella* is atypical and copper-CueP is a major periplasmic metal complex. *The Journal of Biological Chemistry*, *285*, 25259–25268.

Outten, F. W., Huffman, D. L., Hale, J. A., & O'Halloran, T. V. (2001). The independent *cue* and *cus* systems confer copper tolerance during aerobic and anaerobic growth in *Escherichia coli*. *The Journal of Biological Chemistry*, *276*, 30670–30677.

Outten, C. E., & O'Halloran, T. V. (2001). Femtomolar sensitivity of metalloregulatory proteins controlling zinc homeostasis. *Science*, *292*, 2488–2492.

Padilla-Benavides, T., Long, J. E., Raimunda, D., Sassetti, C. M., & Arguello, J. M. (2013). A novel P_{1B}-type Mn^{2+}-transporting ATPase is required for secreted protein metallation in mycobacteria. *The Journal of Biological Chemistry*, *288*, 11334–11347.

Patterson, C. J., Pernil, R., Dainty, S. J., Chakrabarti, B., Henry, C. E., Money, V. A., et al. (2013). Co(ll)-detection does not follow $K_{Co(II)}$ gradient: Channelling in Co(ll)-sensing. *Metallomics*, *5*, 352–362.

Patzer, S. I., & Hantke, K. (1998). The ZnuABC high-affinity zinc uptake system and its regulator Zur in *Escherichia coli*. *Molecular Microbiology*, *28*, 1199–1210.

Percival, S. S. (1998). Copper and immunity. *American Journal of Clinical Nutrition*, *67*, 1064S–1068S.

Petrarca, P., Ammendola, S., Pasquali, P., & Battistoni, A. (2010). The Zur-regulated ZinT protein is an auxiliary component of the high-affinity ZnuABC zinc transporter that facilitates metal recruitment during severe zinc shortage. *Journal of Bacteriology*, *192*, 1553–1564.

Poyart, C., Abachin, E., Razafimanantsoa, I., & Berche, P. (1993). The zinc metalloprotease of *Listeria monocytogenes* is required for maturation of phosphatidylcholine phospholipase C: Direct evidence obtained by gene complementation. *Infection and Immunity*, *61*, 1576–1580.

Prasad, A. S. (2013). Discovery of human zinc deficiency: Its impact on human health and disease. *Advances in Nutrition*, *4*, 176–190.

Prohaska, J. R., & Lukasewycz, O. A. (1981). Copper deficiency suppresses the immune response of mice. *Science*, *213*, 559–561.
Radford, D. S., Kihlken, M. A., Borrelly, G. P., Harwood, C. R., Le Brun, N. E., & Cavet, J. S. (2003). CopZ from *Bacillus subtilis* interacts *in vivo* with a copper exporting CPx-type ATPase CopA. *FEMS Microbiology Letters*, *220*, 105–112.
Raimunda, D., Gonzalez-Guerrero, M., Leeber, B. W., 3rd., & Arguello, J. M. (2011). The transport mechanism of bacterial Cu^{+}-ATPases: Distinct efflux rates adapted to different function. *Biometals*, *24*, 467–475.
Robinson, N. J., & Winge, D. R. (2010). Copper metallochaperones. *Annual Review of Biochemistry*, *79*, 537–562.
Rodrigues, C. M., Takita, M. A., Coletta-Filho, H. D., Olivato, J. C., Caserta, R., Machado, M. A., et al. (2008). Copper resistance of biofilm cells of the plant pathogen *Xylella fastidiosa*. *Applied Microbiology and Biotechnology*, 77, 1145–1157.
Rosadini, C. V., Gawronski, J. D., Raimunda, D., Arguello, J. M., & Akerley, B. J. (2011). A novel zinc binding system, ZevAB, is critical for survival of nontypeable *Haemophilus influenzae* in a murine lung infection model. *Infection and Immunity*, *79*, 3366–3376.
Rowe, J. L., Starnes, G. L., & Chivers, P. T. (2005). Complex transcriptional control links NikABCDE-dependent nickel transport with hydrogenase expression in *Escherichia coli*. *Journal of Bacteriology*, *187*, 6317–6323.
Sabri, M., Houle, S., & Dozois, C. M. (2009). Roles of the extraintestinal pathogenic *Escherichia coli* ZnuACB and ZupT zinc transporters during urinary tract infection. *Infection and Immunity*, 77, 1155–1164.
Schaible, U. E., & Kaufmann, S. H. (2005). A nutritive view on the host-pathogen interplay. *Trends in Microbiology*, *13*, 373–380.
Schlech, W. F., 3rd., Lavigne, P. M., Bortolussi, R. A., Allen, A. C., Haldane, E. V., Wort, A. J., et al. (1983). Epidemic listeriosis—Evidence for transmission by food. *The New England Journal of Medicine*, *308*, 203–206.
Schuchat, A., Swaminathan, B., & Broome, C. V. (1991). Epidemiology of human listeriosis. *Clinical Microbiology Reviews*, *4*, 169–183.
Schuppler, M., & Loessner, M. J. (2010). The opportunistic pathogen *Listeria monocytogenes*: Pathogenicity and interaction with the mucosal immune system. *International Journal of Inflammation*, *2010*, 704321.
Seiler, C., & Berendonk, T. U. (2012). Heavy metal driven co-selection of antibiotic resistance in soil and water bodies impacted by agriculture and aquaculture. *Frontiers in Microbiology*, *3*, 399.
Sekler, I., Sensi, S. L., Hershfinkel, M., & Silverman, W. F. (2007). Mechanism and regulation of cellular zinc transport. *Molecular Medicine*, *13*, 337–343.
Serbina, N. V., Shi, C., & Pamer, E. G. (2012). Monocyte-mediated immune defense against murine *Listeria monocytogenes* infection. *Advances in Immunology*, *113*, 119–134.
Shen, Y., Naujokas, M., Park, M., & Ireton, K. (2000). InlB-dependent internalization of *Listeria* is mediated by the Met receptor tyrosine kinase. *Cell*, *103*, 501–510.
Shi, X., Festa, R. A., Ioerger, T. R., Butler-Wu, S., Sacchettini, J. C., Darwin, K. H., et al. (2014). The copper-responsive RicR regulon contributes to *Mycobacterium tuberculosis* virulence. *mBio*, *5*(1), pii: e00876-13. http://dx.doi.org/10.1128/mBio.00876-13.
Sigdel, T. K., Cilliers, R., Gursahaney, P. R., Thompson, P., Easton, J. A., & Crowder, M. W. (2006). Probing the adaptive response of *Escherichia coli* to extracellular Zn(II). *Biometals*, *19*, 461–471.
Subramanian Vignesh, K., Landero Figueroa, J. A., Porollo, A., Caruso, J. A., & Deepe, G. S., Jr. (2013). Granulocyte macrophage-colony stimulating factor induced Zn sequestration enhances macrophage superoxide and limits intracellular pathogen survival. *Immunity*, *39*, 697–710.

Suttle, N. F., & Jones, D. G. (1989). Recent developments in trace element metabolism and function: Trace elements, disease resistance and immune responsiveness in ruminants. *Journal of Nutrition*, *119*, 1055–1061.

Tang, D. J., Li, X. J., He, Y. Q., Feng, J. X., Chen, B., & Tang, J. L. (2005). The zinc uptake regulator Zur is essential for the full virulence of *Xanthomonas campestris* pv. *campestris*. *Molecular Plant-Microbe Interactions*, *18*, 652–658.

Taudte, N., & Grass, G. (2010). Point mutations change specificity and kinetics of metal uptake by ZupT from *Escherichia coli*. *Biometals*, *23*, 643–656.

Tilney, L. G., & Portnoy, D. A. (1989). Actin filaments and the growth, movement, and spread of the intracellular bacterial parasite, *Listeria monocytogenes*. *The Journal of Cell Biology*, *109*, 1597–1608.

Tottey, S., Patterson, C. J., Banci, L., Bertini, I., Felli, I. C., Pavelkova, A., et al. (2012). Cyanobacterial metallochaperone inhibits deleterious side reactions of copper. *Proceedings of the National Academy of Sciences of the United States of America*, *109*, 95–100.

Tottey, S., Rich, P. R., Rondet, S. A. M., & Robinson, N. J. (2001). Two Menkes-type ATPases supply copper for photosynthesis in *Synechocystis* PCC 6803. *The Journal of Biological Chemistry*, *276*, 19999–20004.

Tufft, L. S., Nockels, C. F., & Fettman, M. J. (1988). Effects of Escherichia coli on iron, copper, and zinc metabolism in chicks. *Avian Diseases*, *32*, 779–786.

Urban, C. F., Ermert, D., Schmid, M., Abu-Abed, U., Goosmann, C., Nacken, W., et al. (2009). Neutrophil extracellular traps contain calprotectin, a cytosolic protein complex involved in host defense against *Candida albicans*. *PLoS Pathogens*, *5*, e1000639.

Varghese, S., Wu, A., Park, S., Imlay, K. R., & Imlay, J. A. (2007). Submicromolar hydrogen peroxide disrupts the ability of Fur protein to control free-iron levels in *Escherichia coli*. *Molecular Microbiology*, *64*, 822–830.

Vazquez-Boland, J. A., Dominguez, L., Fernandez-Garayzabal, J. F., & Suarez, G. (1992). *Listeria monocytogenes* CAMP reaction. *Clinical Microbiology Reviews*, *5*, 343.

Waldron, K. J., & Robinson, N. J. (2009). How do bacterial cells ensure that metalloproteins get the correct metal? *Nature Reviews Microbiology*, *7*, 25–35.

Waldron, K. J., Rutherford, J. C., Ford, D., & Robinson, N. J. (2009). Metalloproteins and metal sensing. *Nature*, *460*, 823–830.

Ward, S. K., Abomoelak, B., Hoye, E. A., Steinberg, H., & Talaat, A. M. (2010). CtpV: A putative copper exporter required for full virulence of *Mycobacterium tuberculosis*. *Molecular Microbiology*, 77, 1096–1110.

Weinberg, E. D. (2009). Iron availability and infection. *Biochimica et Biophysica Acta*, *1790*, 600–605.

Wessling-Resnick, M. (2010). Iron homeostasis and the inflammatory response. *Annual Review of Nutrition*, *30*, 105–122.

Weston, B. F., Brenot, A., & Caparon, M. G. (2009). The metal homeostasis protein, Lsp, of *Streptococcus pyogenes* is necessary for acquisition of zinc and virulence. *Infection and Immunity*, 77, 2840–2848.

White, C., Lee, J., Kambe, T., Fritsche, K., & Petris, M. J. (2009). A role for the ATP7A copper-transporting ATPase in macrophage bactericidal activity. *The Journal of Biological Chemistry*, *284*, 33949–33956.

Witte, C. E., Archer, K. A., Rae, C. S., Sauer, J. D., Woodward, J. J., & Portnoy, D. A. (2012). Innate immune pathways triggered by *Listeria monocytogenes* and their role in the induction of cell-mediated immunity. *Advances in Immunology*, *113*, 135–156.

Wolschendorf, F., Ackart, D., Shrestha, T. B., Hascall-Dove, L., Nolan, S., Lamichhane, G., et al. (2011). Copper resistance is essential for virulence of *Mycobacterium tuberculosis*. *Proceedings of the National Academy of Sciences of the United States of America*, *108*, 1621–1626.

Wunderli-Ye, H., & Solioz, M. (2001). Purification and functional analysis of the copper ATPase CopA of *Enterococcus hirae*. *Biochemical and Biophysical Research Communications*, *280*, 713–719.

Yang, W., Liu, Y., Chen, L., Gao, T., Hu, B., Zhang, D., et al. (2007). Zinc uptake regulator (*zur*) gene involved in zinc homeostasis and virulence of *Xanthomonas oryzae* pv. *oryzae* in rice. *Current Microbiology, 54*, 307–314.

Yoshikawa, Y., Ogawa, M., Hain, T., Chakraborty, T., & Sasakawa, C. (2009). *Listeria monocytogenes* ActA is a key player in evading autophagic recognition. *Autophagy, 5*, 1220–1221.

Zaharik, M. L., Cullen, V. L., Fung, A. M., Libby, S. J., Kujat Choy, S. L., Coburn, B., et al. (2004). The *Salmonella enterica* serovar Typhimurium divalent cation transport systems MntH and SitABCD are essential for virulence in an Nramp1G169 murine typhoid model. *Infection and Immunity, 72*, 5522–5525.

Zaia, A. A., Sappington, K. J., Nisapakultorn, K., Chazin, W. J., Dietrich, E. A., Ross, K. F., et al. (2009). Subversion of antimicrobial calprotectin (S100A8/S100A9 complex) in the cytoplasm of TR146 epithelial cells after invasion by *Listeria monocytogenes*. *Mucosal Immunology, 2*, 43–53.

Zalewski, P., Truong-Tran, A., Lincoln, S., Ward, D., Shankar, A., Coyle, P., et al. (2006). Use of a zinc fluorophore to measure labile pools of zinc in body fluids and cell-conditioned media. *Biotechniques, 40*, 509–520.

Zhu, Y. Q., Zhu, D. Y., Lu, H. X., Yang, N., Li, G. P., & Wang, D. C. (2005). Purification and preliminary crystallographic studies of CutC, a novel copper homeostasis protein from *Shigella flexneri*. *Protein and Peptide Letters, 12*, 823–826.

CHAPTER FOUR

The Role of Macrophages in the Innate Immune Response to *Streptococcus pneumoniae* and *Staphylococcus aureus*: Mechanisms and Contrasts

Joby Cole, Jody Aberdein, Jamil Jubrail, David H. Dockrell[1]
Department of Infection and Immunity, University of Sheffield Medical School and Sheffield Teaching Hospitals, Sheffield, United Kingdom
[1]Corresponding author: e-mail address: d.h.dockrell@sheffield.ac.uk

Contents

Abstract

Macrophages are critical mediators of innate immune responses against bacteria. The Gram-positive bacteria *Streptococcus pneumoniae* and *Staphylococcus aureus* express a range of virulence factors, which challenge macrophages' immune competence. We review how macrophages respond to this challenge. Macrophages employ a range of strategies to phagocytose and kill each pathogen. When the macrophages capacity

Advances in Microbial Physiology
ISSN 0065-2911
http://dx.doi.org/10.1016/bs.ampbs.2014.08.004

to clear bacteria is overwhelmed macrophages play important roles in orchestrating the inflammatory response through pattern recognition receptor-mediated responses. Macrophages also ensure the inflammatory response is tightly constrained, to avoid tissue damage, and play an important role in downregulating the inflammatory response once initial bacterial replication is controlled.

1. INTRODUCTION

Streptococcus pneumoniae (the pneumococcus) is the leading cause of community-acquired pneumonia (CAP). It causes approximately 2 million deaths per year from invasive disease (defined as isolation of the pneumococcus from a normally sterile site such as blood or cerebrospinal fluid), half of these in individuals under 5 years of age, but is likely to cause many more due to pneumonia and other lung disease (O'Brien et al., 2009). Its contribution to the latter is hard to define precisely but since pneumonia accounts for millions of deaths each year and the pneumococcus accounts for up to 40% of these cases the burden of pneumococcal disease is potentially huge (De Roux et al., 2004). It is estimated the pneumococcus accounts for approximately 20 million episodes of pneumonia per year in children under 4 years of age (Walker et al., 2013). Along with malaria, HIV infection and tuberculosis the pneumococcus is one of the four leading causes of infection-related death and the leading cause of death due to an acute bacterial infection. The pneumococcus is also the leading cause of bacterial meningitis in older adults, causing substantial morbidity and mortality (Thigpen et al., 2011). Other common clinical syndromes associated with the pneumococcus include otitis media, sinusitis, bronchitis, empyema, and occult bacteraemia. In addition, it causes several rarer syndromes such as peritonitis, septic arthritis, and endocarditis.

Despite its clinical significance, clinical disease with the pneumococcus is rare in comparison with colonisation. Rates of colonisation vary. In adults up to 10% of the population may be colonised, but rates of >40% are found in young infants, particularly, those exposed in day care settings (Bogaert, De Groot, & Hermans, 2004). Globally, rates of colonisation are even higher with rates of over 70% observed in some studies in sub-Saharan Africa and India. Thus, it becomes apparent that effective immune responses must have developed to protect most individuals against clinical syndromes such as pneumonia and invasive pneumococcal disease. Understanding why most individuals mount effective responses to this pathogen but in a sub-set of individuals these fail is a major unmet medical need. Although protein conjugate vaccines have reduced the burden of invasive pneumococcal disease, the plasticity of the pneumococcal genome means that the pathogen has the potential to adapt to the selective pressure of vaccines. New strains, such as serotype 19A and 22F, emerge and replace previous colonising strains under this selective pressure (Miller, Andrews, Waight, Slack, & George, 2011). Moreover, the periodic spread of antimicrobial-resistant strains, as exemplified by the spread of penicillin resistance in the 1990s, means we cannot guarantee our sustained capacity to effectively manage this pathogen with antimicrobials in the future (Tomasz, 1997). Equally worryingly antimicrobials have made only a limited impact on overall mortality, suggesting we need more effective approaches to identify patients at risk of serious infection and to modulate their inflammatory responses if we are to successfully combat this pathogen.

Staphylococcus aureus represents a separate but over-lapping set of clinical challenges. *S. aureus* has emerged not only as a significant cause of community-acquired infection at all ages but also as a major cause of hospital-acquired infection. It is estimated that *S. aureus* accounts for over 100,000 cases of bloodstream infection every year in Europe, costing health services tens of millions of Euros per annum (De Kraker, Davey, Grundmann, & BURDEN Study Group, 2011). It directly accounted for over 10,000 deaths per year in the United States, but in 2005 approximately 30,000 deaths had an *S. aureus*-related discharge code, suggesting a contributatory role in additional cases (Klein, Smith, & Laxminarayan, 2007). This adds substantially to the costs associated with surgery and critical care. The hallmark of *S. aureus* clinically is the propensity to cause bacteraemia and metastatic infection, with seeding of secondary sites remote from the initial site of infection (Corey, 2009). *S. aureus* colonises the nares, in approximately one third of the population, and spread from the nares to the skin is assumed as a major mechanism of bacteraemia pathogenesis (Kuehnert et al., 2006; Miller et al., 2009; Von Eiff, Becker, Machka, Stammer, & Peters, 2001). In the case of hospitalised patients, any form of instrumentation, such as intravenous cannulae or surgical excision, can facilitate invasion and *S. aureus* is a common cause of post-operative infection. The clinical manifestations of *S. aureus* are protean (John, 2004). Clinical syndromes include; skin and soft tissue infections (SSTI); bone and joint infections, including osteomyelitis and septic arthritis; endovascular infection, including endocarditis and infection of endovascular materials such as vascular grafts; pneumonia, including community-acquired, hospital-acquired, and ventilator-associated forms; central nervous system infections, including brain abscesses, neurosurgical infections, and epidural abscesses; endopthalmitis; post-operative infections including wound infections, intra-abdominal infections, infection of prosthetic materials, dialysis catheter infections, infections of the urinary tract related to catheterisation and instrumentation, and septic shock syndromes in association with production of toxins that function as superantigens (see below).

There are no vaccines currently licenced for *Staphylococcus aureus*. Although initially susceptible to penicillin, the pathogen rapidly acquired a plasmid-encoded beta-lactamase, necessitating treatment with anti-staphylococcal penicillins such as methicillin (Kirby, 1944). However, methicillin resistance emerged in the 1960s and methicillin-resistant *S. aureus* (MRSA) have become endemic in health care settings, resulting in increased mortality rates with bacteraemia (Cosgrove et al., 2003). The emergence of more virulent community-acquired MRSA strains has compounded this problem, particularly over the last 15 years, causing infections in otherwise healthy individuals (Deleo, Otto, Kreiswirth, & Chambers, 2010). Glycopeptides, such as vancomycin, have been a key component of the treatment of MRSA. However, vancomycin-intermediate and vancomycin-resistant strains of *S. aureus* have also emerged (Chang et al., 2003; Hiramatsu et al., 1997).

Since a large proportion of the population may be colonised with these micro-organisms yet only a minority will develop clinical disease it becomes apparent that well developed immune responses must protect most individuals. Yet why these responses fail in a sub-section of the population is unclear. Both pathogens possess a broad array of virulence factors, which help subvert host defence. These will be briefly described below but many other reviews cover these in greater depth (Chambers & Deleo, 2009; Gordon & Lowy, 2008; Jedrzejas, 2001; Kadioglu, Weiser, Paton, & Andrew, 2008; Liu, 2009; Preston & Dockrell, 2008). Our focus will instead be on how the interaction of both pathogens with the tissue macrophages influences outcome. As these cells represent the resident phagocyte in tissue, they play an important role in the early stages of bacterial clearance. In addition, they are involved in the decision to escalate the inflammatory response and play essential roles regulating the inflammatory response; influencing both the recruitment of inflammatory cells and the subsequent resolution of the inflammatory response. As such, macrophage interactions with both pathogens are likely to exert a major influence on the success of the host response and the pathogenesis of clinical disease. Understanding the core elements of the optimal macrophage response to both micro-organisms is therefore essential in order to understand the basis of susceptibility in those who develop disease in response to each infection.

2. *S. pneumoniae* VIRULENCE FACTORS IMPEDE PHAGOCYTOSIS AND ITS CONSEQUENCES

S. pneumoniae represents a significant challenge to phagocytes since the polysaccharide capsule represents a major barrier to phagocytosis and necessitates opsonisation of the bacterium with both immunoglobulin and complement to facilitate ingestion (Ali et al., 2003; Gordon, Irving, Lawson, Lee, & Read, 2000; Janoff et al., 1999; Jonsson, Musher, Chapman, Goree, & Lawrence, 1985). More than 90 capsular variants exist and variability in capsule explains much of the variation in biological behaviour (Dockrell, Whyte, & Mitchell, 2012). Some serotypes, such as serotypes 6A/B, 19F, and 23F, are more efficient colonisers but less likely to cause invasive disease, while others, such as serotypes 1, 4, 7F, and 14, are less likely to colonise but are more likely to cause invasive disease (Sjostrom et al., 2006; Sleeman et al., 2006). The more invasive strains like serotype 1 and 7F behave more like primary pathogens causing disease in previously healthy individuals with lower mortality, while serotype 3 and related strains behave more like opportunistic infections causing disease in individuals with medical co-morbidities associated with higher mortality (Sjostrom et al., 2006; Sleeman et al., 2006). Capsular switch studies, where different capsules are expressed by the same strain, confirm that much of this variation is directly attributable to the capsule. Colonising strains have a capsule, which is more resistant to non-opsonic phagocytosis by neutrophils, a key factor in controlling colonisation (Weinberger et al., 2009; Zhang, Clarke, & Weiser, 2009). Conversely, these strains are more readily ingested by alveolar macrophages when they reach the lower airway. The so called colonising strains have capsules which prevent adhesins, including Pneumococcal surface protein A (PspA), Choline-binding protein A (CbpA, also known as PspC), and Pneumococcal serine-rich repeat protein (PsrP), being accessible to airway epithelial cells, resulting in less efficient adhesion to the airway, a prerequisite for invasive disease (Sanchez et al., 2011). They also have capsules which are more amenable to complement deposition and so these strains are more likely to be phagocytosed by alveolar macrophages in the airspace or cleared from the bloodstream once bacteraemia is established (Hyams, Yuste, et al., 2010). However, capsule alone does not explain all the variation in complement deposition seen between serotypes (Hyams et al., 2011).

Overall the capsule can modify non-opsonic uptake, complement-dependent, and complement-independent uptake of bacteria by phagocytes (Hyams, Camberlein, Cohen, Bax, & Brown, 2010).

In view of the effect of capsule, complement deposition is important to ensure maximal phagocytosis of pneumococci. The classical pathway determines the proportion of bacteria that bind C3 complement components, while the alternative pathway determines the intensity of complement binding (Brown et al., 2002). The classical pathway can be activated by serotype specific antibody following adaptive immune responses, by naturally occurring IgM or by C-reactive protein during innate immune responses but *S. pneumoniae* use a variety of strategies to prevent complement deposition on its surface. Capsular polysaccharide prevents serotype specific antibody, naturally occurring IgM, and CRP binding to the surface of the bacterium, thus limiting complement deposition (Hyams, Camberlein, et al., 2010). The extent of this inhibition varies with capsular serotype and other unidentified genetic factors (Hyams, Yuste, et al., 2010). The toxin pneumolysin prevents classical complement deposition by binding IgG and CRP, while PspA inhibits alternative pathway complement activation to a greater extent than classical pathway activation (Yuste, Botto, Paton, Holden, & Brown, 2005). Another surface protein CbpA/PspC binds the inhibitor of complement factor H and promotes the cleavage of C3b into iC3b, which inhibits the formation of the alternative complement C3 convertase C3b-Bb (Jarva et al., 2002; Lu, Ma, & Zhang, 2006). As mentioned, antibody binding is impeded by pneumolysin-binding IgG, but also by IgA protease-induced cleavage of IgA, which not only limits opsonophagocytosis but also enhances adherence of the bacteria (Kilian, Mestecky, Kulhavy, Tomana, & Butler, 1980; Marriott, Mitchell, & Dockrell, 2008; Weiser et al., 2003).

Another mechanism by which the pneumococcus withstands phagocytosis is through biofilm formation. Although not as extensively studied as in some other micro-organisms biofilm formation has been described for the pneumococcus and may be quite important in certain environments such as the middle ear; the adhesin PsrP, encoded by a pathogenicity island, regulates the later stages of biofilm formation and in its absence bacteria were less able to aggregate or adhere to respiratory epithelium in a murine model (Sanchez et al., 2010). Thus, this may be an important mechanism preventing the phagocytosis of bacteria in the airway and potentially other locations.

Despite the challenges, the pneumococcus presents to effective opsonisation and phagocytosis once ingested *S. pneumoniae* is killed by phagocytes and is not capable of prolonged intracellular survival. That is not to say that the bacterium has not developed an array of methods to limit the effectiveness of specific antimicrobial strategies used by the host. As *S. pneumoniae* uses hydrogen peroxide to kill other bacteria in the nasopharynx and establish its ecological niche, it has developed multiple mechanisms to withstand oxidative stress; SodA, a superoxide dismutase degrades superoxide; a *nox* gene-encoded nicotinamide adenine dinucleotide phosphate (NADPH)-oxidase system converts oxygen to water, thus limiting its availability for superoxide generation; SpxB a pyruvate oxidase limits ATP depletion during oxidative stress; PsaA and PsaD limit hydrogen peroxide production; AdhC, an alcohol dehydrogenase acts to provide the glutathione needed for the glutathione reductase activity of PsaD; AhpD, an alkylhydroperoxidase, and TlpA, a thiredoxin-like protein, both inhibit hydrogen peroxide, Etrx1 and 2, surface exposed thioredoxin-family lipoproteins, act together to inhibit hydrogen peroxide-mediated killing in macrophages; SpMsrAB2, the redox partner of the Etrx proteins, aids resistance to hydrogen peroxide-mediated killing and a heat shock-induced protease HtrA helps the bacterium withstand oxidative stress (Andisi et al., 2012; Ibrahim, Kerr, McCluskey, & Mitchell, 2004; McAllister et al., 2004; Paterson, Blue, & Mitchell, 2006; Pericone, Park, Imlay, & Weiser, 2003; Potter, Kidd, McEwan, & Paton, 2010; Saleh et al., 2013; Tseng, McEwan, Paton, & Jennings, 2002; Yesilkaya et al., 2000; Yu et al., 2001). These factors combine to reduce the efficacy of reactive oxygen species (ROS) as antimicrobicidal molecules against *S. pneumoniae*. Furthermore, PspC impedes nitric oxide (NO) generation, AdhC also functions as a *S*-nitrosoglutathione (GSNO) reductase, while the ClpP protease, which also helps withstand oxidative stress helps pneumococci tolerate the reactive nitrogen species (RNS) generated by the reaction of ROS with NO (Park et al., 2010; Peppoloni et al., 2006; Robertson, Ng, Foley, Gilmour, & Winkler, 2002; Stroeher et al., 2007).

Neutrophil extracellular traps (NETs) have emerged as an important mechanism of killing extracellular bacteria and cells other than neutrophils, such as monocytes may also utilise similar strategies (Brinkmann et al., 2004; Webster et al., 2010). NETs contain a network of DNA, histones, and granule proteases. Pneumococcal capsule can help avoid binding to NETs but endonuclease A (EndA), whose expression varies amongst strains allows the pneumococcus to degrade NETs by causing digestion of the constituent DNA (Beiter et al., 2006; Wartha et al., 2007).

S. pneumoniae does not possess as diverse a range of toxins as *S. aureus*. It does, however, have a significant toxin influencing virulence in the shape of pneumolysin (Kadioglu et al., 2008; Marriott, Mitchell, & Dockrell, 2008). Pneumolysin is a cholesterol-dependent cytolysin, which is not actively secreted and requires bacterial lysis for its release; mechanisms of release can involve autolysin or be autolysin-independent and one autolysin-dependent mechanism involves competent bacteria inducing programmed cell death in non-competent bacteria utilising the two-peptide bacteroicin CibAB (Guiral, Mitchell, Martin, & Claverys, 2005). Pneumolysin alters the cell transcriptome; in one microarray analysis 142 of 4133 genes were differentially regulated in a monocytic cell line by pneumolysin (Rogers et al., 2003). Pneumolysin can also mediate dephosphorylation of serine at position 10 of the H3 histone, suggesting it could modify gene transcription though epigenetic mechanisms, as has been functionally demonostrated for a related toxin listeriolysin produced by *Listeria monocytogenes* (Hamon et al., 2007).

Pneumolysin has cytolytic activity, mediated by formation of membrane pores, and non-cytolytic activity (Marriott, Mitchell, & Dockrell, 2008). Pneumolysin is toxic to a range of cells and causes direct lung injury when instilled into mice lungs. The direct lung injury is not mediated by alveolar macrophages or recruited inflammatory cells but appears to be due to the direct cytolytic effects of the toxin on lung cells (Maus et al., 2004). The deubiquitinase enzyme cylindromatosis (CYLD) is required to mediate acute lung injury in murine models and negatively regulates plasminogen activator inhibitor-1, resulting in pulmonary haemorrhage (Lim et al., 2007). However, CYLD also results in the deubiquitination and therefore activation of receptor-interacting serine/threonine-protein kinase (RIP)1, a kinase which along with RIP3 forms a macromolecular complex the necrosome, which mediates programmed cell necrosis or necroptosis (Moquin, Mcquade, & Chan, 2013). Therefore, it is equally plausible that pneumolysin exerts some of its cytotoxicity through direct activation of necroptosis. Pneumolysin can induce direct toxicity to a range of cells including alveolar macrophages, dendritic cells, and T-cells (Colino & Snapper, 2003; Daigneault et al., 2012; Maus et al., 2007). Generation of ROS during host responses to live bacteria may, however, help protect phagocytes against pneumolysin-induced cytotoxicity (Clark, 1986). Non-cytolytic effects of pneumolysin include complement activation (Jounblat, Kadioglu, Mitchell, & Andrew, 2003) and engagement of a range of pattern recognition receptors, including Toll-like receptor (TLR) 4 and the nucleotide-binding domain and leucine-rich repeat containing gene family, pyrin domain-containing 3 (NLRP3) inflammasome, leading to pro-inflammatory cytokine responses (Fang et al., 2011; Malley et al.,

2003; McNeela et al., 2010; Witzenrath et al., 2011). Thus, pneumolysin contributes to pathogenesis through direct cytotoxicity and also by favouring excessive inflammatory responses that lead to tissue injury and allow bacterial tissue invasion (Rubins et al., 1995). Conversely, it is also a focus of innate host responses through pattern receptor recognition and therefore if pneumococci lack pneumolysin they may not induce as vigorous a host response. This is illustrated by the observation that in mouse models, strains lacking pneumolysin induce chronic bacteraemia with lower cytokine responses (Benton, Everson, & Briles, 1995). It is also supported by observations that some invasive strains express modified pneumolysin and may owe some of their success as invasive pathogens to their ability to not fully engage the innate response (Witzenrath et al., 2011).

Clearly pneumococci possess many other virulence factors (see Fig. 1). These are involved in colonisation (e.g., neuraminidase A and B (NanA and B) and pneumococcal adherence and virulence factor A (PavA)), survival in the nasopharynx (e.g., bacteriocins), induction of inflammatory responses (e.g., pilus), and degradation of the extracellular matrix (e.g., hyaluraonate lyase, Hyl). Collectively they contribute to pathogenesis but further discussion is beyond the scope of the current review, since their primary mechanisms of actions do not typically involve macrophages (Kadioglu et al., 2008; Preston & Dockrell, 2008).

3. *S. aureus* VIRULENCE FACTORS SUBVERT MULTIPLE INNATE IMMUNE RESPONSES INCLUDING INTRACELLULAR KILLING

S. aureus possesses multiple virulence factors that can subvert host defence and individual factors can have several distinct roles in pathogenesis (see Fig. 2). Expression of these factors varies between strains. Virulence factors may be structural or secreted. Colonisation is facilitated by expression of surface proteins termed microbial surface components recognising adhesive matrix molecules (MSCRAMMs) that aid adherence to a range of surfaces including endothelium and prosthetic materials (Gordon & Lowy, 2008). These allow binding to extracellular molecules including collagen, fibronectin, and fibrinogen. Examples of MSCRAMMs include collagen-binding protein (Cna), fibronectin-binding proteins (FnbA and FnbB), and clumping factors (ClfA and ClfB). Presence of a microcapsule, formation of biofilms, and binding of immunoglobulin by MSCRAMM protein A impede phagocytosis (Deisenhofer, 1981; Donlan & Costerton, 2002;

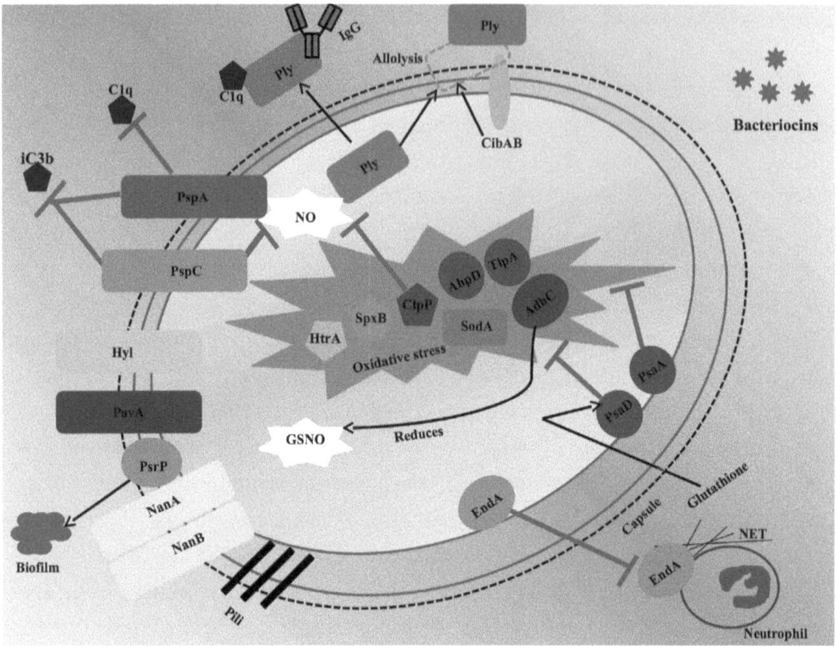

Figure 1 *Streptococcus pneumoniae* virulence factors that resist phagocytosis and antimicrobial killing. *Streptococcus pneumoniae* (Spn) has a polysaccharide capsule that helps it resist phagocytosis. Pneumococcal surface protein A (PspA) and Pneumococcal surface protein C (PspC) inhibit complement activation. Pneumococcal serine-rich repeat protein (PsrP) regulates biofilm formation. Spn synthesises virulence factors to resists oxidative and nitrosative stress. Superoxide dismutase A (SodA) allows Spn to resist superoxide ($O_2^{\cdot -}$). High temperature requirement A (HtrA), alkylhydroperoxidase A (AhpA), caseinolytic peptidase P (ClpP), thioredoxin-like protein A (TlpA), and SpxB (a pyruvate oxidase) help Spn resists hydrogen peroxide (H_2O_2). Pneumococcal surface adhesin A (PsaA) and Pneumococcal surface adhesin D (PsaD) also contribute to resistance to oxidative stress. Alcohol dehydrogenase C (AdhC) provides reduced glutathione required by PsaD. ClpP and PspC help Spn resist nitrosative stress. Endonuclease A (EndA) degrades neutrophil extracellular traps (NETs). Pneumolysin (Ply) prevents complement activation. Ply interacts with CibAB to induce allolysis. Neuraminidase A (NanA) and Neuraminidase B (NanB), Hyaluronate lyase (Hyl), Pneumococcal adherence and virulence factor A (PavA), and the pilus are involved in colonisation. Bacteriocins help survival in the nasopharynx. (See the color plate.)

O'Riordan & Lee, 2004). Several proteins prevent opsonisation with complement by inhibiting the C3 convertase. These include the staphylococcal complement inhibitor (SCIN) and two homologues SCIN-B and SCIN-C that inhibit bacterial phagocytosis (Jongerius et al., 2007; Rooijakkers et al., 2005; Van Wamel, Rooijakkers, Ruyken, Van Kessel, & Van Strijp, 2006).

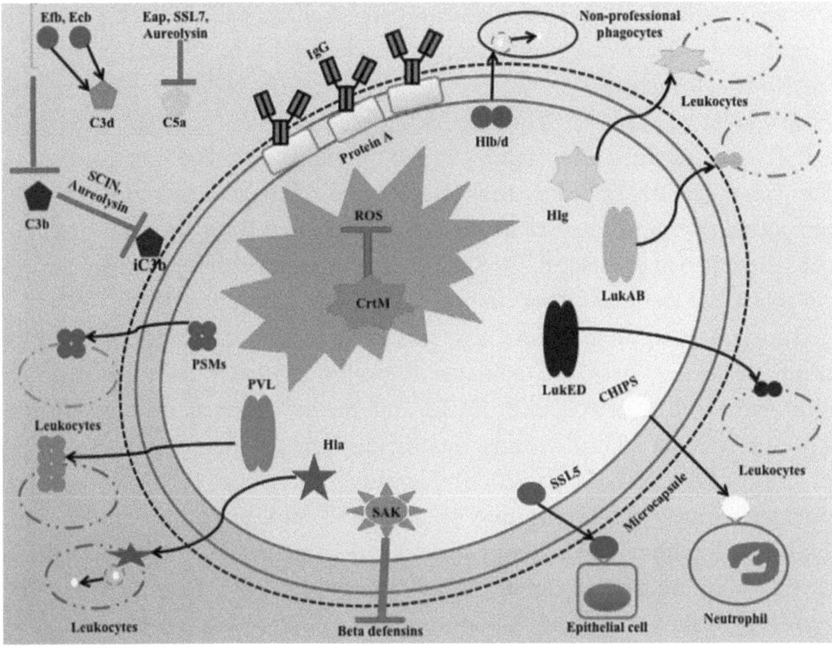

Figure 2 *Staphylococcus aureus* virulence factors that resist phagocytosis and antimicrobial killing. *S. aureus* expresses a range of virulence factors to inhibit phagocytosis and intracellular killing. Some strains express a microcapsule to impede phagocytosis. Surface expressed protein A (Spa) binds the Fc region of IgG inhibiting phagocytosis. The extracellular fibrinogen-binding protein (Efb) binds C3d blocking C3b containing convertases and binds fibrinogen preventing neutrophil activation. The extracellular complement-binding protein (Ecb) also binds C3d and blocks C3b containing convertases and affects C5 convertases. The staphylococcal complement inhibitor (SCIN) and aureolysin prevent complement activation. Superantigen-like protein 7 (SSL7) prevents complement activation by binding C5a. The chemotaxis inhibitory protein of *S. aureus* (CHIPs) prevents neutrophil chemotaxis. Superantigen-like protein 5 (SSL5) binds P-selectin glycoprotein on epithelial cell surfaces preventing neutrophil rolling. *S. aureus* synthesises a range of pore forming toxins to damage host cells. These include Panton–Valentine Leukocidin (PVL), alpha haemolysin (Hla), Phenol Soluble Modulins (PSMs), Leukocidin AB (LukAB), Leukocidin ED (LukED), and gamma haemolysin (Hlg). Expression of Hla of beta haemolysin (Hlb) and delta haemolysin (Hld) contribute to *S. aureus* escape from phagosomes. Carotenoid M (CrtM) helps *S. aureus* survive oxidative stress and staphylokinase (SAK) neutralises defensins. (See the color plate.)

The metalloprotease aureolysin also cleaves C3 and limits C3b deposition on the staphylococcal cell wall (Laarman et al., 2011).

The chemotaxis inhibitory protein of *S. aureus* (CHIPS) binds the C5a and formylated peptide receptors, preventing neutrophil chemotaxis (de Haas et al., 2004). SCINs also prevent chemotaxis by inhibiting C5a activation while extracellular fibrinogen-binding protein (Efb) and its homologue extracellular complement-binding protein (Ecb) have similar functions (Jongerius et al., 2007). Aureolysin reduces C5a release (Laarman et al., 2011). The staphylococcal superantigen-like protein 7 (SSL7) also inhibits C5a, while it also prevents IgA binding to FcαR1 (Langley et al., 2005). SSL5 binds to epithelial cell P-glycoprotein to inhibit neutrophil rolling, an earlier step in neutrophil recruitment to sites of infection (Bestebroer et al., 2007). The extracellular adherence proteins (Eap)s can inactivate neutrophil serine proteases (Stapels et al., 2012).

In addition to these factors, a range of leukocidins can cause destruction of phagocytes. There are epidemiological links between expression of Panton–Valentine leukocidin (PVL) and the emergence of community-acquired MRSA strains causing necrotising pneumonia and invasive skin infections (Gordon & Lowy, 2008). However, the nature of this relationship has been controversial. Four units of LukS-PVL and four units of LukF-PVL can come together to form an octomeric β-barrel pore in cell membranes capable of lysing leukocytes, erythrocytes, and skin cells. Despite this, some studies have shown PVL-deficient mutants of community-acquired MRSA (CA-MRSA) strains USA 300 and USA 400 are no less virulent than wild-type strains in murine models of sepsis or abscess formation and do not alter the capacity of these strains to induce neutrophil lysis (Voyich et al., 2006), leading some to question whether PVL may be a marker rather than the causal factor explaining the virulence of CA-MRSA strains. Some studies have linked PVL to development of necrotising pneumonitis, while others suggest α-hemolysin (Hla) is the major virulence determinant in its pathogenesis (Bubeck Wardenburg, Bae, Otto, Deleo, & Schneewind, 2007; Labandeira-Rey et al., 2007). Phenol soluble modulins (PSMs) are upregualted in CA-MRSA strains, and are capable of inducing neutrophil lysis, particularly the α3 PSM (Wang et al., 2007). They have emerged as an important factor contributing to the virulence of CA-MRSA strains.

Other secreted factors, which may contribute to cell lysis of immune cells include Hla, a 34-kDa pore forming toxin, which causes a cytolytic cell death of mononuclear cells, with morphological features of necroptosis, although neutrophils may be more resistant to Hla-induced cell death (Essmann et al., 2003; Nygaard et al., 2012). Hla also activates the NLRP3 inflammasome, facilitating IL-1β processing and inflammation (Kebaier et al., 2012). The γ-hemolysin (Hlg) also contributes to neutrophil lysis and does so under specific *in vitro* conditions, when PVL does not appear to make a significant contribution to neutrophil lysis (Malachowa et al.,

2011). Other luekocidins may also play a role in leukocyte lysis. These include the leukotoxin AB (LukAB), which lyses neutrophils, macrophages and dendritic cells (Dumont et al., 2011). Another leukotoxin, LukED, can lyse neutrophils and monocytes, via a pathway mediated through LukED binding to the chemokine receptors CXCR1 and CXCR2 (Reyes-Robles et al., 2013). In addition, through binding of another chemokine receptor CCR5, LukED also enables killing of effector memory T-cells (including Th1 and Th17 effectors), as well as macrophages and dendritic cells (Alonzo et al., 2012, 2013).

S. aureus uses the *agr* quorum sensing system to enable phagosomal escape (Giese et al., 2011; Kubica et al., 2008). In non-professional phagocytes, such as endothelial and epithelial cells, the δ hemolysin (Hld) and sphingomyelinase β toxin (Hlb) act synergistically to mediate the phagosomal escape, though whether this mechanism also plays a role in phagocytes is unknown (Giese et al., 2011). In primary human macrophages, Hla has also been implicated in phagosomal escape, which precedes cytolytic death (Kubica et al., 2008).

In addition to mechanisms to prevent phagocyte recruitment or survival at sites of infection, *S. aureus* also uses diverse mechanisms to withstand antimicrobial molecules. The dehydrosqualene synthase CrtM, encodes a carotenoid, which enables resistance against oxidative stress (Liu et al., 2005). USA 300 has a genomic island that encodes an arginine deaminase and since *Streptococcus pyogenes* uses an arginine deaminase to withstand nitrosative stress, *S. aureus* may utilise a similar strategy (Degnan et al., 1998; Diep et al., 2006), though the functional significance of the arginine deaminase in USA 300 remains to be established. In addition, as described above, the pathogen may use strategies to resist antimicrobial proteases (Stapels et al., 2012).

To add to the challenges provided by factors that circumvent the innate immune response, the host must also prevent excessive immune activation. Peptidoglycans, lipopeptides, and other cell wall constituents may activate a diverse range of pattern recognition receptors. While activation of the innate immune response is important, to allow an appropriate response it must be carefully controlled to avoid excessive inflammation and tissue injury. *S. aureus* for example encodes multiple superantigens which result in T-cell activation and stimulation of high levels of cytokines which result in the development of sepsis and tissue injury (Fraser & Proft, 2008).

In summary, *S. pneumoniae* and *S. aureus* express distinct virulence factors and represent separate challenges to the immune system. In the case of *S. pneumoniae*, capsule and resistance to phagocytosis are a major challenge.

Once ingested, although, the bacterium possesses strategies to evade ROS-mediated killing in particular, and expresses a potentially cytolytic toxin, the innate immune system, as we will discuss, has developed strategies to clear bacteria. In contrast, *S. aureus* represents different challenges. Although effectively phagocytosed many of the greatest challenges are encountered as a result of the array of leukocidins expressed by *S. aureus*. Furthermore, strategies that circumvent intracellular killing of bacteria, and attributes limiting recruitment of inflammatory cells when resident host defences fail, represent a major challenge to the innate immune response.

4. ORIGINS OF MACROPHAGES

Although prior to the seminal works of Élie Metchnikoff the process of phagocytosis by leukocytes had been described by multiple authors, notably Osler (Ambrose, 2006), it was he who recognised the importance and centrality of the phenomenon, and of the cells which exhibit it to development, tissue regulation, inflammation, and immunity. Metchnikoff was also able, by use of Ehrlich's alanine dyes, to distinguish microphages, or polymorphonuclear neutrophils, from macrophages (Tauber, 2003). Metchnikoff recognised macrophages in diverse tissues and locations throughout the organisms he studied, a concept furthered by Aschoff (1924) with his proposal of the 'reticulo-endothelial system'. Clear differences in morphology and rate of phagocytosis together with the unreliability of dye uptake techniques as a maker of highly phagocytic cells (due to pinocytosis) led to a further revision to the now familiar mononuclear phagocyte system (Van Furth et al., 1972). Here, diverse cells were able to be placed within the same system by virtue of their structural, lineage and functional features, being bone marrow derived, mononuclear, phagocytic, and adherent to glass.

Early, elegant landmark experiments suggested cells of the mononuclear phagocytic system were commonly derived and maintained from bone marrow precursors. Mouse peritoneal monocytes were demonstrated to differentiate morphologically and biochemically into macrophage-like cells *in vitro* (Cohn & Benson, 1965), thus suggesting the monocyte as a macrophage precursor. Dye labelled rabbit monocytes were also directly observed to migrate across the endothelium into tissues, and to differentiate into macrophages (Ebert & Florey, 1939). Furthermore, tritiated thymidine labelling in parabiotic rats demonstrated that peritoneal exudates were of haematogenous origin, and that labelled macrophages in peritoneal exudates had a rapidly dividing bone marrow precursor (Volkman, 1966). Mouse peripheral blood monocytes were demonstrated to be derived from a rapidly dividing bone marrow precursor, termed a pro-monocyte, in further tritiated thymidine kinetic studies. Extant peritoneal monocytes in these studies did not become radio-labelled, however, upon an inflammatory stimulus such labelled monocytes were demonstrated to appear in the peritoneal compartment. These monocytes were found to persist for a number of weeks after the inflammatory insult (van Furth & Cohn, 1968). Splenectomy was found not to reduce the proportion of labelled blood or peritoneal monocytes in mice, whereas total body irradiation led to minimal labelling. Irradiation with partial screening of bone marrow led to continued production of labelled blood and peritoneal monocytes (van Furth & Cohn, 1968).

Aside from dye and radio-labelling techniques, chimeric rats transplanted with donor bone marrow were found to have tissue macrophages purely of the donor type (Godleski & Brain, 1972; Virolainen, 1968). Further evidence from humans receiving gender disparate bone marrow transplantation, preconditioned either with chemotherapy or radiotherapy, demonstrated alveolar macrophages predominantly of the donor type as judged by presence of a Y body, when later the recipients were investigated with bronchoalveolar lavage (Thomas, Ramberg, Sale, Sparkes, & Golde, 1976). Although more modern investigations had to a certain extent recapitulated the initial view that the mononuclear phagocytic system is derived from a single bone marrow progenitor, with circulating monocytes able to enter tissues during inflammation, and replenish resident tissue macrophages, there are certain experimental findings inconsistent with such a model.

Strontium specifically localises to bone and is therefore able to ablate bone marrow-derived monocytes. In such a model alveolar, and peritoneal, macrophage populations are maintained and were seen to be able to proliferate (Sawyer, Strausbauch, & Volkman, 1982). Furthermore in some adoptive transfer experiments lung and liver macrophages were found even at late time points only to partially be of donor origin, with virtually no replacement of microglia (Kennedy & Abkowitz, 1997, 1998). Tissue macrophages have been observed to divide and self-renew *in vitro*, for example with ovarian derived macrophages (Pascual et al., 2005) and in other tissues (Davies et al., 2011; Hashimoto et al., 2013), particularly Langerhan's cells in skin and microglia (Ajami, Bennett, Krieger, Mcnagny, & Rossi, 2011; Chorro et al., 2009). A helminth infection generated Th2 type environment is associated with local tissue macrophage expansion and proliferation, independent of blood monocyte recruitment (Jenkins et al., 2011). Detailed fate mapping studies have shown that tissue resident macrophage populations arise from an embyronic precursor, and are not dependent on circulating monocytes for maintenance (Yona et al., 2013).

Circulating monocytes recruited to sites of active inflammation in the central nervous system are short lived, and do not contribute to the resident microglial population (Ajami et al., 2011), although at other sites, such as the gut, the recruited macrophages do persist and form a proportion of the tissue resident population. Monocytes, many macrophage subsets, and splenic dendritic cells have a common bone marrow precursor: the monocyte and dendritic cell progenitor. These cells are $CD117^+$, $CX3CR1^+$, and $CD3^-$, $CD19^-$, $NK1.1^-$, IAb^-, $CD11c^-$, $B220^-$, TER^-119^-, $Gr1^-$ (collectively Lin^-), and are able to differentiate to splenic white pulp dendritic cells, splenic marginal sinus macrophages, thioglycolate inflammation recruited peritoneal macrophages, and microglia, *in vivo* (Yona et al., 2013).

Thus, whereas originally thought to be a single lineage derived from a bone marrow precursor, and recruited from circulating monocytes, the range of macrophage populations to be found in tissues have a diverse and organ specific origin. Original tissue resident macrophages may be of embryonic origin prior to definitive haematopoiesis, and are able to self-maintain locally. Depending upon site the population may be augmented, or contributed to by circulating monocytes, particularly, if there is inflammatory insult.

5. THE SPECTRUM OF MACROPHAGE ACTIVATION

Macrophages are to be found in almost all tissues, but there are distinct differences in the activation characteristics of macrophage populations. These differences reflect the diverse functions these cells perform: the removal and recycling of cellular debris generated as a part of normal tissue homeostasis, the response to and removal of damaged cells and/or pathogens in the process of inflammation, and also the regulation and resolution of the inflammatory process.

Early studies demonstrated that macrophages from mice previously infected with an intracellular pathogen *Brucella abortus* were better able to clear an unrelated intracellular pathogen *L. monocytogenes* (Mackaness, 1964). Such macrophage activation was demonstrated to be dependent upon interferon (IFN) γ (Buchmeier & Schreiber, 1985), and IFN γ knockout mice were found to have increased susceptibility to intracellular pathogens and have reduced production of macrophage antimicrobial factors, as well as macrophages with reduced expression of major histocompatibility (MHC) class II antigens (Dalton et al., 1993). Such macrophages have since been termed 'classically activated macrophages', or M1 macrophages, and are thought to represent the macrophage in its role as an immune effector during response to many bacterial pathogens. Such classical activation is dependent on IFNγ to prime, and a second stimulus which is either endogenous tumour necrosis factor (TNF) α, or more commonly a pathway that generates TNFα, such as TLR ligation signalling via MyD88 and NFkB (Bosedasgupta & Pieters, 2014). In addition, IFNγ signalling can be dispensed with if a second signal comes via IFN regulatory factor 3 (IRF3; Mosser & Edwards, 2008). Classically activated macrophages express high levels of MHC class II antigens, are able to kill intracellular pathogens, produce high levels of NO, in particular in rodent cells, and express a variety of pro-inflammatory cytokines such as IL-1β, IL-6, IL-12, and IL-23. M1 macrophages do not, however, uniformly increase their phagocytic ability (Mosser & Handman, 1992), and may switch away from receptor-mediated phagocytosis to macropinocytosis (Bosedasgupta & Pieters, 2014). M1 polarised macrophages are well adapted to phagocytose and kill bacteria, as shown by enhanced early killing of intracellular *S. pneumoniae* in M1 polarised monocyte-derived macrophages (MDMs; Fig. 3).

The finding that mannose receptor expression was not only downregulated by IFNγ, but upregulated by IL-4, which also increased endocytosis, and decreased pro-inflammatory cytokine production (Standiford et al., 1990), led to the description of a second, or alternative activation state of macrophages. Such M2 macrophages also increase MHC class II expression (Cao, Wolff, Meltzer, & Crawford, 1989), and show reduced superoxide production (Abramson & Gallin, 1990). These 'alternatively activated', or M2 macrophages formed the second half of a model analogous to the Th1/Th2 paradigm for lymphocytes (Stein, Keshav, Harris, & Gordon, 1992). Interestingly, IL-4 treated macrophages were found to be better able to phagocytose and kill amastigotes of the protozoan *Trypanasoma cruzi* (Wirth, Kierszenbaum, & Zlotnik, 1989), and have since been found to have an important effector role in the Th2 response associated with helminth infections (Kreider, Anthony, Urban, & Gause, 2007). Further studies have identified M2 macrophages and IL-4/IL-13 as enhancing susceptibility to severe infection with another intracellular parasite *Leishmania major* (Raes, Beschin, Ghassabeh, & De Baetselier, 2007), and also with *Cryptococcus neoformans* (Müller et al., 2007), thus illustrating the requirement for subtle

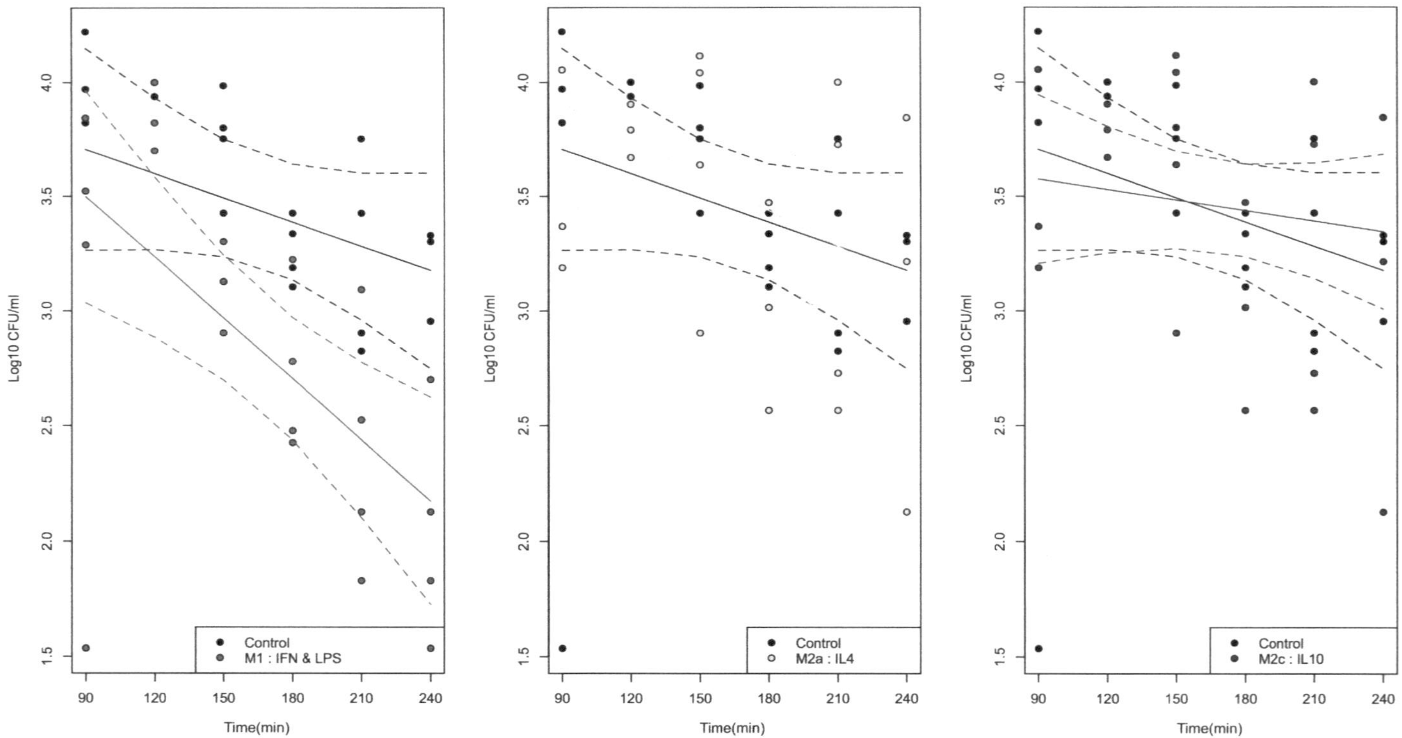

Figure 3 Early intracellular killing of *S. pneumoniae* by polarised human monocyte-derived macrophages (MDMs). Healthy donor monocytes were isolated from peripheral blood mononuclear cells and cultured in 24 well plates as previously described for 14 days to generate MDM and then challenged with opsonised serotype 14 *S. pneumoniae* at a multiplicity of infection of 10 (Dockrell, Lee, Lynch, & Read, 2001). For 72 h prior to infection, they were polarised towards an M1, M2a, or M2c phenotype with 50 ng/ml IFNγ and LPS, IL-4, or IL-10, respectively (Ambarus et al., 2012). Viable intracellular *S. pneumoniae* were quantified at various time points after bacterial challenge using standard gentamicin protection assays and the method of Miles and Misra, as previously described (Dockrell et al., 2001). (See the color plate.)

balance M1 and M2-type cells for effective control of infection (Raes et al., 2007).

A key feature of alternatively activated macrophages as originally described is a change in L-arginine metabolism, with a switch from nitric oxide synthetase (NOS-2) to arginase (Arg-1) caused by IL-4 (Hesse et al., 2001), important for granuloma formation and fibrosis.

Despite the attractive simplicity, and pleasing analogy to the Th1/Th2 paradigm for T-lymphocyte subsets, macrophage activation is more complex and more plastic. In addition to the classical, and the alternatively activated states, it was noted that cross-linking of FcγRI on macrophages inhibited production of the potent pro-inflammatory cytokine IL-12 to normal stimuli such as lipopolysaccharide (LPS; Sutterwala, Noel, Clynes, & Mosser, 1997). Such macrophages were found to produce high levels of the anti-inflammatory cytokine IL-10 (Sutterwala, Noel, Salgame, & Mosser, 1998), which is in turn able to stimulate a Th2 response, and thus induce humoral immunity (Anderson & Mosser, 2002). Furthermore, macrophages in response to glucocorticoid or IL-10 are rather 'deactivated', and secrete transforming growth factor β (TGFβ), and are important in immunoregulation, tissue remodelling and matrix deposition. Thus, the M2 subtype can be further sub-divided into M2a, M2b, and M2c types, respectively (Mantovani et al., 2004). Indeed, rather than distinct subtypes, it may be that macrophages are activated in a large variety of ways, exhibiting greater or lesser degree of characteristics along these broad M1, M2a–c axes (Mosser & Edwards, 2008).

Different macrophage subtypes have been described functionally, and the cytokine milieu resulting in their polarisation has been described. They may also be differentiated on the basis of their cell surface antigens. Classically activated macrophages downregulate CCR2, and also CCR5 and CCR1. Conversely, IL-10 upregulates CCR2 and 5. IL-4 and IL-13 increased expression of CXCR1 and CXCR2. M1 macrophages express high levels of CCD16, CD32, CD64, CD80, and CD86, as well as TLR2 and TLR4. Conversely, M2 macrophages express high levels of the mannose receptor, CD14, CD23, CD163, as well as scavenger receptor (SR) A and SR B (Mantovani et al., 2004).

Profound alterations in gene expression profiles have been identified in differently activated macrophages, with early studies demonstrating that some 25% of the monitored genome of macrophages activated with IFNγ and *Mycobacterium tuberculosis* displayed altered expression (Ehrt et al., 2001). Detailed analysis has led to further insights into the degree of gene expression change upon activation and also has allowed better differentiation of subtypes of activated macrophage, with the identification of at least nine distinct activation programmes (Xue et al., 2014), and leading further insights into the activation states of macrophages in chronic inflammatory diseases such as chronic obstructive pulmonary disease (Shaykhiev et al., 2009). Transcriptomic studies have also added further understanding to the complexity of responses to individual bacterial pathogens. While M1 signatures involving pro-inflammatory cytokines and NO are commonly involved some bacteria may perturb these responses or be associated with M2 polarisation (Benoit, Desnues, & Mege, 2008). Infections such as Whipple's disease and lepromatous leprosy are, for example, associated with M2 macrophage polarisation (Mege, Mehraj, & Capo, 2011). Individual infections may, however, induce their own specific transcriptomic signature with a mixture of both predominant M1 but also some M2-associated transcriptomic responses. For example, although *Streptococcus pyogenes* induce many prototypic M1 cytokine responses in murine macrophages, they also stimulate IL-10 and arginase, responses more typical of M2 polarisation (Goldmann et al., 2007). This emphasises the precarious nature of attempts to characterise macrophage responses globally as merely M1 or M2 polarisation since responses often result from the synthesis of quite specific responses that involve a mixture of polarisation responses.

6. EPIGENETIC REGULATION OF MACROPHAGE FUNCTION

As outlined, macrophages subserve diverse roles to enable the optimal functioning of the innate immune system. In order to perform these roles correctly, the macrophage is dependent on gene transcription. Genetic variation contributes to the intensity of these responses influencing the susceptibility to infectious diseases (Nau et al., 2002), but more recently epigenetic variation has emerged as an additional point of regulation (Jenuwein & Allis, 2001). This allows a mechanism by which the environment can influence the transcriptional response. The epigenome is recognised as making a critical contribution to the control of gene expression in cells.

Epigenetic changes regulate gene transcription through alterations other than changes in the DNA sequence (Berger, 2007; Jenuwein & Allis, 2001). These changes include DNA methylation and modifications to histones. The manner in which DNA is wound around histones and the modelling of the chromatin that results, regulates the availability of DNA sequences to the various transcriptional elements. Chromatin can be densely packed, in which case it is termed heterochromatin; in this state the DNA sequence is so tightly coiled that gene transcription is not possible. Or conversely, in euchromatin, the bound DNA sequence is loosely coiled and genes are available for transcription. The control over chromatin density is thereby in part, responsible for gene control. Chromatin is composed of a series of nucleosomes in turn made of four pairs of histones termed H2A, H2B, H3, and H4. These histones are highly conserved throughout eukaryotic cells. The DNA is coiled around the nucleosomes to form chromatin (Fig. 4).

Modifications of the N-terminal tails of histones influence the availability of the bound DNA sequence by turning euchromatin into heterochromatin and vice versa (Berger, 2007; Jenuwein & Allis, 2001). These changes in the N-terminal of the histone tail, termed posttranslational modifications (PTMs), are formed by chemical alterations, such as phosphorylation, methylation, acetylation, ubiquitination, and sumoylation. These chemical modifications cause adjoining histone tails to be either attracted to each other or repulsed. There are a number of different enzymatic processes involved in the control of these PTM (Jenuwein & Allis, 2001).

DNA methylation refers to the chemical addition of a methyl group to the cytosine base. This process is regulated by DNA methyltransferase (DNMT)1 for maintenance and DNMT3A and DNMT3B for *de novo* methylation in embryonic tissues. The methylation of cytosine bases occurs principally at the cytosine-phosphate-guanine (CpG) dinucleotides. Studies have shown that between 50% and 90% of these are methylated in mammalian tissues (Ehrlich et al., 1982). DNA methylation associated with the promoter region of genes leads to reduced gene expression as the promoter region is not available to initiate transcription. Importantly DNA methylation appears to be conserved over time and is heritable (Smith & Meissner, 2013).

There is a growing body of evidence, albeit not yet in macrophages, that other epigenetic traits can be passed from one generation to the next. In rats, hepatic injury in the parent generation led to a protective effect in the offspring, mediated by modulation in levels trimethylation of Lysine 27 on Histone H3 (H3K27me3; Zeybel et al., 2012). Moreover, it has been recognised for some time that bacteria and viruses can induce chronic changes to the epigenome inducing susceptibility to cancer and the pathogenesis of chronic inflammatory diseases (Esteller, 2007; Paschos & Allday, 2010). These heritable traits are transmitted without a change in DNA sequence, therefore it is likely that the epigenetic make-up of an individual

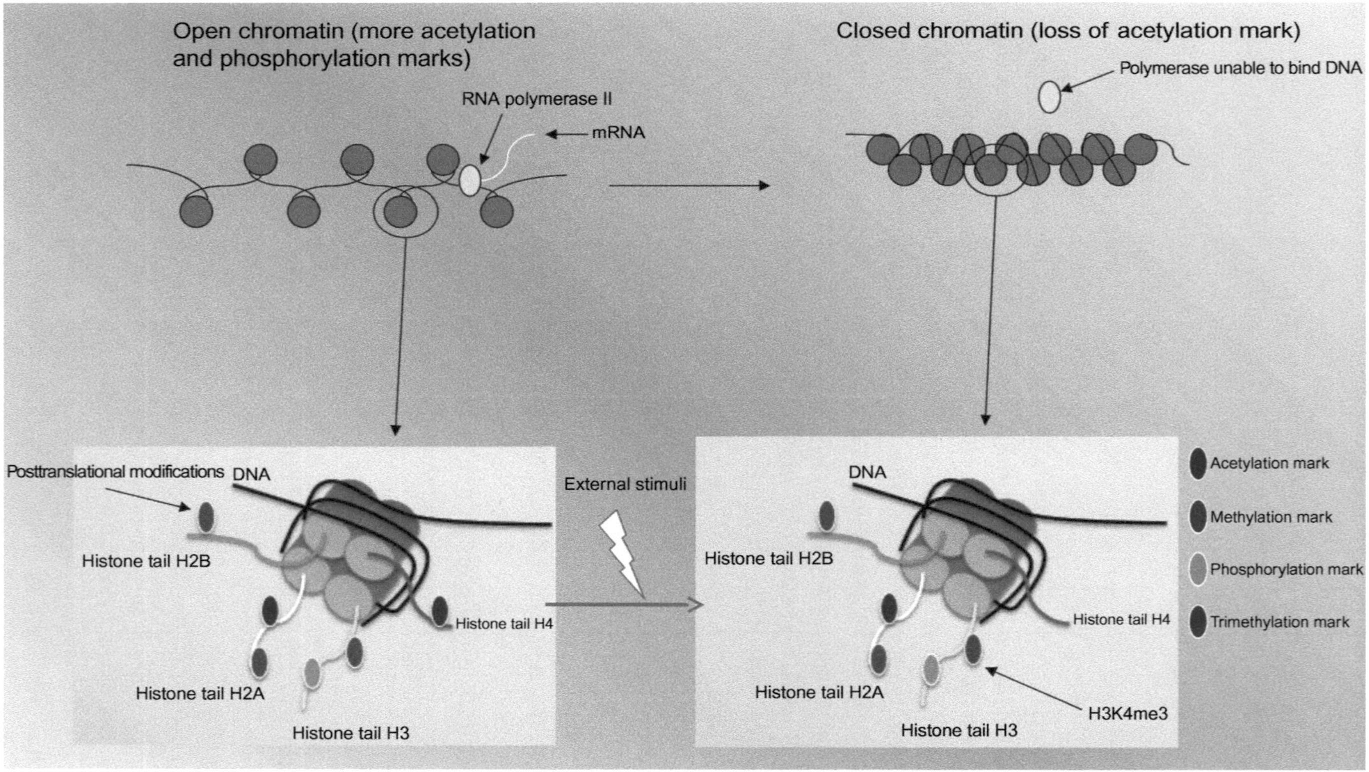

Figure 4 Posttranslational modifications of histone tails regulate macrophage gene transcription. A nucleosome is depicted composed of the four pairs of histones with a schematic representation of some representative posttranslational modifications of the histone tails. Following external stimulation, the levels of H3K4me3 are modified leading to a change in the conformation of the chromatin, in this example from euchromatin to heterochromatin thereby inhibiting access to the DNA sequence wound around it. This in turn leads to decreased transcription of the associated gene, in this example IL-6. (See the color plate.)

may determine their response to future infection and could explain some differences in innate responses between individuals. These epigenetic mechanisms have been shown to play a pivotal role in the regulation of macrophage function in a range of different conditions including smoking-induced inflammation in alveolar macrophages (Philibert et al., 2012) and in the inflammation response to particulate matter, which leads to decreases in DNA methylation through inhibition of DNMT1 activity (Miousse et al., 2014).

The epigenetics of macrophages has mainly been studied in the context of macrophage polarisation and/or differentiation. Both DNA methylation and PTM of histones, as described above, have been shown to play a role in these processes. Epigenetic mechanisms also explain in part the emerging field of trained immunity, sometimes referred to as innate memory. There is mounting evidence that macrophages are capable of developing a form of memory of antecedent events leading to a more vigorous response and a survival advantage that appears to be mediated by epigenetic mechanisms. This is in stark contrast to the traditional description of the innate immune system devoid of any adaptive and memory capacity. The Bacillus Calmette–Guérin (BCG) vaccine has long been used for its protective effect against tuberculosis and leprosy, mediated by T and B lymphocytes. Macrophages from individuals vaccinated with BCG were found to have increased TNF-α and IL-1β following repeat exposure with mycobacteria or fungal elements (Kleinnijenhuis et al., 2012). These experiments demonstrated that vaccination induced a pro-inflammatory circulating monocyte. Subsequently, it has been shown that the PTM involving trimethylation of Lysine 4 on Histone H3 (H3K4me3) was increased following vaccination; furthermore, this PTM was maintained over 3 months. Kleinnijenhuis and colleagues sought to isolate the receptors involved in mediating this and demonstrated that the mechanism of training monocytes with BCG is independent of TLR2 and TLR4, instead it relies on Nucleotide-binding oligomerisation domain-containing protein (NOD)2 receptors. In an attempt to establish functional consequences to training macrophages, this group studied BCG vaccination in a lymphocyte-deficient murine model (using severe combined immunodeficiency (SCID) mice as they are T and B lymphocyte deficient). This showed that BCG vaccination led to a significant survival advantage and decreased fungal burden in kidneys when mice were challenged with a lethal dose of *Candida albicans*, thereby providing *in vivo* evidence of an adaptive form of innate immunity.

Innate memory has also been demonstrated when pre-treating T/B cell-deficient mice, which lack the recombination activating gene (Rag1), with low dose *C. albicans*. Increased survival was subsequently observed when mice were re-infected with a lethal dose of *Candida albicans* and was demonstrated to be monocyte-dependent since it was absent in CCR2-deficient mice (Quintin et al., 2012), emphasising that survival involves training of $CCR2^{+}$ monocytes and not lymphocytes. Genome wide chromatin immunoprecipitation and sequencing (ChIP seq) for the PTM H3K4me3 showed increased levels of this PTM at the promoter regions of genes involved in inflammation, such as TNF-α, IL-6, and IL-18. Thus, providing an epigenetic mechanism underlying the basis for training the innate immune system.

Observations such as these have been made in a number of different models deficient in adaptive immunity, such as plants, where it is termed systemic-acquired resistance (Durrant & Dong, 2004) and in invertebrates. There has also been the historical perspective of crossover protection following vaccination or challenge with one organism conferring survival benefits against another. This led to the term 'trained immunity' being coined in mammalian studies (Netea, Quintin, & van der Meer, 2011). Although this phenomenon has been recognised for a long time, it is only recently that researchers were able to explain the mechanisms underpinning how macrophage training occurs. Indeed, as early as the 1970s observations of increased survival of mice to the parasite *Babesia microti* and *Babesia rodhaini* after BCG vaccination were made (Clark, Allison, & Cox, 1976). Increased survival in mice following pretreatment with low dose *C. albicans* when challenged subsequently with lethal dose *C. albicans* or *S. aureus* was also observed (Bistoni et al., 1986). It has recently been suggested that trained immunity may help to explain the transition of the neonatal immune system from a primarily innate immune system to an adaptive immune system, via trained innate immunity (Levy & Wynn, 2014).

The macrophage response to external stimuli is carefully orchestrated to induce an appropriate level of reaction. While it is important to have the capacity to respond following re-challenge it is equally important to ensure this response does not involve an excessive level of immune response. The mechanisms regulating this have been shown to include epigenetic regulation. One of the principal models used to study inflammation has been that of endotoxic shock. Lypopolyssacharide (LPS) exposure activates TLR4 to induce a number of different pro-inflammatory genes. In contrast, repeated LPS exposure leads to a tolerant, monocyte, which has markedly less production of pro-inflammatory cytokines such as TNF-α, than when first exposed. In murine models, LPS tolerance increases survival following caecal ligation and puncture models of sepsis (Wheeler et al., 2008). The repeated exposure to certain stimuli, such as TLR4 agonists, stimulates tolerance, whereas the response to some other stimuli induces an accentuated pro-inflammatory response, consistent with the model of trained immunity (Ifrim et al., 2014). The development of tolerance helps to decrease the risk of death from an exaggerated pro-inflammatory state, as observed in sepsis. Upon repeated LPS challenge for example the activation of pro-inflammatory cytokine genes such as IL-6 is reduced. However, as the same TLR4 pathway is activated by LPS during initial and repeated responses, a variety of controls are required to induce tolerance. While some of these can involve modification of signal transduction pathways it has become apparent that gene expression can also be regulated by epigenetic changes.

The promoter regions of pro-inflammatory genes during the initial response and on re-challenge in macrophages made tolerant by repeated LPS exposure have been shown to be associated with histone tails with varying levels of H3K4me3 and H4 acetylation (Foster, Hargreaves, & Medzhitov, 2007). The authors demonstrated that on LPS re-challenge the promoter regions of genes associated with LPS tolerance were associated with deacetylated histone H4 tails and were therefore inaccessible for gene transcription, but other genes that continued to be actively transcribed on re-challenge were associated with histone H4 tails that remained acetylated and facilitated promoter availability for gene transcription. Foster and colleagues also showed that the TLR4 response to repeated LPS exposure led to a loss of H3K4me3 (a PTM that when present enables gene transcription) in the histone tails associated with the promoter regions of genes that demonstrated LPS tolerance, while in genes associated with non-tolerance H3K4me3 was maintained in association with sustained transcriptional activity (Foster et al., 2007). Furthermore, when pre-treated with Trichostatin A (a histone deacetylase inhibitor, HDACi) during the initial LPS exposure the pro-inflammatory gene expression of the macrophage remained similar to the naïve macrophage upon re-stimulation, maintaining the secretion of IL-6. This was also seen when pre-treating with pargyline (an inhibitor of H3K4 demethylase) confirming that both H4 acetylation and H3K4me3 marks are important mechanisms underlying the mechanism of regulating pro-inflammatory responses during tolerance (Table 1). Interestingly, the authors observed that the LPS-induced gene responses that did not demonstrate LPS tolerance, included those involved in antimicrobial

Table 1 Summary of various stimuli's influences on histone posttranslational modifications and the outcomes observed

External stimuli	Posttranslational modification	Pathway	Outcome	Comments
BCG	H3K4me3	NOD2	Increase in TNF-α and IL-1β, IL-6, and IL-18	Increased survival in mice with candidaemia
LPS initial challenge	H3K4me3 and H4ac	TLR4		
LPS repeated challenge	Decreased H3K4me3 and H4ac	TLR4	Decrease in IL-6, TNF-α	Increased survival in mice with caecal ligation and puncture
ROM A (*Legionella* species expressed methyltransferase)	Increase in H3K4me		Decrease in IL-6	
LLO (Listeriolysin)	Decrease in H3S10phos and H4ac		Decrease in IL-6	

responses such as the cathelicidin-related antimicrobial peptide, and these genes were not subject to the epigenetic changes associated with tolerance, ensuring these responses were not downregulated on re-challenge (Foster et al., 2007).

DNA methylation has also been shown to regulate inflammatory gene expression and DNA methylation levels are altered in the alveolar macrophages of smokers (Philibert et al., 2012). In particular, smokers were demonstrated to have significantly different methylation status at the CpG sites associated with the promoter regions of genes in alveolar macrophages when compared with non smokers. Furthermore, this correlated with transcriptional changes in the macrophages with increased methylation being associated with decreased mRNA levels and conversely decreased methylation being associated with increased mRNA levels. Finally, GoMiner pathway analysis showed that the majority of the pathways demonstrating differential regulation in smokers involved response to infection, inflammation, and wound healing (Philibert et al., 2012).

Lastly, microRNAs (miRNA) have been demonstrated as a further mechanism by which inflammatory responses may be regulated, for example miR-155 has been found to be increased in response to inflammation (O'Connell, Taganov, Boldin, Cheng, & Baltimore, 2007). There are many other miRNA that are emerging as regulators of inflammatory responses though further discussion of these is beyond the scope of this review.

Macrophage polarisation (as discussed above) has also been shown to be controlled by epigenetic phenomena. In particular, during the induction of an M1 phenotype in macrophages the repressive marks H3K27me3, H3K9me3, and H4K20me3 are removed, this is followed by an increase in the activator marks H3S10phos, H4ac, and H3K4me3 (De Santa et al., 2009; Stender et al., 2012), which in turn leads to activation of genes involved in the acute response to external stimuli associated with the M1 phenotype. This has led to interest in developing novel bromodomain inhibitors to modulate this response (Ivashkiv, 2013). The M2 phenotype is associated with the removal of H3K27me3 repressive mark and is mediated by the H3K27 demethylase Jumonji domain-containing-3 (Jmjd3; Satoh et al., 2010). Satoh and colleagues suggest that Jmjd3 is involved in both the mediation of M1 and M2 phenotypes, but only appears to play a minor role in the M1 response, enabling fine-tuning (De Santa et al., 2009; Satoh et al., 2010). In contrast, it plays a more significant result in M2 polarisation, allowing generation of the transcription factor interferon regulatory factor (IRF) 4 that plays a crucial role in M2 polarisation (Satoh et al., 2010).

Further studies have looked at the location of H4 acetylation, a mark associated with enhanced transcription, in polarised macrophages. The results were mixed but showed that some genes, and, in particular, those mediated by cytokine responses, had increased levels of H4 acetylation associated with the promoter regions (Zhang, Song, Maurer, Bagashev, & Sullivan, 2011). However, not all genes whose expression varied during polarisation showed changes in H4 acetylation levels. The authors suggest that this may be due to the other epigenetic marks that were not studied. This reinforces the view that the epigenome is regulated by a variety of different marks that work in combinations to achieve the fine-tuning of gene regulation. Furthermore, the association of DNA methylation and combinatorial PTMs have not been studied to date.

Micro-organisms have evolved a variety of mechanisms to evade host defences. These include subversion of bacterial killing, inhibition of inflammation, modulation of immune cell chemotaxis and cell lifespan

(Baxt, Garza-Mayers, & Goldberg, 2013; Fernie-King, Seilly, Davies, & Lachmann, 2002). Many of these mechanisms rely on the modulation of host gene expression (Jenner & Young, 2005; Nau et al., 2002). The subversion of the host immune system is one of the key components of bacterial pathogenesis. In effect, microorganisms highjack host gene expression for their own benefit.

In macrophages, histone PTM has been shown to be induced by a number of different bacterial factors including LPS, Listeriolysin and pneumolysin (Hamon et al., 2007). Experiments using both *Legionella pneumophila* and *Listeria monocytogenes* (Eskandarian et al., 2013) have shown that bacterial interactions with the THP-1 monocytic cell line induce histone PTM. We have also demonstrated that the levels of a number of different histone PTM are changed following *S. pneumoniae* infections in both THP-1 and MDMs (Cole & Dockrell, unpublished observations). In the case of *L. pneumophila*, this facilitates intracellular survival of the pathogen. *L. pneumophila* has been shown to use histone PTM to modulate gene expression within the infected cell to aid intracellular replication (Rolando et al., 2013). It secretes RomA, a SET domain-containing methyltransferase, leading to increased H3K14 methylation and decreased levels of acetylation. These histone PTM are associated with the promoter regions of genes, such as IL-6. This shows that *L. penumophilia* modulates host gene expression in particular in genes relating to innate immunity to enhance its intracellular survival. Listeriolysin secreted by *L. monocytogenes* has been shown to cause dephosphorylation at serine 10 of H3 and a reduction in acetylated H4 in THP-1 cells. This correlates with a change in the transcriptional profile in HELA cells which is associated with a decrease in IL-6 and other genes involved in innate immune responses (Hamon et al., 2007). This lends further strength to the theory that bacteria use epigenetic modulation in order to limit the inflammatory response and increase their survival.

The consequences of epigenetic changes during acute bacterial infections are not fully understood but there is growing evidence in the context of certain stimuli, such as LPS exposure, these influence macrophage polarisation and activation. There is mounting interest in the use of histone deacetylase inhibtors (HDACi) in the fields of oncology, HIV, and inflammatory disorders. Their use has been associated with survival benefits in murine models of sepsis. In lethal LPS challenge models, the use of suberoylanilide hydroxamic acid, an HDACi, was associated with increased survival (Li et al., 2010). However, other groups have reported detrimental effects of HDACi on macrophage's killing ability with decreased clearance of bacteria and fungi (Mombelli et al., 2011; Roger et al., 2011). Nevertheless,

Table 2 Posttranslational modifications observed during polarisation in macrophages

	Decreased	Increased
M1 phenotype:	H3K27me3, H3K9me3, and H4K20me3	H3S10phos, H4ac and H3K4me3
M2 phenotype:	H3K27me3	

they also observed increased survival following LPS challenge. Therefore, while it is likely that HDACi could modulate inflammatory responses, in conditions such as sepsis, further studies will be required to optimise the timing of such therapeutic interventions to ensure they do not impair initial macrophage responses to infection. Other therapeutic strategies might utilise specific small molecules targeted at individual modifications, such as the novel jumonji H3K27 demethylase inhibitors, which lead to a modulation in the inflammatory response to LPS (Kruidenier et al., 2012). Furthermore, there is growing interest in developing specific bromodomain inhibitors, which should allow a more targeted approach to the modulation of gene expression.

Epigenetics is a rapidly evolving field and its importance in the control of macrophage function and response to infection is increasingly being recognised. The innate response of macrophages is subject to regulation at multiple levels including the epigenetic changes of histone PTMs and DNA methylation as well as by miRNAs. This complex regulation allows a process that enables both a rapid initial response and protection from a prolonged and excessive inflammatory response. As we have shown, these processes play a role in a variety of processes including immune training, induction of tolerance, and macrophage polarisation (Table 2). Finally, these regulatory mechanisms contribute to the link between inflammatory responses (and by extension those to infection) and the pathogenesis of chronic human diseases.

7. MECHANISMS OF MACROPHAGE PHAGOCYTOSIS

Macrophages internalise particles using one of several processes; pinocytosis, receptor-mediated endocytosis, and phagocytosis (Aderem & Underhill, 1999). Pinocytosis is used for liquids and solutes and is either clathirin dependent or in the case of macropinocytosis of macromolecules may be actin dependent (Swanson & Watts, 1995). Receptor-mediated endocytosis is a clathirin-dependent mechanism, whereas phagocytosis is an actin-dependent mechanism, facilitating the uptake of particles greater than 0.5 μm, such as bacteria (Aderem & Underhill, 1999). Macropinocytosis may also allow internalisation of some bacteria into spacious phagosomes (Alpuche-Aranda, Racoosin, Swanson, & Miller, 1994). Macrophages are regarded as professional phagocytes because of the efficiency with which they internalise particles and differentiated macrophages retain the capacity to ingest particles (Daigneault, Preston, Marriott, Whyte, & Dockrell, 2010). Macrophages are well equipped to phagocytose particles possessing a large surface area, which provides the membrane to form endocytic vacuoles and experimental data suggests the macrophage has the potential to internalise its entire surface membrane as endocytic vacuoles every 30 min (Sokol, Hudson, James, Frost, & Wales, 1987; Steinman, Mellman, Muller, & Cohn, 1983). Phagocytosis involves multiple different receptors and ingestion of bacteria is frequently the net result of internalisation via several receptors as illustrated for *S. pneumoniae* (Ali et al., 2003). Initial receptor interaction with its ligands leads to actin polymerisation and internalisation after which the phagosome matures through a series of microtubule-dependent interactions with endosomes (Aderem & Underhill, 1999).

Opsonised particles are internalised via the Fc receptors or complement receptors. Macrophages express several Fc receptors involved in phagocytosis of opsonised particles. FcγRIIA contain a signalling domain, while FcγRI and RIII require an adaptor protein to signal. FcγRIIA is a low affinity receptor that is the major receptor binding IgG_2 (Parren et al., 1992; Warmerdam, van de Winkel, Vlug, Westerdaal, & Capel, 1991), the subclass of immunoglobulin predominantly produced in response to bacterial polysaccharide capsules in adults (Herrmann et al., 1992), although $Fc\gamma RIIIA_{V158}$ also binds IgG_2 with lower affinity (Bruhns et al., 2009). Both FcγRIIA and FcγRIII are low affinity receptors for IgG_1, the predominant anti-capsular antibody produced in children, and for IgG_3, while FcγRI is a high affinity receptor. Receptor cross-linking activates src family kinases to phosphorylate tyrosine residues in the immunoreceptor tyrosine-based activation motif (ITAM) of the cytoplasmic tail (FcγRIIA) or the adaptor dimer of the receptor. This in turn leads to recruitment of the p72 spleen tyrosine kinase (Syk) and its activation leads to signalling via protein kinase C, phosphoinositide 3 kinase (PI3K), and Rho family GTPases that induce cytoskeletal changes mediating phagocytosis. Macrophages express three complement receptors CR1, CR3, and CR4; CR1 allows particle binding and CR3 and CR4 bind iC3b and facilitate internalisation (Aderem & Underhill, 1999). Unlike the FcγR these receptors require additional stimuli to allow their activation for phagocytosis and phagocytosis by this route is not associated with the production of inflammatory mediators such as ROS (Wright & Silverstein, 1983). In the case of FcγRs Prostaglandin E_2 (PGE_2), downregulates phagocytosis, through cAMP-induced activation of exchange protein directly activated by cAMP (Eap-1) and downstream by activation of Rap-1, a small GTPase, while it also downregulates resultant ROS and TNF-α expression (Aronoff, Canetti, & Peters-Golden, 2004; Aronoff, Canetti, Serezani, Luo, & Peters-Golden, 2005). The Leukotrienes LTB_4 and LTD_4 have the opposite effect and enhance phagocytosis though activation of syk and inhibition of PGE_2 effects on Rap-1 (Canetti et al., 2006; Lee, Serezani, Medeiros, Ballinger, & Peters-Golden, 2009; Mancuso & Peters-Golden, 2000). Certain macrophages, such as alveolar macrophages, utilise leukotriene signalling to enhance signalling cascades that maximise phagocytosis (Lee et al., 2009).

SRs are important for uptake of non-opsonic particles. Macrophages express a range of SR implicated in the recognition and uptake of bacteria. These include SRAI/II, the SRA type macrophage receptor with collagenous structure (MARCO), SRBIII (CD68), CD36 a SRB-like receptor, and lectin-like oxidised low density lipoprotein receptor (LOX-1; Taylor et al., 2005). MARCO is a particularly important receptor for the ingestion of unopsonised bacteria and environmental dusts in the lung (Arredouani et al., 2005). Several receptors recognise carbohydrate moieties and play a role in ingestion of these. The mannose receptor is a member of the C-type lectin superfamily and recognises carbohydrate moieties on a variety of micro-organisms (Allavena, Chieppa, Monti, & Piemonti, 2004), galectin 3 also recognises carbohydrate residues and plays a role in phagocytosis (Jin, Opalek, Marsh, & Wu, 2004), while dectin 1 is an important receptor for the phagocytosis of fungi (Herre et al., 2004). The C-type lectin receptor dendritic cell-specific intracellular adhesion molecule (ICAM)-3 grabbing non-integrin receptor (DC-SIGN) homolog, SIGN related-1 (SIGNR1) is important in recognising neutral microbial polysaccharides and mediating uptake of particles containing these (Geijtenbeek et al., 2002; Taylor et al., 2005).

8. PHAGOCYTOSIS OF *S. pneumoniae* AND *S. aureus*

Pneumococci colonise the upper airway and a major effector of clearance when bacteria reach the distal airway following micro-aspiration is the alveolar macrophage(Dockrell et al., 2012). Mathematical modelling confirms these findings. These models show that alveolar macrophages control initial inocula of bacteria in the lower airway, but have a finite capacity to do this, and with large inocula become overwhelmed (Smith, McCullers, & Adler, 2011). Once pneumonia is established, although neutrophils are key phagocytes for bacterial clearance, recruited inflammatory macrophages are also important, while once bacteraemia is established other mononuclear phagocyte populations, such as monocytes in the blood and splenic macrophages, are important mediators of bacterial clearance through phagocytosis.

As discussed above, the pneumococcal polysaccharide is an important barrier to macrophage phagocytosis, despite the aforementioned strategies to limit opsonisation with complement and antibody. Opsonisation facilitates phagocytosis of pneumococci via FcγRs and CRs (Jonsson et al., 1985). Assays that measure phagocytosis-dependent outputs have illustrated that a range of these receptors including FcγRIIA and FcγRIII, as well as CR1 and CR3 contribute to uptake of opsonised pneumococci (Ali et al., 2003). Some case-control genetic studies suggest a polymorphism of FcγRIIA, the major receptor required to bind bacteria opsonised with IgG_2, the major antibody produced in response to pneumococcal capsule, influence susceptibility to invasive pneumococcal disease, which could suggest it is an important receptor in uptake of opsonised bacteria (Yuan et al., 2008). But this association was not sustained in a meta-analysis (Brouwer et al., 2009). Genetic studies cannot separate the effect of the receptor on macrophages as opposed to neutrophils but absence of a clear signal is consistent with the view that multiple receptors contribute to uptake. FcγRIII is also important for clearance of *S. pneumoniae* when opsonised with specific anti-capsular IgG_1 (Weber, Tian, Van Rooijen, & Pirofski, 2012).

Although both complement and immunoglobulin is required for maximal internalisation by macrophages, immunoglobulin appears to enhance binding to macrophages but once this effect is accounted for rates of phagocytosis appear similar in the presence or absence of immunoglobulin (Ali et al., 2003; Gordon et al., 2000). Anti-pneumococcal IgG is detectable in the bronchoalveolar lavage fluid of many healthy adults, suggesting these *in vitro* findings are relevant in the lung (Eagan et al., 2007). Activation of the classical pathway of complement is particularly important for ensuring clearance of bacteria from the blood suggesting it may be particularly important for phagocytosis of bacteria by neutrophils, monocytes, and splenic macrophages, but since it also influences activation of lung macrophages it is likely it may also contribute to lung clearance (Brown et al., 2002). Opsonic phagocytosis by macrophages is impaired in several clinical conditions; HIV infection, smoking and fasting all impede opsonophagocytosis of pneumococci by macrophages (Eagan et al., 2007; Mancuso, Huffnagle,

Olszewski, Phipps, & Peters-Golden, 2006; Mancuso et al., 2011; Phipps et al., 2010). This provides some basis for the association of these conditions with susceptibility to pneumococcal disease. Uptake of opsonised pneumococci also helps regulate the inflammatory response that results. As outlined above, CRs do not stimulate pro-inflammatory responses, while engagement of the inhibitory FcγRIIB helps downregulate inflammatory responses; mice lacking this receptor may have enhanced clearance of low inocula of bacteria from the peritoneum but have excessive inflammation and worse outcomes when challenged with higher inocula of bacteria (Clatworthy & Smith, 2004). Ingestion of apoptotic cells, a process termed efferocytosis, in the lung increases PGE_2 levels and hence puts another brake on the pro-inflammatory consequence of pneumococcal phagocytosis (Medeiros, Serezani, Lee, & Peters-Golden, 2009). This is advantageous during the resolution phase of inflammation but this mechanism also leaves the host vulnerable to the possibility that during chronic inflammatory conditions enhanced efferocytosis will result in too great a block on inflammation, during the initial stages of the host response, when the inflammatory response requires rapid escalation.

Non-opsonic uptake is also important for phagocytosis of *S. pneumoniae*. MARCO is a critical receptor for uptake of unopsonised pneumococci (Arredouani et al., 2004). MARCO-deficient macrophages phagocytose less effectively and in murine pneumonia models MARCO-deficient mice have reduced early clearance of bacteria in the lung and a greater reliance on neutrophils to control infection. MARCO also emerged as the major mechanism of phagocytosis of unopsonised pneumococci in the nasopharynx of mice, mediating bacterial clearance and TLR2 and nucleotide-binding oligomerisation domain-containing (Nod) 2-mediated nuclear factor kappa B (NFκB) signalling (Dorrington et al., 2013).

Although MARCO has emerged as a major mechanism of uptake of unopsonised bacteria other receptors may also be involved. SRAI/II also contributes to phagocytosis-mediated clearance of pneumococci from the lung (Arredouani et al., 2006), but did not make a contribution to clearance in the nasopharynx (Dorrington et al., 2013). Surfactant protein A in the lung enhances uptake of pneumococci by SRA through upregulation of surface expression of SRA (Kuronuma et al., 2004). CD36 is another SR implicated in phagocytosis of *S. pneumoniae* by alveolar macrophages (Sharif et al., 2013). CD36-deficient alveolar macrophages demonstrate impaired uptake of pneumococci and CD36 downregulates the early inflammatory response during interaction with pneumococci. The authors proposed a model in which CD36 engagement of the phosphocholine residues in the cell wall lipoteichoic acid ensured the inflammatory response was kept in check and not excessive. If these phosphocholine residues were masked with antibody, blocking interaction with CD36, macrophages release increased levels of pro-inflammatory factors such as the chemokine macrophage inflammatory protein (MIP)-2. Although mannose receptors bind *S. pneumoniae* in certain scenarios, as shown for olfactory ensheathing cells in the nose they did not contribute to phagocytosis in the nasopharynx (Dorrington et al., 2013; Macedo-Ramos et al., 2011). The C-type lectin receptor SIGNR1 also plays a role in *S. pneumoniae* phagocytosis. As it is not expressed in alveolar macrophages, but instead in splenic marginal zone macrophages (and also peritoneal macrophages), its primary role in bacterial clearance appears to be in allowing phagocytosis of pneumococci by splenic marginal zone macrophages, enabling them to present antigen. This results in production of natural anti-phospholcholine IgM by marginal zone B cells, aids bacterial clearance, and prevents excessive inflammation (Koppel et al., 2005).

Overall multiple receptors may contribute to the uptake of unopsonised pneumococci. Although MARCO plays a major role other receptors contribute in specific locations. In addition, despite observations that these additional receptors may make only a modest contribution to the overall level of internalisation of bacteria they may still make important contributions through the specific signal transduction pathways they activate with consequences to the overall inflammatory response, which must be tightly regulated within tight boundaries. Phagocytosis of unopsonised *S. pneumoniae* is reduced by concentrated ambient particles, constituents of environmental air pollution, explaining the links between air pollution and lung infection (Zhou & Kobzik, 2007). Some metal ions, such as iron in welding fumes also compromise phagocytosis of unopsonised bacteria, while the more toxic chromium fumes are directly toxic to the macrophages, illustrating that the enhanced susceptibility to pneumococcal disease seen in occupations such as welders is influenced by impairment of these pathways of pneumococcal clearance (Antonini, Lawryk, Murthy, & Brain, 1999; Antonini et al., 2005; Coggon, Inskip, Winter, & Pannett, 1994).

In contrast to the situation observed with *S. pneumoniae*, unopsonised *S. aureus* are avidly phagocytosed by macrophages, since they lack the inhibitory effect of a polysaccharide capsule and adaptations retarding opsonisation as discussed above are less critical (Hof, Repine, Peterson, & Hoidal, 1980; Jonsson et al., 1985). Consistent with these observations phagocytosis of *S. aureus* occurs at rates comparable to unencapsulated

S. pneumoniae (Esposito, Clark, & Poirier, 1988). Surface bound bacteria are efficiently internalised and opsonisation only modestly increases phagocytosis (Lee et al., 1984). Immunoglobulin and complement both opsonise *S. aureus*. In the presence of sufficient antibody, complement does not enhance phagocytosis by macrophages but if immunoglobulin levels are low complement makes a contribution to phagocytosis (Murphey, Root, & Schreiber, 1979). Both the classical and alternative pathway, of complement activation, contribute but the alternative pathway was found to be most important during opsonisation when immunoglobulin was a limiting factor. A range of FcγR and CR are likely to contribute though there is less information on the specific receptors involved for *S. aureus* ingestion than for pneumococci.

Unopsonised *S. aureus* are also phagocytosed via the class A SRs MARCO and SRA I/II (Arredouani et al., 2005; Dunne, Resnick, Greenberg, Krieger, & Joiner, 1994). However, the role of these receptors is likely to vary with strain (which may influence binding to the receptors) and according to the location of the macrophages. One study showed SRs contributed to phagocytosis of heat-killed bacteria but did not contribute to phagocytosis of live bacteria by human MDMs (Deloid, Sulahian, Imrich, & Kobzik, 2009). In contrast, SRAI/II deficiency inhibited phagocytosis and bacterial clearance of *S. aureus* from the peritoneum of mice (Thomas et al., 2000). Other investigators have suggested that in contrast SRs were more important in specific locations such as the lung. SRAI/II, CD36, or MARCO-deficient mice were more susceptible to pulmonary infection, with less bacterial clearance, but more resistant to peritoneal infection (Blanchet, Jouvion, Fitting, Cavaillon, & Adib-Conquy, 2014). Another study though found that CD36 aided bacterial clearance and facilitated TLR activation by *S. aureus*, demonstrating reduced bacterial clearance following intravenous infection in the absence of CD36 (Stuart et al., 2005). The overall differences in these studies make it hard to draw definitive conclusions regarding the role of SRs in the host response against *S. aureus*, but murine studies do suggest they play a role at certain stages of infection, particularly in the lung.

Another receptor that may contribute to phagocytosis of *S. aureus* is the surfactant protein A receptor SP-R210 (Sever-Chroneos et al., 2011). Surfactant protein A can bind to *S. aureus* (Mcneely & Coonrod, 1993), in an interaction dependent on the bacterial adhesin Eap (Sever-Chroneos et al., 2011). SP-R210 exists as two isoforms and in the absence of SP-R210 function, SRAI/II mediates uptake of unopsonised Eap^{+} bacteria (Sever-Chroneos et al., 2011). In the author's proposed model, optimal clearance initially involves SP-R210$_L$-mediated uptake of bacteria opsonised with surfactant protein A, which primes the initial inflammatory response and TNFα production. Subsequently this results in upregulation of SP-R210$_S$. It's interaction with SRAI/II allows a further phase of bacterial phagocytosis that is coupled to generation of anti-inflammatory mediators. This model illustrates some potentially important principles. The consequences of internalisation are often the result of integration of signals generated from multiple receptors and an individual receptor may make a more significant impact to the host–pathogen interaction because of its signalling role, than the extent to which it contributes to the overall rate of phagocytosis would predict it would.

In summary, macrophages are important first effectors of immune responses to both bacteria. Bacterial clearance and co-ordination of immune responses to both bacteria depend on phagocytosis, but these bacteria represent different challenges. *S. pneumoniae* is relatively difficult to phagocytose because of its polysaccharide capsule and the adaptations it has to restrict opsonisation. Conversely, *S. aureus* is more easily ingested but as discussed below it represents different challenges to the macrophage.

9. INTRACELLULAR LOCALISATION OF BACTERIA

Once internalised bacteria are initially localised in a phagosome. This compartment must mature through a series of interactions with other endosomes if the compartment is to effectively kill and degrade the bacteria it contains (Flannagan, Cosio, & Grinstein, 2009). Much of what we know about these events has been garnered from studies of intracellular pathogens, in particular *Mycobacterium tuberculosis* and from pathogens that escape from the phagolysosome like *L. monocytogenes*. Phagosomes initially fuse with early endosomes. The early phagosome contains the early endosome antigen (EEA)1 and fusion with early endosomes is regulated by the small GTPase Rab5; EEA1 is thought to act as a bridge between the early endosome and the phagosome (Bucci et al., 1992; Lawe, Patki, Heller-Harrison, Lambright, & Corvera, 2000). Internalised particles are contained within a tight fitting endosome and membrane fusion events are balanced by fission events and recycling of membranes to the plasma membrane and other compartments. Phagosomal proteins to be removed are ubiquitinated and associated with the endosomal sorting complex required for transport (Lee, Kim, Schreiber, & Grinstein, 2005). The phagosome matures by acquiring lysosomal-associated membrane proteins (LAMP) 1 and 2 and it has been proposed these may facilitate docking with microtubules and centripetal movement of the phagosomes by interaction with the dynein/dynactin motor (Huynh et al., 2007). As the phagosome matures into a late phagolysosome, it acquires Rab7 in place of Rab5, which regulates traffic between the late phagosome and late endosomes or lysosomes (Harrison, Bucci, Vieira, Schroer, & Grinstein, 2003). This enables acquisition of the Rab interacting lysosomal protein (RILP), which allows the centripetal movement of the late phagosome and lysosomes by linking the membrane to the dynein–dynactin motor (Harrison et al., 2003; Jordens et al., 2001). The late phagosome is enriched for LAMP 1 and 2 (Desjardins, Huber, Parton, & Griffiths, 1994). It is also enriched in vacuolar ATPase allowing the luminal pH to drop to 5.5–6. The maturation into a phagolysosome is accompanied by further acidification, facilitated by addition of further vacuolar ATPases, Rab7-mediated lysosomal fusion, and acquisition of lysosomal proteases such as the cathepsins (Flannagan et al., 2009). Ultimately the phagolysosome becomes a compartment equipped to kill and digest bacteria and not surprisingly intracellular bacteria such as *M. tuberculosis*, *Legionella pneumophila,* and *L. monocytogenes* have developed strategies to subvert these steps. Less attention has been focused on how these steps progress following the phagocytosis of extracellular bacteria such as *S. pneumoniae* and *S. aureus.*

S. pneumoniae is internalised into a LAMP-1 positive compartment which suggests it can traffic appropriately to a late phagosome (Gordon et al., 2000). The level of internalisation to this compartment is enhanced by opsonisation. Our unpublished data shows pneumococcal phagolysosomes are appropriately acidified and the majority have a pH < 6, when pneumococci are labelled with pH rhodo and then opsonised to enhance internalisation. These levels are comparable to results with *Escherichia coli* (Jubrail & Dockrell, unpublished observations). After sustained exposure to bacteria, however, the phagolysosomes become permeabilised and the pH rises (Bewley, Marriott, et al., 2011). There is also evidence of activation of cathepsin D, one of the most abundant lysosomal proteases found in macrophages, which is activated by pneumolysin expressed by *S. pneumoniae.* These findings are consistent with a model of effective acidification of the phagolysosome containing pneumococci and its fusion with lysomes to acquire cathepsins, but also suggest that the host–pathogen interaction ultimately triggers permeabilisation of the phagolysosome at later time points, from 10- to 12-h after sustained exposure to bacteria.

A different picture emerges with *S. aureus.* Phagosomes fail to acidify to the same extent as those containing *E. coli* (Sokolovska et al., 2013). The authors provide a model in which both live *S. aureus* and ROS generation produced by NADPH oxidase buffers the luminal pH of the phagolysosome. The author's showed that the early acidification, which does occur (albeit to a lesser extent than with *E. coli*) is the result of a counter-regulatory mechanism that involves activation of the NLRP3 inflammasome and recruitment of active caspase 1 to the phagosome. They show that several components of the NADPH oxidase complex are caspase 1 substrates. In this model, caspase 1-mediated NADPH oxidase inactivation lessens the capacity of NADPH oxidase to raise the luminal pH through production of hydroxyl ions that buffer the influx of hydrogen ions from the vacuolar ATPase. The activation of caspase 1 and the associated effect on luminal pH is associated with increased bacterial killing and generation of IL-6. This clearly demonstrates an important role for NADPH oxidase activation in buffering the phagosomal pH and provides an explanation as to why the *S. aureus* phagolysosome may not fully acidify. The authors also showed that the pH that was generated was dependent on vacuolar ATPase suggesting this system is still active in the *S. aureus* containing phagosome. Important questions remain despite this elegant work. It is unclear if the sustained inability to acidify *S. aureus* containing phagolysosomes is a residual consequence of the earlier buffering capacity of the NADPH oxidase or reflects other activities of the bacterium from within the phagolysosome. We have, however, noted that just as pneumococci traffic to LAMP 1 positive phagolysosomes *S. aureus* also do so, suggesting there is not a failure to fuse with endosomes up to the point of LAMP 1 recruitment (Jubrial & Dockrell, unpublished observations).

10. MICROBIAL KILLING BY MACROPHAGES

The basis of macrophages' antimicrobial activity is still incompletely defined and many of our assumptions are inferred from our knowledge of other phagocytes, such as neutrophils and monocytes. Much attention has focused on ROS on the basis of their assumed role in mediating bacterial killing in neutrophils, although this has been challenged by observations that suggest even here the ROS generated do not directly mediate bacterial killing, but instead it is the serine proteases activated as a consequence of NADPH oxidase activation that actually kill micro-organisms (Reeves et al., 2002). Phagocytes utilise the NOX2 (gp91 phox) NADPH oxidase and transfer electrons from cytosolic NADPH to molecular oxygen, resulting in the generation of superoxide (Quinn & Gauss, 2004).

Superoxide (O_2^-) can dismutate to form hydrogen peroxide (H_2O_2), which in turn can react with superoxide to generate singlet oxygen ($O_2^\bullet$) or hydroxyl radicals (Minakami & Sumimotoa, 2006). How ROS kill bacteria is still an open question; superoxide and hydrogen peroxide can damage biomolecules including enzymes required for metabolism, hydrogen peroxide can react with cysteine residues, while hydroxyl radicals damage DNA (Slauch, 2011). However, many bacteria have adaptations to withstand the hydrogen peroxide which diffuses across their periplasm (Slauch, 2011), including as we have shown *S. pneumoniae* and *S. aureus*. Myeloperoxidase (MPO) converts superoxide into hypochlorous acid and chloramines but differentiated macrophages generally lack MPO and therefore are less able to make these more potent ROS (Shepherd, 1986). Macrophages stop expressing MPO as they differentiate from monocytes (Nakagawara, Nathan, & Cohn, 1981), though macrophages seem capable of expressing MPO under certain circumstances such as in atherosclerotic plaques or in certain locations such as in Kuppfer cells in the liver (Brown, Brunt, & Heinecke, 2001; Sugiyama et al., 2001). Macrophages also need to be activated to produce ROS; cells such as alveolar macrophages are tuned so as not to produce significant levels of ROS constitutively but only to do so after appropriate activation, to ensure they limit the potential for tissue injury. In general, microbicidal production is tightly coupled to activation via pattern recognition receptors and polarisation state (Mantovani et al., 2004; Mosser & Edwards, 2008).

In addition, activation of NADPH oxidase in phagocytes is tightly linked to phagocytosis with maximal production following formation of the phagosome (Deleo, Allen, Apicella, & Nauseef, 1999). Recently an additional source of ROS has emerged with evidence that TLR2 or TLR4 activation can lead to recruitment of mitochondria to the phagolysosome and stimulation of mitochondrial ROS release, which aids killing of bacteria in the phagolysosome (West et al., 2011). This is mediated by the TLR adaptor TNF receptor-associated factor (TRAF)-6 which is recruited to the mitochondria and engages the protein evolutionary conserved signalling intermediate in Toll pathways (ESCIT). ESCIT plays a role in regulating respiratory chain assembly and its TRAF-6-mediated ubiquitination results in increased mitochondrial ROS generation and bacterial killing. In general, however, macrophages lack the capacity to generate the more potent forms of ROS, while their production of ROS is of a lower level than neutrophils and they lack the neutrophil expressed granule-associated serine proteases that have been implicated as mediators of NOX2-dependent killing in neutrophils. In addition, bacteria have many antioxidant defences so it is becoming increasingly clear that the macrophage must utilise other microbicidal strategies in combination with or instead of ROS.

Macrophages generate nitric oxide using inducible nitric oxide synthase, also known as NOS2 (Fang, 2004). Production of nitric oxide is regulated at the level of NOS2 transcription. L-arginine, oxygen and hydrogen ions are used to generate NO and citrulline. NOS2 transcription and NO production is a feature of M1 polarised macrophages (Mosser & Edwards, 2008). NO is produced on the cytoplasmic side of the phagosome but diffuses across the phagosomal membrane and reacts with ROS to generate a series of RNS including peroxynitrite and nitrosothiols (Fang, 2004). These cause protein inactivation, lipid oxidation, and DNA damage in micro-organisms. However, human macrophages appear to generate less NO than rodent species (Jesch et al., 1997). Moreover differentiated macrophages produce less than monocytes (Daigneault et al., 2010). Therefore, the role of NO in host defence in human macrophages has been controversial (Fang, 2004). However, there are multiple reports of human macrophage upregulating NOS2 in response to both intracellular and extracellular bacterial infection and therefore it appears that although tightly regulated, human macrophages can utilise NO or its reaction products in host defence against bacteria (Marriott et al., 2004; Nicholson et al., 1996). Transcription of NOS2 is tightly regulated but induced by antimicrobial products and cytokines such as TNFα and IFNγ (Macmicking, Xie, & Nathan, 1997).

In addition to ROS and NO macrophage, phagolysosomes contain multiple proteins with the potential to contribute to microbicidal killing. These include proteins that limit bacterial growth by removing critical nutrients such as divalent cations like Fe^{2+}, Zn^{2+}, and Mn^{2+}. Natural resistance-associated macrophage protein (NRAMP)1 is an example of this group (Cellier, Courville, & Campion, 2007). It is expressed on late endosomes and lysosomes and is recruited to the phagolysosome where it helps restrict the growth of intracellular bacteria such as *Mycobacterium tuberculosis*. Another antimicrobial strategy involves β-defensins that are produced by macrophages, as well as by neutrophils and epithelial cells, and can permeabilise bacterial membranes by the insertion of voltage-dependent channels (Lehrer, Lichtenstein, & Ganz, 1993). Phagocytes also contain a variety of proteases with the potential to digest bacteria or their products (Flannagan et al., 2009). These include endopeptidases (cysteine and aspartate proteases) and exopeptidases (serine and cysteine proteases) as well as hydrolases that degrade carbohydates, e.g., lysozyme and lipids, e.g., phospholipase A2. Macrophages lack many of the serine proteases used by neutrophils like the granule-associated neutrophil elastase and cathepsin G, although they do possess other serine proteases like tripeptide peptidase I (Jin et al., 2004). Instead, they are enriched in comparison to neutrophils or monocytes, in a range of proteases which possess the potential to digest phagocytosed particles; alveolar macrophages are rich in the aspartate protease cathepsin D and the carboxyproteases cathepsin B and X and the amimopeptidase cathepsin H (Burnett, Crocker, & Stockley, 1983; Jin et al., 2004; Kato, Kojima, & Murachi, 1972). Macrophage elastase (Matrix metalloprotease (MMP) 12) has emerged as an important protease that is delivered to the phagolysosome and contributes to killing of Gram-positive and Gram-negative bacteria (McGarry Houghton, Hartzell, Robbins, Gomis-Rüth, & Shapiro, 2009).

The activity of many of these proteases is dependent on an acidic pH. The vacuolar ATPase is critical to maintaining a low pH, which both inhibits bacterial growth and aids degradation of the ingested bacteria by the proteases contained within the phagolysosome (Huynh & Grinstein, 2007). Other mechanisms which acidify the phagolysosome in macrophages include the activity of Na^+/H^+ exchangers and acid carrier vesicles formed from early endosomes, but which lack vacuolar ATPase (Hackam et al., 1997). The importance of appropriate acidification is demonstrated by the increased bacterial viability and the reduced cytokine responses of macrophages in which phagolysosomal acidification is perturbed (Sokolovska et al., 2013). Moreover successful intracellular pathogens, which are able to survive in macrophages, possess adaptations to perturb phagolysosomal acidification. For example, *M. tuberculosis* expresses a secreted tyrosine phosphatase (PtpA) which binds to the H sub-unit of the vacuolar ATPase, resulting in its extrusion from the phagosome and impaired acidification (Wong, Bach, Sun, Hmama, & Av-Gay, 2011). This is just one example of a series of adaptations successful intracellular pathogens employ to manipulate phagolysosomal maturation, acidification, or membrane integrity to escape the hostile environment of low pH and activated proteases in the phagolysosome (Flannagan et al., 2009).

11. APOPTOSIS-ASSOCIATED KILLING COMPLEMENTS CLEARANCE OF *S. pneumoniae*

As we have discussed, ingested pneumococci are taken up into phagosomes, which mature appropriately to phagolysosomes, and in which pneumococci are efficiently killed (Gordon et al., 2000; Jonsson et al., 1985). The important role of alveolar macrophages in the clearance of pulmonary pneumococci is supported by murine models in which alveolar macrophage depletion prevents efficient clearance of low inocula of pneumococci from the murine lung (Dockrell et al., 2003). The mechanisms of intracellular killing are less well established. ROS alone do not emerge as a major effector of pneumococcal killing by neutrophils in part reflecting the inherent resistance of *S. pneumoniae* to oxidative stress discussed above. Neutrophils appear to be more dependent on the granule-contained serine proteases than ROS itself to kill pneumococci (Standish & Weiser, 2009). It is unclear, however, whether macrophage proteases also contribute to early pneumococcal killing. The gp91 phox-deficient mice also did not have a major defect in pneumococcal clearance in murine models of established pneumonia (Marriott et al., 2008). Similarly, in a murine model of preclinical infection in which macrophages are the major effector of bacterial clearance gp91 phox deficiency did not alter macrophage clearance of bacteria (Marriott, Hellewell, Whyte, & Dockrell, 2007). Consistent with the limited role of ROS alone in early bacterial killing incubation of alveolar macrophages with catalase to degrade hydrogen peroxide did not reduce bacterial killing (Biggar, Buron, & Holmes, 1976). In contrast, human MDM upregulate NOS2 and produce NO after challenge with *S. pneumoniae* and this contributes to bacterial killing (Marriott et al., 2004). NO is required for alveolar macrophages to kill bacteria in the low dose, macrophage-contained, murine models of lung infection (Marriott et al., 2007). However, these studies emphasise that although macrophages are important as first effectors of bacterial clearance they have a finite capacity to kill pneumococci and can be overwhelmed, requiring recruitment of neutrophils and other inflammatory cells to aid bacterial clearance when challenged with large doses of bacteria (Dockrell et al., 2003).

To deal with the challenge of ingested but viable bacteria macrophages utilise, an additional killing mechanism to complement the earlier phase of canonical killing in the phagolysosome. This involves induction of macrophage apoptosis, which is used as a strategy to enable killing at a later stage of the macrophage's interaction with pneumococci, when canonical phagolysosomal killing mechanisms are exhausted (Dockrell et al., 2001). Macrophage apoptosis requires intracellular bacteria to trigger the process and is directly proportional to the level of internalised bacteria (Ali et al., 2003). It is enhanced by opsonisation or by challenge with unencapsulated strains of pneumococci that are phagocytosed in greater numbers. It is dependent on live bacteria and requires pneumolysin to trigger the death pathway, which may reflect the ability of pneumolysin to activate the macrophage through TLR4 (Malley et al., 2003; Marriott et al., 2004; Srivastava et al., 2005). Although pneumolysin may induce apoptosis in other cell types (Bermpohl et al., 2005; Braun et al., 2002) the toxin alone fails to trigger macrophage apoptosis and inhibition of microbial killing reduces apoptosis, instead allowing cells to undergo a necrotic form of cell death (Marriott et al., 2004). This is in keeping with the induction of apoptosis representing a host-mediated response to bacteria rather than a pathogen-mediated means of immune subversion (Aberdein, Cole, Bewley, Marriott, & Dockrell, 2013). It is a phenotype of differentiated macrophages, potentially reflecting the greater need of these cells to complement canonical killing mechanisms (Ali et al., 2003).

The death program is consistent with a classical pathway of apoptosis with caspase induction and nuclear fragmentation (Ali et al., 2003). The cell membrane remains intact till the death program is advanced, indicating the process is not necroptosis (Marriott et al., 2004). Our observations (Bewley et al., 2014) show that the death process is caspase 1 independent and does not have features of pyroptosis. This is consistent with other data showing pneumococci do not induce pyroptosis in monocytes (Webster et al., 2010) and that pneumococci induce chitinase-3-like-1 protein to limit caspase 1 activation (Dela Cruz et al., 2012). It also fits with a view that pneumolysin may preferentially induce a necroptotic form of cell death through CYLD (Lim et al., 2007) and activation of RIP1 as discussed above. Thus, the bacterial subversion of innate immunity through induction of cell death may be via induction of necroptosis and the preferred host response to ensure an effective innate immune response appears to require apoptosis. Our observations that inhibition of microbial killing, or alternatively of steps in the apoptosis pathway, preferentially induce a necrotic form of death are consistent with this. Moreover inhibition of apoptosis in the low dose murine models of pneumococcal lung infection, where macrophages are able to control bacterial replication, illustrate that blocking this response results in increased bacterial replication in the lung and development of bacteraemia (Bewley, Marriott, et al., 2011; Marriott et al., 2005).

The pathway that leads to apoptosis induction stems from the phagolysosome as one would expect of a response to intraphagolysosomal bacteria. The phagolysosome becomes permeabilised and releases activated cathepsin D (Bewley, Marriott, et al., 2011). Cathespin D has a range of effects on the macrophage proteome. It both facilitates apoptosis induction and inhibits competing death processes (Bewley, Pham, et al., 2011). Key to its effect is downregulation of the anti-apoptotic Bcl-2 family member myeloid cell leukaemia sequence 1 protein (Mcl-1). Mcl-1 is very short-lived protein and its expression level regulates the macrophages susceptibility to apoptosis (Liu, Perlman, Pagliari, & Pope, 2001; Schubert & Duronio, 2001). Following macrophages ingestion of *S. pneumoniae* Mcl-1 is initially upregulated at a transcriptional level, to help maintain cell viability during increasing oxidative and nitrosative stress, but with time Mcl-1 is downregualted and a mitochondrial pathway of apoptosis is triggered (Marriott et al., 2005). The onset of Mcl-1 downregulation represents the molecular switch that initiates apoptosis. Cathepsin D activation and phagolysosomal membrane permeabilisation precede Mcl-1 downregualtion (Bewley, Marriott, et al., 2011). Cathepsin D induces the preferential interaction of Mcl-1 with its E3 ubiquitin ligase Mcl-1 ubiquitin ligase (Mule) and the resulting ubiquitination ensures its proteasomal degradation. In addition, cathepsin D downregualtes eukaryotic elongation factor 2 (eEF2; Bewley, Pham, et al., 2011). eEF2 is a critical component of the cells protein translation machinery and a necessary factor to ensure levels of a protein with such a short-lived half-life are maintained.

Once Mcl-1 is downregulated, a mitochondrial pathway of apoptosis is activated with mitochondrial outer membrane permeabilisation, release of cytochrome c and formation of the apoptososome, which results in caspase activation (Marriott et al., 2004). The death pathway does not appear to involve death receptor ligands such as Fas ligand and TNFα (Dockrell et al., 2001). Although the macrophage's own response occurs through an intrinsic pathway recent reports suggest that neutrophils could trigger TNFα-related apoptosis inducing ligand (TRAIL)-dependent apoptosis of macrophages (Steinwede et al., 2012). While this cannot explain the macrophage death process seen in the absence of neutrophils in low dose infection models in the lung or explain the killing of pneumococci by macrophages *in vitro* it remains theoretically possible that other immune cells could interact with macrophages to enhance the induction of apoptosis through this or other pathways. Ultimately the induction of apoptosis appears to mediate killing in a mechanism that requires NO and ROS. Since ROS does not appear to directly kill bacteria it is likely that the mediator of bacterial killing is the RNS produced by their reaction (Bewley, Marriott, et al., 2011). Macrophage apoptosis not only aids bacterial killing but raises the threshold at which neutrophil recruitment is required and helps downregulate the inflammatory response, in essence reducing the inflammatory cost of bacterial clearance (Bewley, Marriott, et al., 2011; Marriott et al., 2006).

12. MACROPHAGE KILLING OF *S. aureus*

In contrast to the situation with pneumococci, NADPH oxidase is required to allow clearance of *S. aureus*. Patients with chronic granulomatous disease who have mutations in their NOX2 are highly susceptible to this micro-organism (Lekstrom-Himes & Gallin, 2000; Reeves et al., 2002). Although alveolar macrophages appear to produce lower levels of ROS than other phagocytes and lack the granule-associated serine proteases (e.g., neutrophil elastase and cathespin G) of neutrophils they still require ROS to clear *S. aureus* (Nibbering, van den Barselaar, van de Gevel, Leijh, & Van Furth, 1989; Watanabe, Ichiki, Shiratsuchi, & Nakanishi, 2007). They do, however, demonstrate a reduced killing capacity compared to neutrophils in keeping with the lower levels or reduced potency of ROS they produce relative to neutrophils (Lee et al., 1984). Further support of the role of NOX2 in *S. aureus* control in mononuclear phagocytes comes from conditional expression in mononuclear phagocytes; macrophage specific reconstitution of NOX2 in p47phox-deficient mice (i.e., mice lacking a functional NOX2) made mice more resistant to the systemic features of *S. aureus* infection, while the mice that lacked NOX2 in all phagocytes were highly susceptible in an arthritis model, although the differences were not reflected in differences in bacterial clearance (Pizzolla et al., 2012). NOS2 also appears to help control *S. aureus*, although as stated its expression depends on appropriate levels of macrophage activation (Mantovani et al., 2004; Sakiniene, Bremell, & Tarkowski, 1997). One model of macrophage activation to upregulate NOS2 and increase NOS2-dependent macrophage killing of *S. aureus* involves IL-33 (Li et al., 2014). IL-33 was upregulated in patients with *S. aureus* skin infection and IL-33 neutralisation inhibited NOS2 expression and bacterial killing in a murine model of *S. aureus* skin infection. Histone deacetylase inhibitors modulated the expression of gp91phox and NOS2, with reduced bacterial clearance in murine macrophages challenged with *S. aureus* infection, emphasising that epigenetic regulation of these microbicidal strategies also plays a part in controlling the host response (Mombelli et al., 2011).

The matrix metalloprotease MMP-12 plays a role in early killing of *S. aureus* in the phagolysome of macrophages (McGarry Houghton et al., 2009). MMP-12 adheres to the bacterial cell wall and disrupts the bacterial cell membrane. It plays a role in protecting against *S. aureus* in murine models of pneumonia and peritonitis. Acidification of the phagolysosome and the resultant activation of lysozyme and serine, cysteine and acid proteases also contributes to bacterial killing (Ip et al., 2010). This is not just because of the direct effects of low pH and proteases discussed above but also because the digestion of bacteria releases microbial factors that trigger TLR-mediated signalling. This then results in the production of pro-inflammatory cytokines that amongst other effects will also help prime the macrophages microbicidal effects. Cathepsin L has recently been identified as contributing to non-oxidative killing of *S. aureus* in macrophages (Muller et al., 2014). In addition, cathepsin K emerged as an important factor in priming certain MyD88-dependent TLR responses that resulted in IL-6 expression following *S. aureus* phagocytosis. This illustrates that phagolysosomal proteases' roles extend beyond direct killing, but through release of microbial factors that can activate pattern recognition receptor-mediated signalling can extend to indirect effects on killing though macrophage activation and pro-inflammatory cytokine production, which may have paracrine effects on other macrophages. These responses must, however, be tightly regulated. Under some circumstances TLR signalling can actually reduce killing of *S. aureus*. In one study, TLR2-mediated activation of c-Jun N-terminal kinases (JNK), a member of the mitogen-activated protein kinase (MAPK) family, actually inhibited ROS production, and *S. aureus* killing in murine macrophages (Watanabe et al., 2007). Once again, this emphasises, the control of microbicidal killing must be tightly regulated with both amplification and inhibitory circuits needing to be optimally controlled.

S. aureus appears to be capable of escaping from the phagolysosome, a feature more reminiscent of intracellular bacteria associated with long term survival in macrophages (Kubica et al., 2008). This will enable the pathogen to escape from the compartment in which it should be killed and in which its degradation products should activate pattern recognition receptors (Ip et al., 2010). Escape from the phagosome was dependent on expression of the accessory gene regulator (agr) and on α-hemolysin (Hla) whose expression is dependent on agr (Kubica et al., 2008). Intriguingly, α-hemolysin had previously been implicated in *S. aureus* escape from the phagosomes of a cystic fibrosis epithelial cell line (Jarry, Memmi, & Cheung, 2008). PSM α also helped MRSA and methicillin sensitive strains escape the phagosomes of a monocytic cell line (Grosz et al., 2014). Of note MDMs challenged with *S. aureus* did not activate a program of apoptosis and showed sustained upregulation of Mcl-1 and a further anti-apoptotic factor Bcl-2 (Koziel et al., 2009). We have also observed that *S. aureus* fails to activate cathepsin D a key factor required to make the macrophage susceptible to apoptosis

(Bewley, Marriott, et al., 2011; Bewley, Pham, et al., 2011). Taken together these findings suggest that *S. aureus* may circumvent the macrophages ability to engage apoptosis-associated killing. Since this is a critical mechanism to remove viable intracellular bacteria and prevent chronic intracellular infection, this may allow *S. aureus* to survive intracellularly in macrophages for extended time periods.

In light of all the barriers, *S. aureus* presents to intracellular killing macrophages may have to engage other killing processes. One proposed mechanism is that macrophages could release extracellular traps, analogous to the NETs (Brinkmann et al., 2004). Murine macrophages including thioglycolate elicited peritoneal macrophages were found to release extracellular traps in response to exposure to the cholesterol lowering agent mevastatin and these statin-induced extracellular traps were released in a murine model of *S. aureus* pneumonia (Chow et al., 2010). These traps, which the authors proposed were produced by neutrophils and macrophages, were enriched with the murine antimicrobial peptide cathelicidin-related antimicrobial peptide (CRAMP) and the authors suggested these could explain some of the potential beneficial effects of stains during severe bacterial infection. However, recent reports suggest *S. aureus* may also have adaptations against extracellular traps (Thammavongsa, Missiakas, & Schneewind, 2013). *S. aureus* was found to secrete the enzymes nuclease and adenosine synthase (AdsA) to degrade NETs and to release deoxyadnosine, which in turn resulted in the caspase 3-mediated death of macrophages. While this work refers to degradation of NETs and it remains to be established how the putative extracellular traps produced by macrophages differ from these, observations suggest *S. aureus* may have evolved strategies for subverting this aspect of host defence and in turn use it as another target to disarm the innate immune response through induction of macrophage apoptosis. In this example, macrophages are restricted from the immune cell cuff that surrounds *S. aureus* abscesses by their rapid removal through apoptosis induced by the degradation of NETs.

13. MACROPHAGE ORCHESTRATION OF THE INFLAMMATORY RESPONSE

As discussed, macrophages are the resident phagocytes of tissue and are the first phagocytes to encounter bacteria that reach sterile locations such as the alveolus. Their ability to phagocytose and kill bacteria means they can clear bacteria efficiently, but as we have discussed their capacity to clear bacteria is finite. With large inocula or more virulent bacteria their capacity to clear bacteria on their own becomes overwhelmed and their role shifts to that of the orchestrators of a co-ordinated inflammatory response involving recruitment of neutrophils and other inflammatory cells (Dockrell et al., 2003). Macrophages form part of a cellular network to mediate these responses, often inducing their effects indirectly by releasing cytokines that act on epithelial cells or other cells which facilitate chemokine release and inflammatory cell recruitment (Krakauer, 2002; Morris et al., 2005, 2006; Standiford, Kunkel, Phan, Rollins, & Strieter, 1991). An example of this is seen during pneumococcal infection where macrophages produce relatively modest levels of the neutrophil chemoattractant CXCL8 but still make an important contribution by releasing IL-1β, which stimulates CXL8 production by epithelial cells and IL-1-dependent CXC chemokine-mediated neutrophil recruitment in murine models of pneumonia (Marriott et al., 2011). Macrophage depletion in lung explants also confirmed an important role for macrophages in the cellular network that produces pro-inflammatory cytokines in response to *S. pneumoniae* (Xu et al., 2008). Host defence against pneumococci in the murine lung requires an early cytokine response involving IL-1 and TNFα, which is required for NF-κB activation in epithelial cells and neutrophil recruitment (Quinton et al., 2007). Macrophage derived cytokines are an important contributor to the NF-κB activation in epithelial cells (Lentsch, Czermak, Bless, Van Rooijen, & Ward, 1999).

In other settings, macrophages themselves may be important sources of neutrophil chemoattractants; perivascular macrophages play a key role in recruitment of neutrophils to the skin during *S. aureus* infection and α-hemolysin-induced macrophage lysis may help the pathogen subvert this host response (Abtin et al., 2014). It is also important that macrophages ensure rapid development of an early inflammatory response because if bacteria are cleared when bacterial burdens are lower the inflammatory response required may not need to be as great and therefore the potential for unwanted tissue injury induced by the inflammatory cells may be less (Dockrell et al., 2012). At later stages, macrophages also must play an important role in downregulating the inflammatory response and ensuring clearance of effete inflammatory cells, through induction of apoptosis in neutrophils but also in monocytes and lymphocytes (Daigneault et al., 2012; Dockrell et al., 2001). The ingestion of these apoptotic cells by efferocytosis not only clears the cells but also helps re-orientate the cytokine network to a less inflammatory profile (Marriott et al., 2006). Monocytes and macrophages may also play important roles in ensuring the death process is apoptosis. For example, pneumolysin can preferentially induce a necrotic cell death in lymphocytes but monocytes/macrophages can ensure lymphocytes undergo death by apoptosis rather than necrosis (Daigneault et al., 2012). During pneumococcal infection, this monocyte/macrophage driven process is ensured by a Fas ligand-mediated program of apoptosis that can remove highly activated cells and prevent their death by necrosis (Daigneault et al., 2012; Marriott et al., 2012). This helps reduce inflammatory injury and ensures better outcomes in murine pneumonia models.

To ensure macrophages respond appropriately to bacteria, they have developed a series of overlapping recognition systems for extracellular and intracellular microbial recognition. These systems of pattern recognition receptors include the TLR family, the nucleotide-binding oligomerisation domain (Nod)-like receptor (NLR) family and the retinoid acid inducible gene-1 system (Chen, Shaw, Kim, & Nunez, 2009; Sabroe et al., 2003). The latter are primarily involved in viral recognition. The TLR family processes an extramembrane domain involved in ligand engagement, either on the extracellular surface or for some TLR in the endosome, and a cytoplasmic signalling domain. TLRs can signal via their Toll–IL-1R (TIR) domain and the adaptor protein MyD88 and TIR containing adaptor protein (TIRAP) to activate NF-κB and MAPKs, resulting in cytokine expression (Sabroe et al., 2003). For many TLRs, the signalling complex involves IL-1R-associated kinases and or TRAF-6. MyD88-independent signalling from some TLR including TLR4, TLR3, and TLR9 results in activation of IRF3 and generation of type 1 interferons. Toll/IL-1R domain-containing adaptor-inducing interferon beta (TRIF) is a key adaptor in TLR-mediated interferon signalling. In addition to cytokine production, TLR signalling activates antimicrobial host defence involving both NO-dependent and -independent responses (Thoma-Uszynski et al., 2001). TLR will often co-operate to mediate their effects, for example TLR2 and TLR4 synergise to maximise their impact (Sato et al., 2000). Whole bacteria will therefore increase signalling by activating multiple pattern recognition receptors.

There are at least 22 human NLRs but many more have been described in some primitive organisms so it is likely additional human NLR will be identified. The NLR proteins Nod1 (NLRC1) recognises D-γ-glutyamyl-meso-DAP (iE-DAP), a constituent of most Gram-negative and select Gram-positive bacteria (e.g., *L. monocytogenes*) and Nod2 (NLRC2) binds muramyl dipeptide, present in the peptidoglycans of Gram-positive and -negative bacteria, (Correa, Milutinovic, & Reed,

2012). Nods signal via MAPK, NF-κB, and IRFs to induce cytokine responses but have also been implicated in the regulation of autophagy and apoptosis (Chen et al., 2009; Correa et al., 2012).

Certain NLR (e.g., NLRP1, NLRP3, and NLRC4) recruit an adaptor molecule, apoptosis associated speck-like protein containing a caspase recruitment domain (ASC) to activate caspase-1 (Chen et al., 2009). These NLR form a multi-molecular complex called the inflammasome, which is required to process IL-1 family members to allow their release. In macrophages, IL-1 processing is not constitutive, unlike in monocytes, rather it requires an additional signal after TLR activation. It has been demonstrated that this second signal is provided by adenosine triphosphate (ATP), which results in activation of the macrophage inflammasome (Netea et al., 2009). The requirement for ATP has been suggested to be because its engagement with the $P2X_7$ receptor results in opening of a pore by Pannexin-1 (Pelegrin & Surprenant, 2006). It is suggested that this can facilitate translocation of bacterial products into the cytosol where they can interact with NLR. Bacterial pore forming toxins and secretion systems may also facilitate this translocation (Chen et al., 2009). The formation of pores may have an additional consequence required for activation of NLRP3. These pores also allow potassium efflux, which is also required to activate the NLRP3 inflammasome (Franchi, Kanneganti, Dubyak, & Nunez, 2007; Petrilli et al., 2007). Caspase 1 generation from the NLRP3 inflammasome not only plays a role in release of IL-1 family cytokines but also results in induction of pyroptosis a form of programmed cell death involving early rupture of the cell membrane, a process distinct from apoptosis that induces more tissue inflammation (Chen et al., 2009). This form of death has been particularly implicated in responses to several intracellular bacteria including *Salmonella enterica* serovar *typhimurium* sp., *Shigella fexneri*, *Yesrsinia pseudotuberculosis*, *Fransicella tularensis*, *Legionella pneumophila*, and *Bacillus anthracis*.

14. PATTERN RECOGNITION RECEPTORS IN THE RECOGNITION OF *S. pneumoniae* AND *S. aureus*

S. pneumoniae activates a range of TLR, Nod 2, and some other NLRs and the spectrum of receptors activated in *S. aureus* infection is similar (Dockrell et al., 2012). The importance of TLR signalling pathways is illustrated by the fact that MyD88 or IRAK 4 deficiency contributes to susceptibility to infection with both organisms (Picard et al., 2003; Von Bernuth et al., 2008), while a TIRAP polymorphism was associated with susceptibility to invasive pneumococcal disease (Khor et al., 2007). In addition, mutations in NF-κB essential modulator/IkappaB kinase gamma (NEMO), a factor required for NF-κB activation (Ku et al., 2007; Smahi et al., 2000) or in IkappaB an inhibitor of NF-κB activation were associated with invasive pneumococcal disease (Chapman et al., 2007).

In murine models, TLR2 deficiency did not influence overall bacteria clearance but reduced expression of some cytokines (Knapp et al., 2004). TLR4-deficient mice had worse outcomes with pneumococcal infection in another study (Malley et al., 2003), but a further study suggested TLR4 only influenced responses to low doses of pneumococci (Branger et al., 2004). TLR9 is the receptor for the unmethylated CpG dinucleotides found in prokaryotic DNA and TLR9 deficiency was associated with a greater defect in host defence against pneumococci than either TLR2 or TLR4 deficiency (Albiger et al., 2007). Overall though, the effects of single TLR knock-outs are modest and restricted to early stages of the host response (Albiger et al., 2007; Branger et al., 2004; Knapp et al., 2004). The combination of TLR9 with either TLR2 or TLR4 deficiency or the deletion of the shared adaptor MyD88 is associated with greater defects (Lee, Scanga, Bachelder, Chen, & Snapper, 2007), but there is clearly considerable redundancy in TLR responses. Nod1 and Nod 2 were upregulated by pneumococci and Nod2 was required for NF-κB activation in a further study (Opitz et al., 2004). The importance of this pathway was illustrated by a model in which lysozyme M-induced peptidoglycan release, and digestion into Nod2 ligands, from internalised pneumococci in macrophages. The release of Nod2 ligands resulted in release of the macrophage chemokine CCL2, which facilitated macrophage recruitment to the upper airway during pneumococcal colonisation in mice (Davis, Nakamura, & Weiser, 2011). This pathway also required the pore forming toxin pneumolysin. TLR2-dependent responses also played a role in the clearance of colonising pneumococci in mice previously naïve to colonisation, through the induction of protective Th17 responses and recruitment of macrophages which cleared bacteria (Zhang et al., 2009). Bacterial DNA is also recognised by the cytoplasmic pattern recognition receptor DNA-dependent activator of interferon regulatory factors and this may contribute further to macrophage recognition of pneumococci (Parker et al., 2011).

Pneumolysin activates TLR4 (Malley et al., 2003). However, pneumolysin has been shown to stimulate TLR-independent responses mediated by NLRP3 that synergise with TLR-dependent responses to stimulate cytokine responses (McNeela et al., 2010). Endotoxin free pneumolysin did not stimulate cytokine production on its own, emphasising the need for stimulation of both TLR and NLRP3. In their model, pneumolysin-induced pore formation and lysosomal permeabilisation. Cathepsin B activation and potassium efflux contributed to inflammasome activation and IL-1β release. This response was required for optimal bacterial clearance in the mouse lung since NLRP3-deficient mice had increased bacterial colony counts. These findings have been extended by another group who suggested that the success of some epidemic strains of *S. pneumoniae* (e.g., serotype 1 strain 4496 (ST3018, lineage C)) arises because they express pneumolysin of reduced pore forming activity (Harvey, Ogunniyi, Chen, & Paton, 2011; Witzenrath et al., 2011). In murine models, strains of reduced haemolytic potential were more likely to invade and cause significant bacteraemia, although a potential consequence of their lower induction of IL-1-dependent responses may then be that they are associated with lower mortality. This again emphasises how a vigorous innate response can favour bacterial clearance but at the cost of inflammation-induced tissue injury. Other NLR may also contribute to the recognition of pneumolysin, including the absent in melanoma 2 NLR (Fang et al., 2011).

Recognition of *S. aureus* shows many similarities. Although TLR2, TLR2/4, or TLR9 deficiency did not modify the response of peritoneal macrophages to *S. aureus* there was a reduced response in MyD88-deficient cells with reduced expression of cytokines such as TNFα (Kapetanovic et al., 2007). The same group also found a role for Nod2 in *S. aureus* recognition, though this response was seen in monocytes and peritoneal macrophages but not in alveolar macrophages (Kapetanovic et al., 2011). Degradation of *S. aureus* in the phagolysosome by lysozyme is required for optimal TLR stimulation (Wolf et al., 2011). Initial macrophage stimulation involved cell surface TLR2 engagement while the response was amplified by TLR2 and TLR9 responses in association with phagolysosomal degradation of bacteria, which enhanced responsiveness to small numbers of bacteria. The importance of TLR or Nod to *S. aureus*-induced pro-inflammatory cytokine production is also supported by the observation that IRAK4-deficient macrophages have defective cytokine responses (Pennini, Perkins, Salazar, Lipsky, & Vogel, 2013). *S. aureus* also produces factors which inhibit the ability of the macrophage to respond to *S. aureus* through TLR signalling; staphylococcal superantigen-like protein 3 bound to TLR2 and inhibited pro-inflammatory cytokine responses to heat-killed bacteria or lipopeptides (Yokoyama et al., 2012). They also can engage the inhibitory pattern recognition receptor paired immunoglobulin-like receptor B (PIR-B) to suppress extracellular signal-regulated kinase (ERK) 1/2 activation, generation of IL-1β or IL-6 and inflammasome activation via lipoteichoic acid binding to PIR-B (Nakayama et al., 2012). Finally, *S. aureus* may subvert macrophage polarisation to compromise the inflammatory response and resulting innate responses; *S. aureus* biofilms can drive M2 polarisation and reduce the macrophage's capacity to control biofilm growth (Hanke, Heim, Angle, Sanderson, & Kielian, 2013).

However, inflammatory responses need tight control. While, as discussed above, the NLRP3 inflammasome was found to aid the host response against *S. pneumoniae* it was found to enhance pathogenicity in *S. aureus* pneumonia. α-Hemolysin activated the NLRP3 inflammasome causing IL-1β release, although its activation had no effect on bacterial clearance from the murine lung. Instead, inflammasome activation was associated with induction of necroptosis and lung injury (Kebaier et al., 2012). Although α hemolysin stimulates a range of caspase 1-dependent outputs, such as IL-1β release from macrophages in a pathway dependent on potassium efflux (Walev, Reske, Palmer, Valeva, & Bhakdi, 1995), it appeared that necroptosis was not the result of IL-1β production or a direct effect of α hemolysin, rather the consequence of inappropriate caspase 1 activation. This once more influences that host responses must be tightly regulated to prevent harmful effects. PVL also activates the NLRP3 inflammasome in macrophages enhancing IL-1β release and inducing necroptosis in a process also dependent on potassium efflux and cathepsin B activation (Holzinger et al., 2012). Increased inflammasome activation and autophagy are reciprocally regulated during optimal functioning of innate immunity (Rodgers, Bowman, Liang, & Jung, 2014). Inflammasome activation and IL-1 signalling may contribute to a defect in autophagy, in select settings such as during defective NADPH oxidase signalling (De Luca et al., 2014). Therefore, excessive activation of this pathway may have other consequences for host defence against *S. aureus*, such as through inhibiting autophagy. Whether this is the case in macrophages during *S. aureus* infection remains to be established.

On the one hand, the inflammatory response produced by pattern recognition receptors is designed to protect the host but on the other it has the potential to cause tissue injury and contribute to pathogenesis, particularly when subverted by pathogen virulence factors. It therefore becomes apparent that the host must carefully regulate the response and an ideal response is initiated early so that pathogen clearance can be achieved with lower levels of an inflammatory response to avoid unwanted tissue damage. If this could be achieved therapeutically, this might enhance outcomes during infection. One way to achieve this could be to maximise the macrophage's bacterial clearance capacity. Mice lacking Phosphatase and tensin homolog (PTEN) demonstrated enhanced PI3K signalling and hence macrophage phagocytosis and clearance of pneumococci in the mouse lung, with reduced neutrophilic inflammation and improved survival (Schabbauer et al., 2010). Engagement of triggering receptor expressed on myeloid cells-1 (TREM-1) amplifies early inflammatory responses but accelerates their resolution in models of pneumococcal pneumonia, in part through upregulation of IRAK-M, which forms part of an inhibitory regulatory loop downregulating inflammatory responses (Lagler et al., 2009). Increased understanding of how the host regulates key responses can provide insights into pathways that could be exploited. The surface receptor CD353 (Slamf8) appears to negatively regulate NOX2 in macrophages during *S. aureus* infection (Wang et al., 2012), while as discussed above caspase 1 can also downregulate NOX2 (Sokolovska et al., 2013). In contrast, the observations that in CGD patients, absence of NOX2 function and unopposed inflammasome function combined with IL-1β production leads to a defect in autophagy, which can be corrected by antagonism of IL-1β, are intriguing. This raises questions about how selective resetting of inflammatory responses could be achieved in other settings to improve the response to these and other infections (De Luca et al., 2014).

15. CONCLUSION

S. pneumoniae and *S. aureus* possess multiple attributes which challenge the macrophage's ability to deliver effective innate immunity. Fine-tuning the host responses requires optimisation of bacterial phagocytosis and killing. Strategies to enhance these responses in macrophages can help improve disease outcomes. In addition, targeting the macrophages role in pathogen recognition and co-ordination of the inflammatory response holds promise as an approach to control dysfunctional inflammatory responses. Investigating approaches that refocus the host response to these pathogens merits attention in order to ensure better therapeutic responses and focusing on the macrophages role in the host–pathogen interaction is likely to provide an informative approach.

CONFLICT OF INTEREST

The authors have no conflicts of interest to declare.

REFERENCES

Aberdein, J. D., Cole, J., Bewley, M. A., Marriott, H. M., & Dockrell, D. H. (2013). Alveolar macrophages in pulmonary host defence the unrecognized role of apoptosis as a mechanism of intracellular bacterial killing. *Clinical and Experimental Immunology*, *174*(2), 193–202. http://dx.doi.org/10.1111/cei.12170.

Abramson, S. L., & Gallin, J. I. (1990). IL-4 inhibits superoxide production by human mononuclear phagocytes. *The Journal of Immunology*, *144*(2), 625–630.

Abtin, A., Jain, R., Mitchell, A. J., Roediger, B., Brzoska, A. J., Tikoo, S., et al. (2014). Perivascular macrophages mediate neutrophil recruitment during bacterial skin infection. *Nature Immunology*, *15*(1), 45–53. http://dx.doi.org/10.1038/ni.2769.

Aderem, A., & Underhill, D. M. (1999). Mechanisms of phagocytosis in macrophages. *Annual Review of Immunology*, *17*, 593–623. http://dx.doi.org/10.1146/annurev.immunol.17.1.593.

Ajami, B., Bennett, J. L., Krieger, C., Mcnagny, K. M., & Rossi, F. M. V. (2011). Infiltrating monocytes trigger EAE progression, but do not contribute to the resident microglia pool. *Nature Neuroscience*, *14*(9), 1142–1149. http://dx.doi.org/10.1038/nn.2887.

Albiger, B., Dahlberg, S., Sandgren, A., Wartha, F., Beiter, K., Katsuragi, H., et al. (2007). Toll-like receptor 9 acts at an early stage in host defence against pneumococcal infection. *Cellular Microbiology*, *9*(3), 633–644. http://dx.doi.org/10.1111/j.1462-5822.2006.00814.x.

Ali, F., Lee, M. E., Iannelli, F., Pozzi, G., Mitchell, T. J., Read, R. C., et al. (2003). Streptococcus pneumoniae-associated human macrophage apoptosis after bacterial internalization via complement and Fcgamma receptors correlates with intracellular bacterial load. *The Journal of Infectious Diseases*, *188*(8), 1119–1131. http://dx.doi.org/10.1086/378675.

Allavena, P., Chieppa, M., Monti, P., & Piemonti, L. (2004). From pattern recognition receptor to regulator of homeostasis: The double-faced macrophage mannose receptor. *Critical Reviews in Immunology*, *24*(3), 179–192.

Alonzo, F., 3rd., Benson, M. A., Chen, J., Novick, R. P., Shopsin, B., & Torres, V. J. (2012). Staphylococcus aureus leucocidin ED contributes to systemic infection by targeting neutrophils and promoting bacterial growth in vivo. *Molecular Microbiology*, *83*(2), 423–435. http://dx.doi.org/10.1111/j.1365-2958.2011.07942.x.

Alonzo, F., 3rd., Kozhaya, L., Rawlings, S. A., Reyes-Robles, T., Dumont, A. L., Myszka, D. G., et al. (2013). CCR5 is a receptor for Staphylococcus aureus leukotoxin ED. *Nature*, *493*(7430), 51–55. http://dx.doi.org/10.1038/nature11724.

Alpuche-Aranda, C. M., Racoosin, E. L., Swanson, J. A., & Miller, S. I. (1994). Salmonella stimulate macrophage macropinocytosis and persist within spacious phagosomes. *The Journal of Experimental Medicine*, *179*(2), 601–608.

Ambarus, C., Krausz, S., Van Eijk, M., Hamann, J., Radstake, T., Reedquist, K., et al. (2012). Systematic validation of specific phenotypic markers for in vitro polarized human macrophages. *Journal of Immunological Methods*, *375*(1), 196–206.

Ambrose, C. T. (2006). The Osler slide, a demonstration of phagocytosis from 1876: Reports of phagocytosis before Metchnikoff's 1880 paper. *Cellular Immunology*, *240*(1), 1–4. http://dx.doi.org/10.1016/j.cellimm.2006.05.008.

Anderson, C. F., & Mosser, D. M. (2002). A novel phenotype for an activated macrophage: The type 2 activated macrophage. *Journal of Leukocyte Biology*, *72*(1), 101–106.

Andisi, V. F., Hinojosa, C. A., De Jong, A., Kuipers, O. P., Orihuela, C. J., & Bijlsma, J. J. (2012). Pneumococcal gene complex involved in resistance to extracellular oxidative stress. *Infection and Immunity*, *80*(3), 1037–1049. http://dx.doi.org/10.1128/IAI.05563-11.

Antonini, J. M., Lawryk, N. J., Murthy, G. G., & Brain, J. D. (1999). Effect of welding fume solubility on lung macrophage viability and function in vitro. *Journal of Toxicology and Environmental Health Part A*, *58*(6), 343–363.

Antonini, J. M., Leonard, S. S., Roberts, J. R., Solano-Lopez, C., Young, S. H., Shi, X., et al. (2005). Effect of stainless steel manual metal arc welding fume on free radical production, DNA damage, and apoptosis induction. *Molecular and Cellular Biochemistry*, *279*(1–2), 17–23. http://dx.doi.org/10.1007/s11010-005-8211-6.
Aronoff, D. M., Canetti, C., & Peters-Golden, M. (2004). Prostaglandin E2 inhibits alveolar macrophage phagocytosis through an E-prostanoid 2 receptor-mediated increase in intracellular cyclic AMP. *Journal of Immunology*, *173*(1), 559–565.
Aronoff, D. M., Canetti, C., Serezani, C. H., Luo, M., & Peters-Golden, M. (2005). Cutting edge: Macrophage inhibition by cyclic AMP (cAMP): Differential roles of protein kinase A and exchange protein directly activated by cAMP-1. *Journal of Immunology*, *174*(2), 595–599.
Arredouani, M. S., Palecanda, A., Koziel, H., Huang, Y. C., Imrich, A., Sulahian, T. H., et al. (2005). MARCO is the major binding receptor for unopsonized particles and bacteria on human alveolar macrophages. *Journal of Immunology*, *175*(9), 6058–6064.
Arredouani, M. S., Yang, Z., Imrich, A., Ning, Y., Qin, G., & Kobzik, L. (2006). The macrophage scavenger receptor SR-AI/II and lung defense against pneumococci and particles. *American Journal of Respiratory Cell and Molecular Biology*, *35*(4), 474–478. http://dx.doi.org/10.1165/rcmb.2006-0128OC.
Arredouani, M., Yang, Z., Ning, Y., Qin, G., Soininen, R., Tryggvason, K., et al. (2004). The scavenger receptor MARCO is required for lung defense against pneumococcal pneumonia and inhaled particles. *The Journal of Experimental Medicine*, *200*(2), 267–272. http://dx.doi.org/10.1084/jem.20040731.
Aschoff, L. (1924). Das reticulo-endotheliale system. In F. Kraus, E. Meyer, O. Minkowski, F. Müller, H. Sahli, A. Schittenhelm, A. Czerny, O. Heubner, & L. Langstein (Eds.), *Ergebnisse der Inneren Medizin und Kinderheilkunde* (pp. 1–118). Berlin: Springer.
Baxt, L. A., Garza-Mayers, A. C., & Goldberg, M. B. (2013). Bacterial subversion of host innate immune pathways. *Science*, *340*(6133), 697–701. http://dx.doi.org/10.1126/science.1235771.
Beiter, K., Wartha, F., Albiger, B., Normark, S., Zychlinsky, A., & Henriques-Normark, B. (2006). An endonuclease allows Streptococcus pneumoniae to escape from neutrophil extracellular traps. *Current Biology*, *16*(4), 401–407. http://dx.doi.org/10.1016/j.cub.2006.01.056.
Benoit, M., Desnues, B., & Mege, J. L. (2008). Macrophage polarization in bacterial infections. *The Journal of Immunology*, *181*(6), 3733–3739.
Benton, K. A., Everson, M. P., & Briles, D. E. (1995). A pneumolysin-negative mutant of Streptococcus pneumoniae causes chronic bacteremia rather than acute sepsis in mice. *Infection and Immunity*, *63*(2), 448–455.
Berger, S. L. (2007). The complex language of chromatin regulation during transcription. *Nature*, *447*(7143), 407–412.
Bermpohl, D., Halle, A., Freyer, D., Dagand, E., Braun, J. S., Bechmann, I., et al. (2005). Bacterial programmed cell death of cerebral endothelial cells involves dual death pathways. *The Journal of Clinical Investigation*, *115*(6), 1607–1615. http://dx.doi.org/10.1172/JCI23223.
Bestebroer, J., Poppelier, M. J., Ulfman, L. H., Lenting, P. J., Denis, C. V., Van Kessel, K. P., et al. (2007). Staphylococcal superantigen-like 5 binds PSGL-1 and inhibits P-selectin-mediated neutrophil rolling. *Blood*, *109*(7), 2936–2943. http://dx.doi.org/10.1182/blood-2006-06-015461.
Bewley, M. A., Marriott, H. M., Tulone, C., Francis, S. E., Mitchell, T. J., Read, R. C., et al. (2011). A cardinal role for cathepsin d in co-ordinating the host-mediated apoptosis of macrophages and killing of pneumococci. *PLoS Pathogens*, 7(1), e1001262. http://dx.doi.org/10.1371/journal.ppat.1001262.
Bewley, M. A., Naughton, M., Preston, J., Mitchell, A., Holmes, A., Marriott, H. M., et al. (2014). Pneumolysin activates macrophage lysosomal membrane permeabilization and

executes apoptosis by distinct mechanisms without membrane pore formation. *MBio*, in press.

Bewley, M. A., Pham, T. K., Marriott, H. M., Noirel, J., Chu, H. P., Ow, S. Y., et al. (2011). Proteomic evaluation and validation of cathepsin D regulated proteins in macrophages exposed to Streptococcus pneumoniae. *Molecular & Cellular Proteomics*. *10*(6). http://dx.doi.org/10.1074/mcp.M111.008193, M111 008193.

Biggar, W. D., Buron, S., & Holmes, B. (1976). Bactericidal mechanisms in rabbit alveolar macrophages: Evidence against peroxidase and hydrogen peroxide bactericidal mechanisms. *Infection and Immunity*, *14*(1), 6–10.

Bistoni, F., Vecchiarelli, A., Cenci, E., Puccetti, P., Marconi, P., & Cassone, A. (1986). Evidence for macrophage-mediated protection against lethal Candida albicans infection. *Infection and Immunity*, *51*(2), 668–674.

Blanchet, C., Jouvion, G., Fitting, C., Cavaillon, J. M., & Adib-Conquy, M. (2014). Protective or deleterious role of scavenger receptors SR-A and CD36 on host resistance to Staphylococcus aureus depends on the site of infection. *PLoS One*, *9*(1), e87927. http://dx.doi.org/10.1371/journal.pone.0087927.

Bogaert, D., De Groot, R., & Hermans, P. W. (2004). Streptococcus pneumoniae colonisation: The key to pneumococcal disease. *The Lancet Infectious Diseases*, *4*(3), 144–154. http://dx.doi.org/10.1016/S1473-3099(04)00938-7.

Bosedasgupta, S., & Pieters, J. (2014). Inflammatory stimuli reprogram macrophage phagocytosis to macropinocytosis for the rapid elimination of pathogens. *PLoS Pathogens*. *10*(1). http://dx.doi.org/10.1371/journal.ppat.1003879.

Branger, J., Knapp, S., Weijer, S., Leemans, J. C., Pater, J. M., Speelman, P., et al. (2004). Role of Toll-like receptor 4 in gram-positive and gram-negative pneumonia in mice. *Infection and Immunity*, *72*(2), 788–794.

Braun, J. S., Sublett, J. E., Freyer, D., Mitchell, T. J., Cleveland, J. L., Tuomanen, E. I., et al. (2002). Pneumococcal pneumolysin and H(2)O(2) mediate brain cell apoptosis during meningitis. *The Journal of Clinical Investigation*, *109*(1), 19–27. http://dx.doi.org/10.1172/JCI12035.

Brinkmann, V., Reichard, U., Goosmann, C., Fauler, B., Uhlemann, Y., Weiss, D. S., et al. (2004). Neutrophil extracellular traps kill bacteria. *Science*, *303*(5663), 1532–1535. http://dx.doi.org/10.1126/science.1092385.

Brouwer, M. C., De Gans, J., Heckenberg, S. G., Zwinderman, A. H., Van Der Poll, T., & Van De Beek, D. (2009). Host genetic susceptibility to pneumococcal and meningococcal disease: A systematic review and meta-analysis. *The Lancet Infectious Diseases*, *9*(1), 31–44. http://dx.doi.org/10.1016/S1473-3099(08)70261-5.

Brown, K. E., Brunt, E. M., & Heinecke, J. W. (2001). Immunohistochemical detection of myeloperoxidase and its oxidation products in Kupffer cells of human liver. *The American Journal of Pathology*, *159*(6), 2081–2088. http://dx.doi.org/10.1016/S0002-9440(10)63059-3.

Brown, J. S., Hussell, T., Gilliland, S. M., Holden, D. W., Paton, J. C., Ehrenstein, M. R., et al. (2002). The classical pathway is the dominant complement pathway required for innate immunity to Streptococcus pneumoniae infection in mice. *Proceedings of the National Academy of Sciences of the United States of America*, *99*(26), 16969–16974. http://dx.doi.org/10.1073/pnas.012669199.

Bruhns, P., Iannascoli, B., England, P., Mancardi, D. A., Fernandez, N., Jorieux, S., et al. (2009). Specificity and affinity of human Fcgamma receptors and their polymorphic variants for human IgG subclasses. *Blood*, *113*(16), 3716–3725. http://dx.doi.org/10.1182/blood-2008-09-179754.

Bubeck Wardenburg, J., Bae, T., Otto, M., Deleo, F. R., & Schneewind, O. (2007). Poring over pores: Alpha-hemolysin and Panton–Valentine leukocidin in Staphylococcus aureus pneumonia. *Nature Medicine*, *13*(12), 1405–1406. http://dx.doi.org/10.1038/nm1207-1405.

Bucci, C., Parton, R. G., Mather, I. H., Stunnenberg, H., Simons, K., Hoflack, B., et al. (1992). The small GTPase rab5 functions as a regulatory factor in the early endocytic pathway. *Cell*, *70*(5), 715–728.

Buchmeier, N. A., & Schreiber, R. D. (1985). Requirement of endogenous interferon-gamma production for resolution of Listeria monocytogenes infection. *Proceedings of the National Academy of Sciences*, *82*(21), 7404–7408.

Burnett, D., Crocker, J., & Stockley, R. A. (1983). Cathepsin B-like cysteine proteinase activity in sputum and immunohistologic identification of cathepsin B in alveolar macrophages. *The American Review of Respiratory Disease*, *128*(5), 915–919.

Canetti, C., Aronoff, D. M., Choe, M., Flamand, N., Wettlaufer, S., Toews, G. B., et al. (2006). Differential regulation by leukotrienes and calcium of Fc gamma receptor-induced phagocytosis and Syk activation in dendritic cells versus macrophages. *Journal of Leukocyte Biology*, *79*(6), 1234–1241. http://dx.doi.org/10.1189/jlb.0705374.

Cao, H., Wolff, R. G., Meltzer, M. S., & Crawford, R. M. (1989). Differential regulation of class II MHC determinants on macrophages by IFN-gamma and IL-4. *The Journal of Immunology*, *143*(11), 3524–3531.

Cellier, M. F., Courville, P., & Campion, C. (2007). Nramp1 phagocyte intracellular metal withdrawal defense. *Microbes and Infection*, *9*(14–15), 1662–1670. http://dx.doi.org/10.1016/j.micinf.2007.09.006.

Chambers, H. F., & Deleo, F. R. (2009). Waves of resistance: Staphylococcus aureus in the antibiotic era. *Nature Reviews Microbiology*, 7(9), 629–641. http://dx.doi.org/10.1038/nrmicro2200.

Chang, S., Sievert, D. M., Hageman, J. C., Boulton, M. L., Tenover, F. C., Downes, F. P., et al. (2003). Infection with vancomycin-resistant Staphylococcus aureus containing the vanA resistance gene. *The New England Journal of Medicine*, *348*(14), 1342–1347. http://dx.doi.org/10.1056/NEJMoa025025.

Chapman, S. J., Khor, C. C., Vannberg, F. O., Frodsham, A., Walley, A., Maskell, N. A., et al. (2007). IkappaB genetic polymorphisms and invasive pneumococcal disease. *American Journal of Respiratory and Critical Care Medicine*, *176*(2), 181–187. http://dx.doi.org/10.1164/rccm.200702-169OC.

Chen, G., Shaw, M. H., Kim, Y. G., & Nunez, G. (2009). NOD-like receptors: Role in innate immunity and inflammatory disease. *Annual Review of Pathology*, *4*, 365–398. http://dx.doi.org/10.1146/annurev.pathol.4.110807.092239.

Chorro, L., Sarde, A., Li, M., Woollard, K. J., Chambon, P., Malissen, B., et al. (2009). Langerhans cell (LC) proliferation mediates neonatal development, homeostasis, and inflammation-associated expansion of the epidermal LC network. *The Journal of Experimental Medicine*, *206*(13), 3089–3100. http://dx.doi.org/10.1084/jem.20091586.

Chow, O. A., Von Kockritz-Blickwede, M., Bright, A. T., Hensler, M. E., Zinkernagel, A. S., Cogen, A. L., et al. (2010). Statins enhance formation of phagocyte extracellular traps. *Cell Host & Microbe*, *8*(5), 445–454. http://dx.doi.org/10.1016/j.chom.2010.10.005.

Clark, R. A. (1986). Oxidative inactivation of pneumolysin by the myeloperoxidase system and stimulated human neutrophils. *The Journal of Immunology*, *136*(12), 4617–4622.

Clark, I. A., Allison, A. C., & Cox, F. E. (1976). Protection of mice against babesia, and plasmodium with BCG. *Nature*, *259*(5541), 309–311. http://dx.doi.org/10.1038/259309a0.

Clatworthy, M. R., & Smith, K. G. (2004). FcgammaRIIb balances efficient pathogen clearance and the cytokine-mediated consequences of sepsis. *The Journal of Experimental Medicine*, *199*(5), 717–723. http://dx.doi.org/10.1084/jem.20032197.

Coggon, D., Inskip, H., Winter, P., & Pannett, B. (1994). Lobar pneumonia: An occupational disease in welders. *Lancet*, *344*(8914), 41–43.

Cohn, Z. A., & Benson, B. (1965). The differentiation of mononuclear phagocytes morphology, cytochemistry, and biochemistry. *The Journal of Experimental Medicine*, *121*(1), 153–170. http://dx.doi.org/10.1084/jem.121.1.153.

Colino, J., & Snapper, C. M. (2003). Two distinct mechanisms for induction of dendritic cell apoptosis in response to intact Streptococcus pneumoniae. *The Journal of Immunology*, *171*(5), 2354–2365.

Corey, G. R. (2009). Staphylococcus aureus bloodstream infections: Definitions and treatment. *Clinical Infectious Diseases*, *48*(Suppl. 4), S254–S259. http://dx.doi.org/10.1086/598186.

Correa, R. G., Milutinovic, S., & Reed, J. C. (2012). Roles of NOD1 (NLRC1) and NOD2 (NLRC2) in innate immunity and inflammatory diseases. *Bioscience Reports*, *32*(6), 597–608. http://dx.doi.org/10.1042/BSR20120055.

Cosgrove, S. E., Sakoulas, G., Perencevich, E. N., Schwaber, M. J., Karchmer, A. W., & Carmeli, Y. (2003). Comparison of mortality associated with methicillin-resistant and methicillin-susceptible Staphylococcus aureus bacteremia: A meta-analysis. *Clinical Infectious Diseases*, *36*(1), 53–59. http://dx.doi.org/10.1086/345476.

Daigneault, M., De Silva, T. I., Bewley, M. A., Preston, J. A., Marriott, H. M., Mitchell, A. M., et al. (2012). Monocytes regulate the mechanism of T-cell death by inducing Fas-mediated apoptosis during bacterial infection. *PLoS Pathogens*, *8*(7), e1002814. http://dx.doi.org/10.1371/journal.ppat.1002814.

Daigneault, M., Preston, J. A., Marriott, H. M., Whyte, M. K., & Dockrell, D. H. (2010). The identification of markers of macrophage differentiation in PMA-stimulated THP-1 cells and monocyte-derived macrophages. *PLoS One*, *5*(1), e8668. http://dx.doi.org/10.1371/journal.pone.0008668.

Dalton, D. K., Pitts-Meek, S., Keshav, S., Figari, I. S., Bradley, A., & Stewart, T. A. (1993). Multiple defects of immune cell function in mice with disrupted interferon-gamma genes. *Science*, *259*(5102), 1739–1742.

Davies, L. C., Rosas, M., Smith, P. J., Fraser, D. J., Jones, S. A., & Taylor, P. R. (2011). A quantifiable proliferative burst of tissue macrophages restores homeostatic macrophage populations after acute inflammation. *European Journal of Immunology*, *41*(8), 2155–2164. http://dx.doi.org/10.1002/eji.201141817.

Davis, K. M., Nakamura, S., & Weiser, J. N. (2011). Nod2 sensing of lysozyme-digested peptidoglycan promotes macrophage recruitment and clearance of *S. pneumoniae* colonization in mice. *The Journal of Clinical Investigation*, *121*(9), 3666–3676. http://dx.doi.org/10.1172/JCI57761.

Degnan, B. A., Palmer, J. M., Robson, T., Jones, C. E., Fischer, M., Glanville, M., et al. (1998). Inhibition of human peripheral blood mononuclear cell proliferation by Streptococcus pyogenes cell extract is associated with arginine deiminase activity. *Infection and Immunity*, *66*(7), 3050–3058.

de Haas, C. J., Veldkamp, K. E., Peschel, A., Weerkamp, F., Van Wamel, W. J., Heezius, E. C., et al. (2004). Chemotaxis inhibitory protein of Staphylococcus aureus, a bacterial antiinflammatory agent. *The Journal of Experimental Medicine*, *199*(5), 687–695. http://dx.doi.org/10.1084/jem.20031636.

Deisenhofer, J. (1981). Crystallographic refinement and atomic models of a human Fc fragment and its complex with fragment B of protein A from *Staphylococcus aureus* at 2.9- and 2.8-A resolution. *Biochemistry*, *20*(9), 2361–2370.

de Kraker, M. E., Davey, P. G., & Grundmann, H.BURDEN Study Group. (2011). Mortality and hospital stay associated with resistant Staphylococcus aureus and Escherichia coli bacteremia: Estimating the burden of antibiotic resistance in Europe. *PLoS Medicine*, *8*(10), e1001104. http://dx.doi.org/10.1371/journal.pmed.1001104.

Dela Cruz, C. S., Liu, W., He, C. H., Jacoby, A., Gornitzky, A., Ma, B., et al. (2012). Chitinase 3-like-1 promotes Streptococcus pneumoniae killing and augments host tolerance to lung antibacterial responses. *Cell Host & Microbe*, *12*(1), 34–46. http://dx.doi.org/10.1016/j.chom.2012.05.017.

Deleo, F. R., Allen, L. A., Apicella, M., & Nauseef, W. M. (1999). NADPH oxidase activation and assembly during phagocytosis. *The Journal of Immunology*, *163*(12), 6732–6740.

Deleo, F. R., Otto, M., Kreiswirth, B. N., & Chambers, H. F. (2010). Community-associated meticillin-resistant Staphylococcus aureus. *Lancet*, *375*(9725), 1557–1568. http://dx.doi.org/10.1016/S0140-6736(09)61999-1.

Deloid, G. M., Sulahian, T. H., Imrich, A., & Kobzik, L. (2009). Heterogeneity in macrophage phagocytosis of Staphylococcus aureus strains: High-throughput scanning cytometry-based analysis. *PLoS One*, *4*(7), e6209. http://dx.doi.org/10.1371/journal.pone.0006209.

de Luca, A., Smeekens, S. P., Casagrande, A., Iannitti, R., Conway, K. L., Gresnigt, M. S., et al. (2014). IL-1 receptor blockade restores autophagy and reduces inflammation in chronic granulomatous disease in mice and in humans. *Proceedings of the National Academy of Sciences of the United States of America*, *111*(9), 3526–3531. http://dx.doi.org/10.1073/pnas.1322831111.

de Roux, A., Marcos, M. A., Garcia, E., Mensa, J., Ewig, S., Lode, H., et al. (2004). Viral community-acquired pneumonia in nonimmunocompromised adults. *Chest*, *125*(4), 1343–1351.

De Santa, F., Narang, V., Yap, Z. H., Tusi, B. K., Burgold, T., Austenaa, L., et al. (2009). Jmjd3 contributes to the control of gene expression in LPS-activated macrophages. *The EMBO Journal*, *28*(21), 3341–3352. http://dx.doi.org/10.1038/emboj.2009.271.

Desjardins, M., Huber, L. A., Parton, R. G., & Griffiths, G. (1994). Biogenesis of phagolysosomes proceeds through a sequential series of interactions with the endocytic apparatus. *The Journal of Cell Biology*, *124*(5), 677–688.

Diep, B. A., Gill, S. R., Chang, R. F., Phan, T. H., Chen, J. H., Davidson, M. G., et al. (2006). Complete genome sequence of USA300, an epidemic clone of community-acquired meticillin-resistant Staphylococcus aureus. *Lancet*, *367*(9512), 731–739. http://dx.doi.org/10.1016/S0140-6736(06)68231-7.

Dockrell, D. H., Lee, M., Lynch, D. H., & Read, R. C. (2001). Immune-mediated phagocytosis and killing of Streptococcus pneumoniae are associated with direct and bystander macrophage apoptosis. *The Journal of Infectious Diseases*, *184*(6), 713–722. http://dx.doi.org/10.1086/323084.

Dockrell, D. H., Marriott, H. M., Prince, L. R., Ridger, V. C., Ince, P. G., Hellewell, P. G., et al. (2003). Alveolar macrophage apoptosis contributes to pneumococcal clearance in a resolving model of pulmonary infection. *Journal of Immunology*, *171*(10), 5380–5388.

Dockrell, D. H., Whyte, M. K., & Mitchell, T. J. (2012). Pneumococcal pneumonia: Mechanisms of infection and resolution. *Chest*, *142*(2), 482–491. http://dx.doi.org/10.1378/chest.12-0210.

Donlan, R. M., & Costerton, J. W. (2002). Biofilms: Survival mechanisms of clinically relevant microorganisms. *Clinical Microbiology Reviews*, *15*(2), 167–193.

Dorrington, M. G., Roche, A. M., Chauvin, S. E., Tu, Z., Mossman, K. L., Weiser, J. N., et al. (2013). MARCO is required for TLR2- and Nod2-mediated responses to Streptococcus pneumoniae and clearance of pneumococcal colonization in the murine nasopharynx. *The Journal of Immunology*, *190*(1), 250–258. http://dx.doi.org/10.4049/jimmunol.1202113.

Dumont, A. L., Nygaard, T. K., Watkins, R. L., Smith, A., Kozhaya, L., Kreiswirth, B. N., et al. (2011). Characterization of a new cytotoxin that contributes to Staphylococcus aureus pathogenesis. *Molecular Microbiology*, *79*(3), 814–825. http://dx.doi.org/10.1111/j.1365-2958.2010.07490.x.

Dunne, D. W., Resnick, D., Greenberg, J., Krieger, M., & Joiner, K. A. (1994). The type I macrophage scavenger receptor binds to gram-positive bacteria and recognizes

lipoteichoic acid. *Proceedings of the National Academy of Sciences of the United States of America, 91*(5), 1863–1867.
Durrant, W. E., & Dong, X. (2004). Systemic acquired resistance. *Annual Review of Phytopathology, 42*(1), 185–209. http://dx.doi.org/10.1146/annurev.phyto.42.040803.140421.
Eagan, R., Twigg, H. L., 3rd., French, N., Musaya, J., Day, R. B., Zijlstra, E. E., et al. (2007). Lung fluid immunoglobulin from HIV-infected subjects has impaired opsonic function against pneumococci. *Clinical Infectious Diseases, 44*(12), 1632–1638. http://dx.doi.org/10.1086/518133.
Ebert, R. H., & Florey, H. W. (1939). The extravascular development of the monocyte observed in vivo. *British Journal of Experimental Pathology, 20*(4), 342–356.
Ehrlich, M., Gama-Sosa, M. A., Huang, L. H., Midgett, R. M., Kuo, K. C., McCune, R. A., et al. (1982). Amount and distribution of 5-methylcytosine in human DNA from different types of tissues of cells. *Nucleic Acids Research, 10*(8), 2709–2721.
Ehrt, S., Schnappinger, D., Bekiranov, S., Drenkow, J., Shi, S., Gingeras, T. R., et al. (2001). Reprogramming of the macrophage transcriptome in response to interferon-γ and mycobacterium tuberculosis signaling roles of nitric oxide synthase-2 and phagocyte oxidase. *The Journal of Experimental Medicine, 194*(8), 1123–1140. http://dx.doi.org/10.1084/jem.194.8.1123.
Eskandarian, H. A., Impens, F., Nahori, M.-A., Soubigou, G., Coppée, J.-Y., Cossart, P., et al. (2013). A role for SIRT2-dependent histone H3K18 deacetylation in bacterial infection. *Science (New York, N.Y.), 341*(6145), 1238858. http://dx.doi.org/10.1126/science.1238858.
Esposito, A. L., Clark, C. A., & Poirier, W. J. (1988). An assessment of the respiratory burst and bactericidal activity of alveolar macrophages from adult and senescent mice. *Journal of Leukocyte Biology, 43*(5), 445–454.
Essmann, F., Bantel, H., Totzke, G., Engels, I. H., Sinha, B., Schulze-Osthoff, K., et al. (2003). Staphylococcus aureus alpha-toxin-induced cell death: Predominant necrosis despite apoptotic caspase activation. *Cell Death and Differentiation, 10*(11), 1260–1272. http://dx.doi.org/10.1038/sj.cdd.4401301.
Esteller, M. (2007). Cancer epigenomics: DNA methylomes and histone-modification maps. *Nature Reviews. Genetics, 8*(4), 286–298. http://dx.doi.org/10.1038/nrg2005.
Fang, F. C. (2004). Antimicrobial reactive oxygen and nitrogen species: Concepts and controversies. *Nature Reviews. Microbiology, 2*(10), 820–832. http://dx.doi.org/10.1038/nrmicro1004.
Fang, R., Tsuchiya, K., Kawamura, I., Shen, Y., Hara, H., Sakai, S., et al. (2011). Critical roles of ASC inflammasomes in caspase-1 activation and host innate resistance to Streptococcus pneumoniae infection. *Journal of Immunology, 187*(9), 4890–4899. http://dx.doi.org/10.4049/jimmunol.1100381.
Fernie-King, B., Seilly, D., Davies, A., & Lachmann, P. (2002). Subversion of the innate immune response by micro-organisms. *Annals of the Rheumatic Diseases, 61*(Suppl. 2), ii8–ii12. http://dx.doi.org/10.1136/ard.61.suppl_2.ii8.
Flannagan, R. S., Cosio, G., & Grinstein, S. (2009). Antimicrobial mechanisms of phagocytes and bacterial evasion strategies. *Nature Reviews. Microbiology,* 7(5), 355–366. http://dx.doi.org/10.1038/nrmicro2128.
Foster, S. L., Hargreaves, D. C., & Medzhitov, R. (2007). Gene-specific control of inflammation by TLR-induced chromatin modifications. *Nature, 447*(7147), 972–978. http://dx.doi.org/10.1038/nature05836.
Franchi, L., Kanneganti, T. D., Dubyak, G. R., & Nunez, G. (2007). Differential requirement of P2X7 receptor and intracellular K+ for caspase-1 activation induced by intracellular and extracellular bacteria. *The Journal of Biological Chemistry, 282*(26), 18810–18818. http://dx.doi.org/10.1074/jbc.M610762200.

Fraser, J. D., & Proft, T. (2008). The bacterial superantigen and superantigen-like proteins. *Immunological Reviews*, *225*, 226–243. http://dx.doi.org/10.1111/j.1600-065X.2008.00681.x.
Geijtenbeek, T. B., Groot, P. C., Nolte, M. A., Van Vliet, S. J., Gangaram-Panday, S. T., Van Duijnhoven, G. C., et al. (2002). Marginal zone macrophages express a murine homologue of DC-SIGN that captures blood-borne antigens in vivo. *Blood*, *100*(8), 2908–2916. http://dx.doi.org/10.1182/blood-2002-04-1044.
Giese, B., Glowinski, F., Paprotka, K., Dittmann, S., Steiner, T., Sinha, B., et al. (2011). Expression of delta-toxin by Staphylococcus aureus mediates escape from phago-endosomes of human epithelial and endothelial cells in the presence of beta-toxin. *Cellular Microbiology*, *13*(2), 316–329. http://dx.doi.org/10.1111/j.1462-5822.2010.01538.x.
Godleski, J. J., & Brain, J. D. (1972). The origin of alveolar macrophages in mouse radiation chimeras. *The Journal of Experimental Medicine*, *136*(3), 630–643. http://dx.doi.org/10.1084/jem.136.3.630.
Goldmann, O., Von Kockritz-Blickwede, M., Holtje, C., Chhatwal, G. S., Geffers, R., & Medina, E. (2007). Transcriptome analysis of murine macrophages in response to infection with Streptococcus pyogenes reveals an unusual activation program. *Infection and Immunity*, *75*(8), 4148–4157. http://dx.doi.org/10.1128/IAI.00181-07.
Gordon, S. B., Irving, G. R., Lawson, R. A., Lee, M. E., & Read, R. C. (2000). Intracellular trafficking and killing of Streptococcus pneumoniae by human alveolar macrophages are influenced by opsonins. *Infection and Immunity*, *68*(4), 2286–2293.
Gordon, R. J., & Lowy, F. D. (2008). Pathogenesis of methicillin-resistant Staphylococcus aureus infection. *Clinical Infectious Diseases*, *46*(Suppl. 5), S350–S359. http://dx.doi.org/10.1086/533591.
Grosz, M., Kolter, J., Paprotka, K., Winkler, A. C., Schafer, D., Chatterjee, S. S., et al. (2014). Cytoplasmic replication of Staphylococcus aureus upon phagosomal escape triggered by phenol-soluble modulin alpha. *Cellular Microbiology*, *16*(4), 451–465. http://dx.doi.org/10.1111/cmi.12233.
Guiral, S., Mitchell, T. J., Martin, B., & Claverys, J. P. (2005). Competence-programmed predation of noncompetent cells in the human pathogen Streptococcus pneumoniae: Genetic requirements. *Proceedings of the National Academy of Sciences of the United States of America*, *102*(24), 8710–8715. http://dx.doi.org/10.1073/pnas.0500879102.
Hackam, D. J., Rotstein, O. D., Zhang, W. J., Demaurex, N., Woodside, M., Tsai, O., et al. (1997). Regulation of phagosomal acidification. Differential targeting of Na+/H+ exchangers, Na+/K+-ATPases, and vacuolar-type H+-atpases. *The Journal of Biological Chemistry*, *272*(47), 29810–29820.
Hamon, M. A., Batsche, E., Regnault, B., Tham, T. N., Seveau, S., Muchardt, C., et al. (2007). Histone modifications induced by a family of bacterial toxins. *Proceedings of the National Academy of Sciences of the United States of America*, *104*(33), 13467–13472. http://dx.doi.org/10.1073/pnas.0702729104.
Hanke, M. L., Heim, C. E., Angle, A., Sanderson, S. D., & Kielian, T. (2013). Targeting macrophage activation for the prevention and treatment of Staphylococcus aureus biofilm infections. *The Journal of Immunology*, *190*(5), 2159–2168. http://dx.doi.org/10.4049/jimmunol.1202348.
Harrison, R. E., Bucci, C., Vieira, O. V., Schroer, T. A., & Grinstein, S. (2003). Phagosomes fuse with late endosomes and/or lysosomes by extension of membrane protrusions along microtubules: Role of Rab7 and RILP. *Molecular and Cellular Biology*, *23*(18), 6494–6506.
Harvey, R. M., Ogunniyi, A. D., Chen, A. Y., & Paton, J. C. (2011). Pneumolysin with low hemolytic activity confers an early growth advantage to Streptococcus pneumoniae in the blood. *Infection and Immunity*, *79*(10), 4122–4130. http://dx.doi.org/10.1128/IAI.05418-11.

Hashimoto, D., Chow, A., Noizat, C., Teo, P., Beasley, M. B., Leboeuf, M., et al. (2013). Tissue-resident macrophages self-maintain locally throughout adult life with minimal contribution from circulating monocytes. *Immunity*, *38*(4), 792–804. http://dx.doi.org/10.1016/j.immuni.2013.04.004.

Herre, J., Marshall, A. S., Caron, E., Edwards, A. D., Williams, D. L., Schweighoffer, E., et al. (2004). Dectin-1 uses novel mechanisms for yeast phagocytosis in macrophages. *Blood*, *104*(13), 4038–4045. http://dx.doi.org/10.1182/blood-2004-03-1140.

Herrmann, D. J., Hamilton, R. G., Barington, T., Frasch, C. E., Arakere, G., Makela, O., et al. (1992). Quantitation of human IgG subclass antibodies to Haemophilus influenzae type b capsular polysaccharide. Results of an international collaborative study using enzyme immunoassay methodology. *Journal of Immunological Methods*, *148*(1–2), 101–114.

Hesse, M., Modolell, M., Flamme, A. C. L., Schito, M., Fuentes, J. M., Cheever, A. W., et al. (2001). Differential regulation of nitric oxide synthase-2 and arginase-1 by type 1/type 2 cytokines in vivo: Granulomatous pathology is shaped by the pattern of L-arginine metabolism. *The Journal of Immunology*, *167*(11), 6533–6544.

Hiramatsu, K., Hanaki, H., Ino, T., Yabuta, K., Oguri, T., & Tenover, F. C. (1997). Methicillin-resistant Staphylococcus aureus clinical strain with reduced vancomycin susceptibility. *The Journal of Antimicrobial Chemotherapy*, *40*(1), 135–136.

Hof, D. G., Repine, J. E., Peterson, P. K., & Hoidal, J. R. (1980). Phagocytosis by human alveolar macrophages and neutrophils: Qualitative differences in the opsonic requirements for uptake of Staphylococcus aureus and Streptococcus pneumoniae in vitro. *The American Review of Respiratory Disease*, *121*(1), 65–71.

Holzinger, D., Gieldon, L., Mysore, V., Nippe, N., Taxman, D. J., Duncan, J. A., et al. (2012). Staphylococcus aureus Panton–Valentine leukocidin induces an inflammatory response in human phagocytes via the NLRP3 inflammasome. *Journal of Leukocyte Biology*, *92*(5), 1069–1081. http://dx.doi.org/10.1189/jlb.0112014.

Huynh, K. K., Eskelinen, E. L., Scott, C. C., Malevanets, A., Saftig, P., & Grinstein, S. (2007). LAMP proteins are required for fusion of lysosomes with phagosomes. *The EMBO Journal*, *26*(2), 313–324. http://dx.doi.org/10.1038/sj.emboj.7601511.

Huynh, K. K., & Grinstein, S. (2007). Regulation of vacuolar pH and its modulation by some microbial species. *Microbiology and Molecular Biology Reviews*, *71*(3), 452–462. http://dx.doi.org/10.1128/MMBR.00003-07.

Hyams, C., Camberlein, E., Cohen, J. M., Bax, K., & Brown, J. S. (2010). The Streptococcus pneumoniae capsule inhibits complement activity and neutrophil phagocytosis by multiple mechanisms. *Infection and Immunity*, *78*(2), 704–715. http://dx.doi.org/10.1128/IAI.00881-09.

Hyams, C., Opel, S., Hanage, W., Yuste, J., Bax, K., Henriques-Normark, B., et al. (2011). Effects of Streptococcus pneumoniae strain background on complement resistance. *PLoS One*, *6*(10), e24581. http://dx.doi.org/10.1371/journal.pone.0024581.

Hyams, C., Yuste, J., Bax, K., Camberlein, E., Weiser, J. N., & Brown, J. S. (2010). Streptococcus pneumoniae resistance to complement-mediated immunity is dependent on the capsular serotype. *Infection and Immunity*, *78*(2), 716–725. http://dx.doi.org/10.1128/IAI.01056-09.

Ibrahim, Y. M., Kerr, A. R., McCluskey, J., & Mitchell, T. J. (2004). Role of HtrA in the virulence and competence of Streptococcus pneumoniae. *Infection and Immunity*, *72*(6), 3584–3591. http://dx.doi.org/10.1128/IAI.72.6.3584-3591.2004.

Ifrim, D. C., Quintin, J., Joosten, L. A., Jacobs, C., Jansen, T., Jacobs, L., et al. (2014). Trained immunity or tolerance: Opposing functional programs induced in human monocytes after engagement of various pattern recognition receptors. *Clinical and Vaccine Immunology*, *21*(4), 534–545. http://dx.doi.org/10.1128/cvi.00688-13.

Ip, W. K., Sokolovska, A., Charriere, G. M., Boyer, L., Dejardin, S., Cappillino, M. P., et al. (2010). Phagocytosis and phagosome acidification are required for pathogen processing and MyD88-dependent responses to Staphylococcus aureus. *Journal of Immunology, 184*(12), 7071–7081. http://dx.doi.org/10.4049/jimmunol.1000110.

Ivashkiv, L. B. (2013). Epigenetic regulation of macrophage polarization and function. *Trends in Immunology, 34*(5), 216–223. http://dx.doi.org/10.1016/j.it.2012.11.001.

Janoff, E. N., Fasching, C., Orenstein, J. M., Rubins, J. B., Opstad, N. L., & Dalmasso, A. P. (1999). Killing of Streptococcus pneumoniae by capsular polysaccharide-specific polymeric IgA, complement, and phagocytes. *The Journal of Clinical Investigation, 104*(8), 1139–1147. http://dx.doi.org/10.1172/JCI6310.

Jarry, T. M., Memmi, G., & Cheung, A. L. (2008). The expression of alpha-haemolysin is required for Staphylococcus aureus phagosomal escape after internalization in CFT-1 cells. *Cellular Microbiology, 10*(9), 1801–1814. http://dx.doi.org/10.1111/j.1462-5822.2008.01166.x.

Jarva, H., Janulczyk, R., Hellwage, J., Zipfel, P. F., Bjorck, L., & Meri, S. (2002). Streptococcus pneumoniae evades complement attack and opsonophagocytosis by expressing the pspC locus-encoded Hic protein that binds to short consensus repeats 8-11 of factor H. *The Journal of Immunology, 168*(4), 1886–1894.

Jedrzejas, M. J. (2001). Pneumococcal virulence factors: Structure and function. *Microbiology and Molecular Biology Reviews, 65*(2), 187–207. http://dx.doi.org/10.1128/MMBR.65.2.187-207.2001, first page, table of contents.

Jenkins, S. J., Ruckerl, D., Cook, P. C., Jones, L. H., Finkelman, F. D., Van Rooijen, N., et al. (2011). Local macrophage proliferation, rather than recruitment from the blood, is a signature of TH2 inflammation. *Science, 332*(6035), 1284–1288. http://dx.doi.org/10.1126/science.1204351.

Jenner, R. G., & Young, R. A. (2005). Insights into host responses against pathogens from transcriptional profiling. *Nature Reviews. Microbiology, 3*(4), 281–294. http://dx.doi.org/10.1038/nrmicro1126.

Jenuwein, T., & Allis, C. D. (2001). Translating the histone code. *Science (New York, N.Y.), 293*(5532), 1074–1080. http://dx.doi.org/10.1126/science.1063127.

Jesch, N. K., Dorger, M., Enders, G., Rieder, G., Vogelmeier, C., Messmer, K., et al. (1997). Expression of inducible nitric oxide synthase and formation of nitric oxide by alveolar macrophages: An interspecies comparison. *Environmental Health Perspectives, 105*(Suppl. 5), 1297–1300.

Jin, M., Opalek, J. M., Marsh, C. B., & Wu, H. M. (2004). Proteome comparison of alveolar macrophages with monocytes reveals distinct protein characteristics. *American Journal of Respiratory Cell and Molecular Biology, 31*(3), 322–329. http://dx.doi.org/10.1165/rcmb.2004-0080OC.

John, J. F. J. (2004). Staphylococcal infection: Emerging clinical syndromes. In D. A. Aldeen, & K. Hiramatsu (Eds.), Staphylococcus aureus*: Molecular and clinical aspects* (pp. 1–30). United Kingdom: Horwood Publishing Limited, Reprinted from: 2009.

Jongerius, I., Kohl, J., Pandey, M. K., Ruyken, M., Van Kessel, K. P., Van Strijp, J. A., et al. (2007). Staphylococcal complement evasion by various convertase-blocking molecules. *The Journal of Experimental Medicine, 204*(10), 2461–2471. http://dx.doi.org/10.1084/jem.20070818.

Jonsson, S., Musher, D. M., Chapman, A., Goree, A., & Lawrence, E. C. (1985). Phagocytosis and killing of common bacterial pathogens of the lung by human alveolar macrophages. *The Journal of Infectious Diseases, 152*(1), 4–13.

Jordens, I., Fernandez-Borja, M., Marsman, M., Dusseljee, S., Janssen, L., Calafat, J., et al. (2001). The Rab7 effector protein RILP controls lysosomal transport by inducing the recruitment of dynein-dynactin motors. *Current Biology, 11*(21), 1680–1685.

Jounblat, R., Kadioglu, A., Mitchell, T. J., & Andrew, P. W. (2003). Pneumococcal behavior and host responses during bronchopneumonia are affected differently by the cytolytic and complement-activating activities of pneumolysin. *Infection and Immunity*, *71*(4), 1813–1819.

Kadioglu, A., Weiser, J. N., Paton, J. C., & Andrew, P. W. (2008). The role of Streptococcus pneumoniae virulence factors in host respiratory colonization and disease. *Nature Reviews Microbiology*, *6*(4), 288–301. http://dx.doi.org/10.1038/nrmicro1871.

Kapetanovic, R., Nahori, M. A., Balloy, V., Fitting, C., Philpott, D. J., Cavaillon, J. M., et al. (2007). Contribution of phagocytosis and intracellular sensing for cytokine production by Staphylococcus aureus-activated macrophages. *Infection and Immunity*, *75*(2), 830–837. http://dx.doi.org/10.1128/IAI.01199-06.

Kapetanovic, R., Parlato, M., Fitting, C., Quesniaux, V., Cavaillon, J. M., & Adib-Conquy, M. (2011). Mechanisms of TNF induction by heat-killed Staphylococcus aureus differ upon the origin of mononuclear phagocytes. *American Journal of Physiology Cell Physiology*, *300*(4), C850–C859. http://dx.doi.org/10.1152/ajpcell.00187.2010.

Kato, T., Kojima, K., & Murachi, T. (1972). Proteases of macrophages in rat peritoneal exudate, with special reference to the effects of actinomycete protease inhibitors. *Biochimica et Biophysica Acta*, *289*(1), 187–193.

Kebaier, C., Chamberland, R. R., Allen, I. C., Gao, X., Broglie, P. M., Hall, J. D., et al. (2012). Staphylococcus aureus alpha-hemolysin mediates virulence in a murine model of severe pneumonia through activation of the NLRP3 inflammasome. *The Journal of Infectious Diseases*, *205*(5), 807–817. http://dx.doi.org/10.1093/infdis/jir846.

Kennedy, D. W., & Abkowitz, J. L. (1997). Kinetics of central nervous system microglial and macrophage engraftment: Analysis using a transgenic bone marrow transplantation model. *Blood*, *90*(3), 986–993.

Kennedy, D. W., & Abkowitz, J. L. (1998). Mature monocytic cells enter tissues and engraft. *Proceedings of the National Academy of Sciences*, *95*(25), 14944–14949.

Khor, C. C., Chapman, S. J., Vannberg, F. O., Dunne, A., Murphy, C., Ling, E. Y., et al. (2007). A Mal functional variant is associated with protection against invasive pneumococcal disease, bacteremia, malaria and tuberculosis. *Nature Genetics*, *39*(4), 523–528. http://dx.doi.org/10.1038/ng1976.

Kilian, M., Mestecky, J., Kulhavy, R., Tomana, M., & Butler, W. T. (1980). IgA1 proteases from Haemophilus influenzae, Streptococcus pneumoniae, Neisseria meningitidis, and Streptococcus sanguis: Comparative immunochemical studies. *The Journal of Immunology*, *124*(6), 2596–2600.

Kirby, W. M. (1944). Extraction of a highly potent penicillin inactivator from penicillin resistant Staphylococci. *Science*, *99*(2579), 452–453. http://dx.doi.org/10.1126/science.99.2579.452.

Klein, E., Smith, D. L., & Laxminarayan, R. (2007). Hospitalizations and deaths caused by methicillin-resistant Staphylococcus aureus, United States, 1999–2005. *Emerging Infectious Diseases*, *13*, 1840–1846.

Kleinnijenhuis, J., Quintin, J., Preijers, F., Joosten, L. A. B., Ifrim, D. C., Saeed, S., et al. (2012). Bacille Calmette–Guerin induces NOD2-dependent nonspecific protection from reinfection via epigenetic reprogramming of monocytes. *Proceedings of the National Academy of Sciences of the United States of America*, *109*(43), 17537–17542. http://dx.doi.org/10.1073/pnas.1202870109.

Knapp, S., Wieland, C. W., van'T Veer, C., Takeuchi, O., Akira, S., Florquin, S., et al. (2004). Toll-like receptor 2 plays a role in the early inflammatory response to murine pneumococcal pneumonia but does not contribute to antibacterial defense. *Journal of Immunology*, *172*(5), 3132–3138.

Koppel, E. A., Wieland, C. W., Van Den Berg, V. C., Litjens, M., Florquin, S., Van Kooyk, Y., et al. (2005). Specific ICAM-3 grabbing nonintegrin-related 1 (SIGNR1) expressed by marginal zone macrophages is essential for defense against pulmonary Streptococcus pneumoniae infection. *European Journal of Immunology, 35*(10), 2962–2969. http://dx.doi.org/10.1002/eji.200526216.

Koziel, J., Maciag-Gudowska, A., Mikolajczyk, T., Bzowska, M., Sturdevant, D. E., Whitney, A. R., et al. (2009). Phagocytosis of Staphylococcus aureus by macrophages exerts cytoprotective effects manifested by the upregulation of antiapoptotic factors. *PLoS One, 4*(4), e5210. http://dx.doi.org/10.1371/journal.pone.0005210.

Krakauer, T. (2002). Stimulant-dependent modulation of cytokines and chemokines by airway epithelial cells: Cross talk between pulmonary epithelial and peripheral blood mononuclear cells. *Clinical and Diagnostic Laboratory Immunology, 9*(1), 126–131.

Kreider, T., Anthony, R. M., Urban, J. F., Jr., & Gause, W. C. (2007). Alternatively activated macrophages in helminth infections. *Current Opinion in Immunology, 19*(4), 448–453. http://dx.doi.org/10.1016/j.coi.2007.07.002.

Kruidenier, L., Chung, C.-w., Cheng, Z., Liddle, J., Che, K., Joberty, G., et al. (2012). A selective jumonji H3K27 demethylase inhibitor modulates the proinflammatory macrophage response. *Nature, 488*(7411), 404–408. http://dx.doi.org/10.1038/nature11262.

Ku, C. L., Picard, C., Erdos, M., Jeurissen, A., Bustamante, J., Puel, A., et al. (2007). IRAK4 and NEMO mutations in otherwise healthy children with recurrent invasive pneumococcal disease. *Journal of Medical Genetics, 44*(1), 16–23. http://dx.doi.org/10.1136/jmg.2006.044446.

Kubica, M., Guzik, K., Koziel, J., Zarebski, M., Richter, W., Gajkowska, B., et al. (2008). A potential new pathway for *Staphylococcus aureus* dissemination: The silent survival of *S. aureus* phagocytosed by human monocyte-derived macrophages. *PLoS One, 3*(1), e1409. http://dx.doi.org/10.1371/journal.pone.0001409.

Kuehnert, M. J., Kruszon-Moran, D., Hill, H. A., Mcquillan, G., McAllister, S. K., Fosheim, G., et al. (2006). Prevalence of Staphylococcus aureus nasal colonization in the United States, 2001–2002. *The Journal of Infectious Diseases, 193*(2), 172–179. http://dx.doi.org/10.1086/499632.

Kuronuma, K., Sano, H., Kato, K., Kudo, K., Hyakushima, N., Yokota, S., et al. (2004). Pulmonary surfactant protein A augments the phagocytosis of Streptococcus pneumoniae by alveolar macrophages through a casein kinase 2-dependent increase of cell surface localization of scavenger receptor A. *The Journal of Biological Chemistry, 279*(20), 21421–21430. http://dx.doi.org/10.1074/jbc.M312490200.

Laarman, A. J., Ruyken, M., Malone, C. L., Van Strijp, J. A., Horswill, A. R., & Rooijakkers, S. H. (2011). Staphylococcus aureus metalloprotease aureolysin cleaves complement C3 to mediate immune evasion. *Journal of Immunology, 186*(11), 6445–6453. http://dx.doi.org/10.4049/jimmunol.1002948.

Labandeira-Rey, M., Couzon, F., Boisset, S., Brown, E. L., Bes, M., Benito, Y., et al. (2007). Staphylococcus aureus Panton–Valentine leukocidin causes necrotizing pneumonia. *Science, 315*(5815), 1130–1133. http://dx.doi.org/10.1126/science.1137165.

Lagler, H., Sharif, O., Haslinger, I., Matt, U., Stich, K., Furtner, T., et al. (2009). TREM-1 activation alters the dynamics of pulmonary IRAK-M expression in vivo and improves host defense during pneumococcal pneumonia. *Journal of Immunology, 183*(3), 2027–2036. http://dx.doi.org/10.4049/jimmunol.0803862.

Langley, R., Wines, B., Willoughby, N., Basu, I., Proft, T., & Fraser, J. D. (2005). The staphylococcal superantigen-like protein 7 binds IgA and complement C5 and inhibits IgA-Fc alpha RI binding and serum killing of bacteria. *The Journal of Immunology, 174*(5), 2926–2933.

Lawe, D. C., Patki, V., Heller-Harrison, R., Lambright, D., & Corvera, S. (2000). The FYVE domain of early endosome antigen 1 is required for both phosphatidylinositol

3-phosphate and Rab5 binding. Critical role of this dual interaction for endosomal localization. *The Journal of Biological Chemistry, 275*(5), 3699–3705.
Lee, D. A., Hoidal, J. R., Garlich, D. J., Clawson, C. C., Quie, P. G., & Peterson, P. K. (1984). Opsonin-independent phagocytosis of surface-adherent bacteria by human alveolar macrophages. *Journal of Leukocyte Biology, 36*(6), 689–701.
Lee, W. L., Kim, M. K., Schreiber, A. D., & Grinstein, S. (2005). Role of ubiquitin and proteasomes in phagosome maturation. *Molecular Biology of the Cell, 16*(4), 2077–2090. http://dx.doi.org/10.1091/mbc.E04-06-0464.
Lee, K. S., Scanga, C. A., Bachelder, E. M., Chen, Q., & Snapper, C. M. (2007). TLR2 synergizes with both TLR4 and TLR9 for induction of the MyD88-dependent splenic cytokine and chemokine response to Streptococcus pneumoniae. *Cellular Immunology, 245*(2), 103–110. http://dx.doi.org/10.1016/j.cellimm.2007.04.003.
Lee, S. P., Serezani, C. H., Medeiros, A. I., Ballinger, M. N., & Peters-Golden, M. (2009). Crosstalk between prostaglandin E2 and leukotriene B4 regulates phagocytosis in alveolar macrophages via combinatorial effects on cyclic AMP. *Journal of Immunology, 182*(1), 530–537.
Lehrer, R. I., Lichtenstein, A. K., & Ganz, T. (1993). Defensins: Antimicrobial and cytotoxic peptides of mammalian cells. *Annual Review of Immunology, 11*, 105–128. http://dx.doi.org/10.1146/annurev.iy.11.040193.000541.
Lekstrom-Himes, J. A., & Gallin, J. I. (2000). Immunodeficiency diseases caused by defects in phagocytes. *The New England Journal of Medicine, 343*(23), 1703–1714.
Lentsch, A. B., Czermak, B. J., Bless, N. M., Van Rooijen, N., & Ward, P. A. (1999). Essential role of alveolar macrophages in intrapulmonary activation of NF-kappaB. *American Journal of Respiratory Cell and Molecular Biology, 20*(4), 692–698. http://dx.doi.org/10.1165/ajrcmb.20.4.3414.
Levy, O., & Wynn, J. L. (2014). A prime time for trained immunity: Innate immune memory in newborns and infants. *Neonatology, 105*(2), 136–141. http://dx.doi.org/10.1159/000356035.
Li, C., Li, H., Jiang, Z., Zhang, T., Wang, Y., Li, Z., et al. (2014). Interleukin-33 increases antibacterial defense by activation of inducible nitric oxide synthase in skin. *PLoS Pathogens, 10*(2), e1003918. http://dx.doi.org/10.1371/journal.ppat.1003918.
Li, Y., Liu, B., Fukudome, E. Y., Kochanek, A. R., Finkelstein, R. A., Chong, W., et al. (2010). Surviving lethal septic shock without fluid resuscitation in a rodent model. *Surgery, 148*(2), 246–254. http://dx.doi.org/10.1016/j.surg.2010.05.003.
Lim, J. H., Stirling, B., Derry, J., Koga, T., Jono, H., Woo, C. H., et al. (2007). Tumor suppressor CYLD regulates acute lung injury in lethal Streptococcus pneumoniae infections. *Immunity, 27*(2), 349–360. http://dx.doi.org/10.1016/j.immuni.2007.07.011.
Liu, G. Y. (2009). Molecular pathogenesis of Staphylococcus aureus infection. *Pediatric Research, 65*(5 Pt 2), 71R–77R. http://dx.doi.org/10.1203/PDR.0b013e31819dc44d.
Liu, G. Y., Essex, A., Buchanan, J. T., Datta, V., Hoffman, H. M., Bastian, J. F., et al. (2005). Staphylococcus aureus golden pigment impairs neutrophil killing and promotes virulence through its antioxidant activity. *The Journal of Experimental Medicine, 202*(2), 209–215. http://dx.doi.org/10.1084/jem.20050846.
Liu, H., Perlman, H., Pagliari, L. J., & Pope, R. M. (2001). Constitutively activated Akt-1 is vital for the survival of human monocyte-differentiated macrophages. Role of Mcl-1, independent of nuclear factor (NF)-kappaB, Bad, or caspase activation. *The Journal of Experimental Medicine, 194*(2), 113–126.
Lu, L., Ma, Y., & Zhang, J. R. (2006). Streptococcus pneumoniae recruits complement factor H through the amino terminus of CbpA. *The Journal of Biological Chemistry, 281*(22), 15464–15474. http://dx.doi.org/10.1074/jbc.M602404200.
Macedo-Ramos, H., Campos, F. S., Carvalho, L. A., Ramos, I. B., Teixeira, L. M., De Souza, W., et al. (2011). Olfactory ensheathing cells as putative host cells for

Streptococcus pneumoniae: Evidence of bacterial invasion via mannose receptor-mediated endocytosis. *Neurosciences Research*, *69*(4), 308–313. http://dx.doi.org/10.1016/j.neures.2010.12.015.

Mackaness, G. B. (1964). The immunological basis of acquired cellular resistance. *The Journal of Experimental Medicine*, *120*(1), 105–120. http://dx.doi.org/10.1084/jem.120.1.105.

Macmicking, J., Xie, Q. W., & Nathan, C. (1997). Nitric oxide and macrophage function. *Annual Review of Immunology*, *15*, 323–350. http://dx.doi.org/10.1146/annurev.immunol.15.1.323.

Malachowa, N., Whitney, A. R., Kobayashi, S. D., Sturdevant, D. E., Kennedy, A. D., Braughton, K. R., et al. (2011). Global changes in Staphylococcus aureus gene expression in human blood. *PLoS One*, *6*(4), e18617. http://dx.doi.org/10.1371/journal.pone.0018617.

Malley, R., Henneke, P., Morse, S. C., Cieslewicz, M. J., Lipsitch, M., Thompson, C. M., et al. (2003). Recognition of pneumolysin by Toll-like receptor 4 confers resistance to pneumococcal infection. *Proceedings of the National Academy of Sciences of the United States of America*, *100*(4), 1966–1971. http://dx.doi.org/10.1073/pnas.0435928100.

Mancuso, P., Huffnagle, G. B., Olszewski, M. A., Phipps, J., & Peters-Golden, M. (2006). Leptin corrects host defense defects after acute starvation in murine pneumococcal pneumonia. *American Journal of Respiratory and Critical Care Medicine*, *173*(2), 212–218. http://dx.doi.org/10.1164/rccm.200506-909OC.

Mancuso, P., & Peters-Golden, M. (2000). Modulation of alveolar macrophage phagocytosis by leukotrienes is Fc receptor-mediated and protein kinase C-dependent. *American Journal of Respiratory Cell and Molecular Biology*, *23*(6), 727–733.

Mancuso, P., Peters-Golden, M., Goel, D., Goldberg, J., Brock, T. G., Greenwald-Yarnell, M., et al. (2011). Disruption of leptin receptor-STAT3 signaling enhances leukotriene production and pulmonary host defense against pneumococcal pneumonia. *Journal of Immunology*, *186*(2), 1081–1090. http://dx.doi.org/10.4049/jimmunol.1001470.

Mantovani, A., Sica, A., Sozzani, S., Allavena, P., Vecchi, A., & Locati, M. (2004). The chemokine system in diverse forms of macrophage activation and polarization. *Trends in Immunology*, *25*(12), 677–686. http://dx.doi.org/10.1016/j.it.2004.09.015.

Marriott, H. M., Ali, F., Read, R. C., Mitchell, T. J., Whyte, M. K., & Dockrell, D. H. (2004). Nitric oxide levels regulate macrophage commitment to apoptosis or necrosis during pneumococcal infection. *The FASEB Journal*, *18*(10), 1126–1128. http://dx.doi.org/10.1096/fj.03-1450fje.

Marriott, H. M., Bingle, C. D., Read, R. C., Braley, K. E., Kroemer, G., Hellewell, P. G., et al. (2005). Dynamic changes in Mcl-1 expression regulate macrophage viability or commitment to apoptosis during bacterial clearance. *The Journal of Clinical Investigation*, *115*(2), 359–368. http://dx.doi.org/10.1172/JCI21766.

Marriott, H. M., Daigneault, M., Thompson, A. A., Walmsley, S. R., Gill, S. K., Witcher, D. R., et al. (2012). A decoy receptor 3 analogue reduces localised defects in phagocyte function in pneumococcal pneumonia. *Thorax*, *67*(11), 985–992. http://dx.doi.org/10.1136/thoraxjnl-2012-201591.

Marriott, H. M., Gascoyne, K. A., Gowda, R., Geary, I., Nicklin, M. J., Iannelli, F., et al. (2011). IL-1beta regulates CXCL8 release and influences disease outcome in response to Streptococcus pneumoniae, defining intracellular cooperation between pulmonary epithelial cells and macrophages. *Infection and Immunity*, *80*(3), 1140–1149. http://dx.doi.org/10.1128/IAI.05697-11.

Marriott, H. M., Hellewell, P. G., Cross, S. S., Ince, P. G., Whyte, M. K., & Dockrell, D. H. (2006). Decreased alveolar macrophage apoptosis is associated with increased pulmonary inflammation in a murine model of pneumococcal pneumonia. *Journal of Immunology*, *177*(9), 6480–6488.

Marriott, H. M., Hellewell, P. G., Whyte, M. K., & Dockrell, D. H. (2007). Contrasting roles for reactive oxygen species and nitric oxide in the innate response to pulmonary infection with Streptococcus pneumoniae. *Vaccine*, *25*(13), 2485–2490. http://dx.doi.org/10.1016/j.vaccine.2006.09.024.

Marriott, H. M., Jackson, L. E., Wilkinson, T. S., Simpson, A. J., Mitchell, T. J., Buttle, D. J., et al. (2008). Reactive oxygen species regulate neutrophil recruitment and survival in pneumococcal pneumonia. *American Journal of Respiratory and Critical Care Medicine*, *177*(8), 887–895. http://dx.doi.org/10.1164/rccm.200707-990OC.

Marriott, H. M., Mitchell, T. J., & Dockrell, D. H. (2008). Pneumolysin: A double-edged sword during the host–pathogen interaction. *Current Molecular Medicine*, *8*(6), 497–509.

Maus, U. A., Backi, M., Winter, C., Srivastava, M., Schwarz, M. K., Ruckle, T., et al. (2007). Importance of phosphoinositide 3-kinase gamma in the host defense against pneumococcal infection. *American Journal of Respiratory and Critical Care Medicine*, *175*(9), 958–966. http://dx.doi.org/10.1164/rccm.200610-1533OC.

Maus, U. A., Srivastava, M., Paton, J. C., Mack, M., Everhart, M. B., Blackwell, T. S., et al. (2004). Pneumolysin-induced lung injury is independent of leukocyte trafficking into the alveolar space. *The Journal of Immunology*, *173*(2), 1307–1312.

McAllister, L. J., Tseng, H. J., Ogunniyi, A. D., Jennings, M. P., McEwan, A. G., & Paton, J. C. (2004). Molecular analysis of the psa permease complex of Streptococcus pneumoniae. *Molecular Microbiology*, *53*(3), 889–901. http://dx.doi.org/10.1111/j.1365-2958.2004.04164.x.

McGarry Houghton, A., Hartzell, W. O., Robbins, C. S., Gomis-Rüth, F. X., & Shapiro, S. D. (2009). Macrophage elastase kills bacteria within murine macrophages. *Nature*, *460*, 637–641. http://dx.doi.org/10.1038/nature08181.

McNeela, E. A., Burke, A., Neill, D. R., Baxter, C., Fernandes, V. E., Ferreira, D., et al. (2010). Pneumolysin activates the NLRP3 inflammasome and promotes proinflammatory cytokines independently of TLR4. *PLoS Pathogens*, *6*(11), e1001191. http://dx.doi.org/10.1371/journal.ppat.1001191.

Mcneely, T. B., & Coonrod, J. D. (1993). Comparison of the opsonic activity of human surfactant protein A for Staphylococcus aureus and Streptococcus pneumoniae with rabbit and human macrophages. *The Journal of Infectious Diseases*, *167*(1), 91–97.

Medeiros, A. I., Serezani, C. H., Lee, S. P., & Peters-Golden, M. (2009). Efferocytosis impairs pulmonary macrophage and lung antibacterial function via PGE2/EP2 signaling. *The Journal of Experimental Medicine*, *206*(1), 61–68. http://dx.doi.org/10.1084/jem.20082058.

Mege, J. L., Mehraj, V., & Capo, C. (2011). Macrophage polarization and bacterial infections. *Current Opinion in Infectious Diseases*, *24*(3), 230–234. http://dx.doi.org/10.1097/QCO.0b013e328344b73e.

Miller, E., Andrews, N. J., Waight, P. A., Slack, M. P., & George, R. C. (2011). Herd immunity and serotype replacement 4 years after seven-valent pneumococcal conjugate vaccination in England and Wales: An observational cohort study. *The Lancet Infectious Diseases*, *11*(10), 760–768. http://dx.doi.org/10.1016/S1473-3099(11)70090-1.

Miller, M., Cook, H. A., Furuya, E. Y., Bhat, M., Lee, M. H., Vavagiakis, P., et al. (2009). Staphylococcus aureus in the community: Colonization versus infection. *PLoS One*, *4*(8), e6708. http://dx.doi.org/10.1371/journal.pone.0006708.

Minakami, R., & Sumimotoa, H. (2006). Phagocytosis-coupled activation of the superoxide-producing phagocyte oxidase, a member of the NADPH oxidase (nox) family. *International Journal of Hematology*, *84*(3), 193–198. http://dx.doi.org/10.1532/IJH97.06133.

Miousse, I. R., Chalbot, M.-C. G., Aykin-Burns, N., Wang, X., Basnakian, A., Kavouras, I. G., et al. (2014). Epigenetic alterations induced by ambient particulate matter in mouse macrophages. *Environmental and Molecular Mutagenesis*, *55*(5), 428–435. http://dx.doi.org/10.1002/em.21855.

Mombelli, M., Lugrin, J., Rubino, I., Chanson, A. L., Giddey, M., Calandra, T., et al. (2011). Histone deacetylase inhibitors impair antibacterial defenses of macrophages. *The Journal of Infectious Diseases*, *204*(9), 1367–1374. http://dx.doi.org/10.1093/infdis/jir553.

Moquin, D. M., Mcquade, T., & Chan, F. K. (2013). CYLD deubiquitinates RIP1 in the TNFalpha-induced necrosome to facilitate kinase activation and programmed necrosis. *PLoS One*, *8*(10), e76841. http://dx.doi.org/10.1371/journal.pone.0076841.

Morris, G. E., Parker, L. C., Ward, J. R., Jones, E. C., Whyte, M. K., Brightling, C. E., et al. (2006). Cooperative molecular and cellular networks regulate Toll-like receptor-dependent inflammatory responses. *The FASEB Journal*, *20*(12), 2153–2155. http://dx.doi.org/10.1096/fj.06-5910fje.

Morris, G. E., Whyte, M. K., Martin, G. F., Jose, P. J., Dower, S. K., & Sabroe, I. (2005). Agonists of toll-like receptors 2 and 4 activate airway smooth muscle via mononuclear leukocytes. *American Journal of Respiratory and Critical Care Medicine*, *171*(8), 814–822. http://dx.doi.org/10.1164/rccm.200403-406OC.

Mosser, D. M., & Edwards, J. P. (2008). Exploring the full spectrum of macrophage activation. *Nature Reviews Immunology*, *8*(12), 958–969. http://dx.doi.org/10.1038/nri2448.

Mosser, D. M., & Handman, E. (1992). Treatment of murine macrophages with interferon-gamma inhibits their ability to bind leishmania promastigotes. *Journal of Leukocyte Biology*, *52*(4), 369–376.

Muller, S., Faulhaber, A., Sieber, C., Pfeifer, D., Hochberg, T., Gansz, M., et al. (2014). The endolysosomal cysteine cathepsins L and K are involved in macrophage-mediated clearance of Staphylococcus aureus and the concomitant cytokine induction. *The FASEB Journal*, *28*(1), 162–175. http://dx.doi.org/10.1096/fj.13-232272.

Müller, U., Stenzel, W., Köhler, G., Werner, C., Polte, T., Hansen, G., et al. (2007). IL-13 induces disease-promoting type 2 cytokines, alternatively activated macrophages and allergic inflammation during pulmonary infection of mice with Cryptococcus neoformans. *The Journal of Immunology*, *179*(8), 5367–5377.

Murphey, S. A., Root, R. K., & Schreiber, A. D. (1979). The role of antibody and complement in phagocytosis by rabbit alveolar macrophages. *The Journal of Infectious Diseases*, *140*(6), 896–903.

Nakagawara, A., Nathan, C. F., & Cohn, Z. A. (1981). Hydrogen peroxide metabolism in human monocytes during differentiation in vitro. *The Journal of Clinical Investigation*, *68*(5), 1243–1252.

Nakayama, M., Kurokawa, K., Nakamura, K., Lee, B. L., Sekimizu, K., Kubagawa, H., et al. (2012). Inhibitory receptor paired Ig-like receptor B is exploited by Staphylococcus aureus for virulence. *The Journal of Immunology*, *189*(12), 5903–5911. http://dx.doi.org/10.4049/jimmunol.1201940.

Nau, G. J., Richmond, J. F. L., Schlesinger, A., Jennings, E. G., Lander, E. S., & Young, R. A. (2002). Human macrophage activation programs induced by bacterial pathogens. *Proceedings of the National Academy of Sciences*, *99*(3), 1503–1508. http://dx.doi.org/10.1073/pnas.022649799.

Netea, M. G., Nold-Petry, C. A., Nold, M. F., Joosten, L. A., Opitz, B., Van Der Meer, J. H., et al. (2009). Differential requirement for the activation of the inflammasome for processing and release of IL-1beta in monocytes and macrophages. *Blood*, *113*(10), 2324–2335. http://dx.doi.org/10.1182/blood-2008-03-146720.

Netea, M. G., Quintin, J., & Van Der Meer, J. W. M. (2011). Trained immunity: A memory for innate host defense. *Cell Host & Microbe*, *9*(5), 355–361. http://dx.doi.org/10.1016/j.chom.2011.04.006.

Nibbering, P. H., van den Barselaar, M. T., van de Gevel, J. S., Leijh, P. C., & van Furth, R. (1989). Deficient intracellular killing of bacteria by murine alveolar macrophages. *American Journal of Respiratory Cell and Molecular Biology*, *1*(5), 417–422.

Nicholson, S., Bonecini-Almeida Mda, G., Lapa E Silva, J. R., Nathan, C., Xie, Q. W., Mumford, R., et al. (1996). Inducible nitric oxide synthase in pulmonary alveolar macrophages from patients with tuberculosis. *The Journal of Experimental Medicine, 183*(5), 2293–2302.

Nygaard, T. K., Pallister, K. B., Dumont, A. L., Dewald, M., Watkins, R. L., Pallister, E. Q., et al. (2012). Alpha-toxin induces programmed cell death of human T cells, B cells, and monocytes during USA300 infection. *PLoS One*, 7(5), e36532. http://dx.doi.org/10.1371/journal.pone.0036532.

O'Brien, K. L., Wolfson, L. J., Watt, J. P., Henkle, E., Deloria-Knoll, M., McCall, N., et al. (2009). Burden of disease caused by Streptococcus pneumoniae in children younger than 5 years: Global estimates. *Lancet, 374*(9693), 893–902. http://dx.doi.org/10.1016/S0140-6736(09)61204-6.

O'Connell, R. M., Taganov, K. D., Boldin, M. P., Cheng, G., & Baltimore, D. (2007). MicroRNA-155 is induced during the macrophage inflammatory response. *Proceedings of the National Academy of Sciences, 104*(5), 1604–1609. http://dx.doi.org/10.1073/pnas.0610731104.

O'Riordan, K., & Lee, J. C. (2004). Staphylococcus aureus capsular polysaccharides. *Clinical Microbiology Reviews, 17*(1), 218–234.

Opitz, B., Puschel, A., Schmeck, B., Hocke, A. C., Rosseau, S., Hammerschmidt, S., et al. (2004). Nucleotide-binding oligomerization domain proteins are innate immune receptors for internalized Streptococcus pneumoniae. *The Journal of Biological Chemistry, 279*(35), 36426–36432. http://dx.doi.org/10.1074/jbc.M403861200.

Park, C. Y., Kim, E. H., Choi, S. Y., Tran, T. D., Kim, I. H., Kim, S. N., et al. (2010). Virulence attenuation of Streptococcus pneumoniae clpP mutant by sensitivity to oxidative stress in macrophages via an NO-mediated pathway. *Journal of Microbiology, 48*(2), 229–235. http://dx.doi.org/10.1007/s12275-010-9300-0.

Parker, D., Martin, F. J., Soong, G., Harfenist, B. S., Aguilar, J. L., Ratner, A. J., et al. (2011). Streptococcus pneumoniae DNA initiates type I interferon signaling in the respiratory tract. *mBio. 2*(3). http://dx.doi.org/10.1128/mBio.00016-11, e00016-00011.

Parren, P. W., Warmerdam, P. A., Boeije, L. C., Arts, J., Westerdaal, N. A., Vlug, A., et al. (1992). On the interaction of IgG subclasses with the low affinity Fc gamma RIIa (CD32) on human monocytes, neutrophils, and platelets. Analysis of a functional polymorphism to human IgG2. *The Journal of Clinical Investigation, 90*(4), 1537–1546. http://dx.doi.org/10.1172/JCI116022.

Paschos, K., & Allday, M. J. (2010). Epigenetic reprogramming of host genes in viral and microbial pathogenesis. *Trends in Microbiology, 18*(10), 439–447. http://dx.doi.org/10.1016/j.tim.2010.07.003.

Pascual, C. J., Sanberg, P. R., Chamizo, W., Haraguchi, S., Lerner, D., Baldwin, M., et al. (2005). Ovarian monocyte progenitor cells: Phenotypic and functional characterization. *Stem Cells and Development, 14*(2), 173–180. http://dx.doi.org/10.1089/scd.2005.14.173.

Paterson, G. K., Blue, C. E., & Mitchell, T. J. (2006). An operon in Streptococcus pneumoniae containing a putative alkylhydroperoxidase D homologue contributes to virulence and the response to oxidative stress. *Microbial Pathogenesis, 40*(4), 152–160. http://dx.doi.org/10.1016/j.micpath.2005.12.003.

Pelegrin, P., & Surprenant, A. (2006). Pannexin-1 mediates large pore formation and interleukin-1beta release by the ATP-gated P2X7 receptor. *The EMBO Journal, 25*(21), 5071–5082. http://dx.doi.org/10.1038/sj.emboj.7601378.

Pennini, M. E., Perkins, D. J., Salazar, A. M., Lipsky, M., & Vogel, S. N. (2013). Complete dependence on IRAK4 kinase activity in TLR2, but not TLR4, signaling pathways underlies decreased cytokine production and increased susceptibility to Streptococcus pneumoniae infection in IRAK4 kinase-inactive mice. *The Journal of Immunology, 190*(1), 307–316. http://dx.doi.org/10.4049/jimmunol.1201644.

Peppoloni, S., Colombari, B., Neglia, R., Quaglino, D., Iannelli, F., Oggioni, M. R., et al. (2006). The lack of Pneumococcal surface protein C (PspC) increases the susceptibility of Streptococcus pneumoniae to the killing by microglia. *Medical Microbiology and Immunology, 195*(1), 21–28. http://dx.doi.org/10.1007/s00430-005-0243-8.

Pericone, C. D., Park, S., Imlay, J. A., & Weiser, J. N. (2003). Factors contributing to hydrogen peroxide resistance in Streptococcus pneumoniae include pyruvate oxidase (SpxB) and avoidance of the toxic effects of the fenton reaction. *Journal of Bacteriology, 185*(23), 6815–6825.

Petrilli, V., Papin, S., Dostert, C., Mayor, A., Martinon, F., & Tschopp, J. (2007). Activation of the NALP3 inflammasome is triggered by low intracellular potassium concentration. *Cell Death and Differentiation, 14*(9), 1583–1589. http://dx.doi.org/10.1038/sj.cdd.4402195.

Philibert, R. A., Sears, R. A., Powers, L. S., Nash, E., Bair, T., Gerke, A. K., et al. (2012). Coordinated DNA methylation and gene expression changes in smoker alveolar macrophages: Specific effects on VEGF receptor 1 expression. *Journal of Leukocyte Biology, 92*(3), 621–631. http://dx.doi.org/10.1189/jlb.1211632.

Phipps, J. C., Aronoff, D. M., Curtis, J. L., Goel, D., O'Brien, E., & Mancuso, P. (2010). Cigarette smoke exposure impairs pulmonary bacterial clearance and alveolar macrophage complement-mediated phagocytosis of Streptococcus pneumoniae. *Infection and Immunity, 78*(3), 1214–1220. http://dx.doi.org/10.1128/IAI.00963-09.

Picard, C., Puel, A., Bonnet, M., Ku, C. L., Bustamante, J., Yang, K., et al. (2003). Pyogenic bacterial infections in humans with IRAK-4 deficiency. *Science, 299*(5615), 2076–2079. http://dx.doi.org/10.1126/science.1081902.

Pizzolla, A., Hultqvist, M., Nilson, B., Grimm, M. J., Eneljung, T., Jonsson, I. M., et al. (2012). Reactive oxygen species produced by the NADPH oxidase 2 complex in monocytes protect mice from bacterial infections. *The Journal of Immunology, 188*(10), 5003–5011. http://dx.doi.org/10.4049/jimmunol.1103430.

Potter, A. J., Kidd, S. P., McEwan, A. G., & Paton, J. C. (2010). The MerR/NmlR family transcription factor of Streptococcus pneumoniae responds to carbonyl stress and modulates hydrogen peroxide production. *Journal of Bacteriology, 192*(15), 4063–4066. http://dx.doi.org/10.1128/JB.00383-10.

Preston, J. A., & Dockrell, D. H. (2008). Virulence factors in pneumococcal respiratory pathogenesis. *Future Microbiology, 3*(2), 205–221. http://dx.doi.org/10.2217/17460913.3.2.205.

Quinn, M. T., & Gauss, K. A. (2004). Structure and regulation of the neutrophil respiratory burst oxidase: Comparison with nonphagocyte oxidases. *Journal of Leukocyte Biology, 76*(4), 760–781. http://dx.doi.org/10.1189/jlb.0404216.

Quintin, J., Saeed, S., Martens, J. H. A., Giamarellos-Bourboulis, E. J., Ifrim, D. C., Logie, C., et al. (2012). Candida albicans infection affords protection against reinfection via functional reprogramming of monocytes. *Cell Host & Microbe. 12*(2). http://dx.doi.org/10.1016/j.chom.2012.06.006.

Quinton, L. J., Jones, M. R., Simms, B. T., Kogan, M. S., Robson, B. E., Skerrett, S. J., et al. (2007). Functions and regulation of NF-kappaB RelA during pneumococcal pneumonia. *Journal of Immunology, 178*(3), 1896–1903. http://dx.doi.org/10.4049/jimmunol.178.3.1896.

Raes, G., Beschin, A., Ghassabeh, G. H., & De Baetselier, P. (2007). Alternatively activated macrophages in protozoan infections. *Current Opinion in Immunology, 19*(4), 454–459. http://dx.doi.org/10.1016/j.coi.2007.05.007.

Reeves, E. P., Lu, H., Jacobs, H. L., Messina, C. G., Bolsover, S., Gabella, G., et al. (2002). Killing activity of neutrophils is mediated through activation of proteases by K+ flux. *Nature, 416*(6878), 291–297.

Reyes-Robles, T., Alonzo, F., 3rd., Kozhaya, L., Lacy, D. B., Unutmaz, D., & Torres, V. J. (2013). Staphylococcus aureus leukotoxin ED targets the chemokine receptors CXCR1 and CXCR2 to kill leukocytes and promote infection. *Cell Host & Microbe*, *14*(4), 453–459. http://dx.doi.org/10.1016/j.chom.2013.09.005.

Robertson, G. T., Ng, W. L., Foley, J., Gilmour, R., & Winkler, M. E. (2002). Global transcriptional analysis of clpP mutations of type 2 Streptococcus pneumoniae and their effects on physiology and virulence. *Journal of Bacteriology*, *184*(13), 3508–3520.

Rodgers, M. A., Bowman, J. W., Liang, Q., & Jung, J. U. (2014). Regulation where autophagy intersects the inflammasome. *Antioxidants & Redox Signaling*, *20*(3), 495–506. http://dx.doi.org/10.1089/ars.2013.5347.

Roger, T., Lugrin, J., Le Roy, D., Goy, G., Mombelli, M., Koessler, T., et al. (2011). Histone deacetylase inhibitors impair innate immune responses to Toll-like receptor agonists and to infection. *Blood*, *117*(4), 1205–1217. http://dx.doi.org/10.1182/blood-2010-05-284711.

Rogers, P. D., Thornton, J., Barker, K. S., McDaniel, D. O., Sacks, G. S., Swiatlo, E., et al. (2003). Pneumolysin-dependent and -independent gene expression identified by cDNA microarray analysis of THP-1 human mononuclear cells stimulated by Streptococcus pneumoniae. *Infection and Immunity*, *71*(4), 2087–2094.

Rolando, M., Sanulli, S., Rusniok, C., Gomez-Valero, L., Bertholet, C., Sahr, T., et al. (2013). Legionella pneumophila effector RomA uniquely modifies host chromatin to repress gene expression and promote intracellular bacterial replication. *Cell Host & Microbe*, *13*(4), 395–405. http://dx.doi.org/10.1016/j.chom.2013.03.004.

Rooijakkers, S. H., Ruyken, M., Roos, A., Daha, M. R., Presanis, J. S., Sim, R. B., et al. (2005). Immune evasion by a staphylococcal complement inhibitor that acts on C3 convertases. *Nature Immunology*, *6*(9), 920–927. http://dx.doi.org/10.1038/ni1235.

Rubins, J. B., Charboneau, D., Paton, J. C., Mitchell, T. J., Andrew, P. W., & Janoff, E. N. (1995). Dual function of pneumolysin in the early pathogenesis of murine pneumococcal pneumonia. *The Journal of Clinical Investigation*, *95*(1), 142–150. http://dx.doi.org/10.1172/JCI117631.

Sabroe, I., Read, R. C., Whyte, M. K., Dockrell, D. H., Vogel, S. N., & Dower, S. K. (2003). Toll-like receptors in health and disease: Complex questions remain. *Journal of Immunology*, *171*(4), 1630–1635.

Sakiniene, E., Bremell, T., & Tarkowski, A. (1997). Inhibition of nitric oxide synthase (NOS) aggravates Staphylococcus aureus septicaemia and septic arthritis. *Clinical and Experimental Immunology*, *110*(3), 370–377.

Saleh, M., Bartual, S. G., Abdullah, M. R., Jensch, I., Asmat, T. M., Petruschka, L., et al. (2013). Molecular architecture of Streptococcus pneumoniae surface thioredoxin-fold lipoproteins crucial for extracellular oxidative stress resistance and maintenance of virulence. *EMBO Molecular Medicine*, *5*(12), 1852–1870. http://dx.doi.org/10.1002/emmm.201202435.

Sanchez, C. J., Hinojosa, C. A., Shivshankar, P., Hyams, C., Camberlein, E., Brown, J. S., et al. (2011). Changes in capsular serotype alter the surface exposure of pneumococcal adhesins and impact virulence. *PLoS One*, *6*(10), e26587. http://dx.doi.org/10.1371/journal.pone.0026587.

Sanchez, C. J., Shivshankar, P., Stol, K., Trakhtenbroit, S., Sullam, P. M., Sauer, K., et al. (2010). The pneumococcal serine-rich repeat protein is an intra-species bacterial adhesin that promotes bacterial aggregation in vivo and in biofilms. *PLoS Pathogens*, *6*(8), e1001044. http://dx.doi.org/10.1371/journal.ppat.1001044.

Sato, S., Nomura, F., Kawai, T., Takeuchi, O., Muhlradt, P. F., Takeda, K., et al. (2000). Synergy and cross-tolerance between toll-like receptor (TLR) 2- and TLR4-mediated signaling pathways. *Journal of Immunology*, *165*(12), 7096–7101.

Satoh, T., Takeuchi, O., Vandenbon, A., Yasuda, K., Tanaka, Y., Kumagai, Y., et al. (2010). The Jmjd3-Irf4 axis regulates M2 macrophage polarization and host responses against helminth infection. *Nature Immunology*, *11*(10), 936–944. http://dx.doi.org/10.1038/ni.1920.

Sawyer, R. T., Strausbauch, P. H., & Volkman, A. (1982). Resident macrophage proliferation in mice depleted of blood monocytes by strontium-89. *Laboratory Investigation, 46*(2), 165–170.

Schabbauer, G., Matt, U., Gunzl, P., Warszawska, J., Furtner, T., Hainzl, E., et al. (2010). Myeloid PTEN promotes inflammation but impairs bactericidal activities during murine pneumococcal pneumonia. *Journal of Immunology, 185*(1), 468–476. http://dx.doi.org/10.4049/jimmunol.0902221.

Schubert, K. M., & Duronio, V. (2001). Distinct roles for extracellular-signal-regulated protein kinase (ERK) mitogen-activated protein kinases and phosphatidylinositol 3-kinase in the regulation of Mcl-1 synthesis. *The Biochemical Journal, 356*(Pt 2), 473–480.

Sever-Chroneos, Z., Krupa, A., Davis, J., Hasan, M., Yang, C. H., Szeliga, J., et al. (2011). Surfactant protein A (SP-A)-mediated clearance of Staphylococcus aureus involves binding of SP-A to the staphylococcal adhesin eap and the macrophage receptors SP-A receptor 210 and scavenger receptor class A. *The Journal of Biological Chemistry, 286*(6), 4854–4870. http://dx.doi.org/10.1074/jbc.M110.125567.

Sharif, O., Matt, U., Saluzzo, S., Lakovits, K., Haslinger, I., Furtner, T., et al. (2013). The scavenger receptor CD36 downmodulates the early inflammatory response while enhancing bacterial phagocytosis during pneumococcal pneumonia. *The Journal of Immunology, 190*(11), 5640–5648. http://dx.doi.org/10.4049/jimmunol.1202270.

Shaykhiev, R., Krause, A., Salit, J., Strulovici-Barel, Y., Harvey, B. G., O'Connor, T. P., et al. (2009). Smoking-dependent reprogramming of alveolar macrophage polarization: Implication for pathogenesis of chronic obstructive pulmonary disease. *Journal of Immunology, 183*(4), 2867–2883. http://dx.doi.org/10.4049/jimmunol.0900473.

Shepherd, V. L. (1986). The role of the respiratory burst of phagocytes in host defense. *Seminars in Respiratory Infections, 1*(2), 99–106.

Sjostrom, K., Spindler, C., Ortqvist, A., Kalin, M., Sandgren, A., Kuhlmann-Berenzon, S., et al. (2006). Clonal and capsular types decide whether pneumococci will act as a primary or opportunistic pathogen. *Clinical Infectious Diseases, 42*(4), 451–459. http://dx.doi.org/10.1086/499242.

Slauch, J. M. (2011). How does the oxidative burst of macrophages kill bacteria? Still an open question. *Molecular Microbiology, 80*(3), 580–583. http://dx.doi.org/10.1111/j.1365-2958.2011.07612.x.

Sleeman, K. L., Griffiths, D., Shackley, F., Diggle, L., Gupta, S., Maiden, M. C., et al. (2006). Capsular serotype-specific attack rates and duration of carriage of Streptococcus pneumoniae in a population of children. *The Journal of Infectious Diseases, 194*(5), 682–688. http://dx.doi.org/10.1086/505710.

Smahi, A., Courtois, G., Vabres, P., Yamaoka, S., Heuertz, S., Munnich, A., et al. (2000). Genomic rearrangement in NEMO impairs NF-kappaB activation and is a cause of incontinentia pigmenti. The International Incontinentia Pigmenti (IP) Consortium. *Nature, 405*(6785), 466–472. http://dx.doi.org/10.1038/35013114.

Smith, A. M., McCullers, J. A., & Adler, F. R. (2011). Mathematical model of a three-stage innate immune response to a pneumococcal lung infection. *Journal of Theoretical Biology, 276*(1), 106–116. http://dx.doi.org/10.1016/j.jtbi.2011.01.052.

Smith, Z. D., & Meissner, A. (2013). DNA methylation: Roles in mammalian development. *Nature Reviews Genetics, 14*(3), 204–220. http://dx.doi.org/10.1038/nrg3354.

Sokol, R. J., Hudson, G., James, N. T., Frost, I. J., & Wales, J. (1987). Human macrophage development: A morphometric study. *Journal of Anatomy, 151*, 27–35.

Sokolovska, A., Becker, C. E., Ip, W. K., Rathinam, V. A., Brudner, M., Paquette, N., et al. (2013). Activation of caspase-1 by the NLRP3 inflammasome regulates the NADPH oxidase NOX2 to control phagosome function. *Nature Immunology, 14*(6), 543–553. http://dx.doi.org/10.1038/ni.2595.

Srivastava, A., Henneke, P., Visintin, A., Morse, S. C., Martin, V., Watkins, C., et al. (2005). The apoptotic response to pneumolysin is Toll-like receptor 4 dependent and protects against pneumococcal disease. *Infection and Immunity*, *73*(10), 6479–6487.

Standiford, T. J., Kunkel, S. L., Phan, S. H., Rollins, B. J., & Strieter, R. M. (1991). Alveolar macrophage-derived cytokines induce monocyte chemoattractant protein-1 expression from human pulmonary type II-like epithelial cells. *Journal of Biological Chemistry*, *266*(15), 9912–9918.

Standiford, T. J., Strieter, R. M., Chensue, S. W., Westwick, J., Kasahara, K., & Kunkel, S. L. (1990). IL-4 inhibits the expression of IL-8 from stimulated human monocytes. *The Journal of Immunology*, *145*(5), 1435–1439.

Standish, A. J., & Weiser, J. N. (2009). Human neutrophils kill Streptococcus pneumoniae via serine proteases. *Journal of Immunology*, *183*(4), 2602–2609. http://dx.doi.org/10.4049/jimmunol.0900688.

Stapels, D. A. C., Ramyar, K. X., Ricklin, D., Milder, F. J. J., Bischoff, M., Hermann, M., et al. (2012). Extracellular adherence protein (Eap) of Staphylococcus aureus evades innate immunity by inhibiting complement activation and neutrophil elastase. *Immunobiology*, *217*(11), 1170.

Stein, M., Keshav, S., Harris, N., & Gordon, S. (1992). Interleukin 4 potently enhances murine macrophage mannose receptor activity: A marker of alternative immunologic macrophage activation. *The Journal of Experimental Medicine*, *176*(1), 287–292. http://dx.doi.org/10.1084/jem.176.1.287.

Steinman, R. M., Mellman, I. S., Muller, W. A., & Cohn, Z. A. (1983). Endocytosis and the recycling of plasma membrane. *The Journal of Cell Biology*, *96*(1), 1–27.

Steinwede, K., Henken, S., Bohling, J., Maus, R., Ueberberg, B., Brumshagen, C., et al. (2012). TNF-related apoptosis-inducing ligand (TRAIL) exerts therapeutic efficacy for the treatment of pneumococcal pneumonia in mice. *The Journal of Experimental Medicine*, *209*(11), 1937–1952. http://dx.doi.org/10.1084/jem.20120983.

Stender, J. D., Pascual, G., Liu, W., Kaikkonen, M. U., Do, K., Spann, N. J., et al. (2012). Control of proinflammatory gene programs by regulated trimethylation and demethylation of histone H4K20. *Molecular Cell*, *48*(1), 28–38. http://dx.doi.org/10.1016/j.molcel.2012.07.020.

Stroeher, U. H., Kidd, S. P., Stafford, S. L., Jennings, M. P., Paton, J. C., & McEwan, A. G. (2007). A pneumococcal MerR-like regulator and S-nitrosoglutathione reductase are required for systemic virulence. *The Journal of Infectious Diseases*, *196*(12), 1820–1826. http://dx.doi.org/10.1086/523107.

Stuart, L. M., Deng, J., Silver, J. M., Takahashi, K., Tseng, A. A., Hennessy, E. J., et al. (2005). Response to Staphylococcus aureus requires CD36-mediated phagocytosis triggered by the COOH-terminal cytoplasmic domain. *The Journal of Cell Biology*, *170*(3), 477–485. http://dx.doi.org/10.1083/jcb.200501113.

Sugiyama, S., Okada, Y., Sukhova, G. K., Virmani, R., Heinecke, J. W., & Libby, P. (2001). Macrophage myeloperoxidase regulation by granulocyte macrophage colony-stimulating factor in human atherosclerosis and implications in acute coronary syndromes. *The American Journal of Pathology*, *158*(3), 879–891. http://dx.doi.org/10.1016/S0002-9440(10)64036-9.

Sutterwala, F. S., Noel, G. J., Clynes, R., & Mosser, D. M. (1997). Selective suppression of interleukin-12 induction after macrophage receptor ligation. *The Journal of Experimental Medicine*, *185*(11), 1977–1985. http://dx.doi.org/10.1084/jem.185.11.1977.

Sutterwala, F. S., Noel, G. J., Salgame, P., & Mosser, D. M. (1998). Reversal of proinflammatory responses by ligating the macrophage Fcγ receptor type I. *The Journal of Experimental Medicine*, *188*(1), 217–222. http://dx.doi.org/10.1084/jem.188.1.217.

Swanson, J. A., & Watts, C. (1995). Macropinocytosis. *Trends in Cell Biology*, *5*(11), 424–428.

Tauber, A. I. (2003). Metchnikoff and the phagocytosis theory. *Nature Reviews Molecular Cell Biology*, *4*(11), 897–901. http://dx.doi.org/10.1038/nrm1244.
Taylor, P. R., Martinez-Pomares, L., Stacey, M., Lin, H. H., Brown, G. D., & Gordon, S. (2005). Macrophage receptors and immune recognition. *Annual Review of Immunology*, *23*, 901–944. http://dx.doi.org/10.1146/annurev.immunol.23.021704.115816.
Thammavongsa, V., Missiakas, D. M., & Schneewind, O. (2013). Staphylococcus aureus degrades neutrophil extracellular traps to promote immune cell death. *Science*, *342*(6160), 863–866. http://dx.doi.org/10.1126/science.1242255.
Thigpen, M. C., Whitney, C. G., Messonnier, N. E., Zell, E. R., Lynfield, R., Hadler, J. L., et al. (2011). Bacterial meningitis in the United States, 1998–2007. *The New England Journal of Medicine*, *364*(21), 2016–2025. http://dx.doi.org/10.1056/NEJMoa1005384.
Thomas, C. A., Li, Y., Kodama, T., Suzuki, H., Silverstein, S. C., & El Khoury, J. (2000). Protection from lethal gram-positive infection by macrophage scavenger receptor-dependent phagocytosis. *The Journal of Experimental Medicine*, *191*(1), 147–156.
Thomas, E. D., Ramberg, R. E., Sale, G. E., Sparkes, R. S., & Golde, D. W. (1976). Direct evidence for a bone marrow origin of the alveolar macrophage in man. *Science*, *192*(4243), 1016–1018.
Thoma-Uszynski, S., Stenger, S., Takeuchi, O., Ochoa, M. T., Engele, M., Sieling, P. A., et al. (2001). Induction of direct antimicrobial activity through mammalian toll-like receptors. *Science*, *291*(5508), 1544–1547. http://dx.doi.org/10.1126/science.291.5508.1544.
Tomasz, A. (1997). Antibiotic resistance in Streptococcus pneumoniae. *Clinical Infectious Diseases*, *24*(Suppl. 1), S85–S88.
Tseng, H. J., McEwan, A. G., Paton, J. C., & Jennings, M. P. (2002). Virulence of Streptococcus pneumoniae: PsaA mutants are hypersensitive to oxidative stress. *Infection and Immunity*, *70*(3), 1635–1639.
van Furth, R., & Cohn, Z. A. (1968). The origin and kinetics of mononuclear phagocytes. *The Journal of Experimental Medicine*, *128*(3), 415–435.
van Furth, R., Cohn, Z. A., Hirsch, J. G., Humphrey, J. H., Spector, W. G., & Langevoort, H. L. (1972). The mononuclear phagocyte system: A new classification of macrophages, monocytes, and their precursor cells. *Bulletin of the World Health Organization*, *46*(6), 845–852.
van Wamel, W. J., Rooijakkers, S. H., Ruyken, M., Van Kessel, K. P., & Van Strijp, J. A. (2006). The innate immune modulators staphylococcal complement inhibitor and chemotaxis inhibitory protein of Staphylococcus aureus are located on beta-hemolysin-converting bacteriophages. *Journal of Bacteriology*, *188*(4), 1310–1315. http://dx.doi.org/10.1128/JB.188.4.1310-1315.2006.
Virolainen, M. (1968). Hematopoietic origin of macrophages as studied by chromosome markers in mice. *The Journal of Experimental Medicine*, *127*(5), 943–952. http://dx.doi.org/10.1084/jem.127.5.943.
Volkman, A. (1966). The origin and turnover of mononuclear cells in peritoneal exudates in rats. *The Journal of Experimental Medicine*, *124*(2), 241–254. http://dx.doi.org/10.1084/jem.124.2.241.
von Bernuth, H., Picard, C., Jin, Z., Pankla, R., Xiao, H., Ku, C. L., et al. (2008). Pyogenic bacterial infections in humans with MyD88 deficiency. *Science*, *321*(5889), 691–696. http://dx.doi.org/10.1126/science.1158298.
von Eiff, C., Becker, K., Machka, K., Stammer, H., & Peters, G. (2001). Nasal carriage as a source of *Staphylococcus aureus* bacteremia. Study Group. *The New England Journal of Medicine*, *344*(1), 11–16. http://dx.doi.org/10.1056/NEJM200101043440102.
Voyich, J. M., Otto, M., Mathema, B., Braughton, K. R., Whitney, A. R., Welty, D., et al. (2006). Is Panton–Valentine leukocidin the major virulence determinant in community-associated methicillin-resistant Staphylococcus aureus disease? *The Journal of Infectious Diseases*, *194*(12), 1761–1770. http://dx.doi.org/10.1086/509506.

Walev, I., Reske, K., Palmer, M., Valeva, A., & Bhakdi, S. (1995). Potassium-inhibited processing of IL-1 beta in human monocytes. *The EMBO Journal*, *14*(8), 1607–1614.

Walker, C. L., Rudan, I., Liu, L., Nair, H., Theodoratou, E., Bhutta, Z. A., et al. (2013). Global burden of childhood pneumonia and diarrhoea. *Lancet*, *381*(9875), 1405–1416. http://dx.doi.org/10.1016/S0140-6736(13)60222-6.

Wang, G., Abadia-Molina, A. C., Berger, S. B., Romero, X., O'Keeffe, M. S., Rojas-Barros, D. I., et al. (2012). Cutting edge: Slamf8 is a negative regulator of Nox2 activity in macrophages. *The Journal of Immunology*, *188*(12), 5829–5832. http://dx.doi.org/10.4049/jimmunol.1102620.

Wang, R., Braughton, K. R., Kretschmer, D., Bach, T. H., Queck, S. Y., Li, M., et al. (2007). Identification of novel cytolytic peptides as key virulence determinants for community-associated MRSA. *Nature Medicine*, *13*(12), 1510–1514. http://dx.doi.org/10.1038/nm1656.

Warmerdam, P. A., van de Winkel, J. G., Vlug, A., Westerdaal, N. A., & Capel, P. J. (1991). A single amino acid in the second Ig-like domain of the human Fc gamma receptor II is critical for human IgG2 binding. *The Journal of Immunology*, *147*(4), 1338–1343.

Wartha, F., Beiter, K., Albiger, B., Fernebro, J., Zychlinsky, A., Normark, S., et al. (2007). Capsule and D-alanylated lipoteichoic acids protect Streptococcus pneumoniae against neutrophil extracellular traps. *Cellular Microbiology*, *9*(5), 1162–1171. http://dx.doi.org/10.1111/j.1462-5822.2006.00857.x.

Watanabe, I., Ichiki, M., Shiratsuchi, A., & Nakanishi, Y. (2007). TLR2-mediated survival of Staphylococcus aureus in macrophages: A novel bacterial strategy against host innate immunity. *Journal of Immunology*, *178*(8), 4917–4925.

Weber, S., Tian, H., Van Rooijen, N., & Pirofski, L. A. (2012). A serotype 3 pneumococcal capsular polysaccharide-specific monoclonal antibody requires Fcgamma receptor III and macrophages to mediate protection against pneumococcal pneumonia in mice. *Infection and Immunity*, *80*(4), 1314–1322. http://dx.doi.org/10.1128/IAI.06081-11.

Webster, S. J., Daigneault, M., Bewley, M. A., Preston, J. A., Marriott, H. M., Walmsley, S. R., et al. (2010). Distinct cell death programs in monocytes regulate innate responses following challenge with common causes of invasive bacterial disease. *The Journal of Immunology*, *185*(5), 2968–2979. http://dx.doi.org/10.4049/jimmunol.1000805.

Weinberger, D. M., Trzcinski, K., Lu, Y. J., Bogaert, D., Brandes, A., Galagan, J., et al. (2009). Pneumococcal capsular polysaccharide structure predicts serotype prevalence. *PLoS Pathogens*, *5*(6), e1000476. http://dx.doi.org/10.1371/journal.ppat.1000476.

Weiser, J. N., Bae, D., Fasching, C., Scamurra, R. W., Ratner, A. J., & Janoff, E. N. (2003). Antibody-enhanced pneumococcal adherence requires IgA1 protease. *Proceedings of the National Academy of Sciences of the United States of America*, *100*(7), 4215–4220. http://dx.doi.org/10.1073/pnas.0637469100.

West, A. P., Brodsky, I. E., Rahner, C., Woo, D. K., Erdjument-Bromage, H., Tempst, P., et al. (2011). TLR signalling augments macrophage bactericidal activity through mitochondrial ROS. *Nature*, *472*(7344), 476–480. http://dx.doi.org/10.1038/nature09973.

Wheeler, D. S., Lahni, P. M., Denenberg, A. G., Poynter, S. E., Wong, H. R., Cook, J. A., et al. (2008). Induction of endotoxin tolerance enhances bacterial clearance and survival in murine polymicrobial sepsis. *Shock*, *30*(3), 267–273. http://dx.doi.org/10.1097/shk.0b013e318162c190.

Wirth, J. J., Kierszenbaum, F., & Zlotnik, A. (1989). Effects of IL-4 on macrophage functions: Increased uptake and killing of a protozoan parasite (Trypanosoma cruzi). *Immunology*, *66*(2), 296–301.

Witzenrath, M., Pache, F., Lorenz, D., Koppe, U., Gutbier, B., Tabeling, C., et al. (2011). The NLRP3 inflammasome is differentially activated by pneumolysin variants and contributes to host defense in pneumococcal pneumonia. *Journal of Immunology*, *187*(1), 434–440. http://dx.doi.org/10.4049/jimmunol.1003143.

Wolf, A. J., Arruda, A., Reyes, C. N., Kaplan, A. T., Shimada, T., Shimada, K., et al. (2011). Phagosomal degradation increases TLR access to bacterial ligands and enhances macrophage sensitivity to bacteria. *The Journal of Immunology*, *187*(11), 6002–6010. http://dx.doi.org/10.4049/jimmunol.1100232.

Wong, D., Bach, H., Sun, J., Hmama, Z., & Av-Gay, Y. (2011). Mycobacterium tuberculosis protein tyrosine phosphatase (PtpA) excludes host vacuolar-H+-ATPase to inhibit phagosome acidification. *Proceedings of the National Academy of Sciences of the United States of America*, *108*(48), 19371–19376. http://dx.doi.org/10.1073/pnas.1109201108.

Wright, S. D., & Silverstein, S. C. (1983). Receptors for C3b and C3bi promote phagocytosis but not the release of toxic oxygen from human phagocytes. *The Journal of Experimental Medicine*, *158*(6), 2016–2023.

Xu, F., Droemann, D., Rupp, J., Shen, H., Wu, X., Goldmann, T., et al. (2008). Modulation of the inflammatory response to Streptococcus pneumoniae in a model of acute lung tissue infection. *American Journal of Respiratory Cell and Molecular Biology*, *39*(5), 522–529. http://dx.doi.org/10.1165/rcmb.2007-0328OC.

Xue, J., Schmidt, S. V., Sander, J., Draffehn, A., Krebs, W., Quester, I., et al. (2014). Transcriptome-based network analysis reveals a spectrum model of human macrophage activation. *Immunity*, *40*(2), 274–288. http://dx.doi.org/10.1016/j.immuni.2014.01.006.

Yesilkaya, H., Kadioglu, A., Gingles, N., Alexander, J. E., Mitchell, T. J., & Andrew, P. W. (2000). Role of manganese-containing superoxide dismutase in oxidative stress and virulence of Streptococcus pneumoniae. *Infection and Immunity*, *68*(5), 2819–2826.

Yokoyama, R., Itoh, S., Kamoshida, G., Takii, T., Fujii, S., Tsuji, T., et al. (2012). Staphylococcal superantigen-like protein 3 binds to the Toll-like receptor 2 extracellular domain and inhibits cytokine production induced by Staphylococcus aureus, cell wall component, or lipopeptides in murine macrophages. *Infection and Immunity*, *80*(8), 2816–2825. http://dx.doi.org/10.1128/IAI.00399-12.

Yona, S., Kim, K.-W., Wolf, Y., Mildner, A., Varol, D., Breker, M., et al. (2013). Fate mapping reveals origins and dynamics of monocytes and tissue macrophages under homeostasis. *Immunity*, *38*(1), 79–91. http://dx.doi.org/10.1016/j.immuni.2012.12.001.

Yu, J., Bryant, A. P., Marra, A., Lonetto, M. A., Ingraham, K. A., Chalker, A. F., et al. (2001). Characterization of the Streptococcus pneumoniae NADH oxidase that is required for infection. *Microbiology*, *147*(Pt 2), 431–438.

Yuan, F. F., Marks, K., Wong, M., Watson, S., De Leon, E., McIntyre, P. B., et al. (2008). Clinical relevance of TLR2, TLR4, CD14 and FcgammaRIIA gene polymorphisms in Streptococcus pneumoniae infection. *Immunology and Cell Biology*, *86*(3), 268–270. http://dx.doi.org/10.1038/sj.icb.7100155.

Yuste, J., Botto, M., Paton, J. C., Holden, D. W., & Brown, J. S. (2005). Additive inhibition of complement deposition by pneumolysin and PspA facilitates Streptococcus pneumoniae septicemia. *The Journal of Immunology*, *175*(3), 1813–1819.

Zeybel, M., Hardy, T., Wong, Y. K., Mathers, J. C., Fox, C. R., Gackowska, A., et al. (2012). Multigenerational epigenetic adaptation of the hepatic wound-healing response. *Nature Medicine*, *18*(9), 1369–1377. http://dx.doi.org/10.1038/nm.2893.

Zhang, Z., Clarke, T. B., & Weiser, J. N. (2009). Cellular effectors mediating Th17-dependent clearance of pneumococcal colonization in mice. *The Journal of Clinical Investigation*, *119*(7), 1899–1909. http://dx.doi.org/10.1172/JCI36731.

Zhang, Z., Song, L., Maurer, K., Bagashev, A., & Sullivan, K. E. (2011). Monocyte polarization: The relationship of genome-wide changes in H4 acetylation with polarization. *Genes and Immunity*, *12*(6), 445–456. http://dx.doi.org/10.1038/gene.2011.17.

Zhou, H., & Kobzik, L. (2007). Effect of concentrated ambient particles on macrophage phagocytosis and killing of Streptococcus pneumoniae. *American Journal of Respiratory Cell and Molecular Biology*, *36*(4), 460–465. http://dx.doi.org/10.1165/rcmb.2006-0293OC.

CHAPTER FIVE

Aeromonas Flagella and Colonisation Mechanisms

Rebecca Lowry*, Sabela Balboa*,†, Jennifer L. Parker*, Jonathan G. Shaw*,[1]

*Department of Infection and Immunity, University of Sheffield, Sheffield, United Kingdom
†Departamento de Microbiología y Parasitología, Universidad de Santiago de Compostela, Santiago de Compostela, Spain
[1]Corresponding author: e-mail address: j.g.shaw@sheffield.ac.uk

Contents

Abstract

Aeromonas species are inhabitants of aquatic environments and are able to cause disease in humans and fish among other animals. In aquaculture, they are responsible for the economically important diseases of furunculosis and motile *Aeromonas* septicaemia (MAS). Whereas gastroenteritis and wound infections are the major human diseases associated with the genus. As they inhabit and survive in diverse environments, aeromonads possess a wide range of colonisation factors. The motile species are able to swim in liquid environments through the action of a single polar flagellum, the flagellin subunits of which are glycosylated; although essential for function the biological role of

Advances in Microbial Physiology
ISSN 0065-2911
http://dx.doi.org/10.1016/bs.ampbs.2014.08.007

glycan addition is yet to be determined. Approximately 60% of aeromonads possess a second lateral flagella system that is expressed in viscous environments for swarming over surfaces; both flagellar systems have been shown to be important in the initial colonisation of surfaces. Subsequently, other non-flagellar colonisation factors are employed; these can be both filamentous and non-filamentous. The aeromonads possess a number of fimbrial systems with the bundle-forming MSHA type IV pilus system, having a major role in human cell adherence. Furthermore, a series of outer-membrane proteins have also been implicated in the aeromonad adhesion process. A number of strains are also capable of cell invasion and that maybe linked with the more invasive diseases of bacteraemia or wound infections. These strains employ cell surface factors that allow the colonisation of these niches that protect them from the host's immune system such as S-layers, capsules or particular lipopolysaccharides.

1. INTRODUCTION

Bacteria of the genus *Aeromonas* are Gram negative, rod-shaped, facultative anaerobes that along with the genera *Oceanimonas* and *Tolumonas* make up the family *Aeromonadaceae* within the Gammaproteobacteria (Martin-Carnahan & Joseph, 2005). The genus *Aeromonas* is taxonomically complex but is divided into two main groups, the first is the non-motile psychrophilic aeromonads, with optimal growth temperatures of 22–28 °C represented by *Aeromonas salmonicida*. The second much larger group contains the mesophilic motile aeromonads that have optimal growth temperatures of 30–37 °C. Recently with the application of molecular methods supplementing biochemical analysis, the number of species within the genus has expanded rapidly and currently the genus contains 25 accepted and validated species (Beaz-Hidalgo & Figueras, 2013).

Aeromonads are found ubiquitously in the environment, but are mainly associated with fresh or estuarine brackish water. They have been isolated from vegetables, meat, seafood or processed foods. Some strains are able to grow in 0–6% salt, whereas others can grow at refrigerator temperatures or at pH ranges of 4–10 and are therefore thought to be a potential food contamination problem (Janda & Abbott, 2010; Kirov, 1997).

Aeromonas are able to cause disease in many animal species but are mainly associated with infections in aquatic poikilothermic animals. The psychrophilic species *A. salmonicida* subsp. *salmonicida* is a problem in aquaculture causing furunculosis, an infective bacteraemic and ulcerative disease of salmonid fish (salmon, trout) as well as several other economically important fish species such as turbot, or cod (Beaz-Hidalgo & Figueras, 2013). Motile

Aeromonas septicaemia (MAS) is predominantly caused by *A. hydrophila* and a number of other motile aeromonad species (Beaz-Hidalgo & Figueras, 2013); it is another economically important fish disease that mainly affects fresh water species such as, tilapia, perch and carp; and it has caused extensive losses to commercial catfish operations in the southern USA (Hossain et al., 2013). Aeromonads can cause a series of others infections in both reptiles and amphibians, *A. hydrophila* can cause the internal haemorrhaging and often fatal disease of *Xenopus* frogs through a disease called Red Leg, whereas they are also linked to ulcerative stomatitis in reptiles (Densmore & Green, 2007; Janda & Abbott, 2010).

In humans, the mesophilic *Aeromonas* are mainly linked to gastroenteritis and wound infections, with more than 85% of human clinical cases being caused by three species, *A. hydrophila*, *A. caviae* and *A. veronii* biovar sobria. Other extra-intestinal clinical infections include septicaemia in immunocompromised patients that usually have an underlying condition such as impaired hepatobiliary function, cases of meningitis, peritonitis, epiglottitis and pneumonia from near-drowning experiences have also been recorded (Parker & Shaw, 2011). The most common clinical disease associated with *Aeromonas* is gastroenteritis; this can manifest itself in three ways, mild self-limiting watery diarrhoea, a bloody dysentery-like illness or chronic diarrhoea. *Aeromonas* gastroenteritis can be problematic as certain aeromonad strains can carry Shiga-like toxins that can result in the rare complication of haemolytic uremic syndrome (Robson, Leung, & Trevenen, 1992). The second most common aeromonad clinical manifestations are wound infections. These usually occur in healthy individuals following trauma in an aquatic environment, and can present as a mild cellulitis, through to myonecrosis or even necrotising fasciitis that has a high mortality (Parker & Shaw, 2011). Zoonotic aeromonad wound infections have also been reported in patients that have survived bites from bears, tigers, alligators, snakes and other reptiles (Abrahamian & Goldstein, 2011; Easow & Tuladhar, 2007).

Certain *Aeromonas* strains are also major components of the animal microbiota. Strains of *A. veronii* along with *Rikenella* species are the major symbionts of the medicinal leech *Hirudo verbana* (Silver, Rabinowitz, Küffer, & Graf, 2007) possibly helping the digestion of the blood meal. *Aeromonas* species mostly *A. veronii* are also known to be members of the mosquito microbiota; however, a recent paper using deep sequencing of wild mosquitoes from Kenya has revealed that *Aeromonas* spp. are the commonest member of the mosquito mid-gut microbiota making up nearly 40% of the bacterial strains isolated (Osei-Poku, Mbogo, Palmer, & Jiggins, 2012).

Therefore, with their ability to live in a number of environments and their diverse series of both warm-blooded and cold-blooded hosts *Aeromonas* utilise a number of different colonisation mechanisms, the major ones of which are the subject of this review and are described below.

2. FLAGELLA

Bacterial flagella are complex nano-machines that comprise a 10–15 μm helical filament that protrudes out from the cell surface that is anchored to a rotating basal body that spans the bacterial cell envelope (Terashima, Kojima, & Homma, 2008). The main function of flagella is motility, in liquid environments and on solid surfaces. This allows the bacteria to move towards nutrients and more favourable environments and away from less promising locations, a process known as chemotaxis. As well as motility, the flagellum is also an important antigenic structure involved in successful adhesion and colonisation. Flagellin, the protein that polymerises to make up the filament, is a direct target of the innate immune system through Toll-like receptor 5 (TLR-5) leading to increased production of interleukin-6 in mammals (Hayashi et al., 2001). Approximately 60% of mesophilic *Aeromonas* species possess dual flagella systems, including the clinically important species *A. caviae* and *A. hydrophila*. In these strains, a polar flagellum is constitutively expressed for motility in liquid environments, as well as a completely distinct inducible lateral flagellar system for movement across solid surfaces or through viscous environments (Gavin et al., 2002; Kirov et al., 2002; Rabaan, Gryllos, Tomás, & Shaw, 2001). Here, we describe in detail the structure and regulation of the polar and lateral flagella systems in mesophilic aeromonads and also discuss the distinct differences between these systems and that of psychrophilic aeromonads such as *A. salmonicida*.

2.1. Polar flagella

The mesophilic aeromonads such as *A. caviae* and *A. hydrophila* are motile in liquid environments (in broth culture) by the action of a single unsheathed polar flagellum. This polar flagellum is constitutively expressed and is encoded by over 50 genes located within five major gene clusters on the chromosome. The helical filament of the polar flagellum of *Aeromonas* is a polymer of two repeating flagellin subunits known as FlaA and FlaB (Canals, Altarriba, et al., 2006; Canals, Jimenez, et al., 2006; Canals, Ramirez, et al., 2006; Rabaan et al., 2001).

Other components of the aeromonad polar flagellum of interest are the flagellar motor components. The motor is composed of the rotor and stator. The rotary component (rotor) comprises the proteins FliG, FliM and FliN that make up the C-ring, and in *Escherichia coli* and *Salmonella*, the stationary component (stator) is made up of the proteins MotA and MotB. In the *E. coli* and *Salmonella* model of motility, MotA and MotB couple the flow of protons down an ion gradient for flagella rotation. In contrast to the *E. coli* and *Salmonella* systems, the stator complex for the polar flagellum in *Aeromonas* is more complex and similar to those of *Vibrio parahaemolyticus* and *Shewanella oneidensis*. In aeromonads, the stator complex is driven by sodium ions rather than protons and requires four proteins, namely, PomA, PomB, MotX and MotY, where PomA and PomB are the paralogs of the proton-driven MotA and MotB (Molero, Wilhelms, Infanzón, Tomás, & Merino, 2011; Wilhelms et al., 2009). MotX and MotY are absent in the *E. coli* flagella system, but in the *Vibrio* polar flagellum they form the T-ring component of the basal body. Furthermore, aeromonads have two sets of the Pom-motor proteins designated PomA and PomB and $PomA_2$ and $PomB_2$ (Wilhelms et al., 2009). Both sets are sodium-coupled stator complexes and interestingly neither set is required for full motility; however, a recent study has shown that mutation of either set of Pom motors does result in variations in their sensitivities to sodium ion concentrations (Wilhelms et al., 2009).

Most work on the *Aeromonas* flagellar systems has concentrated on the "motile" mesophilic species. In these systems, the helical filament is composed of repeating subunits known as flagellins. The aeromonad polar flagellum has two flagellin subunits (FlaA and FlaB) that are approximately 32 kDa in size and are exported as unfolded monomers via a dedicated type III protein secretion system (T3SS). Following secretion, the flagellin monomers are folded and able to polymerise assembling into filaments which requires the presence of the filament capping protein FlaH. The cap helps with the formation of the filament which elongates as monomers of flagellin continue to be secreted and polymerised, resulting in the formation of a full-length flagellum (Rabaan et al., 2001) (Fig. 1).

Flagellin proteins are composed of three main domains that are required for function (Fig. 1A); the N-terminal region (approximately amino acids 1–50) of the protein which contains a putative signal sequence required for export of the protein via the dedicated T3SS; the variable central D2/D3 antigenic domain (approximately amino acids 146–231) which is modified with the addition of sugar residues in a large number of bacteria including *A. caviae* and *A. hydrophila*; and the C-terminal region that is

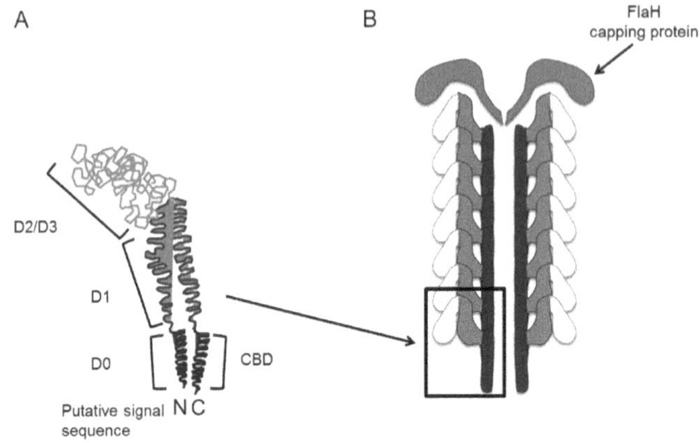

Figure 1 Schematic diagram depicting the cross-section of a flagella filament and flagellin subunit. (A) A single flagellin subunit, showing the conserved D0 and D1 domains located at the C- and N-termini of the flagellin; this is the area recognised by the innate immune systems Toll-like receptor TLR-5. The variable D2/D3 domain is also indicated that in the *Aeromonas* is glycosylated with pseudaminic acid. The N-terminal D0 regions contain the putative signal sequence for flagellin secretion, whereas the C-terminal final 45 amino acids comprises the chaperone (FlaJ) binding domain (CBD). (B) Cross-section of part of a flagella filament showing a number of individual flagellin subunits polymerised into a filament showing the hollow tube through which flagellin subunits are secreted. The capping proteins (FlaH) help the flagellins polymerise into the filament at the distal end of the growing flagellum. The conserved parts of the flagellins the D0 and D1 domains are located inside the filament, whereas the variable glycosylated D2/D3 domain of the flagellin are exposed on the filament exterior. (See the color plate.)

(approximately amino acids 262-end), required for recognition by the dedicated flagellin-specific chaperone (annotated in aeromonads as FlaJ). The conserved regions at the C- and N-termini are also known as the D0 domains, and these are the sections of flagellin that are recognised by the innate immune system receptor TLR-5. They are linked to the central D2/D3 domain via two further conserved stretches of the protein known as the D1 regions. The flagellin monomers fold in such a way that the less conserved or variable central D2/D3 domain that is modified through glycosylation (see below) is exposed to the outer surface of the filament while concealing the parts of the flagellin that are recognised by the innate immune system inside the structure of the polymerised flagellar filament (Smith et al., 2003) (Fig. 1B).

The acknowledged discrepancies regarding *Aeromonas* taxonomy issues has been largely avoided with *A. salmonicida* due to its distinct role as a fish pathogen and its apparent lack of motility making it simpler to identify than other species. It has been previously defined as both aflagellated and non-motile. However, in 1991 McIntosh and Austin demonstrated that *A. salmonicida* was capable of polar flagella-mediated motility when grown at temperatures higher than optimal (30–37 °C) (McIntosh & Austin, 1991). The first genetic evidence that *A. salmonicida* were in fact capable of producing flagella was provided by Umelo and Trust in 1997. In this study, the flagellin genes (*flaA* and *flaB*) of *A. salmonicida* A449 were identified, cloned and expressed in *E. coli*. The study was extended through electron microscopy analysis of *A. salmonicida* A449 which confirmed that it expresses an unsheathed polar flagella at extremely low frequencies when grown in liquid culture under normal laboratory growth conditions (Umelo & Trust, 1997), providing an explanation as to why it had been previously undiscovered. Genome sequencing of *A. salmonicida* A449 demonstrated the presence of all the genes required for the biosynthesis of a polar flagellum as well as lateral flagella. However, several of the genes contained disruptions through nucleotide deletions (*flrA* and *maf1*) or nucleotide duplications (*flgL*), resulting in abrogation of the synthesis of a functional flagellum (Reith et al., 2008).

2.2. Flagellin glycosylation

O-linked glycosylation of flagellins, the major subunits of the flagellar filament, has been described for a number of bacteria (Josenhans, Vossebein, Friedrich, & Suerbaum, 2002; LeClerc, Wang, & Ely, 1998; Schirm et al., 2004, 2003; Thibault et al., 2001; Verma et al., 2006) and was first described in *A. caviae* in 2001 (Rabaan et al., 2001), where it was observed that the polar flagellins FlaA and FlaB (32 kDa) showed aberrant migration (35–36 kDa) when analysed by SDS-PAGE. In a later study, mass spectrometry analysis of purified flagellin of *A. caviae* indicated that FlaA and FlaB of *A. caviae* Sch3 are homogeneously glycosylated with between six and eight pseudaminic acid (Pse5Ac7Ac) moieties of 317 Da that are linked on to serine and threonine residues in the central immunogenic D2/D3 domains of the flagellin via O-linked glycosylation (Tabei et al., 2009). Pseudaminic acid is a nine-carbon nonulosonic acid sugar related to sialic acid; Pse5Ac7Ac also decorates the flagellins of other pathogenic bacteria such as *Campylobacter jejuni* and *Helicobacter pylori*. Transposon mutagenesis initially discovered the *A. caviae* flagellin glycosylation locus (Gryllos, Shaw, Gavin, Merino, & Tomás, 2001). Tabei and coworkers subsequently demonstrated that five of the genes in this locus *flmA*, *flmB*, *neuB*, *flmD* and *neuA* were required for the synthesis and activation of Pse5Ac7Ac (Tabei et al.,

2009) and that these genes were essential for motility through the production of flagella. This locus appears to be the smallest flagellar glycosylation locus so far discovered in bacteria. In order for Pse5Ac7Ac to be transferred onto the flagellin, it has to be activated with the addition of cytidine-monophosphate (CMP) by NeuA (also called PseF) to give Pse5Ac7Ac-CMP. There is now strong evidence to suggest that the family of motility accessory factor (Maf) proteins performs transfer of activated pseudaminic acid on to the flagellin monomers. The paralogous family of Maf proteins were first identified in the flagellar glycosylation islands of *C. jejuni* (Karlyshev, Linton, Gregson, & Wren, 2002). The genes encoding Maf proteins are present in the flagellar glycosylation loci of bacteria that glycosylate their flagellins with nonulosonic acids (Parker, Day-Williams, Tomás, Stafford, & Shaw, 2012). Putative *maf* homologues have been identified in a number of *Aeromonas* species, including the clinically important *A. caviae* (Parker et al., 2012), *A. hydrophila* (Canals et al., 2007) and *A. veronii* (J. Shaw unpublished), as well as *A. diversa* and *A. dhakensis* (Shaw unpublished). In addition, *maf* homologues have also been identified in the psychrophilic fish pathogens *A. salmonicida* and *A. molluscorum* (Reith et al., 2008; Shaw unpublished). Genes encoding putative Maf proteins have been shown to be required for flagellin glycosylation and the formation of a fully functional flagellum in *A. caviae* and *A. hydrophila* (Canals et al., 2007; Parker et al., 2014). More recently, direct *in vivo* interactions between Maf1 and the flagellin proteins FlaA and FlaB in *A. caviae* have been observed (Parker et al., 2014) (Fig. 2). The genome of *A. caviae* encodes a single *maf* homologue (Parker et al., 2012) and its flagellin has been shown to be homogenously glycosylated (Tabei et al., 2009), making it an ideal model organism to use for the study of flagellin glycosylation.

In contrast to the glycosylated polar flagellin of *A. caviae*, the polar flagellin of *A. hydrophila* AH-3 appears to be more complicated as they are modified with a heterogeneous glycan that consists of a heptasaccharide linked to a pseudaminic acid derivative of 376 Da (Wilhelms, Fulton, Twine, Tomás, & Merino, 2012). In addition the flagellin glycosylation island of *A. hydrophila*, AH-3 is more complex with the presence of several extra genes that could possibly account for the heterogeneous glycan or the modified pseudaminic acid derivative, as well as the presence of two *maf* gene homologues that may be required for transferring specific sugars on to the flagellin monomers (Canals et al., 2007). A recent study also implicates a lipid carrier protein WecX in flagellin glycosylation in *A. hydrophila* (Merino et al., 2014). For *A. hydrophila*, the authors suggest that pseudaminic

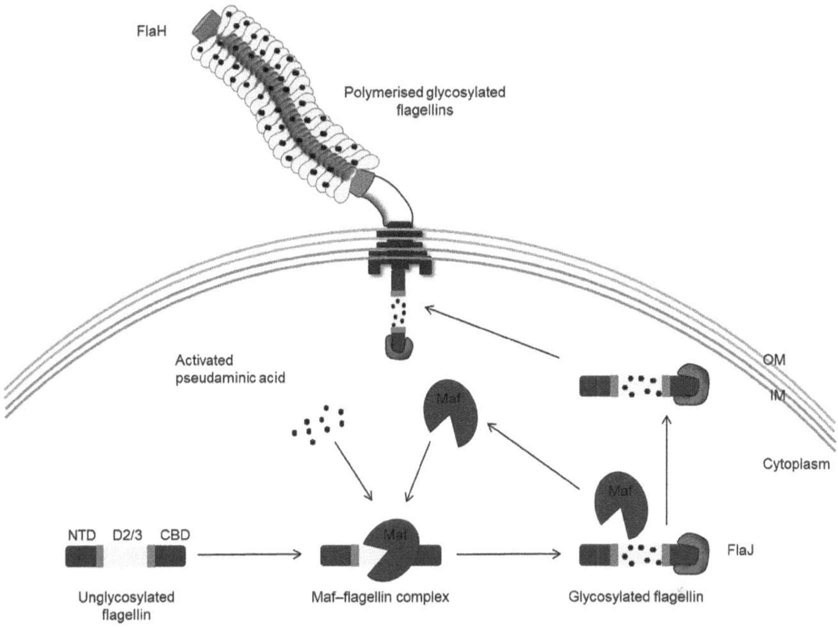

Figure 2 Flagellin glycosylation model derived from studies in *A. caviae.* Synthesised flagellin (FlaA or FlaB) is initially bound by the glycosyl-transferase Maf-1 in a chaperone (FlaJ) independent manner. Maf-1 then glycosylates the flagellins D2/D3 domains with six to eight residues to pseudaminic acid (Pse5Ac7Ac) using activated pseudaminic acid (CMP-Pse5Ac7Ac) as a donor. Once glycosylated, the flagellin-specific chaperone FlaJ binds the flagellins C-terminal chaperone-binding domain (CBD) causing the release of Maf-1 from the flagellin. The glycosylated chaperone-bound flagellin is then guided to the flagellar basal body where the putative N-terminal signal sequence (in the N-terminal domain, NTD) is recognised by the flagellar-specific type three secretion system (T3SS) and the flagellin is secreted through the flagellar hook to polymerise and create the growing flagella filament.

acid is transferred on to the lipid carrier undecaprenyl phosphate (Und-P) and that the heptasaccharide is built upon this initial sugar similar to LPS O-antigen biosynthesis, before a flagellin-specific transferase transfers the heptasaccharide on the serine and threonine residues on the flagellin in an O-linked manner (Merino et al., 2014).

A recent study from our laboratory has started to elucidate the flagellin glycosylation pathway (Fig. 2) and has indicated that Maf-dependent flagellin glycosylation occurs in the cytoplasm rather than at the cell surface. In this study, the flagellin-specific chaperone FlaJ was found to preferentially bind glycosylated flagellin over the unglycosylated version indicating that glycosylation occurs prior to chaperone binding and that this system has evolved to favour secretion of the polymerisation competent glycosylated form (Parker et al., 2014).

Currently, the exact cellular function of flagellin glycosylation is unknown although some studies suggest that it is necessary for flagella filament assembly via its dedicated T3SS apparatus with unglycosylated flagellin exported less efficiently than the glycosylated form leading to accumulation of flagellin in the cytoplasm (Josenhans et al., 2002), or not forming functional filaments (Parker et al., 2014). Because unglycosylated flagellins are not exported (or for *A. caviae* Sch3 only exported at a considerably lower efficiency) and do not produce a functional filament, it is difficult to determine whether or not the modifications themselves have a precise role in pathogenicity or infection. Additional hypotheses for the role of flagellin glycosylation include the idea that glycosylation acts as a signal for secretion or increases the solubility of the flagellin. A recent study in our laboratory has shown that very low levels of unglycosylated flagellin are able to be secreted to the culture supernatant (Parker et al., 2014) and therefore the glycosylation moieties on the flagellin do not function as a secretion signal for its dedicated T3SS. It is likely that these very low levels of export of unglycosylated flagellin are rather linked to the FlaJ chaperone's preferential binding of glycosylated flagellin and subsequent efficient export (Parker et al., 2014). Further hypotheses suggest that flagellin glycosylation may be required to evade the immune system by preventing (or decreasing) recognition by TLR-5 (Logan, 2006). However, glycosylation occurs on the flagellin D2/D3 domain and not the flagellin D0 domain recognised by TLR-5. Evidence from *A. caviae* Sch3 suggests that the glycosylated flagellin is recognised better by antisera than the unglycosylated form (Parker et al., 2012).

2.3. Lateral flagella

An early study by Shimada and coworkers using electron microscopy reported that mesophilic aeromonads expressed a single polar flagellum when grown in liquid medium, but when grown on surfaces, 20–30% of the strains tested produced peritrichous flagella (Shimada, Sakazaki, & Suzuki, 1985). Subsequent work has demonstrated that approximately 60% of mesophilic aeromonads produce and use this entirely distinct peritrichous lateral flagella system for movement across a solid surface (known as swarming motility) (Kirov et al., 2002). In *A. hydrophila*, the complete gene set of 38 genes required for lateral flagella biosynthesis are found in a single chromosomal region and are known as the *laf* genes (Canals, Altarriba, et al.,

2006; Canals, Jimenez, et al., 2006; Canals, Ramirez, et al., 2006). In 2002, a comprehensive study showed that 27 out of 55 (49%) clinical isolates tested contained *laf* genes and the strains containing these genes could swarm successfully over the surface of solid media (Kirov et al., 2002). Electron microscopy analysis of *Aeromonas* grown under laboratory conditions showed that >60% of an *A. caviae* Sch3 population expressed lateral flagella when grown on solid medium at 37 °C for 6 h (Kirov et al., 2002). The lateral flagella of a number of *Aeromonas* species were investigated further under varying growth conditions and found that no lateral flagella could be identified in *A. caviae*, *A. veronii* biovar sobria and *A. hydrophila* when they were grown in liquid medium (Kirov et al., 2002).

Lateral flagella filaments on aeromonads are composed of repeating lateral flagellin known as LafA, which form unsheathed flagella filaments that are generated via a dedicated T3SS in the same way as the polar flagella. Swarming motility over solid surfaces mediated by the lateral flagella is powered using by the stator complex that comprises the proteins LafT and LafU which is thought to be proton driven (Gavin et al., 2002; Wilhelms et al., 2009), unlike the sodium-driven polar flagella stator.

Most of the genome-sequenced mesophilic aeromonads that contain lateral flagella encode a single lateral flagellin gene (*lafA*). However, the genome of *A. caviae* Sch3 as well as a small number of other aeromonad strains encode two lateral flagellin genes (*lafA1* and *lafA2*) with studies showing that the lateral flagella filament of *A. caviae* Sch3 comprises two distinct lateral flagellins known as LafA1 and LafA2 (Gavin et al., 2002). In addition to the mesophilic aeromonads, this single gene cluster containing the complete set of genes for lateral flagella biosynthesis is also present in *A. salmonicida*. In this system, *lafA* is disrupted by the insertion sequence *IS3* and the *lfgD* gene has a nucleotide deletion resulting in the lateral flagella negative phenotype of this bacterium (Merino, Gavin, Vilches, Shaw, & Tomás, 2003; Reith et al., 2008).

Glycosylation of lateral flagellin has not been studied to the extent of that of the polar flagellin. It was previously observed that the lateral flagellins of *A. caviae* and *A. hydrophila* exhibited aberrant migration when analysed by SDS-PAGE (Gavin et al., 2002). More recently, findings have shown that this aberrant migration is due to the modification of lateral flagellin of *A. hydrophila* via O-linked glycosylation on to serine and threonine residues. *A. hydrophila* lateral flagellins are glycosylated exclusively with a 376 Da derivative of pseudaminic acid that is also used to modify the polar flagellin (Wilhelms et al., 2012), although it is not known whether the two flagella systems (polar and lateral), share glycosylation machinery. It appears that glycosylation of the lateral flagellin in this strain of *A. hydrophila* is essential for swarming motility over surfaces. As there is a gene encoding a homologue of the Maf family of proteins in the *A. hydrophila* AH-3 lateral flagella gene cluster, this is designated *maf-5*. Mutation of *maf-5* resulted in the loss of lateral flagellin similar to the *maf* mutants of the polar system that caused loss of polar flagellin (Canals, Altarriba, et al., 2006; Canals, Jimenez, et al., 2006; Canals, Ramirez, et al., 2006).

2.4. Genetic organisation and regulation of the *Aeromonas* flagella systems

Bacterial motility and concomitant flagella synthesis is metabolically very expensive, requiring the expression of many genes. For this reason, flagella genes are transcribed in a hierarchical manner and are tightly regulated, being expressed in a sequential process according to the temporal requirements of the assembly process (Aldrige & Hughes, 2002; MacNab, 1996). As described above, up to 60% of mesophilic *Aeromonas* possess dual flagella systems; this gives added levels of complexity for the coordination of gene expression and regulation. This is an area of study that is only now being addressed (Gavin et al., 2002; Kirov et al., 2002; Merino, Shaw, & Tomás, 2006; Rabaan et al., 2001). The expression of two flagella systems has been described in other bacterial species such as *V. parahaemolyticus* (McCarter & Silverman, 1990), *Azospirillum brasilense* (Moens, Michiels, Keijer, Van Leuven, & Vanderleyden, 1995), *Rhodospirillum centenum* (Jiang & Bauer, 1997), *Helicobacter mustelae* (O'Rouke, Lee, & Fox, 1992) and *Plesiomonas shigelloides* (Inoue, Kosako, Suzuky, & Shimada, 1991); however, it is still a rare feature of motile bacteria.

The best studied *Aeromonas* flagella system are those of *A. hydrophila* AH-3 where the genetic regulation of the polar (Canals, Ramirez, et al., 2006; Wilhelms, Molero, Shaw, Tomás, & Merino, 2011) and the lateral (Canals, Altarriba, et al., 2006; Wilhelms, Gonzalez, Tomás, & Merino, 2013) flagella systems have been studied in detail and show similarities to those described for *V. parahaemolyticus* (Stewart & McCarter, 2003; Yun-Kyeong & McCarter, 2000).

2.4.1 Polar flagella regulation

In *Aeromonas*, the complete and functional single polar flagellum is encoded by more than 50 genes including the regulatory proteins necessary for the biogenesis. These genes are distributed in five chromosomal regions, although most of the genes are contained in two loci (Canals, Ramirez, et al., 2006). The gene organisation of the polar flagella seems to be highly conserved among different species of *Aeromonas* and also with *Vibrio* and *Pseudomonas*, although the distribution in the chromosomal regions in the latter species is different (Canals, Ramirez, et al., 2006; McCarter, 2001).

The aeromonad genetic region 1 is similar to that of *Vibrio* spp. It comprises three gene clusters, all transcribed in the same direction with the exception of that of *flgNMA*. In this region are encoded the chemotaxis factors (CheVR) and genes encoding structural proteins that form the L and P-ring and the rod (*flgBCDEFGHIJKL*). Composition and distribution of genes in the aeromonad region 2 is identical between *A. hydrophila* AH-3 and *A. caviae* Sch3, and it is highly conserved among other *Aeromonas* species (Canals, Ramirez, et al., 2006; Rabaan et al., 2001). It contains six ORFs transcribed in the same direction some of which are under the regulation of either sigma factor σ^{54} or σ^{28}, suggesting differential regulation (Rabaan et al., 2001). The flagellins (FlaAB) are encoded in this locus as well as the filament length control protein (FlaG), the capping protein (FlaH), the flagellin-specific chaperone (FlaJ) and the Maf-1. This region shows similar organisation in *V. parahaemolyticus*, although in the *Vibrio* genome there is one more flagellin gene (*flaF*) and a second putative chaperone encoding gene (*flaI*) as well as the *fli* genes that in *Aeromonas* are located in region 3. On the other hand, the *maf*-1 gene is absent in *Vibrio* but it is present in *H. pylori*, *Clostridium acetobutylicum* and *C. jejuni*. The aeromonad region 3 is composed by 29 ORFs, all of them transcribed in the same direction and encoding proteins involved in chemotaxis and flagella biogenesis: *fliEF* (M/S ring), *fliGHIJKLM* (the switch and export/assembly machinery, hook length), *fliOPQR* and *flhAB* (export and assembly), *flhFG* (regulation and positioning), as well as the σ^{28} factor (*fliA*) and the stator of the motor (*pomAB*). This region shows similar organisation as with region 2 in *Vibrio*, with the exception of the genes for the motor proteins *pomA* and *pomB*. Region 4 includes in both *Vibrio* and *Aeromonas* the sodium-driven motor protein MotX that together with PomAB from region 3 constitutes the motor of the polar flagella. Region 5 in *Aeromonas* contains the genes encoding master regulatory proteins, FlrA, FlrB and FlrC.

2.4.2 Lateral flagella

In *Aeromonas*, the lateral flagella is encoded by 38 genes, all of them arranged in a single chromosomal region (Canals, Altarriba, et al., 2006), this organisation also occurs in *E. coli* 042 (Ren, Beatson, Parkhill, & Pallen, 2005). The genes are distributed in six operons all transcribed in the same direction with the exception of the one that contains the anti-σ^{28} factor (*flgNMA*$_L$). This cluster, together with the *flgBCDEFGHIJKL*$_L$, encode for the hook, rod, peptidoglycan hydrolase and hook-associated proteins; these are located in region one of *V. parahaemolyticus*. In *V. parahaemolyticus*, the lateral flagella genes are arranged in two distinct chromosomal regions, while in *Aeromonas* and *E. coli* 042 they occur in a single chromosomal locus (Merino et al., 2006). The other five operons, *flgBCDEFGHIJKL*$_L$, *lafKfliEFGHIJ*$_L$, *fliMNPQR*$_L$*flhBA*$_L$, *maf-5* and *lafABCXEFSTUpilZ* are transcribed in the same direction in *Aeromonas* in contrast with their counterparts in *V. parahaemolyticus* and *E. coli* that are transcribed divergently. In addition, *V. parahaemolyticus* also contains the *mot*Y gene in this cluster which encodes an outer-membrane motor protein similar to MotYp in *V. alginolyticus* (Okabe, Yakushi, Kojima, & Homma, 2002); this gene is not present in *Aeromonas*.

All *Aeromonas* species sequenced so far that contain lateral flagella exhibit the same genetic organisation with some variation occurring around the flagellin genes. Most of the lateral flagella gene clusters of *Aeromonas*, *V. parahaemolyticus* and *E. coli* 042 show a unique flagellin gene (*lafA*) that is located upstream of the gene encoding the capping protein LafB. In *A. caviae* Sch3, *A. hydrophila* SSU and AH-1, two flagellin subunit genes have been described, *lafA1* and *lafA2*, both transcribed in the same direction. In *A. caviae* Sch3 and *A. hydrophila* SSU, upstream of the flagellin genes is a divergently transcribed gene encoding a L-lysine methylase (*fliU*) involved in flagella biosynthesis (Gavin et al., 2002). This variable flagellin encoding region also contains *maf-5* in *A. hydrophila* strain AH-3, a gene possibly involved in lateral flagellin glycosylation (Canals, Altarriba, et al., 2006; Canals, Jimenez, et al., 2006; Canals, Ramirez, et al., 2006). The lateral flagella gene cluster is located in the same chromosomal position in all lateral flagella positive strains of *Aeromonas* that have been genome sequenced. This is between the genes equivalent to the *A. hydrophila* type strain ATCC7966^T genes *aha3893* and *aha3896* (Seshadri et al., 2006).

2.4.3 Regulation of flagella expression

Either in *Aeromonas* or *Vibrio*, the polar and lateral flagella do not share structural or regulatory genes (Kirov et al., 2002; McCarter, 2004) although some sharing does occur in other species with dual flagellar systems such as *Azospirillum* or *R. centenum* (Jiang, Rushing, Bai, & Gest, 1998; Kalir, 2001).

In the regulation cascade of flagella systems, the genes can be divided into early, middle and late genes and each tier of the cascade is sensitive to the previous assembly stage (Aldrige & Hughes, 2002). The early genes control the entire regulon and are mainly composed of regulatory proteins. The middle genes include structural components of the hook–basal body, the secretion system and those regulatory proteins coupled with the late gene expression. Late genes include filament, motor force generators and chemosensory machinery. The expression is controlled by a group of transcriptional regulators including alternative sigma factors and anti-sigma factors (Chilcott & Hughes, 2000). The transcriptional hierarchies of flagellar biosynthesis include three different promoter classes (Chilcott & Hughes, 2000; Jenal, 2000; Kim & McCarter, 2000; McCarter, 2001). The regulatory flagellar hierarchies for both polar and lateral flagella have been investigated in *A. hydrophila* AH-3 (Canals, Altarriba, et al., 2006; Canals, Jimenez, et al., 2006; Canals, Ramirez, et al., 2006; Wilhelms et al., 2011, 2013).

The transcriptional hierarchy of the genes required for polar flagellum biogenesis and its regulation have been described in *A. hydrophila* AH-3 by Wilhelms et al. (2011). They described four transcriptional levels (classes I–IV) where each level contains the activator for the subsequent transcriptional level in the hierarchy (Fig. 3). The class I genes encode the sigma 54 (σ^{54}) dependent enhancer-binding regulatory protein FlrA whose expression is σ^{28} and σ^{54} independent, suggesting that FlrA is the master regulator of the polar flagella regulon. Evidence of its expression by σ^{70} RNA polymerase have been published, as it was expected considering that the polar flagella in *Aeromonas* is constitutively expressed. The *flrA* mutants do not produce polar flagella, however, lateral flagella formation is unaffected (Canals, Altarriba, et al., 2006). Furthermore, the major regulator of the lateral flagellar system LafK is not able to compensate for the disruption of FlrA as occurs in *V. parahaemolyticus*. Even though both FlrA and LafK proteins share a high degree of sequence similarity at the amino acid level, it has been hypothesised that the C-terminal domain of those two proteins might bind different DNA-binding sites and therefore they are unable to compensate for each other (Canals, Altarriba, et al., 2006; Wilhelms et al., 2011).

Aeromonas polar flagellar class II genes are FlrA and σ^{54} dependent and include the majority of the structural proteins such as the basal body components, the L and P rings, hook formation and the export/assembly machinery. In this class is included the genes for the next regulatory step in the hierarchy that encode for the enhancer-binding protein FlrC required

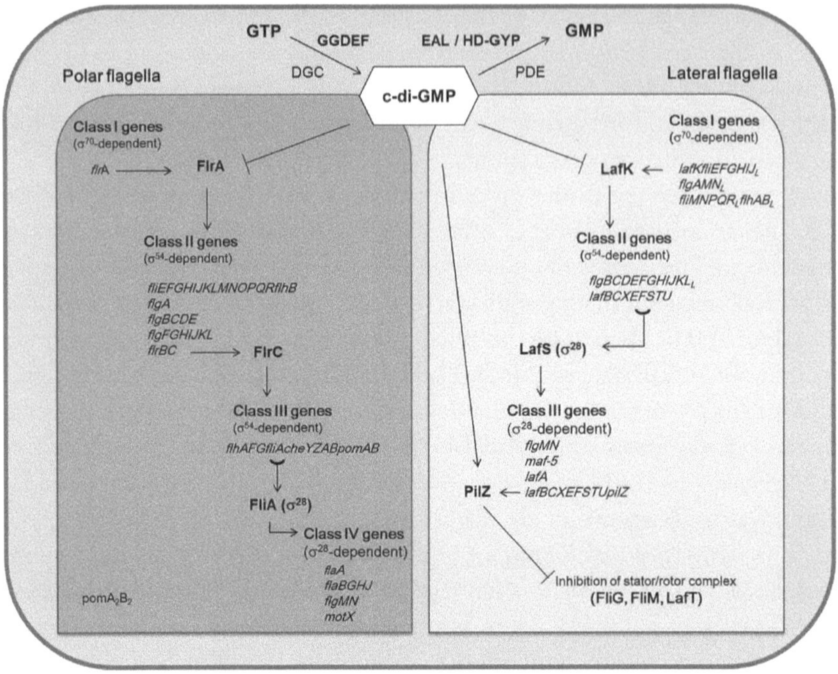

Figure 3 The predicted role of cyclic-di-GMP on the previously determined polar and lateral flagella regulatory cascades in *Aeromonas* species. C-di-GMP is produced by GGDEF domain containing diguanylate cyclase (DGC) enzymes and is broken down by EAL or HD-GYP domain containing phosphodiesterases (PDE) proteins. The rise in the intracellular concentration of c-di-GMP in *Aeromonas* species has been demonstrated to inhibit both swimming (polar flagella dependent) and swarming (lateral flagella dependent) motility, while favouring sessile activity such as biofilm formation (Kozlova, Khajanchi, Sha, & Chopra, 2011). In related bacteria (*Vibrio*, *Pseudomonas*), c-di-GMP has been shown to bind to the enhancer-binding protein-response regulators FlrA and FleQ, respectively. These proteins are homologs of the aeromonad FlrA and LafK proteins that control the expression of the polar and lateral flagella gene expression cascades, respectively (Wilhelms et al., 2011, 2013). PilZ the c-di-GMP binding protein encoded within the lateral flagella gene cluster is postulated to modulate lateral flagella rotation through interaction with the rotor/stator complex analogous to YcgR of *E. coli*. The genes encoding the extra set of stator proteins (*pomA*$_2$*B*$_2$) are transcribed independently. (See the color plate.)

for class III expression (Fig. 3) and its cognate kinase FlrB. Similar proteins are found in the polar flagella systems of *V. cholerae*, *V. parahaemolyticus* and *Pseudomonas* (FlrBC, FlaLM and FleSR, respectively). FlrC together with σ^{54} activates the class III genes that encode, among others, the polar flagellar-specific sigma factor FliA (σ^{28}) that activates the class IV genes. Class

IV make up the late flagellar genes that include the filament proteins flagellins (*flaA* and *flaB*), filament cap (*flaH*), flagellin-specific chaperone (*flaJ*), the anti-sigma factor FlgM and the motor components (Wilhelms et al., 2011). FlgM negatively regulates the class IV promoters and is transcribed from class II genes. The FlgM protein accumulates preventing the expression of the late genes products by binding to the sigma factor FliA. FlgM is eventually exported along with the hook and basal body components, and therefore frees FliA (σ^{28}) to initiate the transcription of the class IV promoters (Chadsey, Karlinsey, & Hughes, 1998). *A. hydrophila* possesses a redundant set of motor proteins in the polar flagellum, PomAB and $PomA_2B_2$, *pomAB* is transcribed from a class III promoter and is σ^{54} and FlrC dependent, while $pomA_2B_2$ is transcribed independently from the flagella hierarchy (Wilhelms et al., 2009).

In contrast to constitutively expressed polar flagella genes, those required for lateral flagella biosynthesis are expressed only under certain conditions and are environmentally regulated (Merino et al., 2006). Expression of the lateral flagella system is mainly induced when bacteria are grown on solid or in highly viscous environments although other secondary environmental signals have been described such as iron-depleted media in *V. parahaemolyticus* (McCarter & Silverman, 1989) or static liquid growth in *A. brasilense* (Madi, Kessel, Sadovnik, & Henis, 1988).

The regulation of lateral flagella gene expression is more complicated than polar flagella, with different promoters acting in the same gene class. In general terms, in *A. hydrophila* AH-3 genes are categorised into early, middle and late, with the enhancer-binding protein LafK acting as a major regulator (Fig. 3). Class I early genes are σ^{70} dependent and include the major regulator of the system LafK and some structural components, such as the proteins forming the C and P rings. The expression of the middle genes is controlled by σ^{54} and LafK that controls the transcription of the sigma factor LafS (σ^{28}) and the rest of the structural genes. In most cases, late gene expression is controlled by the lateral flagellar-specific sigma factor LafS (σ^{28}) and its cognate anti-σ factor FlgM (Canals, Altarriba, et al., 2006). Class I genes are σ^{70} dependent and are transcribed in both liquid and viscous media. Studies performed with RT-PCR and Western blot demonstrated that these lateral flagella genes are transcribed in liquid, viscous and solid media but are only translated in viscous or solid environments. The fact that class I genes are σ^{70} dependent support the initial constitutive expression of this system (Wilhelms et al., 2013). Other genes regulated by the σ^{70} factor are $flgAMN_L$, and the structural and motor proteins encoded by *fliMNPQR$_L$flhAB$_L$* and *lafTUpilZ*. Class III genes are expressed from LafS σ^{28}-dependent promoters and include *flgMN*, *maf-5* and *lafA* (Fig. 3).

2.4.4 Regulation of the flagella systems by c-di-GMP

The signal that activates LafK and how *Aeromonas* senses a surface is currently unknown; however, it appears to be different from the mechanism described for *V. parahaemolyticus*. This bacterium uses its polar flagellum as a mechanosensor that measures the viscosity of the media and transduces the signal to control lateral flagella expression. Mutations in the polar flagellum of *V. parahaemolyticus* cause constitutive expression of the lateral flagella system (Merino et al., 2006) but that is not the case in *Aeromonas*, as mutations in the polar flagellum do not affect lateral flagella expression (Gavin et al., 2002; Merino et al., 2006). Similarly, the signal that represses the expression of FlrA, the master regulator of the polar flagella is still unknown although this protein is very similar to its counterparts in *V. cholerae* and *P. aeruginosa* (FlrA, FleQ) and therefore is possibly regulated in a similar manner (Fig. 3).

Cyclic-di-GMP is made by diguanylate cyclase (DGC) proteins that have GGDEF domains and it is broken down by phosphodiesterase proteins (PDE) that have EAL or HD-GYP domains. The intracellular concentration of c-di-GMP controls the change between sessile and motile bacterial lifestyles. Therefore, the link between c-di-GMP and motility in bacteria and with exo-polysaccharides production (and thus, biofilm formation) is well documented (Hickman & Harwood, 2008; Römling et al., 2013; Srivastava, Hsieh, Khataokar, Neiditch, & Waters, 2013). Evidence of the regulation of motility by c-di-GMP has been demonstrated in bacteria with dual flagellar systems, such as *V. parahaemolyticus*. At high c-di-GMP concentrations, biofilm formation is increased and lateral flagella expression is impaired while low concentrations promote lateral flagella biogenesis and therefore swarming motility. Although the exact mechanism of how c-di-GMP regulates lateral flagella expression in *V. parahaemolyticus* is unknown, it has been demonstrated that the protein ScrC plays an essential role in this regulation. ScrC is a protein that contains both GGDEF (DGC) and EAL (PDE) domains (Kim & McCarter, 2007).

Initial studies demonstrated that c-di-GMP regulates motility in *Aeromonas*. Rahman et al. (2007) showed that the overexpression of AdrA a GGDEF domain containing protein from *Salmonella* in *A. veronii* biovar sobria increased biofilm formation but repressed motility. Similarly, the overexpression of an EAL domain protein, YhjH, impaired biofilm formation and increased motility (Rahman et al., 2007). The genome sequences of

Aeromonas species have shown the presence of many genes encoding GGDEF or EAL domain proteins (Seshadri et al., 2006). The complete unpublished draft genome of *A. caviae* Sch3 showed a number of genes that encoded proteins that contained GGDEF domains (31), EAL or HD-GYP domains (12) and dual GGDEF/EAL (14) domains. The overexpression of one of the dual GGDEF/EAL domain proteins, designated AHA0383-MshH, in *A. caviae* Sch3 completely inhibited swarming and expression of the lateral flagella (Shaw unpublished). This protein is conserved among other species of *Aeromonas*, *Vibrio*, *Shewanella* and *Photobacterium* (Hadi et al., 2012).

A. hydrophila possesses three quorum-sensing (QS) mechanisms, the auto inducer I (AI-1) *N*-acyl homoserine lactone (AHL) based system designated AhyRI, the AI-2-based system LuxS and the AI-3-based QseBC system AI-3 (Kozlova, Khajanchi, Popov, Wen, & Chopra, 2012). Mutation of *ahyRI* does not affect aeromonad motility, whereas mutation of *luxS* does (Kozlova et al., 2008, 2011). The third QseBC AI-3-based system positively regulates swimming and swarming in *A. hydrophila* (Kozlova et al., 2012). Genetically linked to *luxS* on the aeromonad chromosome is a gene (*AHA0701h*) encoding a GGDEF domain protein. The overexpression of this protein in *Aeromonas* leads to a dramatic increase in biofilm formation and reduction in motility thought to be due to the concomitant increase in c-di-GMP (Kozlova et al., 2012). In *A. hydrophila* SSU, it has been demonstrated that c-di-GMP plays a major role inter-connecting the QS mechanisms with that of biofilm formation and motility (Kozlova et al., 2012). The overexpression of AHA0701h in the null mutants for both QS systems showed that only LuxS has a direct effect on the motility, while the AhyRI system is able to prevent the effect of high levels of c-di-GMP in the cell (Kozlova et al., 2011). These two QS systems also affect the transcription of the master regulator of the aeromonad polar flagella system FlrA (termed FleQ in these studies). The over-production of the GGDEF domain protein AHA0701 in *A. caviae* Sch3 demonstrated that increased levels of c-di-GMP impair the swarming motility of this strain (Shaw unpublished).

Comparative amino acid sequence analysis of the major regulators of the aeromonad polar (FlrA) and lateral (LafK) flagella systems showed that they belong to a family of transcription factors that regulates the activity of σ^{54}-containing RNA polymerase. These regulators are similar to those found in other bacteria with polar flagellation, namely, FlrA and FlrC in *V. cholerae* and FleQ and FleR in *P. aeruginosa*. They are members of the enhancer-binding protein family consisting in an N-terminal receiver domain that usually requires phosphorylation, a σ^{54}-interacting AAA+ domain and a

C-terminal DNA-binding helix-turn-helix (Hickman & Harwood, 2008; Srivastava et al., 2013). The σ^{54}-type regulatory protein FlrC in the aeromonad polar flagella system and the corresponding FlrC in *V. cholerae* and FleR proteins in *P. aeruginosa* posses the conserved amino acid signature of a phosphorylation-dependent response regulator and are associated with a cognate kinase (FlrB); however, the aeromonad and *Vibrio* polar flagella FlrA and lateral flagella LafK proteins do not have a complete phosphorylation motif, nor do they have a cognate sensor kinase partner and are therefore believed not to be phosphorylated for their activation. In *P. aeruginosa*, the activity of FleQ has been demonstrated to be modulated not by phosphorylation but by the action of an anti-activator protein FleN (Dasgupta & Ramphal, 2001; Dasgupta, Ferrell, Kanack, West, & Ramphal, 2002), a similar protein called FlhG is present in *Vibrio* species (Kusumoto et al., 2008) and *Aeromonas* (Canals, Altarriba, et al., 2006, Canals, Jimenez, et al., 2006, Canals, Ramirez, et al., 2006). Recently, it has been shown that *Pseudomonas* FleQ and *V. cholerae* FlrA are directly regulated by the c-di-GMP. The direct binding of c-di-GMP to FlrA prevents the subsequent attachment of the master regulator to the promoter region of the *flrBC* genes; therefore, the expression of these important genes encoding subsequent activators of the polar flagellar system is abrogated, resulting in a non-motile phenotype (Srivastava et al., 2013). *Pseudomonas* FleQ also binds c-di-GMP, and this binding of c-di-GMP likewise affects flagella expression (Hickman & Harwood, 2008). As c-di-GMP affects *Aeromonas* motility and aeromonads possess homologous proteins (FlrA, LafK) to both of these *Pseudomonas* and *Vibrio* proteins, they could possibly be regulated by c-di-GMP in a similar manner (Fig. 3).

In addition to the possible c-di-GMP modulation of flagellar master regulator activity in *Aeromonas*, there is another level of potential regulation by this molecule. During the annotation of the complete draft genome sequence of the *A. caviae* strain Sch3 (Shaw unpublished), the last gene in the lateral flagella gene cluster, was identified as encoding a PilZ containing domain protein that contains a classic c-di-GMP receptor domain (Fig. 3). The gene for this protein is immediately downstream of the gene encoding the motor protein LafU and they are transcriptionally coupled (Shaw unpublished). In *E. coli*, a similar PilZ domain protein, YcgR, acts as a brake of the flagellar motor upon binding with c-di-GMP by hindering the rotation of the motor (Boehm et al., 2010; Paul, Nieto, Carlquist, Blair, & Harshey, 2010). Recently, a similar PilZ domain protein, FlgZ, has been identified in *Pseudomonas* and has been shown to be involved in the regulation of the swimming motility of *P. putida* and *P. fluorescens* in response to high concentrations of c-di-GMP (Martínez-Granero et al., 2014).

3. LIPOPOLYSACCHARIDE AND CAPSULES

3.1. Lipopolysaccharide

In the Gram negative bacterial cell envelope, the exterior leaflet of the outer membrane is composed of lipopolysaccharide (LPS); this is also true in the aeromonads. LPS is separated into three domains: the extremely variable O-antigen polysaccharide which is linked to the core oligosaccharide that is in turn linked to the conserved and toxic lipid A component. Aeromonad lipid A and the core oligosaccharide have been recently described in an excellent review by Juan Tomás (Tomás, 2013). O-antigens are composed of oligosaccharide polymers of repeating units of 1–6 sugars, these variable units give rise to the wide variety of O-antigen groups (sergroups) that have been demonstrated in Gram negative bacteria. Originally 44 O-antigen serogroups were detected among 307 aeromonad strains (Sakazaki & Shimada, 1984), a subsequent study reported a further 52 additional serogroups among 1255 *Aeromonas* strains, thus showing the highly variable nature of *Aeromonas* serogroups. However, even the combination of these two schemes still appears to be a underestimate of the number of serogroups present in the genus *Aeromonas*, as many strains do not belong to a particular known serogroup, suggesting there are yet more to be discovered (Thomas, Gross, Cheasty, & Rowe, 1990). However, among *Aeromonas* strains isolated from human infections, the serogroups O:3, O:11, O14, O:16 and O:34 predominate (Sakazaki & Shimada, 1984; Thomas et al., 1990). Initial electrophoretic studies on aeromonad LPS demonstrated three main patterns, the first being a major ladder-like pattern that is often seen in the enterobacteria and is associated with smooth (S) LPS. The second is characterised by the presence of some O-antigen chains demonstrating a homogeneous pattern and the third a rough (R) or semi-rough pattern in which no O-antigen is present and there is usually one main band present representing the core (Khashe, Hill, & Janda, 1996; Kokka, Vedros, & Janda, 1990; Tso & Dooley, 1995). A number of aeromonad LPS O-antigen serogroups have been shown to be regulated by heat, these being O:13, O:33, O:34 and O:44. With the smooth or S-LPS being prevalent at 20 °C and higher osmolarity, whereas the rough or R-LPS is more common at 37 °C and low osmolarity. This LPS thermoregulation has been linked to colonisation, as S-LPS has been associated with adherence to HEp-2 cells by

A. hydrophila. In addition when strains of *A. hydrophila* O:34 were grown at 20 °C, the presence of their S-LPS was correlated with higher virulence in fish and mice than their R-LPS counterparts grown at higher temperatures (Merino, Camprubi, & Tomás, 1992).

O-antigens are synthesised through the linkage of a sugar to a lipid carrier Und-P. In *A. hydrophila* AH-3, the WecP protein transfers *N*-acetyl-galactosamine from UDP-*N*-acetyl-galactosamine to Und-P (Merino et al., 2011). Other reported strains use WecA that transfers *N*-acetyl-glucosamine-1-phosphate (Tomás, 2013). This initial transfer reaction is then followed by specific glycosyltransferases transferring particular sugars onto the Und-P sugar to create an O-antigen unit. As O-antigens are highly variable, their chemical composition is therefore diverse, but recently a number of aeromonad O-antigen structures or compositions have been determined by chemical analysis or inferred from bioinformatics analysis of genome sequences. The O-antigen of a diarrhoeal isolate of *A. caviae* was one of the first to be elucidated; this was composed of a branched pentasaccharide of L-rhamnose, D-galactose, *N*-acetyl-D-mannosamine and *N*-acetyl-D-galactosamine (Linnerborg et al., 1996). *A. hydrophila* AH-3 contains an O:34 LPS and has been extensively investigated by the Tomás and Merino laboratory in Barcelona over a number of years. They have shown that aeromonad LPS biosynthesis is complex with the requirement of over 40 genes. The O:34 structure has been shown to contain D-mannosamine, *N*-acetyl-D-galactosamine and 6-deoxy-L-talose (Knirel, Shaskov, Senchenkova, Merino, & Tomás, 2002). The O:34 $wb^*_{O:34}$ O-antigen gene cluster is encoded within a 18.1 kb genetic island on the chromosome of *A. hydrophila* AH-3 (Jimenez, Canals, Saló, et al., 2008). The $wb^*_{O:34}$ gene cluster has a G+C content of 44.3% that is significantly lower than the aeromonad average of around 60% and the locus encodes 17 open reading frames (ORFs) required for the biosynthesis of D-mannose, L-rhamnose and 6-deoxy-L-talose. Also present in the $wb^*_{O:34}$ locus are *wzy* and *wzx* genes encoding the O-antigen polymerase and the flippase, respectively, suggesting that the O:34 antigen is exported in a Wzy-dependent manner. Correlating with their previous findings on LPS thermoregulation in *A. hydrophila* AH-3, the gene that encodes the polysaccharide chain length protein Wzz in the $wb^*_{O:34}$ gene cluster, was demonstrated to have a much higher transcription at 20 °C in comparison to 37 °C (Aguilar, Merino, Rubires, & Tomàs, 1997; Jimenez, Canals, Saló, et al., 2008).

The *A. hydrophila* AH-3 O:34 LPS core structure has also been determined (Knirel, Vinogradov, Jimenez, Merino, & Tomás, 2004); the inner core was shown to be conserved as it was highly similar to that of *A. salmonicida* A450 (Wang, Li, Vingradov, & Altman, 2006). Three gene clusters associated with core biosynthesis in *A. hydrophila* strain AH-3 have been isolated (Jimenez, Canals, Lacasta, et al., 2008). Seven genes make up the first region; there are four genes in region 2 and two in the third. The specific roles of these genes in LPS core biosynthesis were elucidated by applying bioinformatic data to the core structural information (Jimenez, Canals, Lacasta, et al., 2008).

Recent genome analysis has provided more information on LPS in the aeromonads. Workers investigating the emergence of epidemic *A. hydrophila* strains that are the causative agents of MAS in catfish have sequenced the genomes of 11 *A. hydrophila* isolates, which included five reference (RAh) strains and six epidemic strains termed VAh (Hossain et al., 2013). The O-antigen gene clusters of all strains were compared and four different clusters were identified; however, all epidemic and one reference isolate shared the same novel LPS locus. The epidemic VAh *A. hydrophila* strains O-antigen gene clusters were larger at 26.5 kb than those previously reported for *A. hydrophila* strains AH-3 and PPD134/91 that were 18–19 kb in size (Hossain et al., 2013). The MAS epidemic strains locus contained 25 ORFs that encoded genes for the putative biosynthesis D-rhamnose, D-mannose, GDP-L-Fucose and 3-acetamido-3,6-dideoxy-D-galactose (D-Fuc*p*3NAc) as well as genes for glycosyltransferases and O-antigen processing. As the genes for Fuc*p*3NAc biosynthesis were present in all the MAS epidemic strains, but not in the reference or previously sequenced strains, it was suggested that this sugar is an important component of the epidemic strain O-antigen. The VAh strains contained the *wecA* gene that encodes the protein that transfers *N*-acetyl-glucosamine-1-phosphate to the lipid carrier Und-P, to which the O-antigen unit is built by specific glycosyltransferases. The single oligosaccharide unit is then predicted to be flipped to the periplasm by a Wzx flippase polymerised in a Wzy-dependent manner (Hossain et al., 2013). Analysis of the O-antigen gene loci in two of the reference strains RAh showed the presence of two additional gene clusters required for S-layer protein (SLP) synthesis (see below) and its dedicated T2SS. The authors suggest that the S-layer of these two strains is anchored to the cell envelope by this particular O-antigen (Hossain et al., 2013). For the reader interested in more structural information on aeromonad O-antigens and core-oligosaccharides, a recent comprehensive review has been published (Nazarenko, Crawford, & Ivanova, 2011).

The *A. caviae* Sch3 flagellin glycosylation locus has also been demonstrated to have a role in LPS O-antigen biogenesis and therefore has a dual role (Tabei et al., 2009). Transposon mutants isolated in *A. caviae* Sch3 were shown to be located in a locus that affected motility, adherence and caused the loss of the LPS O-antigen (Gryllos et al., 2001). The transposon insertions mapped to a locus that encoded seven genes (Gryllos et al., 2001; Tabei et al., 2009). Five of these genes (*flmA*, *flmB*, *neuB*, *flmD* and *neuA*) encode proteins involved in biosynthesis of the sialic acid related nine-carbon sugar pseudaminic acid (Pse5Ac7Ac) (Tabei et al., 2009). As outlined elsewhere in this review, pseudaminic acid also decorates the polar flagellins of *A. caviae* and is essential for flagellar function. The other two genes in the mutated locus *lst* and *lsg* encoded a putative LPS pseudamyl-transferase and a Wzx-like LPS O-antigen flippase, respectively. Mutation of either *lst* or *lsg* only caused loss of the LPS O-antigen and did not affect flagellin glycosylation or motility demonstrating that they were LPS specific (Tabei et al., 2009). Subsequent analysis by mass spectrometry demonstrated the presence of at least two pseudaminic acid residues in the LPS O-antigen of this strain along with two unidentified sugars. Therefore, this gene cluster is necessary for the biosynthesis of the sugar pseudaminic acid that is required for flagellin glycosylation and for the O-antigen in this particular *A. caviae* strain (Tabei et al., 2009).

3.2. Capsules

The bacterial capsule is usually a hydrated polysaccharide structure that covers the outer layer of the cell wall, and in most bacteria it is composed of monosaccharides linked together via glycosidic bonds. However, amino acid (peptide) and protein–carbohydrate capsules have also been described. Due to its exposure to the environment capsules are thought to have roles in the virulence and colonisation of a number of pathogenic bacteria. Capsules can help to prevent bacterial desiccation, foil phagocytosis by host cells or reduce complement-mediated lysis. Although aeromonads are generally described as non-capsulate, capsules have been identified in both motile and non-motile *Aeromonas* with the first capsule being described in *A. salmonicida* when grown in glucose-rich media *in vitro* (Garrote, Bonet, Merino, Simonpujol, & Congregado, 1992) and *in vivo* in Atlantic salmon (Garduno & Kay, 1995). After initial reports in psychrophilic aeromonads, capsules were then also described in mesophilic *A. hydrophila* (serotypes O:11 and O:34) when grown on glucose-rich media (Martinez et al., 1995). Although glucose is the predominant sugar of many aeromonad capsules, the composition of the capsule has been found to vary. Studies with *A. hydrophila* strains from serogroups O:11 and O:34 discovered a capsule composed of hetero-polymers of D-glucose, L-rhamnose, D-mannose, D-mannuronic acid and acetic acid (Martinez et al., 1995). Garrote et al. (1992) also described hetero-polymers forming the *A. salmonicida* capsule with the addition of *N*-acetyl-mannosamine; however, in contrast to *A. hydrophila*, acetic acid was not present.

Capsules are produced in response to environmental conditions, such as in environments with high glucose concentrations and a temperature of 37 °C. The capsule is therefore likely to have a role in the virulence of *Aeromonas* species capable of causing bacteraemia, with expression being induced upon entry of the host blood stream where the environment is glucose rich. Studies have demonstrated that the capsule increases invasiveness of both psychrophilic and mesophilic aeromonads in fish cell lines (Merino, Aguilar, et al., 1996; Merino, Rubires, Aguilar, Alberti, et al., 1996; Merino, Rubires, Aguillar, Guillot, & Tomas, 1996; Merino, Rubires, Aguilar, & Tomás, 1997) and is able to protect the bacterium from host immune responses by preventing complement-mediated killing and macrophage opsonisation. Aguilar, Merino, Nogueras, Regue, and Tomas (1999) identified two genes in *A. hydrophila* 1051–88 (serogroup O:34) that allowed usually uncapsulated *E. coli* to produce a colonic acid capsule and in doing so conferred serum resistance to this strain. These genes initially identified in the *A. hydrophila* (O:34) capsule gene cluster were *orf1* and *wcaJ*, with *wcaJ* mutants being unable to form a capsule under inducing conditions (Aguilar et al., 1999).

A gene cluster for capsule production has now been described in depth for *A. hydrophila* PPD134/91 where a 17.5 kb locus has been found to contain 13 genes organised into three regions (I, II, III), the gene organisation is very similar to group II capsules of *E. coli* (Zhang, Arakawa, & Leung, 2002). Genes from regions I and III are involved in capsule export, whereas region II genes are involved in capsule biosynthesis (Zhang et al., 2002; Zhang, Lau, Arakawa, & Leung, 2003). In addition, capsule polysaccharides from virulent *A. hydrophila* PPD134/91 were able to protect an avirulent strain (PPD35/85) from naive tilapia serum (Zhang et al., 2002). Zhang et al. (2003) carried out a study to determine the occurrence of group II capsules among *A. hydrophila* strains. 33 random *A. hydrophila* strains were analysed and 10 discovered to possess a group II capsule. Further analysis enabled group II aeromonad capsules to be subdivided into two groups due to the organisation of their region II genes involved in capsule biosynthesis. These subgroups were named group IIA and IIB, with group IIA being present in O:18 and O:34 serogroups and group IIB in O:21 and O:27 serogroups (Zhang et al., 2003). This study also demonstrated that the presence of a group II capsules correlated with the ability of *A. hydrophila* to evade host immune responses (i.e. serum and phagocytosis resistance) (Zhang et al., 2003). With the recent increase in aeromonad genome sequences being available, more capsule information should be forthcoming.

4. PILI

Pili are filamentous, extracellular structures that allow bacteria to adhere to biotic and abiotic surfaces. A variety of pili have been identified on the cell surface of *Aeromonas* spp. with early studies initially describing two distinct pili types due to their structural differences: short-rigid (S/R) and long-wavy (L/W) pili (Carrello, Silburn, Budden, & Chang, 1988; Ho, Mietzner, Smith, & Schoolnik, 1990). *Aeromonas* strains collected from environmental samples were reported to preferentially display large numbers of S/R pili, whereas isolates from clinical samples were more likely to produce small numbers of L/W pili (Carrello et al., 1988); proposing L/W pili to be involved in virulence. In Gram negative bacteria, there are four main groups of bacterial pili: pili assembled by the 'chaperone-usher pathway', type IV pili, pili assembled by the extracellular nucleation pathway and pili assembled by the 'alternative chaperone-usher pathway'. *Aeromonas* spp. have been found to express type I pili assembled by the 'chaperone-usher pathway' and type IV pili. Ultra-structural studies carried out by Ho et al. (1990) on the S/R pilus of *A. hydrophila* AH26 revealed the pilin (major repeating unit of the pilus) to have homology with type I and Pap (pyelonephritis-associated pili) pili from *E. coli* (Ho et al., 1990), and belong to the group of pili assembled by the chaperone-usher pathway. Type I pili from mesophilic *Aeromonas* species have been associated with autoagglutination of cells but not with pathogenicity. Adherence studies from Honma and Nakasone (1990) showed that although their *A. hydrophila* strain (Ae6) was highly adherent to rabbit and human intestines, antibodies raised to the S/R pili did not block adherence to cells and sheared off S/R pili also had no affinity to host cells. The genes encoding type I pili have now been located in the genomes of a number of *Aeromonas* species including some mesophilic species and the psychrophilic aeromonad, *A. salmonicida* subsp *salmonicida* (Reith et al., 2008; Seshadri et al., 2006). Unlike mesophilic aeromonads, the type I pilus of *A. salmonicida* does appear to have a role in colonisation of Atlantic salmon. Mutational studies carried out by Dacanay and coworkers (2010) demonstrated that *A. salmonicida* lacking the type I pilus operon have a reduced ability to adhere to the gastrointestinal tract of Atlantic salmon. However, once adhered, mutant ability to invade the host was comparable to the wild type, suggesting type I pili to be involved in the initial stages of colonisation only (Dacanay et al., 2010).

Considerable amounts of work carried out on aeromonad adhesion have concentrated on the L/W pili. Adherence studies with a variety of mesophilic strains in the 1990s highlighted the importance of these appendages in the colonisation of host tissues. Mechanical removal of L/W pili from the bacterial cell surface, or pre-treatment of bacterial cells with an anti-pilin antibody, was shown to block bacterial adherence to a variety of animal and human cell lines (Barnett & Kirov, 1999; Hokama, Honma, & Nakasone, 1990; Honma & Nakasone, 1990; Iwanaga & Hokama, 1992; Kirov, Hayward, & Nerrie, 1995; Nakasone, Iwanaga, Yamashiro, Nakashima, & Albert, 1996). L/W pilins generally have a molecular weight between 19 and 23 kDa and are antigenically diverse, with only the *N*-terminal amino acid sequence showing homology to type IV pilins (Kirov & Sanderson, 1996). Some type IV pili are able to form filamentous networks or bundles (Kirov, Hayward, et al., 1995) and it is these bundle-forming pili (Bfp), in particular, that are considered major colonisation factors in mesophilic *Aeromonas* spp. The first Bfp isolated from a species of *Aeromonas* was from *A. veronii* biovar sobria (strain BC88) (Kirov & Sanderson, 1996). The N-terminal sequence of the major pilin from this Bfp displayed sequence homology with the mannose-sensitive haemagglutinin (MSHA) type IV pilin from *Vibrio cholerae* (Kirov & Sanderson, 1996) which belongs to the 'classical' type IVa family of pilins compared to the type IVb family that Bfp usually belong. The *A. veronii* bv. sobria bundle-forming pilus was subsequently shown by genetic means to be a member of the MSHA pilus family (Hadi et al., 2012). Expression of the MSHA pilus in these studies appeared to be environmentally regulated as *A. veronii* biovar sobria showed increased expression when grown at 22 °C in liquid culture, compared to 37 °C on solid media (Kirov & Sanderson, 1996). These results also agreed with earlier studies from Kirov, Hayward, et al. (1995) where *A. veronii* bv. sobria isolates displayed optimal expression of filamentous structures at 22 °C, with a few environmental strains showing optimal expression at 7 °C. Further investigations of the type IV pili present on *A. veronii* bv. sobria resulted in the discovery of a new type IV pilus system, the Tap pilus, which was found to differ significantly from the Bfp (Kirov & Sanderson, 1996). Shortly after its initial detection in *A. veronii* bv. sobria, Tap pili biogenesis genes were identified in *A. hydrophila* (Ah65) (*tapABCD*) and found to share homology with type IV pilus genes in *P. aeruginosa* (*pilABCD*); the genes from *A. hydrophila* were able to complement corresponding mutations in *P. aeruginosa* (Pepe, Eklund, & Strom, 1996). Although Tap pili are present in all *Aeromonas* strains whose genomes have been so far sequenced, very little is known about their function. Deletion of the Tap pilin gene (*tapA*), and therefore deletion of the Tap pilus filament, had no effect on the ability of clinical *A. veronii* bv. sobria isolates to adhere to HEp-2 cells, adhere to human intestinal cells or colonise a young mouse model; similar findings were also reported with a fish isolate of *A. hydrophila* (Kirov, Barnett, Pepe, Strom, & Albert, 2000). The Tap pilus does contribute to the pathogenicity of the psychrophilic aeromonad *A. salmonicida* subsp. *salmonicida*, however, when infecting Atlantic salmon; although it is not essential for virulence as *tapA* mutants retain much of their ability to colonise the host (Boyd et al., 2008). Gene expression analysis in *A. salmonicida* has demonstrated that the Tap pilus is constitutively expressed and together with *tapABCD*, the Tap pilus system is also encoded on another 18 genes scattered around the genome (Boyd et al., 2008), similar to the type IV pili genes of *Pseudomonas* and *Neisseria* spp. (Mattick, 2002). While the role of the Tap pilus remains to be elucidated, its biogenesis has strong implications in virulence in other ways. TapD, a prepilin peptidase (Pepe et al., 1996) is not only required for the processing of the prepilins (from multiple pilus systems) before they are incorporated into the pilus filament but also for the processing of the pseudopilins of the T2SS, that are essential for the secretion of many virulence factors, such as proteases, haemolysin and DNase (Hadi et al., 2012; Kirov et al., 2000).

Along with the MSHA and Tap pilus, a third type IV pilus, belonging to the type IVb class, has also been identified in the *A. salmonicida* subsp. *salmonicida*, named the Flp pilus due to its homology with the Flp system of *Actinobacillus actinomycetemcomitans* (Boyd et al., 2008). In *A. salmonicida*, the genes for this pilus are located in a single operon (*flpA-L*) and its expression is iron regulated, with upregulation of the genes occurring under low iron conditions (Boyd et al., 2008). Very little is known about the role of the Flp pilus in *Aeromonas* and despite being a regulated system it does not appear to be involved in *A. salmonicida* virulence as *flp*-deficient strains fully retained their ability to colonise and infect Atlantic salmon (Boyd et al., 2008). The

Flp pilus has been found to be widespread among bacteria and contributes to biofilm formation and autoagglutination (Kachlany et al., 2000; Planet, Kachlany, Fine, DeSalle, & Figurski, 2003); it therefore may be more important in bacteria–bacteria interactions rather than host cell interactions.

Therefore, three distinct type IV pili systems have been described in aeromonads, with most species possessing at least two of these. For mesophilic aeromonads, early adherence studies in the 1990s suggested that the MSHA bundle-forming pilus is the major filamentous appendage responsible for tissue adherence and colonisation, however only recently was this genetically determined by Hadi et al. (2012). A 22 kb Bfp MSHA locus was isolated and characterised from *A. veronii* bv. sobria which was found to consist of 17 genes arranged into two operons, four of which encoded prepilins (one major and three minor) the subunits of the pilus filament (Hadi et al., 2012). It was concluded that all four prepilins are essential for pilus biogenesis with individual mutations in each prepilin gene reducing bacterial adherence to HEp-2 cells by 90% (Hadi et al., 2012). Ability of the mutants to form biofilms was also assayed and demonstrated that the MSHA pilus genes of *A. veronii* bv. sobria are essential for biofilm formation. These studies confirmed that the MSHA pilus is the major adherence system in *A. veronii* bv. sobria and this could also be the case for other mesophilic *Aeromonas* species. Although the MSHA pilus operon is present in *A. salmonicida*, there is a deletion of eight genes within the operon, including the gene encoding the major pilin (Boyd et al., 2008). The MSHA pilus in *A. salmonicida* is therefore not expected to be expressed and so appears to only aid the virulence of mesophilic aeromonads.

5. OUTER-MEMBRANE PROTEINS AND S-LAYER

5.1. Outer-membrane proteins

A number of outer-membrane proteins (OMPs) have been identified in *Aeromonas* spp. as monomeric adhesins. Early studies by Atkinson and Trust (1980) showed that most strains of *A. hydrophila* require pili for haemagglutination and adherence to host cells; however, some species retained these abilities upon pili removal. Proteins were discovered on the bacterial cell surface that were carbohydrate-reactive (crOMPs) and found to be involved in adherence as downregulation of OMPs coincided with decreased ability to haemagglutinate (Atkinson & Trust, 1980; Quinn, Wong, Atkinson, & Flower, 1993). Many more OMPs have now been discovered which vary in size (30–54 kDa) and show expression variation due to environmental changes, such as temperature, osmolarity and iron availability, effecting the adherence of some aeromonads. It has been suggested that OMPs facilitate adherence of non-piliated aeromonads to host cells. For example, a non-piliated *A. caviae* strain (Ae398) showed increased expression of a 43 kDa OMP when grown at 22 °C compared to 37 °C and this correlated with an improved ability to adhere to HEp-2 cells (Rocha-De-Souza et al., 2001). OMP expression correlating with growth temperature is similar to Bfp expression in piliated aeromonads, suggesting that these proteins are involved in bacterial adherence.

OMPs may also act as alternative adherence factors or adhesion enhancers for aeromonads during colonisation. Omp48 (48 kDa) was initially identified on the surface of *A. veronii* and found to share homology with a well characterised OMP of *E. coli* involved in maltose transport, LamB (Vazquez-Juarez, Tutt, & Torres, 2004). It was demonstrated that purified Omp48 has the ability to bind proteins of the extracellular matrix (collagen and fibronectin) and mucus glycoproteins (lactoferrin and mucin); binding was inhibited by sugars indicating that Omp48 interacts with carbohydrate domains on these host proteins (Vazquez-Juarez et al., 2004). Furthermore, *A. veronii* that had been pre-incubated with anti-Omp48 anti-serum showed a decreased ability to bind HeLa cells, confirming that this protein has a role in the *A. veronii* adhesion process (Vazquez-Juarez et al., 2004). Omp48 is regulated by maltose availability which may be of use during infection where maltose is present due to host digestion. This may induce the expression of this OMP and aid the colonisation process.

The published *A. hydrophila* ATCC 7966^{T} genome (Seshadri et al., 2006) and the unpublished *A. caviae* Sch3 genome (Shaw unpublished) have annotated a number of new OMPs as potential invasins. These proteins are encoded by the *AHA1066* gene in *A. hydrophila* ATCC 7966^{T} and *aigA*/*aigB* genes in *A. caviae* Sch3, and are predicted to contain bacterial immunoglobulin-like domains (BIG domains). Similar genes encoding BIG domain proteins are also present in *A. hydrophila* 4AK4 and *A. veronii* AMC34 (Shaw unpublished). Intimin of enteropathogenic *E. coli* (Luo et al., 2000) and the invasin protein of *Yersinia* spp. (Hamburger, Brown, Isberg, & Bjorkman, 1999) are both BIG domain containing OMPs and are involved in adherence and invasion of host tissues, respectively, implicating these BIG domain containing proteins in colonisation. However, the *Aeromonas* BIG-like proteins (AHA1066, AigA and AigB) are thought to be different to intimin or invasin as they do not contain β-barrel domains at their N-terminus which anchor the *E. coli* and *Yersinia* proteins into the outer membrane (Hamburger et al., 1999; Luo et al., 2000). Instead, they contain lipidation sequences at their N-terminus and are therefore likely to be lipoproteins, similar to the Lig lipoproteins of pathogenic *Leptospira* spp. (Cerqueira et al., 2009; Choy et al., 2007; Matsunaga et al., 2003; Palaniappan et al., 2002), to which AigA and AigB of *Aeromonas* show most sequence homology to. The Lig proteins also contain BIG domains and are located on the bacterial cell surface; they are essential for the colonisation and pathogenesis of virulent *Leptospira* and have been shown to bind extracellular matrix proteins such as fibronectin, collagen and laminin (Choy et al., 2007). It is therefore possible that Lig-like proteins are *Aeromonas* virulence factors and may be the reason why some *Aeromonas* strains are highly invasive as the production of Lig proteins in *Leptospira* spp. is characteristic of extremely pathogenic strains (Cerqueira et al., 2009).

As OMPs are located on the bacterial cell surface they are extremely immunogenic, the host immune system recognises them as foreign material, making them attractive vaccine candidates, especially for the protection of fish against aeromonad disease (Rahman & Kawai, 2000). For example, the previously mentioned Omp48 is present in many aeromonads and a recombinant version from *A. hydrophila* has recently been shown to induce a protective response in the Indian major carp, rohu (Khushiramani et al., 2012). This OMP also shares homology with OMPs of other fish pathogens (such as *Edwardsiella tarda*) and so may elicit a multiple protective response (Khushiramani et al., 2012). OmpW is another vaccine candidate; it is widespread among bacteria and has previously shown promise as a vaccine candidate against *Vibrio cholerae* (Das, Chopra, Cantu, & Peterson, 1998). OmpW from *A. hydrophila* is highly immunogenic in common carp and so may be useful for aquaculture disease management (Maiti, Raghunath, & Karunasagar, 2009; Maiti, Shetty, Shekar, Karunasagar, & Karunasagar, 2012). In addition, the immunogenic response to OmpG (another highly conserved OMP in aeromonads) in European eels was found to provide some protection against *A. hydrophila* and *A. sobria* (Guan, Xiong, Huang, & Guo, 2011). With respect to OMPs in psychrophilic aeromonads, iron-regulated proteins (IR-OMPs) present on the *A. salmonicida* cell surface that are thought to be potentially involved in virulence (Ebanks, Dacanay, Goguen, Pinto, & Ross, 2004; Ebanks, Goguen, McKinnon, Pinto, & Ross, 2005). Although the function of these IR-OMPs is currently unknown, they stimulate an immunogenic response in Atlantic salmon and could therefore be of use in vaccine development (Ebanks et al., 2005).

As many OMPs are conserved in aeromonads, a number of studies have highlighted the usefulness of OMPs as a tool for the typing of *Aeromonas* isolates. For example, Omp48 and OmpW can be used for *Aeromonas* detection due to their widespread nature and OmpTS for the identification of virulent strains as it is commonly found decorating clinical isolates (Guan et al., 2011; Khushiramani, Girisha, Bhowmick, Karunasagar, & Karunasagar, 2008).

5.2. S-layer

Surface layers or S-layers are a bi-dimensional crystalline protein array that cover the entire organism they are found in both Gram positive and Gram negative bacteria, in addition to *Archaea*. Bacterial S-layers are composed of proteins or glycoproteins of between 40 and 200 kDa in size, the so-called SLPs that constitutes one of the most abundant proteins in the cell. Once secreted SLPs auto-assemble to form a 5–10 nm thick paracrystalline structure that contains pores (Fagan & Fairweather, 2014). Depending upon bacterial species, S-layers can exhibit different paracrystalline structures that can possess oblique, square or hexagonal lattice symmetry.

In *Aeromonas*, the S-layer is formed by a unique protein, VapA (sometimes called AshA in mesophilic aeromonads), this protein forms a tetragonal array complex on the surface of the strains that contain the *vapA* gene (Chu et al., 1991). The first description of an S-layer (initially called A-layer) in *Aeromonas* was in *A. salmonicida* (Kay et al., 1981) and was recognised as a major virulence factor related with the ability of causing septicaemia in fish. The S-layer is less common in mesophilic *Aeromonas* but has been described in strains of *A. hydrophila* and *A. veronii* biovar sobria of LPS serogroups O:11, O:14 and O:81; however, it has not been demonstrated to be present in *A. caviae* (Kokka et al., 1990; Esteve et al., 2004). Moreover, the role of S-layers in the pathogenicity of mesophilic aeromonads appears to be clouded, as *A. hydrophila* S-layer mutants do not exhibit increased susceptibility to rabbit serum (Noonan & Trust, 1997), or loss of virulence in mice (Kokka, Vedros, & Janda, 1991).

The *vapA* gene encoding for the SLP is usually located within an LPS O-antigen gene cluster along with an operon encoding an S-layer dedicated T2SS. The gene *vapA* encoding the SLP in *A. salmonicida* can be deleted when the bacterium is grown at over 25 °C, resulting in loss of the S-layer (Belland & Trust, 1987). As with other LPS O-antigen-associated gene clusters, the G + C content of this chromosomal region is usually lower than the 60% aeromonad average, suggesting that the genes for the S-layer have been lateral acquired (Reith et al., 2008). Such lateral acquisition has been demonstrated in other bacteria such as *Clostridium difficile*, where the horizontal transfer of the cassette containing the genes *slpA* and *secA2* between different lineages has been demonstrated (Dingle et al., 2013). Genetic linkage between the genes encoding for the SLP and its dedicated T2SS has been demonstrated in *C. difficile*, *Bacillus anthracis*, *Campylobacter fetus* and *Caulobacter crescentus* (Fagan & Fairweather, 2014).

In aeromonads, as S-layers are encoded within LPS gene clusters they are thought to be anchored to the cell wall by the LPS through the tethering onto the O-antigen of a homogeneous length. Mutants of *A. salmonicida* lacking the smooth LPS O-antigen are unable to assemble a functional S-layer, although the SLP VapA is still secreted in to the media. Mutations resulting in partial loss of the LPS O-antigen also result in S-layer disruption (Kokka et al., 1990; Griffiths & Lynch, 1990). The C-terminal domain of the SLP VapA has been shown to be essential for LPS interaction (Thomas, Austin, McCubbin, Kay, & Trust, 1992). In aeromonads, SLPs are usually between 50 and 52 kDa in size (Kokka et al., 1990; Kay et al., 1981), and according to their predicted molecular weights, the SLP of *A. hydrophila*, AshA, appears to be post-translationally modified while the SLP VapA of *A. salmonicida* is not (Noonan & Trust, 1997). The *A. hydrophila* AshA SLP post-translational modifications appear to include phosphorylation of tyrosine residues that decrease the pI of the SLP from 6.7 to 4.6 (Thomas & Trust, 1995).

Bacterial S-layers can represent 10–15% of the total cellular protein, therefore their production and maintenance constitutes a challenge for the bacteria. The SLP must be efficiently transcribed and translated. Until now, two promoters have been identified upstream of the *vapA* gene in *A. salmonicida* (Chu et al., 1991), one of which is most active during the early stages of growth (Noonan & Trust, 1995a).

The SLP VapA contains an N-terminal signal sequence of 19 to 21 amino acids in length that is essential for secretion. The *A. salmonicida vapA* gene lies downstream from an operon encoding a dedicated T2SS (Reith et al., 2008). This T2SS was demonstrated to be specific for the secretion of the VapA protein that forms the S-layer. Mutation of the S-layer T2SS genes *spsD* (a PulD T2SS homologue) or *spsE* (a PulE T2SS ATPase homologue) resulted in the accumulation of the VapA in the bacterial periplasm but did not affect the secretion of other extracellular proteins known to be secreted by the cells general T2SS (Noonan & Trust, 1995b).

In *A. salmonicida* and *A. hydrophila*, S-layers have been associated with the ability to cause disease in fish (Kay et al., 1981; Murray, Dooley,

Whippey, & Trust, 1988) and have been suggested to act by protecting the cells from complement-mediated killing and proteases (Merino, Aguilar, et al., 1996; Merino, Rubires, Aguilar, Alberti, et al., 1996; Merino, Rubires, Aguillar, Guillot, et al., 1996). In both *A. salmonicida* and *A. hydrophila*, the SLPs are resistant to proteolysis inside the phagolysosome of the macrophages (Noonan & Trust, 1997). Although the role of the S-layers of *A. hydrophila* and *A. veronii* biovar sobria have not been fully determined, their importance is not thought to be as great as for *A. salmonicida* in fish disease (Merino, Rubires, Konkel, & Tomás, 1995). In other mesophilic aeromonads, more specifically the mesophilic O:11 serogroup, S-layers are thought to contribute to the colonisation of the intestinal mucosa and facilitate the spread all over the intestinal mucosa by conferring a higher resistance to opsono-phagocytosis (Kay et al., 1981).

6. AEROMONAD COLONISATION AND HOST RESPONSE

6.1. Adherence factors

Adherence to host cells is a crucial step for pathogen colonisation and infection. Aeromonads possess a variety of adhesins on the cell surface, both filamentous and non-filamentous, that contribute to this process. The adherence abilities of aeromonads can also vary due to a number of factors such as, temperature, growth media and growth phase (Ho et al., 1990; Kirov, Hayward, et al., 1995; Kirov, Jacobs, Hayward, & Hapin, 1995). This therefore suggests that the adherence of *Aeromonas* spp. is a complex, multifactorial process with some colonisation factors having several roles.

The flagellar systems of mesophilic aeromonads are not only required for motility of bacterial cells once they have reached the host mucosa but also for adherence to cells; mutational studies have demonstrated that the polar and lateral flagella of *A. caviae* are essential for adherence to HEp-2 cells (Gavin et al., 2002; Rabaan et al., 2001). In addition, flagella have a role in biofilm formation which generally allows disease persistence (Gavin et al., 2002; Rabaan et al., 2001). Flagellins are recognised by the TLR-5 in mammals via their D0 domains (Hayashi et al., 2001) and this induces a proinflammatory response. It has previously been established with a number of bacteria, such as *E. coli* and *S. enterica* serovar. Typhimurium, that flagellins stimulate IL-8 secretion in epithelial cells (Gewirtz et al., 2001; Steiner, Nataro, Poteet-Smith, Smith, & Guerrant, 2000; Zhou et al., 2003) which is a chemoattractant for macrophages and neutrophils. A number of mesophilic aeromonads glycosylate their polar and potentially their lateral flagella

(Tabei et al., 2009) and the reason for this post-translational modification is a mystery, although it has been speculated that glycosylation may mask flagellins from the host immune response by preventing this TLR-5 recognition (Logan, 2006). However, studies by Parker et al. (2012) in *A. caviae* have suggested glycosylated polar flagellins are more strongly recognised by antisera than unglycosylated flagellins. And investigations with *P. aeruginosa* flagellins have demonstrated that the glycosylated form are able to induce a stronger IL-8 response than the unglycosylated (Verma, Arora, Kuravi, & Ramphal, 2005), which may aid disease progression and be the reason why hyper-inflammation is present in respiratory infections caused by this bacterium. Both polar and lateral flagella of mesophilic aeromonads are therefore important virulence factors being involved in multiple steps of the colonisation pathway. The polar flagellum has also been reported as a potential vaccine candidate for the immunisation of channel catfish; recombinant FlgK (hook–filament junction protein) of *A. hydrophila* was shown to strongly react with catfish sera upon experimental infection (Yeh & Klesius, 2011).

Other filamentous appendages on the cell surface are also involved in aeromonad colonisation. As previously mentioned, type IV MSHA pili have a major role in the adherence of mesophilic species to host cells and, similarly to the flagella, are involved in biofilm formation (Hadi et al., 2012). Furthermore, studies in *V. cholerae* have provided evidence that the PilA containing type IV pilus (Tap) is induced by chitin, facilitating chitin colonisation (Meibom et al., 2004). It is therefore possible that the aeromonad Tap pilus system may be induced in a similar way and be of importance to bacterial survival in aquatic environments by aiding adhesion to chitin containing surfaces. Although type I and type IV (Tap) pili from psychrophilic *A. salmonicida* do have a role in adherence, they are not essential for host colonisation and the majority of strains are non-motile, so lack flagella.

LPS is another aeromonad colonisation factor; early studies with serogroup O:34 *Aeromonas spp* described the importance of the LPS in the colonisation of a number of animal models, such as the germ-free chicken (Merino, Rubires, Aguillar, Guillot, et al., 1996) and the medicinal leech (Braschler, Merino, Tomas, & Graf, 2003; Silver et al., 2007). LPS is often referred to as 'smooth' or 'rough', the presence of full-length O-chains (O-antigen) causes LPS to be smooth (S-LPS) and it is this S-LPS version that is implicated in pathogen virulence. Mutational studies in *A. hydrophila* (AH-3) serotype O:34 have demonstrated that strains unable to produce the O:34 antigen of the LPS have a reduced ability in adherence to HEp-2 cells and showed altered bacterial motility, even though both flagella systems were still present (Canals, Altarriba, et al., 2006; Canals, Jimenez, et al., 2006, Canals, Ramirez, et al., 2006; Vilches et al., 2007). Mutants were also less virulent in fish and mouse models of infection compared to the wild-type strain (Canals, Altarriba, et al., 2006; Canals, Jimenez, et al., 2006, Canals, Ramirez, et al., 2006; Vilches et al., 2007) demonstrating the importance of complete LPS (smooth LPS) in bacterial behaviour and virulence. LPS affects the overall 'wettability' of the cell surface, influencing how bacteria interact with their environment and therefore affecting colonisation capabilities. More recently it was discovered that the LPS O-antigen of *A. caviae* (Sch3) is glycosylated with pseudaminic acid, similar to the polar flagellin of this strain (Tabei et al., 2009). Mutation in the *A. caviae* biosynthetic locus for this sugar (*flm* locus) resulted in a loss of the LPS O-antigen and both flagella systems (Tabei et al., 2009). Mutants displayed a reduced ability to adhere to HEp-2 cells (Gryllos et al., 2001; Tabei et al., 2009), further supporting the notion that flagella and the LPS are involved in aeromonad colonisation. In addition, this work also suggests that the post-translational modification of these cell surface structures has important implications in virulence.

LPS of Gram negative bacteria is highly immunogenic and stimulates a proinflammatory response via TLR-4 on host cells which can lead to chronic inflammation and tissue destruction (Lu, Yeh, & Ohashi, 2008). Although it induces a potent immune response the LPS is a protective layer that allows aeromonads to evade complement-mediated lysis by the host (Merino, Alberti, & Tomás, 1994; Merino et al., 1998; Merino, Rubires, Aguilar, Alberti, et al., 1996; Merino, Rubires, Aguillar, Guillot, et al., 1996). Merino et al. (1998) demonstrated that *A. hydrophila* strains possessing an O-antigen were able to mask a 39 kDa OMP with the ability to bind C1q of the classical complement cascade, preventing cascade initiation and providing serum resistance. OMPs are also associated with host cell adherence and are thought to provide protection from the host immune response. The oldest known OMP is the S-layer (A-layer in aeromonads) protein, VapA which has previously been connected to aeromonad colonisation of fish. A study by Han et al. (2011) provided evidence that the production of S-layers in some *A. salmonicida* species is directly linked to the colonisation and infection of black rockfish. These strains were also able to evade the host complement system and macrophage opsonisation (Han et al., 2011) which agreed with previous studies from Merino, Aguilar, et al. (1996) that suggested the S-layer of *A. salmonicida* is involved in immune evasion. For aeromonads that produce an S-layer, VapA makes up around 60% of their OMP content; the abundance of this one protein emphasises its importance to the bacterium. As previously discussed, there are many other OMPs thought to assist in aeromonad colonisation (OMPs), although to what extent is yet to be determined. Also, in the genomes of *A. hydrophila* (ATCC 7966^{T}) (Seshadri et al., 2006) and *A. caviae* (Sch3) (Shaw unpublished) new OMPs have been identified, the Lig-like invasins, and although they have not yet been studied their potential role in *Aeromonas* colonisation is intriguing.

Similar to the S-layer and the LPS, the capsule, possessed by a number of aeromonads, is also involved in bacterial protection by preventing complement-mediated lysis and macrophage opsonisation (Garduno, Thornton, & Kay, 1993; Merino et al., 1994). Additionally, *A. salmonicida* grown under capsule-inducing conditions were more able to invade fish cells lines (Merino, Aguilar, et al., 1996; Merino, Rubires, Aguilar, Alberti, et al., 1996; Merino, Rubires, Aguillar, Guillot, et al., 1996) demonstrating a role for this polysaccharide in colonisation.

6.2. Secreted factors

There are a number of secreted virulence factors also involved in host colonisation and further invasion of host tissues to stabilise infections. Five secretion systems have been identified in aeromonads, out of the six characterised in Gram negative bacteria (type 1, 2, 3, 4 and 6), responsible for the transport of these virulence determinants. Little or no work has been carried out on T1SS, bioinformatics analysis suggests it maybe required for RTX toxin delivery (Shaw unpublished). The T2SS is a general secretion pathway that secretes extracellular factors, some of which are involved in virulence such as DNases, proteases, amylases and the major haemolysin aerolysin. So far, the T2SS has only been studied in *A. hydrophila* (Li, Miller, Bull, & Howard, 2011; Seshadri et al., 2006), however, genes for structural subunits have now been identified in a number of other aeromonad genomes, such as in *A. salmonicida* subsp. *salmonicida* (Reith et al., 2008). As well as transporting virulence determinants into eukaryotic cells, the T4SS is also capable to genetic transfer between bacteria via conjugation (Rangrez et al., 2006). This system is therefore likely to have a role in the transfer of antibiotic resistance genes.

The type 3 and 6 secretion systems (T3SS/T6SS), in particular, are associated with virulence and have been detected in some clinical and environmental aeromonad isolates (Chacon, Soler, Groisman, Guarro, & Figueras,

2004; Suarez et al., 2008). The T3SS is a needle-like structure, often described as an 'injectisome' that can inject effectors directly into the cytoplasm of target cells (Chacon et al., 2004; Vilches et al., 2004). Vilches, Jiménez, Merino, and Tomás (2012) have recognised the importance of this secretion system as a loss of a functional T3SS in *A. hydrophila* corresponds to attenuated virulence in animal models (fish and mice) and in an alternative amoebae model (*Dictyostelium*). So far, five effectors have been described for the aeromonad type III secretion pathway: AexT, AexU, AopP, AopH and AopO. Both AexT and AexU are bifunctional toxins that contain a GTPase-activating domain, allowing the interruption of host cell-signalling pathways and an ADP-ribosylating domain that can disrupt the host cytoskeleton, leading to the induction of apoptosis. AexT is present in *A. salmonicida* and shares homology with ExoT/S of *P. aeruginosa* (Dacanay et al., 2006), whereas AexU (sometimes referred to as AexT) is present in *A. hydrophila* and although it exhibits a similar function to AexT, the ADP-ribosylating domain shows no homology to any bacterial secreted effectors currently known (Sha et al., 2007; Sierra et al., 2007). There is evidence to suggest that both ADP-ribosylating toxins are directly involved in aeromonad virulence. Studies with an *aexT A. salmonicida* subsp. *salmonicida* mutant had no toxic effect on RTG-2 fish cells (Braun et al., 2002) and mice infected with an *aexU A. hydrophila* mutant displayed increased survival rates compared to those infected with the wild type (Sha et al., 2007; Sierra et al., 2007). In addition to AexT, some *A. salmonicida* strains possess the final three effectors known to be secreted by the type III pathway: AopP (NFκB inhibitor), AopH (tyrosine phosphatase) and AopO (serine/threonine kinase), which are able to interrupt host cell signalling pathways and induce toxicity (Dacanay et al., 2006; Reith et al., 2008).

The T6SS also has the ability to translocate virulence determinants directly into the host cell cytoplasm (Suarez et al., 2008, 2010). The T6SS has been analysed from a clinical *A. hydrophila* isolate and studies have demonstrated that mutations in genes encoding components of the T6SS alters bacterial virulence by modifying the ability of the bacterium to be phagocytosed (Suarez et al., 2008). Four effectors of the T6SS have so far been characterised, one of them being the haemolysin co-regulated protein (Hcp) which has been implicated in preventing this bacterial phagocytosis, although the mechanism of action has not been determined (Suarez et al., 2008). The expression of Hcp in HeLa cells also lead to apoptosis, demonstrating the importance of this toxin in bacterial virulence (Suarez et al., 2008). Other effectors that have been identified are Vgr1, Vgr2 and

Vgr3, where it has been established that Vgr1 is an ADP-ribosylating toxin capable of interrupting the host cell cytoskeleton and inducing apoptosis in HeLa cells (Suarez et al., 2008, 2010).

6.3. Host relationships

Aeromonads are ubiquitous in nature and have the ability to colonise a wide range of hosts, whether associations are detrimental to the host (disease in humans and fish) or in some cases beneficial, as aeromonads can exist in a symbiotic nature with their host (medicinal leech and possibly the mosquito).

With respect to human disease, aeromonads generally infect immunocompromised individuals and are therefore usually opportunistic pathogens. Identifying pathogenic and non-pathogenic strains has been a problem for this bacterial genus, however locating the presence of particular virulence determinants via whole-genome sequencing may predict which aeromonads are capable of host colonisation and infection, aiding the future of microbial diagnostics. For example, Grim et al. (2013) highlighted the potential of comparative genomics in the identification of virulent aeromonads during wound infections. They identified an infective *A. hydrophila* subtype that contained a plethora of virulence genes; it was found to be lethal in a mouse model and extremely aggressive during human infection (Grim et al., 2013). Pathogenesis of this strain was attributed to the possession of flagella, a capsule, T3SS and production of haemolysin (Grim et al., 2013). Further comparative studies could therefore help to predict which virulence factors are largely associated with pathogenic *Aeromonas* strains.

Conversely, *Aeromonas* colonisation can be beneficial, where bacterium and host can live in symbiosis. For example, *Aeromonas* is present as a symbiont in the medicinal leech gut with *A. veronii* being the most abundant species (Nelson & Graf, 2012). In addition, *Aeromonas* spp. have previously been isolated from wound infections following leech therapy (Ardehali, Hand, Nduka, Holmes, & Wood, 2006; Lineaweaver et al., 1992), demonstrating that the same *Aeromonas* strain can potentially elicit two outcomes in different hosts as a result of colonisation: either pathogenesis or symbiosis. The purpose of this symbiotic relationship has been investigated and transcriptome analysis by Bomar, Maltz, Colston, and Graf (2011) has suggested that the aeromonads may be able to feed from nutrients (acetate) released by another leech gut symbiont, a *Rikenella*-like bacterium. It is also possible that these nutrients released by symbiotic bacteria may be utilised by the leech, providing optimum nutrition (Bomar et al., 2011). Furthermore, similar to the normal flora of the human gut, symbionts may be present to protect the host organism from pathogenic bacteria, providing 'colonisation resistance'.

A number of factors required for *A. veronii* leech colonisation have been identified that allow persistence in the gut without having negative effects on the host. Mammalian blood ingested by the leech is one factor involved in *A. veronii* colonisation, it contains innate immunity factors (i.e. complement) that remains active in the leech and protect the new host from bacterial infection (Indergand & Graf, 2000). However, many aeromonads are complement resistant due to features such as the LPS; this therefore allows serum resistant *A. veronii* to proliferate in the leech compared to the destruction of serum-sensitive bacteria which could be a threat to the host (Braschler et al., 2003; Indergand & Graf, 2000). Surprisingly, the type 2 and 3 secretion systems, and the export of a number of toxins, are essential for the survival of *A. veronii* in the leech gut also. A study by Maltz and Graf (2011) concluded that the T2SS and its ability to secrete haemolysin is critical for *A. veronii* to associate with the gut wall, suggesting erythrocyte lysis to be involved in the early steps of colonisation. Moreover, an *A. veronii* T3SS mutant analysed by Silver et al. (2007) was found to be targeted by hemocytes (macrophage-like cells of invertebrates) and phagocytosed, whereas the wild type was able to evade this host immune response. The same study also demonstrated that the *A. veronii* mutant was less virulent in a mouse septicaemia model, signifying that secreted factors of the T3SS have a role in aeromonad colonisation in each instance, even though the relationship with each host is drastically different (Silver et al., 2007).

Recently, the analysis of gut microbiota of wild Kenyan mosquitoes has revealed that *Aeromonas* spp. are the most prevalent, making up around a third of bacteria present (Osei-Poku et al., 2012). It has been suggested that bacterial taxa present in the mosquito gut may have an effect on disease transmission rates and studies have already demonstrated the potential of genetically engineered symbiotic bacteria as a tool to fight malaria (Wang et al., 2012). Being the most prevalent gut symbiont, aeromonads may be of use in future attempts to combat malaria.

7. CONCLUSIONS AND OUTLOOK

The recent explosion of aeromonad genome sequences that have become available have been a great asset in revealing the potential colonisation systems possessed within the genus *Aeromonas*. Utilising this data there have been a number of recent manuscripts on comparative genomics of both clinical aeromonad strains involved in wound infections (Grim et al., 2013) and epidemic strains of fish diseases (Hossain et al., 2013) with those of environmental non-pathogenic or non-epidemic origin. However, these investigations did not find the 'ultimate' virulence determinant that separated the pathogenic strains from the environmental strains. Therefore in can be concluded from these studies that pathogenicity of *Aeromonas* is a multi-factorial process that needs the input many virulence determinants, even though some appear to be redundant as for example the possession of multiple pilus systems. However, the possession of multiple systems would be expected from a bacterium that inhabits such a wide series of environments that can be a symbiont in some animals but then colonises and infects other animals that can be either warm blooded or cold blooded. Here, we have outlined the current knowledge of the factors that have been linked to the biotic and abiotic colonisation process in the aeromonads. The major factors we have included in a summary diagram (Fig. 4). In the mesophilic aeromonads, motility plays a great role in their biology with the majority of strains being in the possession of two independent flagella systems. Genomics has demonstrated why certain aeromonad species and strains are motile, while other species are non-motile even though they encode both flagella systems. The structure and the majority of the regulation of the constitutively expressed polar flagellum have been determined. The polar flagella filament has been shown to be glycosylated with the unusual sugar pseudaminic acid, similar to what happens in other bacterial pathogens such as *Helicobacter pylori* or *Campylobacter jejuni*. *A. caviae* Sch3 has a simple flagella glycosylation system that is helping elucidate the molecular mechanisms and pathway of this glycan addition process, with the first purification of a Maf protein from any bacteria and the demonstration of the protein interacting with its flagellin substrate (Parker et al., 2014). Moreover, polar flagellin has been shown to be more complicated in other aeromonad strains such as *A. hydrophila* AH-3, this strain decorates its flagellin with a heptasaccharide and whose glycosylation process involves an Und-P lipid carrier and a specialised transferase WecX. Therefore, there is still much to learn about the intricacies and variations of aeromonad flagellar glycosylation, its complete mechanism, pathway and by far the role of this post-translational modification in the biology of the organisms that posses it. With regard to the lateral flagella system, although it is produced up on contact with solid surfaces or in viscous environments, the actual mechanism of

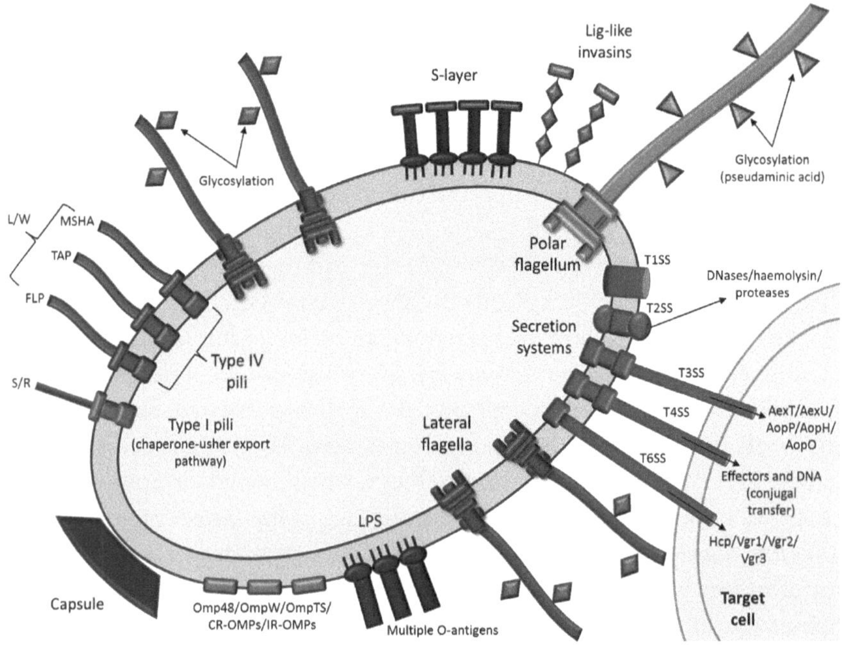

Figure 4 Schematic diagram showing the numerous structures that have been associated or implicated in the aeromonad colonisation process. These include the motility organelles: the polar flagellum glycosylated with pseudaminic acid for swimming motility and lateral flagella for swarming motility over surfaces. For adherence to biotic and abiotic surfaces, type I and type IV pili (Flp, MSHA, Tap), outer-membrane proteins (OMPs, CR-OMPs and IR-OMPs) have been implicated. For protection from the host immune response, lipopolysaccharide (LPS), capsule and S-layers have been shown to be important. For the delivery of effectors that target host or bacterial cells, multiple types of secretion system (T1SS, T2SS, T3SS, T4SS and T6SS) are present.

how this is sensed and transduced to the expression platforms is yet to be elucidated. Initial studies have demonstrated the importance of the signalling molecule c-di-GMP in this process with its potential to interact with FlrA of the polar flagella system and both LafK and PilZ of the lateral flagella system. Investigations currently being undertaken in these areas should now begin to provide answers to these questions.

After motility, aeromonad cells use multiple adhesion systems in order to adhere to biotic or abiotic surfaces. Genomics has shown that aeromonads possess many fimbrial systems, with some strains encoding up to three type IV pilus systems among others. Mutagenesis has complemented previous biochemical findings in demonstrating that the bundle-forming MSHA type IV pilus is one of the most important mesophilic aeromonad adherence systems, being essential for host cell adhesion in addition to biofilm formation on biotic surfaces. However, roles for the Tap and Flp pilus systems have still to be fully determined; although recent findings with the equivalent Tap system of *V. cholerae* may suggest or function in adherence to chitin in the aquatic environment. At present, next to nothing is known about the biology of the other pilus systems present in aeromonad strains.

Great in roads have been made in the field of aeromonad LPS with regard to their structure and genetics. Recent findings from genomic analysis of epidemic *A. hydrophila* strains causing MAS in channel catfish have suggested a particular LPS O-antigen subtype that contains Fuc*p*3NAc having an important role in the pathogenesis of these strains. Other previously important areas of *Aeromonas* study appear to have come to a standstill, with only a small amount of investigation being carried out on S-layers in the past 10 years. This may yet become an important area of study as many S-layers are now known to be glycosylated. Glycosylation is a process that was not thought to take place in bacteria when the original ground breaking work was done on the *Aeromonas* "A-layer" (S-layer) over 20 years ago. Therefore there are still many questions to be answered with regard to S-layers in aeromonad biology and these could provide new and interesting research avenues for the future.

Multiple aeromonad genome analysis has revealed many laterally acquired genes, a number of which have been shown to encode for OMPs. The most potentially interesting of which are the ones that are putative adhesins or invasins of the Lig-like family such as AigA and AigB of *A. caviae* and AHA1066 of *A. hydrophila*, as similar proteins have been demonstrated to be important virulence determinants of the *Leptospira* species. However, many of the OMPs of *Aeromonas* species still need to be studied at the genetic level, with the creation of isogenic mutants to be tested in pathogenicity models.

Analysis of the microbiota of a number of animals including important insect vectors has demonstrated the presence of *Aeromonas*, sometimes being there as the most prevalent species. Currently, very little is known about whether this is a commensal relationship or if it is symbiotic as is found between *Aeromonas* and the medicinal leech. Although the answers and the mechanisms of action of a number of important colonisation factors are being determined for *Aeromonas*, there are still many questions remaining to be solved for the biology of this enigmatic pathogen. Therefore, it is an exciting time to be working with such an interesting bacterium.

REFERENCES

Abrahamian, F. M., & Goldstein, E. J. C. (2011). Microbiology of animal bite wound infections. *Clinical Microbiology Reviews, 24*, 231–246.

Aguilar, A., Merino, S., Nogueras, M. M., Regue, M., & Tomas, J. M. (1999). Two genes from the capsule of *Aeromonas hydrophila* (serogroup 0: 34) confer serum resistance to *Escherichia coli* K12 strains. *Research in Microbiology, 150*, 395–402.

Aguilar, A., Merino, S., Rubires, X., & Tomàs, J. M. (1997). The influence of osmolarity on lipopolysaccharide and virulence of *Aeromonas hydrophila* serotype O:34 strains grown at 37°C. *Infection and Immunity, 65*, 1245–1250.

Aldrige, P., & Hughes, K. T. (2002). Regulation of flagellar assembly. *Current Opinion in Microbiology, 5*, 160–165.

Ardehali, B., Hand, K., Nduka, C., Holmes, A., & Wood, S. (2006). Delayed leech-borne infection with *Aeromonas hydrophila* in escharotic flap wound. *Journal of Plastic Reconstructive and Aesthetic Surgery, 59*, 94–95.

Atkinson, H. M., & Trust, T. J. (1980). Hemagglutination properties and adherence ability of *Aeromonas hydrophila. Infection and Immunity, 27*, 938–946.

Barnett, T. C., & Kirov, S. M. (1999). The type IV *Aeromonas* pilus (Tap) gene cluster is widely conserved in *Aeromonas* species. *Microbial Pathogenesis, 26*, 77–84.

Beaz-Hidalgo, R., & Figueras, M. J. (2013). *Aeromonas* spp. whole genomes and virulence factors implicated in fish disease. *Journal of Fish Diseases, 36*, 371–388.

Belland, R. J., & Trust, T. J. (1987). Cloning of the gene for the surface array protein of Aeromonas salmonicida and evidence linking loss of expression with genetic deletion. *Journal of Bacteriology, 169*, 4086–4091.

Boehm, A., Kaiser, M., Li, H., Spangler, C., Kasper, C. A., Ackermann, M., et al. (2010). Second messenger-mediated adjustment of bacterial swimming velocity. *Cell, 141*, 107–116.

Bomar, L., Maltz, M., Colston, S., & Graf, J. (2011). Directed culturing of microorganisms using metatranscriptomics. *mBio, 2*, e00012–e00011.

Boyd, J. M., Dacanay, A., Knickle, L. C., Touhami, A., Brown, L. L., Jericho, M. H., et al. (2008). Contribution of type IV pili to the virulence of *Aeromonas salmonicida* subsp salmonicida in Atlantic salmon (Salmo salar l.). *Infection and Immunity, 76*, 1445–1455.

Braschler, T. R., Merino, S., Tomas, J. M., & Graf, J. (2003). Complement resistance is essential for colonization of the digestive tract of Hirudo medicinalis by Aeromonas strains. *Applied and Environmental Microbiology, 69*, 4268–4271.

Braun, M., Stuber, K., Schlatter, Y., Wahli, T., Kuhnert, P., & Frey, J. (2002). Characterization of an ADP-ribosyltransferase toxin (AexT) from *Aeromonas salmonicida* subsp salmonicida. *Journal of Bacteriology, 184*, 1851–1858.

Canals, R., Altarriba, M., Vilches, S., Horsburgh, G., Shaw, J. G., Tomás, J. M., et al. (2006). Analysis of the lateral flagellar gene system of *Aeromonas hydrophila* AH-3. *Journal of Bacteriology, 188*, 852–862.

Canals, R., Jimenez, N., Vilches, S., Regue, M., Merino, S., & Tomas, J. M. (2006). The UDP N-acetylgalactosamine 4-epimerase gene is essential for mesophilic *Aeromonas hydrophila* serotype O34 virulence. *Infection and Immunity, 74*, 537–548.

Canals, R., Ramirez, S., Vilches, S., Horsburgh, G., Shaw, J. G., Tomás, J. M., et al. (2006). Polar flagellum biogenesis in *Aeromonas hydrophila. Journal of Bacteriology, 188*, 542–555.

Canals, R., Vilches, S., Wilhelms, M., Shaw, J. G., Merino, S., & Tomás, J. M. (2007). Non-structural flagella genes affecting both polar and lateral flagella-mediated motility in *Aeromonas hydrophila. Microbiology, 153*, 1165–1175.

Carrello, A., Silburn, K. A., Budden, J. R., & Chang, B. J. (1988). Adhesion of clinical and environmental *Aeromonas* isolates to HEp-2 cells. *Journal of Medical Microbiology, 26*, 19–27.

Cerqueira, G. M., McBride, A. J. A., Picardeau, M., Ribeiro, S. G., Moreira, A. N., Morel, V., et al. (2009). Distribution of the leptospiral immunoglobulin-like (lig) genes in pathogenic *Leptospira* species and application of ligB to typing leptospiral isolates. *Journal of Medical Microbiology, 58*, 1173–1181.

Chacon, M. R., Soler, L., Groisman, E. A., Guarro, J., & Figueras, M. J. (2004). Type III secretion system genes in clinical *Aeromonas* isolates. *Journal of Clinical Microbiology, 42*, 1285–1287.

Chadsey, M. S., Karlinsey, J. E., & Hughes, K. T. (1998). The flagellar anti-sigma factor FlgM actively dissociates Salmonella typhimurium sigma 28 RNA polymerase holoenzyme. *Genes and Development, 12*, 3123–3136.

Chilcott, G. S., & Hughes, K. T. (2000). Coupling of flagellar gene expression to flagellar assembly in *Salmonella enterica* serovar Typhimurium and *Escherichia coli*. *Microbiology Molecular Biological Review, 64*, 694–708.

Choy, H. A., Kelley, M. M., Chen, T. L., Moller, A. K., Matsunaga, J., & Haake, D. A. (2007). Physiological osmotic induction of *Leptospira interrogans* adhesion: LigA and LigB bind extracellular matrix proteins and fibrinogen. *Infection and Immunity, 75*, 2441–2450.

Chu, S., Cavaignac, J. Feutrier, Phipps, B. M., Kostrzynska, M., Kay, W. W., & Trust, T. J. (1991). Structure of the tetragonal surface virulence array protein and gene of *Aeromonas salmonicida*. *The Journal of Biological Chemistry, 266*, 15258–15265.

Dacanay, A., Boyd, J. M., Fast, M. D., Knickle, L. C., & Reith, M. E. (2010). Aeromonas salmonicida Type I pilus system contributes to host colonization but not invasion. *Diseases of Aquatic Organisms, 88*, 199–206.

Dacanay, A., Knickle, L., Solanky, K. S., Boyd, J. M., Walter, J. A., Brown, L. L., et al. (2006). Contribution of the type III secretion system (TTSS) to virulence of *Aeromonas salmonicida* subsp *salmonicida*. *Microbiology, 152*, 1847–1856.

Das, M., Chopra, A. K., Cantu, J. R., & Peterson, J. W. (1998). Antisera to selected outer membrane proteins of *Vibrio cholerae* protect against challenge with homologous and heterologous strains of *V. cholerae*. *FEMS Immunology and Medical Microbiology, 22*, 303–308.

Dasgupta, N., Ferrell, E. P., Kanack, K. J., West, S. E. H., & Ramphal, R. (2002). *fleQ*, the gene encoding the major flagellar regulator of *Pseudomonas aeruginosa*, is 70 dependent and is downregulated by Vfr, a homolog of *Escherichia coli* cyclic AMP receptor protein. *Journal of Bacteriology, 184*, 5240–5250.

Dasgupta, N., & Ramphal, R. (2001). Interaction of the antiactivator FleN with the transcriptional activator FleQ regulates flagellar number in *Pseudomonas aeruginosa*. *Journal of Bacteriology, 183*, 6636–6644.

Densmore, C. L., & Green, D. E. (2007). Diseases of amphibians. *ILAR Journal, 48*, 235–254.

Dingle, K. E., Didelot, X., Ansari, M. A., Eyre, D. W., Vaughan, A., Griffiths, D., et al. (2013). Recombinational switching of the *Clostridium difficile* S-layer and a novel glycosylation gene cluster revealed by large scale whole genome sequencing. *Journal of Infectious Diseases, 207*, 675–686.

Easow, J. M., & Tuladhar, R. (2007). *Aeromonas hydrophila* wound infection following a tiger bite in Nepal. *Southeast Asian Journal of Tropical Medicine and Public Health, 38*, 867–870.

Ebanks, R. O., Dacanay, A., Goguen, M., Pinto, D. M., & Ross, N. W. (2004). Differential proteomic analysis of *Aeromonas salmonicida* outer membrane proteins in response to low iron and in vivo growth conditions. *Proteomics, 4*, 1074–1085.

Ebanks, R. O., Goguen, M., McKinnon, S., Pinto, D. M., & Ross, N. W. (2005). Identification of the major outer membrane proteins of *Aeromonas salmonicida*. *Diseases of Aquatic Organisms, 68*, 29–38.

Esteve, C., Alcaide, E., Canals, R., Merino, S., Blasco, D., Figueras, M. J., et al. (2004). Pathogenic *Aeromonas hydrophila* serogroup O: 14 and O: 81 strains with an S layer. *Applied and Environmental Microbiology, 70*, 5898–5904.

Fagan, R. P., & Fairweather, N. F. (2014). Biogenesis and functions of bacterial S-layers. *Nature Reviews Microbiology, 12*, 211–222.

Garduno, R. A., & Kay, W. W. (1995). Capsulated cells of *Aeromonas-salmonicida* grown *in-vitro* have different functional-properties than capsulated cells grown *in-vivo*. *Canadian Journal of Microbiology, 41*, 941–945.
Garduno, R. A., Thornton, J. C., & Kay, W. W. (1993). *Aeromonas salmonicida* grown in-vivo. *Infection and Immunity, 61*, 3854–3862.
Garrote, A., Bonet, R., Merino, S., Simonpujol, M. D., & Congregado, F. (1992). Occurrence of a capsule in *Aeromonas salmonicida*. *FEMS Microbiology Letters, 95*, 127–132.
Gavin, R., Rabaan, A. A., Merino, S., Tomás, J. M., Gryllos, I., & Shaw, J. G. (2002). Lateral flagella of Aeromonas species are essential for epithelial cell adherence and biofilm formation. *Molecular Microbiology, 43*, 383–397.
Gewirtz, A. T., Simon, P. O., Schmitt, C. K., Taylor, L. J., Hagedorn, C. H., O'Brien, A. D., et al. (2001). *Salmonella typhimurium* translocates flagellin across intestinal epithelia, inducing a proinflammatory response. *Journal of Clinical Investigation, 107*, 99–109.
Griffiths, S. G., & Lynch, W. H. (1990). Characterization of *Aeromonas salmonicida* variants with altered cell surfaces and their use in studying surface protein assembly. *Archives Microbiology, 154*, 308–312.
Grim, C. J., Kozlova, E. V., Sha, J., Fitts, E. C., van Lier, C. J., Kirtley, M. L., et al. (2013). Characterization of *Aeromonas hydrophila* wound pathotypes by comparative genomic and functional analyses of virulence genes. *mBio, 4,* e00064–00013.
Gryllos, I., Shaw, J. G., Gavin, R., Merino, S., & Tomás, J. M. (2001). Role of *flm* locus in mesophilic *Aeromonas* adherence. *Infection and Immunity, 69*, 65–74.
Guan, R., Xiong, J., Huang, W., & Guo, S. (2011). Enhancement of protective immunity in European eel (*Anguilla anguilla*) against *Aeromonas hydrophila* and *Aeromonas sobria* by a recombinant *Aeromonas* outer membrane protein. *Acta Biochimica et Biophysica Sinica, 43*, 79–88.
Hadi, N., Yang, Q., Barnett, T. C., Tabei, S. M. B., Kirov, S. M., & Shaw, J. G. (2012). Bundle-forming pilus locus of *Aeromonas veronii* bv sobria. *Infection and Immunity, 80*, 1351–1369.
Hamburger, Z. A., Brown, M. S., Isberg, R. R., & Bjorkman, P. J. (1999). Crystal structure of invasin: A bacterial integrin-binding protein. *Science, 286*, 291–295.
Han, H. J., Kim, D. Y., Kim, W. S., Kim, C. S., Jung, S. J., Oh, M. J., et al. (2011). Atypical *Aeromonas salmonicida* infection in the black rockfish, Sebastes schlegeli Hilgendorf, in Korea. *Journal of Fish Diseases, 34*, 47–55.
Hayashi, F., Smith, K. D., Ozinsky, A., Hawn, T. R., Yi, E. C., Goodlett, D. R., et al. (2001). The innate immune response to bacterial flagellin is mediated by Toll-like receptor 5. *Nature, 410*, 1099–1103.
Hickman, J. W., & Harwood, C. S. (2008). Identification of FleQ from Pseudomonas aeruginosa a c-di-GMP-responsive transcription factor. *Molecular Microbiology, 69*, 376–389.
Ho, A. S. Y., Mietzner, T. A., Smith, A. J., & Schoolnik, G. K. (1990). The pili of *Aeromonas-hydrophila*—Identification of an environmentally regulated mini pilin. *Journal of Experimental Medicine, 172*(3), 795–806.
Hokama, A., Honma, Y., & Nakasone, N. (1990). Pili of an *Aeromonas-hydrophila* strain as a possible colonization factor. *Microbiology and Immunology, 34*, 901–915.
Honma, Y., & Nakasone, N. (1990). Pili of *Aeromonas-hydrophila*—Purification, characterization, and biological role. *Microbiology and Immunology, 34*, 83–98.
Hossain, M. J., Waldbieser, G. C., Sun, D., Capps, N. K., Hemstreet, W. B., Carlisle, K., et al. (2013). Implication of lateral genetic transfer in the emergence of *Aeromonas hydrophila* isolates of epidemic outbreaks in channel catfish. *PLoS One, 8*, 0080943.
Indergand, S., & Graf, J. (2000). Ingested blood contributes to the specificity of the symbiosis of Aeromonas veronii biovar sobria and Hirudo medicinalis, the medicinal leech. *Applied and Environmental Microbiology, 66*, 4735–4741.

Inoue, K., Kosako, Y., Suzuky, K., & Shimada, T. (1991). Peritrichous flagellation in *Plesiomonas shigelloides* strains. *Japanese Journal of Medical Science Biology*, *44*, 141–146.

Iwanaga, M., & Hokama, A. (1992). Characterization of *Aeromonas-sobria* Tap13 pili—A possible new colonization factor. *Journal of General Microbiology*, *138*, 1913–1919.

Janda, J. M., & Abbott, S. L. (2010). The genus *Aeromonas*: Taxonomy, pathogenicity and infection. *Clinical Microbiology Reviews*, *23*, 35–73.

Jenal, U. (2000). Signal transduction mechanisms in *Caulobacter crescentus* development and cell cycle control. *FEMS Microbiology Reviews*, *24*, 177–191.

Jiang, Z. Y., & Bauer, C. E. (1997). Analysis of a chemotaxis operon from *Rhodospirillum centenum*. *Journal of Bacteriology*, *179*, 5712–5719.

Jiang, Z. Y., Rushing, B. G., Bai, Y., & Gest, H. (1998). Isolation of *Rhodospirillum centenum* mutants defective in phototactic colony motility by transposon mutagenesis. *Journal of Bacteriology*, *180*, 1248–1255.

Jimenez, N., Canals, R., Lacasta, A., Kondakova, A. N., Lindner, B., Knirel, Y. A., et al. (2008). Molecular analysis of three *Aeromonas hydrophila* AH-3 (serotype O:34) lipopolysaccharide core biosynthesis gene cluster. *Journal of Bacteriology*, *190*, 3176–3184.

Jimenez, N., Canals, R., Saló, M. T., Vilches, S., Merino, S., & Tomás, J. M. (2008). The *Aeromonas hydrophila wb**O:34 gene cluster: Genetics and temperature regulation. *Journal of Bacteriology*, *190*, 4198–4209.

Josenhans, C., Vossebein, L., Friedrich, S., & Suerbaum, S. (2002). The *neuA/flmD* gene cluster of *Helicobacter pylori* is involved in flagellar biosynthesis and flagellin glycosylation. *FEMS Microbiology Letters*, *210*, 165–172.

Kachlany, S. C., Planet, P. J., Bhattacharjee, M. K., Kollia, E., DeSalle, R., Fine, D. H., et al. (2000). Nonspecific adherence by *Actinobacillus actinouaycetemcomitans* requires genes widespread in Bacteria and Archaea. *Journal of Bacteriology*, *182*, 6169–6176.

Kalir, S. (2001). Ordering genes in a flagella pathway by analysis of expression kinetics from living bacteria. *Science*, *292*, 2080–2083.

Karlyshev, A. V., Linton, D., Gregson, N. A., & Wren, B. W. (2002). A novel paralogous gene family involved in phase variable flagella mediated motility in *Campylobacter jejuni*. *Microbiology*, *148*, 473–480.

Kay, W. W., Buckley, J. T., Ishiguro, E. E., Phipps, B. M., Monette, J. P. L., & Trust, T. J. (1981). Purification and disposition of a surface protein associated with virulence of *Aeromonas salmonicida*. *Journal of Bacteriology*, *147*, 1077–1084.

Khashe, S., Hill, W., & Janda, M. J. (1996). Characterisation of *Aeromonas hydrophila* strains of clinical, animal and environmental origin expressing the O:34 antigen. *Current Microbiology*, *33*, 104–108.

Khushiramani, R., Girisha, S. K., Bhowmick, P. P., Karunasagar, I., & Karunasagar, I. (2008). Prevalence of different outer membrane proteins in isolates of *Aeromonas* species. *World Journal of Microbiology & Biotechnology*, *24*, 2263–2268.

Khushiramani, R. M., Maiti, B., Shekar, M., Girisha, S. K., Akash, N., Deepanjali, A., et al. (2012). Recombinant *Aeromonas hydrophila* outer membrane protein 48 (Omp48) induces a protective immune response against *Aeromonas hydrophila* and *Edwardsiella tarda*. *Research in Microbiology*, *163*, 286–291.

Kim, Y. K., & McCarter, L. L. (2000). Analysis of the polar flagellar gene system of *Vibrio parahaemolyticus*. *Journal of Bacteriology*, *182*, 3693–3704.

Kim, Y. K., & McCarter, L. L. (2007). ScrG, a GGDEF-EAL protein, participates in regulating swarming and sticking in *Vibrio parahaemolyticus*. *Journal of Bacteriology*, *189*, 4094–4107.

Kirov, S. M. (1997). *Aeromonas* and *Plesiomonas* species. In M. P. Doyle, L. R. Beuchat, & MontvilleT. J. (Eds.), *Food microbiology, fundamentals and frontiers* (pp. 265–287). Washington, DC, USA: ASM Press.

Kirov, S. M., Barnett, T. C., Pepe, C. M., Strom, M. S., & Albert, M. J. (2000). Investigation of the role of type IV *Aeromonas* pilus (Tap) in the pathogenesis of *Aeromonas* gastrointestinal infection. *Infection and Immunity*, *68*, 4040–4048.

Kirov, S. M., Hayward, L. J., & Nerrie, M. A. (1995). Adhesion of *Aeromonas sp* to cell lines used as models for intestinal adhesion. *Epidemiology and Infection*, *115*, 465–473.

Kirov, S. M., Jacobs, I., Hayward, L. J., & Hapin, R. H. (1995). Electron-microscopic examination of factors influencing the expression of filamentous surface-structures on clinical and environmental isolates of *Aeromonas-veronii* biotype sobria. *Microbiology and Immunology*, *39*, 329–338.

Kirov, S. M., & Sanderson, K. (1996). Characterization of a type IV bundle-forming pilus (SFP) from a gastroenteritis-associated strain of *Aeromonas veronii* biovar sobria. *Microbial Pathogenesis*, *21*, 23–34.

Kirov, S. M., Tassell, B. C., Semmler, A. B., O'Donovan, L. A., Rabaan, A. A., & Shaw, J. G. (2002). Lateral flagella and swarming motility in *Aeromonas* species. *Journal of Bacteriology*, *184*, 547–555.

Knirel, Y. A., Shaskov, A. S., Senchenkova, S. N., Merino, S., & Tomás, J. M. (2002). Structure of the O-polysaccharide of *Aeromonas hydrophila* O:34: a case of random O-acetylation of 6-deoxy-L-talose. *Carbohydrate Research*, *337*, 1381–1386.

Knirel, Y. A., Vinogradov, E., Jimenez, N., Merino, S., & Tomás, J. M. (2004). Structural studies on the R-type lipopolysaccharide of *Aeromonas hydrohila*. *Carbohydrate Research*, *339*, 787–793.

Kokka, R. P., Vedros, N. A., & Janda, M. J. (1990). Electrophoretic analysis of the surface components of autoagglutinating surface array protein-positive and surface array protein-negative *Aeromonas hydrophila* and *Aeromonas sobria*. *Journal of Clinical Microbiology*, *28*, 2240–2247.

Kokka, R. P., Vedros, N. A., & Janda, M. J. (1991). Characterisation of classic and atypical serogroup O:11 *Aeromonas*: Evidence that the surface array protein is not directly involved in mouse pathogenicity. *Microbial Pathogensis*, *10*, 71–79.

Kozlova, E. V., Khajanchi, B. K., Popov, V. L., Wen, J., & Chopra, A. K. (2012). Impact of QseBC system in c-di-GMP-dependent quorum sensing regulatory network in a clinical isolate SSU of *Aeromonas hydrophila*. *Microbial Pathogenesis*, *53*, 115–124.

Kozlova, E. V., Khajanchi, B. K., Sha, J., & Chopra, A. K. (2011). Quorum sensing and c-di-GMP-dependent alterations in gene transcripts and virulence-associated phenotypes in a clinical isolate of *Aeromonas hydrophila*. *Microbial Pathogenesis*, *50*, 213–223.

Kozlova, E. V., Popov, V. L., Sha, J., Foltz, S. M., Erova, T. E., Agar, S. L., et al. (2008). Mutation in the S-ribosylhomocysteinase (*luxS*) gene involved in quorum sensing affects biofilm formation and virulence in a clinical isolate of *Aeromonas hydrophila*. *Microbial Pathogenesis*, *45*, 343–354.

Kusumoto, A., Shinohara, A., Terashima, H., Kojima, S., Yakushi, T., & Homma, M. (2008). Collaboration of FlhF and FlhG to regulate polar-flagella number and localization in *Vibrio alginolyticus*. *Microbiology*, *154*, 1390–1399.

LeClerc, G., Wang, S., & Ely, B. (1998). A new class of *Caulobacter crescentus* flagellar genes. *Journal of Bacteriology*, *180*, 5010–5019.

Li, G., Miller, A., Bull, H., & Howard, S. P. (2011). Assembly of the Type II secretion system: Identification of ExeA residues critical for peptidoglycan binding and secretin multimerization. *Journal of Bacteriology*, *193*, 197–204.

Lineaweaver, W. C., Hill, M. K., Buncke, G. M., Follansbee, S., Buncke, H. J., Wong, R. K. M., et al. (1992). *Aeromonas-hydrophila* infections following use of medicinal leeches in replantation and flap surgery. *Annals of Plastic Surgery*, *29*, 238–244.

Linnerborg, M., Widmalm, G., Rahman, M. M., Jansson, P., Holme, T., Qadri, F., et al. (1996). Structural studies of the O-antigenic polysaccharide from an *Aeromonas* caviae strain. *Carbohydrate Research*, *291*, 165–174.

Logan, S. M. (2006). Flagellar glycosylation—A new component of the motility repertoire? *Microbiology*, *152*, 1249–1262.

Lu, Y.-C., Yeh, W.-C., & Ohashi, P. S. (2008). LPS/TLR4 signal transduction pathway. *Cytokine*, *42*, 145–151.

Luo, Y., Frey, E. A., Pfuetzner, R. A., Creagh, A. L., Knoechel, D. G., Haynes, C. A., et al. (2000). Crystal structure of enteropathogenic *Escherichia coli* intimin-receptor complex. *Nature*, *405*, 1073–1077.

MacNab, R. M. (1996). Flagella and motility. In F. C. Neidhardt, et al. (Ed.), *Escherichia coli and Salmonella typhimurium:* Cellular and Molecular Biology (pp. 123–145). Washington, DC: ASM Press.

Madi, L., Kessel, M., Sadovnik, E., & Henis, Y. (1988). Electron microscopic studies of aggregation and pellicle formation in *Azospirillum* spp. *Plant and Soil*, *109*, 115–121.

Maiti, B., Raghunath, P., & Karunasagar, I. (2009). Cloning and expression of an outer membrane protein OmpW of *Aeromonas hydrophila* and study of its distribution in *Aeromonas spp. Journal of Applied Microbiology*, *107*, 1157–1167.

Maiti, B., Shetty, M., Shekar, M., Karunasagar, I., & Karunasagar, I. (2012). Evaluation of two outer membrane proteins, Aha1 and OmpW of *Aeromonas hydrophila* as vaccine candidate for common carp. *Veterinary Immunology and Immunopathology*, *149*, 298–301.

Maltz, M., & Graf, J. (2011). The Type II secretion system is essential for erythrocyte lysis and gut colonization by the leech digestive tract symbiont *Aeromonas veronii*. *Applied and Environmental Microbiology*, 77, 597–603.

Martin-Carnahan, A., & Joseph, S. W. (2005). Order XII. Aeromonadales. In D. J. Brenner, N. R. Krieg, & S. T. Staley (Eds.), *Bergey's Manual of Systematic Bacteriology: Vol. 2.* (pp. 556–578). New York, NY, US: Springer, Part B.

Martinez, M. J., Simonpujol, D., Congregado, F., Merino, S., Rubires, X., & Tomás, J. M. (1995). The presence of capsular polysaccharide in mesophilic *Aeromonas hydrophila* serotypes O-11 and O-34. *FEMS Microbiology Letters*, *128*, 69–73.

Martínez-Granero, F., Navazo, A., Barahona, E., Redondo-Nieto, M., González de Heredia, E., Baena, I., et al. (2014). Identification of *flgZ* as a flagellar gene encoding a PilZ domain protein that regulates swimming motility and biofilm formation in pseudomonas. *PLoS One*, *9*, e87608.

Matsunaga, J., Barocchi, M. A., Croda, J., Young, T. A., Sanchez, Y., Siqueira, I., et al. (2003). Pathogenic *Leptospira* species express surface-exposed proteins belonging to the bacterial immunoglobulin superfamily. *Molecular Microbiology*, *49*, 929–945.

Mattick, J. S. (2002). Type IV pili and twitching motility. *Annual Review of Microbiology*, *56*, 289–314.

McCarter, L. L. (2001). Polar flagellar motility of the *Vibrionaceae*. *Microbiology and Molecular Biology Reviews*, *65*, 445–462.

McCarter, L. L. (2004). Dual flagellar systems enable motility under different circumstances. *Journal of Molecular Microbiology and Biotechnology*, *7*, 18–29.

McCarter, L. L., & Silverman, M. (1989). Iron regulation of swarmer cell differentiation of *Vibrio parahaemolyticus*. *Journal of Bacteriology*, *171*, 731–736.

McCarter, L. L., & Silverman, M. (1990). Surface-induced swarmer cell differentiation of *Vibrio parahaemolyticus*. *Molecular Microbiology*, *4*, 1057–1062.

McIntosh, D., & Austin, B. (1991). Atypical characteristics of the salmonid pathogen *Aeromonas salmonicida*. *Journal of General Microbiology*, *137*, 1341–1343.

Meibom, K. L., Li, X. B., Nielsen, A. T., Wu, C.-Y., Roseman, S., & Schoolnik, G. (2004). The *Vibrio cholerae* chitin utilization program. *Proceeding of the National Academy of Sciences of the United States of America*, *101*, 2524–2529.

Merino, S., Aguilar, A., Rubires, X., SimonPujol, D., Congregado, F., & Tomás, J. M. (1996). The role of the capsular polysaccharide of *Aeromonas salmonicida* in the adherence and invasion of fish cell lines. *FEMS Microbiology Letters*, *142*, 185–189.

Merino, S., Alberti, S., & Tomás, J. M. (1994). *Aeromonas-salmonicida* resistance to complement-mediated killing. *Infection and Immunity, 62*, 5483–5490.

Merino, S., Camprubi, S., & Tomás, J. M. (1992). Effect of growth temperature on outer membrane components and virulence of *Aeromonas hydrophila* strains of serotype O:34. *Infection and Immunity, 60*, 4343–4349.

Merino, S., Fulton, K. M., Twine, S. M., Wilhelms, M., Molero, R., & Tomás, J. M. (2014). *Aeromonas hydrophila* flagella glycosylation involvement of a lipid carrier. *PLoS One, 9*, e89630.

Merino, S., Gavin, R., Vilches, S., Shaw, J. G., & Tomás, J. M. (2003). A colonization factor (production of lateral flagella) of mesophilic *aeromonas* spp. is inactive in *Aeromonas salmonicida* strains. *Applied and Environmental Microbiology, 69*, 663–667.

Merino, S., Jimenez, N., Molero, R., Bouamama, L., Regue, M., & Tomás, J. M. (2011). A UDP-HexNAc:polyprenol-P GalNAc-1-P transferase (WecP) representing a new subgroup of the enzyme family. *Journal of Bacteriology, 193*, 1943–1952.

Merino, S., Nogueras, M. M., Aguilar, A., Rubires, X., Alberti, S., Benedi, V. J., et al. (1998). Activation of the complement classical pathway (C1q binding) by mesophilic *Aeromonas hydrophila* outer membrane protein. *Infection and Immunity, 66*, 3825–3831.

Merino, S., Rubires, X., Aguilar, A., Alberti, S., HernandezAlles, S., Benedi, V. J., et al. (1996). Mesophilic *Aeromonas sp* serogroup 0:11 resistance to complement-mediated killing. *Infection and Immunity, 64*, 5302–5309.

Merino, S., Rubires, X., Aguilar, A., & Tomás, J. M. (1997). The role of flagella and motility in the adherence and invasion to fish cell lines by *Aeromonas hydrophila* serogroup O:34 strains. *FEMS Microbiology Letters, 151*, 213–217.

Merino, S., Rubires, X., Aguillar, A., Guillot, J. F., & Tomás, J. M. (1996). The role of the O-antigen lipopolysaccharide on the colonization in vivo of the germfree chicken gut by *Aeromonas hydrophila* serogroup O:34. *Microbial Pathogenesis, 20*, 325–333.

Merino, S., Rubires, X., Konkel, S., & Tomás, J. M. (1995). Emerging pathogens: *Aeromonas* spp. *International Journal of Food Microbiology, 28*, 157–168.

Merino, S., Shaw, J. G., & Tomás, J. M. (2006). Bacterial lateral flagella: An inducible flagella system. *FEMS Microbiology Letters, 263*, 127–135.

Moens, S., Michiels, K., Keijer, V., Van Leuven, F., & Vanderleyden, J. (1995). Cloning, sequencing, and phenotypic analysis of *laf1*, encoding the flagellin of the lateral flagella of *Azospirillum brasilense* Sp7. *Journal of Bacteriology, 177*, 5419–5426.

Molero, R., Wilhelms, M., Infanzón, B., Tomás, J. M., & Merino, S. (2011). *Aeromonas hydrophila motY* is essential for polar flagellum function and requires coordinate expression of *motX* and Pom proteins. *Microbiology, 157*, 2772–2784.

Murray, R. G. E., Dooley, J. S. G., Whippey, P. W., & Trust, T. J. (1988). Structure of an S-layer on a pathogenic strain of *Aeromonas hydrophila*. *Journal of Bacteriology, 170*, 2625–2630.

Nakasone, N., Iwanaga, M., Yamashiro, T., Nakashima, K., & Albert, M. J. (1996). *Aeromonas trota* strains, which agglutinate with *Vibrio cholerae* O139 Bengal antiserum, possess a serologically distinct fimbrial colonization factor. *Microbiology, 142*, 309–313.

Nazarenko, E. L., Crawford, R. J., & Ivanova, E. P. (2011). The structural diversity of carbohydrate antigens of selected Gram-negative marine bacteria. *Marine Drugs, 9*, 1914–1954.

Nelson, M. C., & Graf, J. (2012). Bacterial symbioses of the medicinal leech *Hirudo verbana*. *Gut Microbes, 3*, 322–331.

Noonan, B., & Trust, T. J. (1995a). The leucine zipper of *Aeromonas salmonicida* AbcA is required for the transcriptional activation of the P2 promoter of the surface-layer structural gene, *vapA*, in *Escherichia coli*. *Molecular Microbiology, 17*, 379–386.

Noonan, B., & Trust, T. J. (1995b). Molecular analysis of an A-protein secretion mutant of *Aeromonas salmonicida* reveals a surface layer-specific protein secretion pathway. *Journal of Molecular Biology, 248*, 316–327.

Noonan, B., & Trust, T. J. (1997). The synthesis, secretion and role in virulence of the paracrystalline surface layers of *Aeromonas salmonicida* and *A. hydrophila*. *FEMS Microbiology Letters, 154*, 1–7.

O'Rouke, J., Lee, A., & Fox, J. G. (1992). An ultrastructural study of *Helicobacter mustelae* and evidence of a specific association with gastric mucosa. *Journal of Medical Microbiology, 36*, 420–427.

Okabe, M., Yakushi, T., Kojima, M., & Homma, M. (2002). MotX and MotY, specific components of the sodium-driven flagellar motor, colocalize to the outer membrane in *Vibrio alginolyticus*. *Molecular Microbiology, 46*, 125–134.

Osei-Poku, J., Mbogo, C. M., Palmer, W. J., & Jiggins, F. M. (2012). Deep sequencing reveals extensive variation in the gut microbiota of wild mosquitoes from Kenya. *Molecular Ecology, 21*, 5138–5150.

Palaniappan, R. U. M., Chang, Y. F., Jusuf, S. S. D., Artiushin, S., Timoney, J. F., McDonough, S. P., et al. (2002). Cloning and molecular characterization of an immunogenic LigA protein of *Leptospira interrogans*. *Infection and Immunity, 70*, 5924–5930.

Parker, J. L., Day-Williams, M. J., Tomás, J. M., Stafford, G. P., & Shaw, J. G. (2012). Identification of a putative glycosyltransferase responsible for the transfer of pseudaminic acid onto the polar flagellin of *Aeromonas caviae* Sch3N. *Microbiology Open, 1*, 149–160.

Parker, J. L., Lowry, R., Couto, N. A. S., Wright, P. C., Stafford, G. S., & Shaw, J. G. (2014). Maf dependent bacterial flagellin glycosylation occurs before chaperone binding and flagellar T3SS export. *Molecular Microbiology, 92*, 258–272.

Parker, J. L., & Shaw, J. G. (2011). *Aeromonas* spp. clinical microbiology and disease. *Journal of Infection, 62*, 109–118.

Paul, K., Nieto, V., Carlquist, W. C., Blair, D. F., & Harshey, R. M. (2010). The c-di-GMP binding protein YcgR controls flagellar motor direction and speed to affect chemotaxis by a "backstop brake" mechanism. *Molecular Cell, 38*, 128–139.

Pepe, C. M., Eklund, M. W., & Strom, M. S. (1996). Cloning of an *Aeromonas hydrophila* type IV pilus biogenesis gene cluster: Complementation of pilus assembly functions and characterization of a type IV leader peptidase/N-methyltransferase required for extracellular protein secretion. *Molecular Microbiology, 19*, 857–869.

Planet, P. J., Kachlany, S. C., Fine, D. H., DeSalle, R., & Figurski, D. H. (2003). The widespread colonization island of *Actinobacillus actinomycetemcomitans*. *Nature Genetics, 34*, 193–198.

Quinn, D. M., Wong, C. Y. F., Atkinson, H. M., & Flower, R. L. P. (1993). Isolation of carbohydrate-reactive outer-membrane proteins of *Aeromonas-hydrophila*. *Infection and Immunity, 61*, 371–377.

Rabaan, A. A., Gryllos, I., Tomás, J. M., & Shaw, J. G. (2001). Motility and the polar flagellum are required for *Aeromonas caviae* adherence to HEp-2 cells. *Infection and Immunity, 69*, 4257–4267.

Rahman, M. H., & Kawai, K. (2000). Outer membrane proteins of *Aeromonas hydrophila* induce protective immunity in goldfish. *Fish & Shellfish Immunology, 10*, 379–382.

Rahman, M., Simm, R., Kader, A., Basseres, E., Römling, U., & Möllby, R. (2007). The role of c-di-GMP signaling in an *Aeromonas veronii* biovar sobria strain. *FEMS Microbiology Letters, 273*, 172–179.

Rangrez, A. Y., Dayananda, K. M., Atanur, S., Joshi, R., Patole, M. S., & Shouche, Y. S. (2006). Detection of conjugation related type four secretion machinery in *Aeromonas culicicola*. *PLoS One, 1*, 0000115.

Reith, M. E., Singh, R. K., Curtis, B., Boyd, J. M., Bouevitch, A., Kimball, J., et al. (2008). The genome of *Aeromonas salmonicida* subsp. *salmonicida* A449: Insights into the evolution of a fish pathogen. *BMC Genomics*, *9*, 427–441.

Ren, C.-P., Beatson, S. A., Parkhill, J., & Pallen, M. J. (2005). The flag-2 locus, an ancestral gene cluster, is potentially associated with a novel flagellar system from *Escherichia coli*. *Journal of Bacteriology*, *187*, 1430–1440.

Robson, W. L., Leung, A. K., & Trevenen, C. L. (1992). Haemolytic-uraemic syndrome associated with *Aeromonas hydrophila* enterocolitis. *Pediatric Nephrology*, *6*, 221.

Rocha-De-Souza, C. M., Colombo, A. V., Hirata, R., Mattos-Guaraldi, A. L., Monteiro-Leal, L. H., Previato, J. O., et al. (2001). Identification of a 43-kDa outer-membrane protein as an adhesin in *Aeromonas caviae*. *Journal of Medical Microbiology*, *50*, 313–319.

Römling, U., Galperin, M. Y., & Gomelsky, M. (2013). Cyclic di-GMP: The first 25 years of a universal bacterial second messenger. *Microbiology and Molecular Biology Reviews*, *77*, 1–52.

Sakazaki, R., & Shimada, T. (1984). O-serogrouping scheme for mesophilic *Aeromonas* strains. *Japanese Journal of Medical Science and Biology*, *37*, 247–255.

Schirm, M., Arora, S. K., Verma, A., Vinogradov, E., Thibault, P., Ramphal, R., et al. (2004). Structural and genetic characterization of glycosylation of type a flagellin in *Pseudomonas aeruginosa*. *Journal of Bacteriology*, *186*, 2523–2531.

Schirm, M., Soo, E. C., Aubry, A. J., Austin, J., Thibault, P., & Logan, S. M. (2003). Structural, genetic and functional characterization of the flagellin glycosylation process in *Helicobacter pylori*. *Molecular Microbiology*, *48*, 1579–1592.

Seshadri, R., Joseph, S. W., Chopra, A. K., Sha, J., Shaw, J., Graf, J., et al. (2006). Genome sequence of *Aeromonas hydrophila* ATCC 7966(T): Jack of all trades. *Journal of Bacteriology*, *188*, 8272–8282.

Sha, J., Wang, S. F., Suarez, G., Sierra, J. C., Fadl, A. A., Erova, T. E., et al. (2007). Further characterization of a type III secretion system (T3SS) and of a new effector protein from a clinical isolate of *Aeromonas hydrophila*—Part I. *Microbial Pathogenesis*, *43*, 127–146.

Shimada, T., Sakazaki, R., & Suzuki, K. (1985). Peritrichous flagella in mesophilic strains of *Aeromonas*. *Japanese Journal of Medical Science and Biology*, *38*, 141–145.

Sierra, J. C., Suarez, G., Sha, J., Foltz, S. M., Popov, V. L., Galindo, C. L., et al. (2007). Biological characterization of a new type III secretion system effector from a clinical isolate of *Aeromonas hydrophila*—Part II. *Microbial Pathogenesis*, *43*, 147–160.

Silver, A. C., Rabinowitz, N. M., Küffer, S., & Graf, J. (2007). Identification of *Aeromonas veronii* genes required for colonization of the medicinal leech *Hirudo verbana*. *Journal of Bacteriology*, *189*, 6763–6772.

Smith, K. D., Andersen-Nissen, E., Hayashi, F., Strobe, K., Bergman, M. A., Barrett, S. L., et al. (2003). Toll-like receptor 5 recognizes a conserved site on flagellin required for protofilament formation and bacterial motility. *Nature Immunology*, *4*, 1247–1253.

Srivastava, D., Hsieh, M.-L., Khataokar, A., Neiditch, M. B., & Waters, C. M. (2013). Cyclic di-GMP inhibits Vibrio cholerae motility by repressing induction of transcription and inducing extracellular polysaccharide production. *Molecular Microbiology*, *90*, 1262–1276.

Steiner, T. S., Nataro, J. P., Poteet-Smith, C. E., Smith, J. A., & Guerrant, R. L. (2000). Enteroaggregative *Escherichia coli* expresses a novel flagellin that causes IL-8 release from intestinal epithelial cells. *Journal of Clinical Investigation*, *105*, 1769–1777.

Stewart, B. J., & McCarter, L. L. (2003). Lateral flagellar gene system of *Vibrio parahaemolyticus*. *Journal of Bacteriology*, *185*, 4508–4518.

Suarez, G., Sierra, J. C., Erova, T. E., Sha, J., Horneman, A. J., & Chopra, A. K. (2010). A Type VI secretion system effector protein, VgrG1, from *Aeromonas hydrophila* that induces host cell toxicity by ADP ribosylation of actin. *Journal of Bacteriology*, *192*, 155–168.

Suarez, G., Sierra, J. C., Sha, J., Wang, S., Erova, T. E., Fadl, A. A., et al. (2008). Molecular characterization of a functional type VI secretion system from a clinical isolate of *Aeromonas hydrophila*. *Microbial Pathogenesis*, *44*, 344–361.

Tabei, S. M. B., Hitchen, P. G., Day-Williams, M. J., Merino, S., Vart, R., Pang, P. C., et al. (2009). An *Aeromonas caviae* Genomic Island is required for both O-antigen lipopolysaccharide biosynthesis and flagellin glycosylation. *Journal of Bacteriology*, *191*, 2851–2863.

Terashima, H., Kojima, S., & Homma, M. (2008). Flagellar motility in bacteria structure and function of flagellar motor. *International Review of Cell and Molecular Biology*, *270*, 39–85.

Thibault, P., Logan, S. M., Kelly, J. F., Brisson, J.-R., Ewing, C. P., Trust, T. J., et al. (2001). Identification of the carbohydrate moieties and glycosylation motifs in *Campylobacter jejuni* flagellin. *Journal of Biological Chemistry*, *276*, 34862–34870.

Thomas, S., Austin, J. W., McCubbin, W. D., Kay, C. D., & Trust, T. J. (1992). Roles of structural domains in the morphology and surface anchoring of the tetragonal paracrystalline array of *Aeromonas hydrophila*. *Journal of Molecular Biology*, *228*, 652–661.

Thomas, L. V., Gross, R. J., Cheasty, T., & Rowe, B. (1990). Extended serogrouping scheme for motile mesophilic *Aeromonas* species. *Journal of Clinical Microbiology*, *28*, 980–984.

Thomas, S. R., & Trust, T. J. (1995). A specific PulD homolog is required for the secretion of paracrystalline surface array subunits in *Aeromonas hydrophila*. *Journal of Bacteriology*, *177*, 3932–3939.

Tomás, J. M. (2013). The main Aeromonas pathogenic factors. *ISRN Microbiology*, *2012*, 1–22. http://dx.doi.org/10.5402/2012/256261.

Tso, M. D., & Dooley, J. S. G. (1995). Temperature-dependent protein and lipopolysaccharide expression in clinical *Aeromonas* isolates. *Journal of Medical Microbiology*, *42*, 32–38.

Umelo, E., & Trust, T. J. (1997). Identification and molecular characterization of two tandemly located flagellin genes from *Aeromonas salmonicida* A449. *Journal of Bacteriology*, *179*, 5292–5299.

Vazquez-Juarez, R. C., Tutt, C. B., & Torres, A. G. (2004). A pathoadaptive mutation that mediates adherence of shiga toxin-producing *Escherichia coli* 0111: H8. *Abstracts of the General Meeting of the American Society for Microbiology*, *104*, 51–52.

Verma, A., Arora, S. K., Kuravi, S. K., & Ramphal, R. (2005). Roles of specific amino acids in the N terminus of *Pseudomonas aeruginosa* flagellin and of flagellin glycosylation in the innate immune response. *Infection and Immunity*, *73*, 8237–8246.

Verma, A., Schirm, M., Arora, S. K., Thibault, P., Logan, S. M., & Ramphal, R. (2006). Glycosylation of b-Type flagellin of *Pseudomonas aeruginosa*: Structural and genetic basis. *Journal of Bacteriology*, *188*, 4395–4403.

Vilches, S., Canals, R., Wilhelms, M., Salo, M. T., Knirel, Y. A., Vinogradov, E., et al. (2007). Mesophilic *Aeromonas* UDP-glucose pyrophosphorylase (GalU) mutants show two types of lipopolysaccharide structures and reduced virulence. *Microbiology*, *153*, 2393–2404.

Vilches, S., Jiménez, N., Merino, S., & Tomás, J. M. (2012). The *Aeromonas* dsbA mutation decreased their virulence by triggering type III secretion system but not flagella production. *Microbial Pathogenesis*, *52*, 130–139.

Vilches, S., Urgell, C., Merino, S., Chacon, M. R., Soler, L., Castro-Escarpulli, G., et al. (2004). Complete type III secretion system of a mesophilic *Aeromonas hydrophila* strain. *Applied and Environmental Microbiology*, *70*, 6914–6919.

Wang, S., Ghosh, A. K., Bongio, N., Stebbings, K. A., Lampe, D. J., & Jacobs-Lorena, M. (2012). Fighting malaria with engineered symbiotic bacteria from vector mosquitoes. *Proceedings of the National Academy of Sciences of the United States of America*, *109*, 12734–12739.

Wang, Z., Li, J., Vingradov, E., & Altman, E. (2006). Structural studies of the core region of *Aeromonas salmonicida* subsp. *salmonicida* lipopolysaccharide. *Carbohydrate Research*, *341*, 109–117.

Wilhelms, M., Fulton, K. M., Twine, S. M., Tomás, J. M., & Merino, S. (2012). Differential glycosylation of polar and lateral flagellins in *Aeromonas hydrophila* AH-3. *Journal of Biological Chemistry, 287*, 27851–27862.

Wilhelms, M., Gonzalez, V., Tomás, J. M., & Merino, S. (2013). *Aeromonas hydrophila* lateral flagellar gene transcriptional hierarchy. *Journal of Bacteriology, 195*, 1436–1445.

Wilhelms, M., Molero, R., Shaw, J. G., Tomás, J. M., & Merino, S. (2011). Transcriptional Hierarchy of *Aeromonas hydrophila* polar-flagellum genes. *Journal of Bacteriology, 193*, 5179–5190.

Wilhelms, M., Vilches, S., Molero, R., Shaw, J. G., Tomás, J. M., & Merino, S. (2009). Two redundant sodium-driven stator motor proteins are involved in *Aeromonas hydrophila* polar flagellum rotation. *Journal of Bacteriology, 191*, 2206–2217.

Yeh, H.-Y., & Klesius, P. H. (2011). Over-expression, purification and immune responses to *Aeromonas hydrophila* AL09-73 flagellar proteins. *Fish & Shellfish Immunology, 31*, 1278–1283.

Yun-Kyeong, K., & McCarter, L. L. (2000). Analysis of the polar flagellar gene system of *Vibrio parahaemolyticus*. *Journal of Bacteriology, 182*, 3693–3704.

Zhang, Y. L., Arakawa, E., & Leung, K. Y. (2002). Novel *Aeromonas hydrophila* PPD134/91 genes involved in O-antigen and capsule biosynthesis. *Infection and Immunity, 70*, 2326–2335.

Zhang, Y. L., Lau, Y. L., Arakawa, E., & Leung, K. Y. (2003). Detection and genetic analysis of group II capsules in *Aeromonas hydrophila*. *Microbiology, 149*, 1051–1060.

Zhou, X., Giron, J. A., Torres, A. G., Crawford, J. A., Negrete, E., Vogel, S. N., et al. (2003). Flagellin of enteropathogenic *Escherichia coli* stimulates interleukin-8 production in T84 cells. *Infection and Immunity, 71*, 2120–2129.

CHAPTER SIX

Physiological Adaptations of Key Oral Bacteria

C.W. Ian Douglas, Kathryn Naylor, Chatchawal Phansopa, Andrew M. Frey, Thomas Farmilo, Graham P. Stafford[1]

Integrated BioSciences Group, School of Clinical Dentistry, University of Sheffield, Sheffield, United Kingdom
[1]Corresponding author: e-mail address: g.stafford@sheffield.ac.uk

Contents

Abstract

Oral colonising bacteria are highly adapted to the various environmental niches harboured within the mouth, whether that means while contributing to one of the major oral diseases of caries, pulp infections, or gingival/periodontal disease or as part of a commensal lifestyle. Key to these infections is the ability to adhere to surfaces via a range of specialised adhesins targeted at both salivary and epithelial proteins, their glycans and to form biofilm. They must also resist the various physical stressors they are subjected to, including pH and oxidative stress. Possibly most strikingly, they have

Advances in Microbial Physiology
ISSN 0065-2911
http://dx.doi.org/10.1016/bs.ampbs.2014.08.005

developed the ability to harvest both nutrient sources provided by the diet and those derived from the host, such as protein and surface glycans. We have attempted to review recent developments that have revealed much about the molecular mechanisms at work in shaping the physiology of oral bacteria and how we might use this information to design and implement new treatment strategies.

1. INTRODUCTION

The oral cavity presents a number of challenges that colonising bacteria must meet if they are to survive, prosper,s and in some cases cause infection. In this chapter, we consider how oral dwelling bacteria, including those causing infection, have evolved and tailored their physiology and properties to adapt to this environment and how what we have learnt might provide new approaches for improved treatment development.

We first set the scene by describing the clinical features of the major oral conditions of caries, endodontic/periapical infections and gingival/periodontal disease while also describing the major microbial etiological aspects. The main focus will be on how the causative bacterial agents are well adapted for life in the environmental niches present in the oral cavity and how they use these to survive, prosper and cause disease. We focus on recent developments in the field, enabled by increased understanding of events at the molecular level that have been enabled by improved bioinformatics information based on knowledge of the human oral microbiome and by the improved molecular tools now available for molecular oral microbiology.

2. KEY ORAL ENVIRONMENTAL NICHES

As an introduction, it is first useful to consider the main environmental niches that are pertinent to infectious conditions of the oral cavity. These are illustrated in Fig. 1 in relation to both health and disease.

At birth, the primary ecological niches in the mouth are the epithelial surfaces of the cheeks, gingivae (gums), palate and tongue. Some of these surfaces are keratinised and desquamating, which partly serves to control the buildup of bacterial populations. However, the mucosa covering the anterior two-third of the upper surface of the tongue comprises numerous small projections called the lingual papillae (Du Toit, 2003). This papillary structure provides an extended surface area compared with cheeks and palate and also presents deep crypts and grooves between the papillae, both of which offer opportunities for microbial colonisation. The teeth erupt

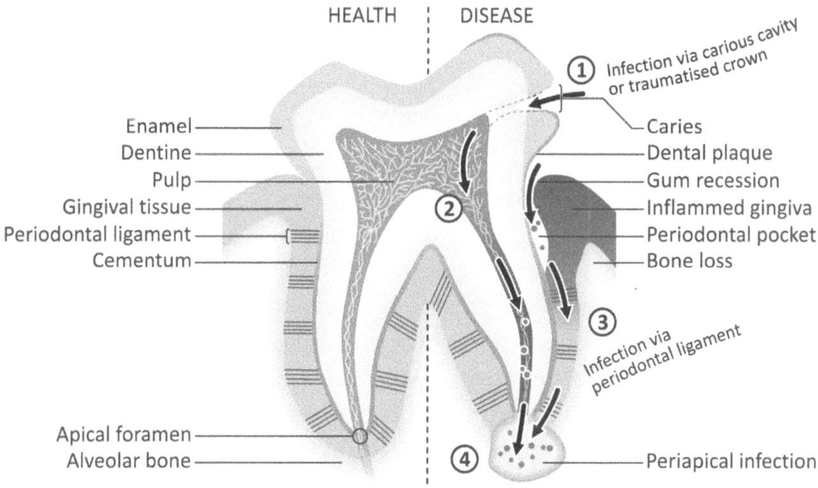

Figure 1 Schematic illustrating the major structures and features of teeth in health and disease, summarising caries, pulp/periapical and periodontal infections.

between 6 and 33 months of age, and these provide a variety of new habitats for commensal bacteria to occupy. In contrast to the majority of mucosal surfaces, which have a high cellular turnover and remove adherent organisms by desquamation, the teeth are nonshedding and stable surfaces for microbial attachment. Also, some teeth (molars) have deep pits and fissures that are relatively protected from the physical forces of salivary flow and abrasive mastication and from which much of the carious lesions arise. Moreover, when two teeth erupt adjacent to each other, a new interdental niche is created, which is also relatively protected from these physical removal forces, resulting in increased dental plaque formation and which, if not removed, leads to local inflammation of the tissues called gingivitis. Finally, the teeth erupt through the gingivae, which creates a so-called gingival crevice between the tooth and the soft tissues. This crevice is also relatively protected from physical removal forces around the teeth and becomes colonised by a diverse population of microorganisms (see below), which in a proportion of individuals leads to destructive periodontitis and ultimately tooth loss.

2.1. External solid surfaces of the tooth

The physical position of teeth in the mouth and their biological function as food processing tools means that they and their resident microbial flora are exposed not only to repeated mechanical stresses, as mentioned above, but also to a variety of environmental perturbations. These environmental perturbations include not only variations in pH due to dietary intake directly but also transient drops in pH from a relatively neutral pH to acidic (pH 5–6) pH in response to the fermentation of dietary carbohydrates by resident acidogenic bacteria (Moynihan, Ferrier, Blomley, Wright, & Russell, 1998; Vroom et al., 1999) (see Section 3.1). In addition, there are obvious fluctuations in temperature experienced by oral bacteria, again due to dietary intake, while the oral cavity in general displays a lower general temperature than core body temperature (i.e. about 1–2 °C lower as compared to sublingual measurements) (Marcotte & Lavoie, 1998).

In addition, in normal circumstances of good oral hygiene residence by bacteria on the nonshedding tooth surfaces is part of a fairly sparse biofilm and as such is exposed to the shear forces from salivary flow and those encountered during mastication of food. Thus, at least in the early stages of colonisation, bacterial attachment can be considered transient. Over time however, and with less than perfect oral hygiene (which is very difficult to achieve), the colonising bacteria grow and multiply to create a complex biofilm containing up to around 500 species (Kistler, Booth, Bradshaw, & Wade, 2013). It is from this biofilm, called dental plaque, that much of dental disease emanates and which involves either growth of the biofilm down the tooth to affect the supporting tissues (periodontitis) or penetration into the underlying dentine and pulp of the tooth, resulting in pulpitits and periapical disease (Fig. 1). Saliva itself presents something of a paradox in terms of tooth colonisation by bacteria. It provides a mechanical removal force due to flow, which varies according to the proximity of the salivary glands, and a delivery function of a range of antimicrobial factors (Oppenheim, Salih, Siqueira, Zhang, & Helmerhorst, 2007). The latter include secretory IgA, lysozyme, lactoferrin, histatins, sialoperoxidase, aggregating factors such as MG1 and gp340 and antimicrobial peptides (Dale & Fredericks, 2004). Of the human β-defensin antimicrobial peptides, hBD1 and -2 have been detected in salivary glands and saliva has the cathelicidin peptide, LL37 (Murakami, Ohtake, Dorschner, & Gallo, 2002). Histatins are largely antifungal, while lactoferrin (and its peptide derivative lactoferricin), lysosozyme and sialoperoxidase are antibacterial by a range of mechanisms (reviewed by Helmerhorst & Oppenheim, 2007; Humphrey & Williamson, 2001). Despite these various defensive factors, the oral biofilm still develops, which it does by pioneer colonising species adhering initially to a film of salivary proteins adsorbed to the enamel surface. This film is called the *acquired pellicle*

(Lee et al., 2013). The paradox here is that certain salivary components have differential interactive effects when either bound to the tooth in the pellicle or when free in salivary solution (Loimaranta et al., 2005). This group of workers recognised three interactive types of streptococci, one which were aggregated by the salivary agglutinin (SAG) gp340 in solution but also attached to it in the pellicle, second which were not aggregated but adhered to the surface-bound form of gp340 and third which were aggregated by gp340 but did not attach to it on the surface. These features varied with strain as well as species and the aggregation interactions were mainly sialidase-sensitive indicating a sugar-dependent mechanism. However, this feature was independent of one of the major sialic acid-binding lectins of these streptococci, Hsa (Yajima et al., 2008) These results indicate the complexity of interaction between bacteria and salivary constituents that ultimately influences the colonisation and removal of bacteria from the oral cavity, and it is this targeting of adhesion to the salivary pellicle that is the basis of the initial development of the dental plaque biofilm. That biofilm is a key factor in the major dental infections and will be a focus of this chapter.

2.2. Internal tooth structures and niches

Teeth are made up of enamel, which overlies the dentine and which in turn surrounds the pulp cavity (Fig. 1). The enamel is 96% mineralised tissue (calcium hydroxyphosphate) which is created by ameloblasts before tooth eruption, but since these cells die when the tooth erupts, there is not intrinsic reparative function to the enamel. Having said that saliva is supersaturated with calcium which can contribute to remineralisation following a period of acid demineralisation. Dentine is formed by odontoblasts which can provide a reparative/defensive function for the tissue. It differs from bone in not being vascularised but it does contain type I collagen and hydroxyapatite and has a series of tubules which connect to the pulp cavity at the centre of the tooth. The pulp cavity extends down through the roots as a so-called root canal and this contains the nerve and blood supply to the tooth providing pain response and host defensive factors against invading microorganisms, respectively.

2.3. Life below the gum line

Of at least equal importance and surface area in the mouth are the less visible tissues and structures that are present below the gum line (subgingivally), i.e. below the gingival margin. In health, a tight junction between the gingiva and the tooth means that the volume of the underlying gingival crevice is small and opportunities for colonisation mostly prevented physically by this junction (see Fig. 1). However, even in health, this area represents a significant environmental niche for bacterial colonisation. The subgingival area is generally speaking exposed to salivary proteins at a much lower concentration and is also not subject to the same physical and environmental fluctuations as the supragingival surfaces. As a result, the available nutrient sources in this area differ with lower concentrations of dietary carbohydrates and other nutrients meaning that bacteria wishing to colonise this niche must be adapted to the use of other nutritional sources such as host-derived glycoproteins on the surface of epithelial cells and those present in the fluid that fills the gingival crevice (namely gingival crevicular fluid, GCF). This GCF is rich in plasma and serum-derived proteins such as albumin, serotransferrin and macroglobulin alongside keratins and actins associated with the high epithelial turnover in this niche (Baliban et al., 2012; Carneiro, Venuleo, Oppenheim, & Salih, 2012).

In terms of physical characteristics, the pH of the GCF in healthy gingival crevices lies around the neutral level that one would expect (Bickel & Cimasoni, 1985). However, within the gingival crevices of inflamed periodontally diseased sites, the pH can vary between pH 7 and 9 with an overall trend for pH to increase at individual subgingival sites displaying gingivitis or periodontal disease (Bickel & Cimasoni, 1985; Eggert, Drewell, Bigelow, Speck, & Goldner, 1991; Galgut, 2001). However, this picture does not seem to be uniform and local variations in pH within plaque biofilms exist (Vroom et al., 1999). There is also evidence that the organisms themselves can influence the pH of the plaque biofilm environment with proteolytic metabolism often resulting in ammonia production that has the dual effect of raising the pH toward the optimum for growth of some pathogenic species, e.g. the periodontal pathogen *Porphyromonas gingivalis* grows optimally at around a pH of 7.5–8 (Takahashi, 2003; Zilm, Mira, Bagley, & Rogers, 2010), and the pH optimum of its main virulence factors—the gingipains which is also in this pH range (Potempa et al., 1998).

These marked differences between the healthy and diseased environments in gingival crevices is further highlighted by the findings over a number of years that local temperature fluctuations occur in the gingival crevice during inflammatory episodes (Fedi & Killoy, 1992). However, on average this temperature is again lower than the sublingual core temperature in the mouth and is raised around 1 °C in inflamed sites (Fedi & Killoy, 1992; Kung, Ochs, & Goodson, 1990).

3. THE MAJOR INFECTIONS OF THE ORAL CAVITY

Most humans suffer from one or more of the major dental diseases at some stage in their life and these are illustrated in Fig. 1. These diseases are, therefore, one of the most common bacterial diseases affecting man and are dental caries, pulp and periapical disease and periodontitis, including gingivitis. All of these conditions are preceded by a polymicrobial colonisation of the relevant hard surfaces of the tooth to form a biofilm referred to as dental plaque. Nutrition for the biofilm comes from degradation of salivary glycoproteins and proteins, cross-feeding between resident species and from host dietary constituents. If the host consumes a diet high in refined carbohydrate over a prolonged period then caries can result, which is due to local bacterial acid production from fermentation of the dietary carbohydrate. The usual consequence of enamel caries is invasion of the underlying dentine of the tooth and unless arrested, ultimately infection of the tooth pulp with the risk of dental abscess formation. Such tissue invasion presents a new set of environment for the bacteria to survive and attempt to thrive in. The purpose of this chapter, therefore, is to consider current knowledge of the physiological adaptations made by bacteria in the oral environment and in relation to these major dental diseases.

3.1. Caries: microbial involvement and aetiology

One of the most common dental diseases in the western world is dental caries. This is characterised by demineralisation of the tooth structure and is due to the production of acids from fermentation of refined carbohydrate in the host's diet. Although oral bacteria hydrolyse and ferment sugars from salivary glycoproteins, such as mucins, the quantity of acid generated is too low to overwhelm the homeostasis of the local oral ecology to cause caries. Indeed the normal resting pH in the oral biofilm is around neutral and many years ago, Yamada, Igarashi, and Mitsutomi (1980) demonstrated by *in vivo* pH telemetry that dietary sugar results in plaque biofilm pH being lowered from 7 to 4 in only 3 min (Yamada et al., 1980).

Another key study conducted some years ago looked at the cultivable plaque microbiota and dietary habits of subjects with active caries or who were caries-free (Hayes, Carter, & Griffiths, 1983). The study revealed that the acidogenic composition of the dental plaque correlated with dietary sugar intake and that a high proportion of acidogenic species was associated with active caries. While several streptococci and *Actinomyces* species were isolated along with *Streptococcus mutans* from all samples, there was no consistent relationship between their relative proportions and caries status, suggesting that the acidogenic population in plaque varies and that although *S. mutans* increases in response to dietary sugar its presence cannot necessarily be correlated with caries. Many other subsequent studies that support the 'ecological hypothesis' for caries put forward by Marsh (2003) states that it is the environment in the plaque biofilm that determines the cariogenic challenge as opposed to a defined cariogenic microbiota. More recently, there have been several studies of the microbiome associated with caries using modern sequencing techniques, and these have indicated that a diverse community of bacteria contributes to the initiation and progression of dental caries (Becker et al., 2002). Nonetheless, the presence of high numbers of *S. mutans* remains strongly associated with caries, and it is still considered a major caries pathogen and is still the subject of much research (Burne et al., 2012). This organism is capable of rapid fermentation of dietary sugars causing a substantial drop in the pH of oral biofilms that it inhabits (Moynihan et al., 1998). The resultant acid generated can lead to marked change in environmental pH. This not only affects the mineralisation of the tooth upon which the biofilm is growing but also causes significant physiological stress on the biofilm community itself, resulting in the selection of those species most able to tolerate low pH, i.e. the 'fittest'. A classic study by Bradshaw and Marsh (1998) using a mixed species population in a chemostat clearly showed that the primary selective pressure under low pH environments is the acid generation itself. They showed that when the pH in the chemostat was allowed to fall to 4.5 or below, two species from an artificial community population of 9 in the culture, namely *S. mutans* and *Lactobacillus rhamnosus*, became dominant, yet they were relatively minor components if the pH was maintained around neutrality by the addition of alkali. These data and that of others around that time indicated that *S. mutans* and *Lactobacillus* species are the most acidogenic species but since then and using more appropriate culture techniques, it became obvious that yeasts, other acidogenic Streptococci, *Actinomyces* species, *Propioniibacterium* and *Bifidobacterium* species contribute to caries and usually outnumber the mutans streptococci (Becker et al., 2002; Beighton, 2005; Nakajo, Takahashi, & Beighton, 2010). Indeed, *Bifidobacteria* were isolated from 95% of caries-active children but only 9% of caries-free children (Kaur, Gilbert, Sheehy, & Beighton, 2013), however, compared with *Bifidobacteria*, *S. mutans* is capable of producing a significantly lower pH from glucose, and more rapidly (Moynihan et al., 1998). Consequently, *S. mutans* is still considered to be a key cariogenic species.

Despite the ongoing development of our understanding of the bacterial population associated with caries, the underlying principle remains that acid generation locally is required for disease initiation and progression. In order to generate acid from sugar fementation, potentially cariogenic bacteria must possess appropriate sugar transport systems. Oral bacteria can transport sugars by three systems, the phosphoenolpyruvate phosphotransferase system (PEP-PTS), the multiple sugar metabolism system (msm) and a glucose permease (Cvitkovitch, Boyd, Thevenot, & Hamilton, 1995; Russell, Aduse-Opoku, Sutcliffe, Tao, & Ferretti, 1992). The PEP-PTS is the primary system and involves phosphorylation of the cytoplasmic enzyme I which in turn phosphorylates a histidine protein (HPr) that then transfers the phosphoryl group to a sugar-specific membrane bound enzyme complex II. In this way, the sugar is phosphorylated upon its entry into the cell using phosphoenolpyruvate as the phosphoryl donor and goes into the Embden–Meyerhof–Parnas glycolytic pathway. Analysis of the genome sequence of *S. mutans* indicates that this bacterium has the capacity to transport and metabolise a wide range of sugars (Ajdić et al., 2002), and 14 sugar-specific PEP-PTSs have been described (Cornejo et al., 2013), some of which are constitutive and some inducible. There is further modulation by the prevailing environmental conditions and these high-affinity transport systems have optimal activity at neutral pH, under carbohydrate limitation and at slow growth rates. In contrast under conditions of low pH and at high sugar concentrations, as would prevail during cariogenic challenge to the tooth, there is an ATP-dependent glucose permease that transports sugar allowing continued metabolism.

The multiple sugar metabolism transporter is a member of the ATP-binding cassette (ABC) superfamily (Russell et al., 1992; Tao, Sutcliffe, Russell, & Ferretti, 1993). It comprises eight contiguous genes involved with the transport of raffinose, melibiose, stachyose, isomaltose and isomaltotriose into the cell. The genes are transcribed as a single operon (McLaughlin & Ferretti, 1996) which is controlled by a positive regulatory gene (*msmR*). The structural genes include α-galactosidase (*aga*), a sugar-binding protein (*msmE*), two membrane proteins (*msmF*, *msmG*), sucrose phosphorylase (*gtfA*), an ATP-binding protein (*msmK*) and dextran glucosidase (*dexB*). Similar systems have been found in other species, such as the raffinose-binding protein RafE of *Streptococcus pneumoniae* that shows 76% similarity to MsmE of *S. mutans*. The ATPase component of Msm is shared with another ABC transporter, MalXFGK, which takes up maltotriose and maltodextrins, indicating that it is central to transport of a range of sugars in these species (Webb, Homer, & Hosie, 2008).

However, acidogenicity alone is not enough to permanently select a shift in the biofilm population because without the ability of the acidogenic species to continue to metabolise at low pH (acid tolerance) the effects of acid production would only be transitory. Most information on the mechanisms by which acid tolerance is achieved with respect to caries relates to *S. mutans*. The major strategy appears to be to maintain a transmembrane pH gradient by keeping the cytoplasm near neutrality while the pH of the extracellular environment falls. *S. mutans* does this by expending energy through increased activity of the proton-translocating F_1F_0-ATPase in order to export H^+ from the cell (Lemos & Burne, 2008; Quivey, Kuhnert, & Hahn, 2001). The *S. mutans* F_1F_0-ATPase has a lower pH optimum than many species (Quivey et al., 2001; Sasaki, Nogami, Maeda, Nakanishi-Matsui, & Iwamoto-Kihara, 2014), allowing it to tolerate lower pHs. However, the *S. mutans* F_1F_0-ATPase has another useful function in that it can also act as an ATP synthase at low pH and under low nutrient conditions (Sheng & Marquis, 2006), so it is able to both provide acid tolerance and generate ATP for growth. The acid tolerance of other species associated with caries (e.g. Bifidobacteria) is similar in that two intracellular subunits of the F_1F_o-ATPase are overproduced in acid conditions (Len, Harty, & Jacques, 2004a) and this presumably accounts for why *Bifidobacteria* exist together with mutans streptococci in caries lesions.

While the F_1F_0-ATPase provides the ecological advantage to *S. mutans* and *Bifidobacteria*, there are other acid stress-induced adaptations (Takahashi & Nyvad, 2011). Protein profiles from several proteomic studies of *S. mutans* where growth at a neutral pH was compared to that at pH 5 or 5.5 reveals a change in the expression of proteins from several pathways that seem to be a key to acid tolerance. These include upregulation of a range of glycolytic enzymes in the Embden–Meyerhoff Parnas glycolytic pathway, F_1F_0-ATPase subunits and branched amino acid production (as a means to reduce formic acid levels and buffer the cytoplasm) while downregulation of several sugar transporters, the protein synthesis machinery and a significant dampening of acid production pathways such as the acetate and ethanol branches of the pyruvate–formate lyase axis were also observed (Len, Harty, & Jacques, 2004b; Wilkins, Homer, & Beighton, 2002). In support of these data, mutants of the *ilvE* gene, which was found to be upregulated in acid stress (Len et al., 2004a) and is involved in the production of branched chain amino acids, displayed reduced acid tolerance and are regulated in response to low pH (Santiago, MacGilvray, Faustoferri, & Quivey, 2012). It was also no surprise that several stress resistance proteins were also differentially regulated between pH 7 and pH 5, with the observation that one of the Clp family of ATP-dependent proteases, ClpL was upregulated 3000-fold in acid-stressed cells (Len et al., 2004b). Some of this work was backed up by an elegant mutagenesis study in which mutation in the associated *clpP* gene was shown to display reduced doubling times at pH 5.5 compared to wildtype (Shibata et al., 2011), while others have shown *clpP* to be responsible for regulating levels of a range of stress resistance genes in *S. mutans* (Chattoraj, Banerjee, Biswas, & Biswas, 2010). In this work, the authors also identify a major role for the *dgk* gene in conferring acid stress resistance, this gene encodes a diacylglycerol kinase that is involved in lipid turnover and which has also been implicated in stress resistance in *Escherichia coli* (Yamashita & Shibata, 2011).

Of note is that both the *clpP* and *dgk* mutants of *S. mutans* display reduced biofilm formation in *in vitro* experiments (Chattoraj et al., 2010; Shibata et al., 2009). These observations are of importance given that the ability to form biofilm is not only considered an essential step in the initiation of oral colonisation but also in caries disease progression while also endowing the bacteria with increased resistance to acid and other environmental stresses, including antibiotic resistance (McNeill & Hamilton, 2003). It is also of note that Dgk inhibitors may have potential as anticaries agents (Shibata et al., 2011). Biofilm formation of *S. mutans* is intimately linked to expression and responses to quorum sensing by the products of the genes *comC, D* and *E* (namely CSP—Competence Stimulating Peptide) that are required for cell-density-dependent generation of genetic competence and control of several other virulence factors including bacteriocin production (Li et al., 2002; Wang, Deutch, Hong, & Kuramitsu, 2011; Yoshida & Kuramitsu, 2002). In keeping with this is the finding that interference with CSP levels by proteases produced by commensal oral Streptococci, in particular *Streptococcus gordonii*, hinders biofilm formation by *S. mutans in vitro* (Jakubovics, 2010; Wang, He, et al., 2011).

Under normal circumstances, the pH of the biofilm gradually rises and returns to near neutrality as long as further excess carbohydrate is not supplied. This process is due to a combination of acid diffusion away from the biofilm, diffusion in of salivary buffering salts and acid neutralisation by the production of alkali within the biofilm. A number of streptococci possess urease and arginine deiminase genes that generate ammonia, CO_2 and amines, and it is assumed that their activity contributes to the raising of pH in the plaque biofilm. *S. mutans* does not have these genes but it is capable of generating some alkali through an agmatine deiminase system

(Griswold, Chen, & Burne, 2004). The mechanisms of the agmatine deiminase of *S. mutans* have been studied and shown to have a pH optimum around 8.0 indicating that this system could only operate if the intracellular pH is maintained significantly above the external environment (Jones, Dreyton, Flick, Causey, & Thompson, 2010).

Colonisation of the tooth surface involves interaction between a range of protein adhesins (e.g. antigen I/II proteins) interacting with salivary proteins adsorbed to the enamel surface (see Table 1 for adhesins involved in colonisation of the oral tissues). However, the mutans streptococci also possess the ability to synthesise a range of extracellular polysaccharides from sucrose by the action of a number of glucosyltransferases (Gtfs) which contribute to the biofilm (reviewed by Bowen & Koo, 2011). In brief, *S. mutans* expresses three Gtf genes which play different roles. GtfB synthesises a glucan that is highly branched and water-insoluble, often called mutan. GtfC synthesises a mixture of soluble and insoluble glucans, and GtfD synthesises water-soluble glucans. GtfC can become adsorbed to enamel within the salivary pellicle, while GtfB binds to bacterial surfaces promoting interbacterial adhesion with the products of GtfD being a source of glucose for metabolism and acting as a primer for the product of GtfB. Glucans synthesised *in situ* on the tooth surface by GtfC provide sites for colonisation by other microorganisms (Van Hijum, Kralj, Ozimek, Dijkhuizen, & van Geel-Schutten, 2006). Evidence for an important *in vivo* role comes from studies in rats in which colonisation of the oral cavity was reduced for mutants of *S. mutans* that lacked the *gtf* genes (Fujiwara, Hoshino, Ooshima, & Hamada, 2002).

3.2. Pulp and periapical infection

Bacterial infection of the pulp (often referred to as endodontic or root canal infections) can result in tooth ache which is surprisingly painful given the small size of the tissue involved; the average volume of a human pulp is only 20 μl. The infection can then progress from the pulp to the apex of the root of the tooth to give rise to an abscess and in some cases, systemic complications arising from these that lead to approximately 1 death per month in the United Kingdom. These infections are most commonly preceded by dental caries, although tooth trauma is also a cause. Infection of the pulp is usually via the dentinal tubules, which are microscopic channels radiating from the pulp through the dentine to the exterior or enamel border and these become exposed during caries or trauma. The number of dentinal tubules per mm^2 varies from 15,000 where the dentine joins the enamel up to 45,000 at the

Table 1 Adhesins of oral bacteria

Organism	Adhesin	Description	Targets	References
Actinomyces oris	FimP, FimQ	Type 1 fimbriae: FimP-shaft, FimQ tip	Proline-rich pellicle proteins	Clark et al. (1989)
	FimA, FimB	Type 2 fimbriae: FimA-shaft, FimB-tip	Asialofetuin, Streptococci, epithelial cells via GalNAcβ1-3Gal or Galβ1-3NAc	Mishra, Das, Cisar, and Ton-That (2007) and Mishra et al. (2010, 2011)
	CafA	Type 2 fimbriae tip protein	*Streptococci*	Reardon-Robinson et al. (2014)
Aggregatibacter actinomycetemcomitans	EmaA	Trimeric autotransporter (T5cSS), Oca	Types I, II, III and V collagen	Mintz (2004)
	Aae	Classical autotransporter (T5aSS)	Epithelial cells	Fine, Velliyagounder, Furgang, and Kaplan (2005), Fine et al. (2010) and Rose, Meyer, and Fives-Taylor (2003)
	Tad, Flp-1, Flp-2	Putative type 4 pilus, suggested T2SS or T4SS	Nonspecific adhesion	Clock, Planet, Perez, and Figurski (2008), Inoue et al. (1998), Kachlany, Planet, Desalle, Fine, and Figurski (2001) and Perez et al. (2006)
	ApiA	Trimeric autotransporter (T5cSS)	Buccal epithelial cells	Yue, Kaplan, Furgang, Mansfield, and Fine (2007)

Continued

Table 1 Adhesins of oral bacteria—cont'd

Organism	Adhesin	Description	Targets	References
Fusobacterium nucleatum	FomA	Porin, OMP	Statherin-derived peptide	Nakagaki et al. (2010) and Puntervoll et al. (2002)
	FadA	Pilus	Host cells, Endothelial-cadherin	Fardini et al. (2011) and Han et al. (2005)
	RadD	OMP, arginine dependent	Streptococci	Kaplan, Lux, Haake, and Shi (2009)
Porphyromonas gingivalis	Major fimbriae-FimA	Unclassified pilus	Epithelial cells, salivary proteins, Streptococci	Maeda et al. (2004) and Umemoto and Hamada (2003)
	Minor Fimbriae-MfaI	Unclassified pilus	Epithelial cells	Umemoto and Hamada (2003)
Strptococcus gordonii	Hsa	Serine-rich repeat	Platelets via gangliosides containing α-2,3 or α-2,5-sia, fibronectin	Jakubovics, Brittan, Dutton, and Jenkinson (2009), Takahashi, Konishi, Cisar, and Yoshikawa (2002) and Takahashi, Yajima, Cisar, and Konishi (2004)
	CshA, CshB	Csh family-form fibrils	Fibronectin, oral bacteria, *Candida*	Holmes, McNab, and Jenkinson (1996), McNab, Holmes, Clarke, Tannock, and Jenkinson (1996) and McNab, Jenkinson, Loach, and Tannock (1994)

	SspA	AgI/II	gp340, Platelets and epithelial cells via β-1 integrin, *Candida*, *Actinomyces*, *P. gingivalis*	Kerrigan et al. (2007), Nobbs, Shearer, Drobni, Jepson, and Jenkinson (2007) and Okuda et al. (2010)
	SspB	AgI/II	gp340, Platelets and epithelial cells via β-1 integrin, *Candida*, *Actinomyces*, *P. gingivalis*	Brooks, Demuth, Gil, and Lamont (1997), Holmes et al. (1996), Jakubovics, Strömberg, van Dolleweerd, Kelly, and Jenkinson (2005), Kerrigan et al. (2007), Nobbs et al. (2007) and Okuda et al. (2010)
	CbdA	Collagen-binding Adhesin	Type I collagen	Moses et al. (2013)
Streptococcus mitis	PblA, PblB	Phage-encoded: homologues of phage tail protein	Platelets via α-2,8-sia	Bensing, Rubens, and Sullam (2001) and Mitchell and Sullam (2009)
Streptococcus mutans	Cbm	MSCRAMM	Type I collagen	Nomura et al. (2012)
	Cnm	MSCRAMM	Type I collagen	Nakano et al. (2010)
	SpaP	AgI/II	gp340 and other salivary glycoproteins, PDLs via α5β1 integrins, type I collagen, laminin	Engels-Deutsch, Rizk, and Haïkel (2011), Okahashi, Sasakawa, and Yoshikawa (1989) and Sciotti, Yamodo, Klein, and Ogier (1997)
	WapA	Sucrose Dependent	Type I collagen, fibronectin	Qian (1993), Han (2006) and Zhu (2006)

Continued

Table 1 Adhesins of oral bacteria—cont'd

Organism	Adhesin	Description	Targets	References
Streptococcus sanguinis	PilA, PilB, PilC	Pilus: PilA-adhesin PilB-major component PilC-anchor	Fibronectin, human cells, salivary amylase	Okahashi et al. (2010, 2011)
Tannerella forsythia	S-layer: TfsA TfsB	Glycosylated surface layer- crystalline lattice of both proteins	Streptococci, *P. gingivalis*, epithelial cells	Lee et al. (2006), Sakakibara et al. (2007), and Shimotahira et al. (2013)
	BspA	OMP, leucine-rich repeat	Epithelial cells, *Fusobacteria*	Sharma et al. (1998), Mishima and Sharma (2011), and Inagaki et al. (2006)
Tannerella denticola	Dentilisin	Chymotrypsin-like protease	Fibrinogen, *P. gingivalis* fimbriae	Bamford, Fenno, Jenkinson, and Dymock (2007) and Hashimoto, Ogawa, Asai, Takai, and Ogawa (2003)
	Msp	Outer sheath protein, porin	Laminin, fibronectin, keratin, type I collagen, fibrinogen	Edwards, Jenkinson, Woodward, and Dymock (2005)
	Lipooligosaccharide	LPS analogue	Laminin, Fibronectin, Type I collagen, *P. micros*, *S. salivarius*, *S. mutans*, *S. mitis*, *A. viscosus*	Grenier (2013)
	FhbB	Factor H-binding protein	Factor H and FHL-1	Mcdowell et al. (2005)
	LrrA	Leucine-rich repeat, OMP	Epithelial cells, *T. forsythia*	Ikegami, Honma, Sharma, and Kuramitsu (2004)

Key: Gal, galactosamine; NAc, *N*-acetyl; sia, sialic acid; T5aSS, classical autotransporter; T5cSS, trimeric adhesin autotransporter type pulp (Garberoglio & Brännström, 1976). It has been known for many years that to reach the pulp oral bacteria gradually grow along these tubules and so it would be expected that species recovered from infected pulps would closely reflect the bacteria that cause demineralisation of enamel and dentine (i.e. caries).

3.2.1 Microbial aetiology of pulp infections

A study by Munson, Banerjee, Watson, and Wade (2004) showed that the advancing caries front comprised 14 major taxa dominated by *Lactobacillus spp.*, *S. mutans*, *Rothia dentocariosa* and *Propionibacterium* spp. (Munson et al., 2004). However, paradoxically the population that dominates in infected pulps is not the same as this. Indeed, a recent microbiome study showed that the infected pulp contains around 180 genera dominated by *Bacteroidetes* (Li et al., 2010), while other studies have shown that *Firmicutes* dominate (Siqueira & Rôças, 2009). It is acknowledged, therefore, that there is considerable variation from subject to subject in the composition of the microbiota associated with these common infections but it is not yet known whether different bacterial dominances contribute to different clinical symptoms or clinical outcomes. Having said that, *P. gingivalis*, which is also strongly implicated in destructive periodontitis (Socransky, Haffajee, Cugini, Smith, & Kent, 1998) (see later), is recovered in low numbers from asymptomatic chronic root canal infections (Le Goff, Bunetel, Mouton, & Bonnaure-Mallet, 1997), but the numbers of *Porphyromonas* and *Prevotella* species increase dramatically when there are signs and symptoms of acute periapical infection (Hashioka, Yamasaki, Nakane, Horiba, & Nakamura, 1992).

Given that dental caries is a prime factor in subsequent pulpal infection, it is of interest to consider how the population shifts from a relatively simple saccharolytic and acidogenic population in early caries to a complex, primarily proteolytic population in pulpal and periapical infection. Work by Love and colleagues has shown that cariogenic oral streptococci adhere to the walls of the dentinal tubules via a well-known family of adhesins known as the antigen I/II family (Love & Jenkinson, 2002; Love, McMillan, Park, & Jenkinson, 2000). In this context, these mediate binding to type-1 collagen lining the tubule. However, these adhesins also mediate 'coaggregation' (interspecies adhesion) with other organisms, such as *Porphyromonas* species, allowing a 'piggy-back' style growth of the pair toward the pulp (see Section 5). Similarly, results of a study by Nagaoka, Liu, Minemoto, and Kawagoe (1995) indicated that invasion of dentinal tubules by *Lactobacillus casei* was enhanced when cultured with *Streptococcus sobrinus* or *Actinomyces naeslundii*, and other coaggregation partnerships are also known among the oral bacteria suggesting that similar mechanisms probably account for the gradual increase in complexity of the population that becomes established in dentinal tubules once the initial barrier is breached by caries or trauma (Hojo, Nagaoka, Ohshima, & Maeda, 2009; Nagaoka et al., 1995).

It is important also to consider what is within the tubules apart from collagen. The dentinal tubules are filled with a combination of odontoblast processes (the cells that manufacture dentine) and a fluid, called dentinal fluid, the composition of which is not entirely clear but is thought to be similar to serum (Knutsson, Jontell, & Bergenholtz, 1994). In the upper parts of dentine, in some cases, there may also be ingress of salivary constituents. The flow of fluid within the tubules is thought to influence the migration of bacteria along them and the proteinaceous *milieu* favours the growth of proteolytic species, particularly in the deeper areas of the dentine, while more superficially, the species that are acidogenic and acid tolerant, such as *S. mutans* predominate. Indeed, this gradation of environment along the tubules is likely to explain the selection of species that thrive in different locations within the dentine.

The *Bacteroidetes* that tend to dominate pulp and periapical infections belong to *Prevotella* and *Porphyromonas* genera and these along with *Fusobacterium* and *Streptococcus* are often thought of as the primary pathogens in these infections (Hsiao et al., 2012; Li et al., 2010). They are also common in periodontitis (gum disease) and periimplantitis, which cause failure of the implant to integrate with bone. The reason these bacteria are common in these different types of infection is because they possess an array of virulence factors and physiological adaptations that are considered in the rest of this chapter.

3.3. Gum disease: gingivitis and periodontitis

Gum disease is a term that arguably encompasses the major inflammatory oral infections encountered in humans. It covers the reversible and less serious transient gingival inflammatory episodes described as gingivitis and the more serious, often prolonged and irreversible condition known as periodontitis.

Gingivitis is estimated to affect up to 90% of the world's population at one time or another (Pihlstrom, Michalowicz, & Johnson, 2005), while periodontitis affects 20% of all 35- to 44-year olds worldwide (WHO:

Fact sheet N°318, 2012), with an estimated 47% of the US population suffering from one form of periodontal disease or another (Eke, Dye, Wei, Thornton-Evans, & Genco, 2012), demonstrating that gingival and periodontal disease is a major health issue. Despite this, our understanding of the mechanisms, health factors and aetiology of these diseases is less well developed than other more high-profile diseases. However, given the increasing links being made between poor gingival and periodontal status and systemic diseases such as rheumatoid arthiritis (Kaur, White, & Bartold, 2013; Moen et al., 2006), cardiovascular disease (Schaefer et al., 2009), diabetes (Borgnakke et al., 2013) and possibly even Alzheimer's disease (Poole, Singhrao, Kesavalu, Curtis, & Crean, 2013), it is more pressing than ever that we understand the biology of these diseases and improve treatment regimens. In addition, there is a significant cost to our health systems from dealing with periodontal disease with an estimated £1.6 billion per annum spent by the NHS in the United Kingdom (El-Kheshen, 2008; Marsh, 2003) and an estimated $14 billion in the United States (Brown, Johns, & Wall, 2002).

Both gingivitis and periodontitis are defined microbiologically as polymicrobial infections caused by population dysbioses characterised by shifts from 'healthy' commensal and largely Gram-positive flora to a dysbiotic population with increased proportions of Gram-negative, anaerobic, virulence factor producing species (Socransky et al., 1998). Periodontitis in particular is a complicated multifactorial microbial disease but has a strong host-defined inflammatory component that is ultimately stimulated by oral bacteria present as part of a dense subgingival plaque biofilm that forms on the root of the tooth surface (Fig. 1). This biofilm is made up of a mixed population of bacteria in combination with typical biofilm matrix components derived from both host (extracellular matrix proteins, salivary proteins and GCF) (Nobbs, Jenkinson, & Jakubovics, 2011) and bacterial origin (exopolysaccharides, proteins, outer membrane vesicles and DNA) (Flemming, Neu, & Wozniak, 2007; Jakubovics, Shields, Rajarajan, & Burgess, 2013).

Periodontitis at the most basic level describes an inflammation of the periodontium, or supporting tissues of the tooth, which causes loss of collagen, cellular damage, loss of alveolar bone and can lead to tooth loss (illustrated graphically in Fig. 1; Di Benedetto, Gigante, Colucci, & Grano, 2013). Periodontitis develops from gingivitis, a mild condition characterised by inflammation of the gingivae often with transient bleeding (Schätzle et al., 2003). However, if this is allowed to persist in a significant proportion of subjects, the resulting expansion of the gingival crevice results in the irreversible formation of a periodontal pocket that presents a protected niche for the further growth of subgingival plaque (Fig. 1; Schätzle et al., 2003). While gingivitis is characterised by influx of neutrophils, periodontal disease is characterised by a marked plasma cell response, a shift in T-cell responses (Myneni et al., 2011; Settem, Honma, Nakajima, et al., 2013; Zeituni, McCaig, Scisci, Thanassi, & Cutler, 2010) and cytokine production all of which influences not only pathogen interactions with epithelial cells but also other important cell types such as osteoblasts and osteoclasts. This influence on cells involved in bone turnover in the periodontium ultimately leads to net bone resorption presenting radiologically as loss of vertical and horizontal tooth-supporting bone and this is often accompanied by the presence of inflammatory exudate and bleeding on probing the site with a blunt needle-like instrument (Di Benedetto et al., 2013; Kim, Obst, Zehaczek, & Geenen, 2008; Offenbacher, 1996).

3.4. Microbial aetiology of gingivitis and periodontitis

Strangely, our first observation of the microbial aetiology of periodontal infections might arguably come from the founder of modern microbiology, Antonie Van Leeuwenhoek, whose famous drawings of 'animalcules' that he observed from his own gingival plaque indicate organisms reminiscent of motile rods (possibly oral *Campylobacter* spp.), needle-like long rods (oral *Fusobacteria* or *Leptospira*) and cocci. However, it was the classic experiments of Loe et al. in the 1960s that began to reveal the microbiological shift that occurs in the onset of gingivitis from a largely Gram-positive, nonmotile population to a more diverse population rich in Gram-negative anaerobes and motile bacteria (Loe, Theilade, & Jensen, 1960). This view of the microbial aetiology of gingivitis has moved on considerably since then with several recent pyrosequencing studies of both clinical and experimental gingivitis confirming these observations but providing deeper understanding of the species involved (Abusleme et al., 2013; Griffen et al., 2012; Kistler et al., 2013). Many of the species observed as significantly enriched in plaque biofilm from gingivitis patients are also enriched in subgingival plaque isolated from periodontal pockets in periodontitis patients.

The classic study of Socransky et al. (1998) based on DNA–DNA hybridisation technology was seminal in the description of the pathogenic consortia by which much evidence supports are strongly associated with periodontal disease (Kamma, Nakou, Gmur, & Baehni, 2004; Socransky et al., 1998). These studies define the coloured complexes that are widely accepted as what we might loosely term periodontal pathogens, these include the red-complex of *P. gingivalis*, *Treponema denticola* and *Tannerella forsythia* (formerly *Bacteroides forsythus*); all Gram-negative obligate anaerobes that have subsequently been shown to possess a range of virulence attributes (see below). However, the complex polymicrobial nature of periodontal disease means we also consider the orange complex organisms of *Prevotella*, *Fusobacteria*, *Campylobacter* spp. and others as important organisms that also contribute to the establishment of a pathogenic community and correlation with more severe symptoms (Kamma et al., 2004). These observations have been largely backed up by more recent pyrosequencing studies that have proliferated with the reduction in cost and improvements in bioinformatics analysis and have revealed the potential importance of new or reconsidered periodontal pathogens such as the Gram-positive anaerobe *Filifactor alocis* (Kistler et al., 2013; Li et al., 2014).

This view of a small number of pathogenic bacteria somehow being more important than the rest of the diverse oral microbial population has recently been backed up by elegant studies showing that low-abundance 'keystone' pathogens influence the composition of the whole microbiota and thus result in the observed pathologies in *in vivo* experimental animal models of periodontitis (Hajishengallis, Darveau, & Curtis, 2012; Hajishengallis & Lamont, 2012; Hajishengallis et al., 2011). In this scenario, the red-complex pathogen *P. gingivalis* seems to influence both the composition and immune reactions to resident bacteria (Curtis, Zenobia, & Darveau, 2011; Hajishengallis et al., 2012). While these studies have focused on the well-characterised and high-profile periodontal pathogen *P. gingivalis*, it is probable that at least some of the other most strongly associated periodontal pathogens might also have this kind of wider community influence. Some evidence for this exists in that *T. forsythia*, much like *P. gingivalis*, is able to manipulate the immune response in terms of dictating T-cell responses that might influence the whole subgingival community (Settem, Honma, Nakajima, et al., 2013; Settem, Honma, Stafford, & Sharma, 2013). This kind of study, where the keystone organism in polymicrobial infections alongside the growing evidence that certain microbial populations can influence gut function and development, points to the possibility that monitoring and controlling our resident microflora or targeting dysbiotic keystone pathogens in disease might help to reduce disease burden (Hajishengallis et al., 2012).

4. BACTERIAL ADAPTATIONS IN THE ORAL CAVITY

While clinically the conditions described above differ in terms of their symptoms, pathologies and main causative agents, many of the key colonisation and virulence factors alongside physiological adaptations are common between caries, pulp and periapical infection and periodontitis. Therefore, we will deal with the adaptations of oral pathogens from a biological and molecular rather than a condition-specific point of view in the remainder of this chapter.

4.1. Nutritional adaptations to life at the host–pathogen interface

4.1.1 Adaptation to a proteolytic lifestyle

As mentioned above, habitation of environments below the gum line and the tooth pulp is often predominantly favoured by bacteria with the ability to degrade and harvest the rich source of proteins available to them. As such this ability to adapt to a proteolytic lifestyle is considered a key to a number of bacteria in these environments. In many cases, the ability to harvest proteins from human cells, such as erythrocytes, is also closely linked to heme/iron acquisition and is a key characteristic of several of the familiar black-pigmented anaerobes (Prevotellae, Porphyromonads) that inhabit these niches whose black pigmentation is due to accumulation of haemoglobin-derived hemin (protoporphyrin IX-dimers) and hematin (protoporphyrin IX-monomers) on the surface of the cells (Lewis, Dawson, Hannis, Macrina, & Muddiman, 1999; Smalley, Silver, Birss, Withnall, & Titler, 2003) (also see Section 4.2). These iron containing compounds ultimately derive from haemoglobin released from erythrocytes by the action of proteases produced by *Porphyromonas* and *Prevotella* spp., and specifically the lysine gingipain Kgp of *Porphyromonas* and the interpain InpA of *Prevotella intermedia* (Byrne, Potempa, Olczak, & Smalley, 2013; Byrne et al., 2010; Lewis et al., 1999). While much attention has focused on the gingipains as virulence factors (reviewed by (Curtis, Aduse-Opoku, & Rangarajan, 2001; Fitzpatrick, Wijeyewickrema, & Pike, 2009) with roles in degradation of complement (Guo, Nguyen, & Potempa, 2010), cytokines (Wang et al., 2010), cellular receptors (Chen, Nakayama, Belliveau, & Duncan, 2001) and matrix proteins (Ruggiero et al., 2013) alongside degradation of intracellular signalling (Stafford et al., 2013) and cytoskeletal proteins (Kinane, Benakanakere, Zhao, Hosur, & Kinane, 2012), they are also key to nutrient acquisition of periodontal pathogens with most knowledge arising from work on *P. gingivalis*. *P. gingivalis* is an asaccharolytic anaerobe whose nutrition seems to derive from the degradation of proteins to oligopeptides and further to dipeptides with a preference for glutamate and aspartate as amino acids of choice (Takahashi, Sato, & Yamada, 2000). The initial degradation of proteins is performed by the arg- and lys-specific gingipains that target these respective exposed residues as recognition sites in target proteins (Pike, Potempa, McGraw, Coetzer, & Travis, 1996). The released oligopeptides are then further degraded to dipeptides and tripeptides by the action of a suite of cell surface associated and released dipeptidyl peptidases (DPPIV, DPP7, DPP11 and DPP5) (Clais et al., 2014; Ohara-Nemoto et al., 2014, 2011) and a prolyl-peptidase (PTP-A) (Oda, Saiki, Tonosaki, Yajima, & Konishi, 2009). The role of both the gingipains and dipeptidyl proteases in nutrition has been well established with mutants of these genes showing retarded growth and presenting targets for inhibitory peptides for treatment (Curtis et al., 2002; Ohara-Nemoto et al., 2014) The involvement of DPP enzymes in proteolytic growth are not exclusive to *Porphyromonas*, however, since strains of *Prevotella endodontalis* (Ohara-Nemoto et al., 2014) (DPP5 paper), *P. intermedia* and *Prevotella nigrescens* also contain DPPases that seem to be present despite the lack of gingipain equivalents (Fujimura, Ueda, Shibata, & Hirai, 2003; Shibata, Miwa, Hirai, & Fujimura, 2003).

While the proteases of the black-pigmented anaerobes have been well studied, with good reason, several other oral pathogens produce potentially key secreted proteases that are now receiving much attention. One prominent example is the novel metalloproteinase karilysin that is produced by *T. forsythia* and which seems to play a role in survival in the oral cavity since it confers complement resistance and also targets human antimicrobial peptides such as LL37 and matrix proteins such as fibronectin (Jusko et al., 2012; Koziel et al., 2010). Indeed *T. forsythia*, which has long been associated with trypsin-like activity, also produces a protease, PrtH, that seems to play a role in virulence *in vivo* but there is no evidence that it has a nutritional role (Schenkein, Fletcher, Bodnar, & Macrina, 1995). A similar story seems to apply for the third red-complex pathogen, the spirochaete *T. denticola*. This organism produces a well-characterised surface-associated chymotrypsin-like protease (CLTP), also known as dentilisin, that contributes to virulence by degrading proteins in serum and the extracellular matrix, among other activities, that are thought to contribute to periodontal disease symptoms and enhance colonisation and persistence in the body (Goetting-Minesky et al., 2013). Among the oral commensal streptococci the presence of Immunoglobulin A degrading proteases has been established for several years (Reinholdt, Tomana, Mortensen, & Kilian, 1990), with their function assumed to aid in the evasion of IgA-dependent clearance and aid in initial colonisation of oral surfaces where secretions are rich in IgA. However, these proteases may also play a role in interspecies competition with recent work showing that another protease, challisin, of the commensal *S. gordonii* (which is not IgA protease positive) inhibits biofilm formation by the more pathogenic *S. mutans* (Wang, He, et al., 2011).

Taken overall, it is clear that in the oral cavity proteases play a key role in a number of processes, both not only as a means to access host nutrients but also as colonisation factors and virulence determinants.

4.1.2 Exploitation of the host glycome as a nutritional interface

When one considers the environmental niche of oral bacteria in the context of nutrition, it is easy to think that freely available dietary sugars might be the most important sources in relation to oral bacteria (see Section 3.1). However, exposure to common dietary sugars such as glucose, sucrose, or even lactose are often transient and are the subject of much competition by cohabiting bacteria. One source of sugar-derived carbon for nutritional purposes by oral bacteria is, like those resident in the human gut, sugars contained as glycoconjugates that form the glycan moieties of human glycoproteins, sometimes called the glycocalyx or the host glycome. Human membrane proteins are well known to be frequently glycosylated while mucosal secretions, such as mucins contain a range of glycan-containing proteins, with saliva and serum being no exception. For example, some salivary mucins are known to contain 85% of their mass in the form of highly branched and concentrated glycan groups (Levine et al., 1987; Zalewska, Zwierz, Zółkowski, & Gindzieński, 2000). These glycans exist as *N*- or *O*-glycosidically linked groups attached at asparagine or serine/threonine residues, respectively (Stanley & Cummings, 2009). The *O*-linked glycans are present on both glycoproteins and mucins and are represented by a range of core glycans that often contain *N*-acetylgalactosamine (GalNAc) as the sugar proximal to the serine/threonine residue which is then linked to GalNAc, galactose, or GlcNAc (*N*-acetyl glucosamine) (Brockhausen, Schachter, & Stanley, 2009; Stanley & Cummings, 2009). It is worth noting here that mucins come in a large variety of forms, with the secreted gel-forming (MG1, also called MUC5B) and small mucins (MG2, also called MUC7) that are present in mucous secretions such as saliva are not the only form. Indeed, many are cell membrane anchored and are present on epithelial and other cell types (including neutrophils) with roles in cell–cell adhesion, tissue coating and antiadhesion among others (Fukuda, 2002; Levine et al., 1987; Tabak, 1995; Vavasseur et al., 1994). These core glycans are then decorated further with glycans that often contain sialic acid as the terminal sugar and commonly contain branched structures that in many cases also include fucose (as in the Lewis antigens). In the case of *N*-glycans, the first sugar proximal to the asparagine residue is GlcNAc connected to a larger number of core sugars (including mannose) before branching and cell type and protein-specific antigens are added (Stanley & Cummings, 2009).

4.1.2.1 Exploitation of host-derived sialic acid

In the oral cavity, several species of pathogenic bacteria have evolved the ability to harvest this rich source of carbon present in the human environment. Indeed, there has been much recent interest in the relationship between human-dwelling bacteria and one common terminal sugar moiety on human glycans, namely sialic acid. This is particularly pertinent in the oral context given that the name sialic acid is in fact derived from the greek sialon, meaning derived from saliva. The most common form of this 9-carbon sugar that is part of a wider family of sialic acids is Neu5Ac, where an acetyl group is present at the 5-carbon position (Neu5Ac is shown with other prominent sialic acids, Neu5Gc and Neu5,9Ac that are relevant in the oral cavity in Fig. 2). Unsurprisingly, oral bacteria (like several other species that colonise humans) have evolved the means to access this sugar via the action of bacterial sialidases, a group of enzymes that are related

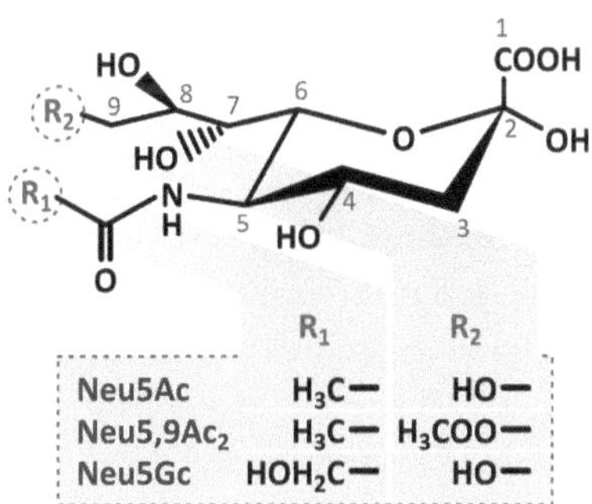

Figure 2 Schematic illustrating the structure of the major forms of sialic acid found in the oral cavity. Neu5Ac contains an acetyl group at the carbon-5 position while the diacetylated Neu5,9Ac contains an extra acetyl group at carbon-9. The non-human form of Neu5Gc contains an extra O atom on the R_1 group.

evolutionarily to both viral and human sialidases (Buschiazzo & Alzari, 2008). They are characterised by the presence of a number of conserved primary sequence motifs, such as the four to five Asp-boxes (Ser/Thr-x-Asp-x-Gly-x-Thr-Trp/Phe; where x represents any amino acid) that are evenly spaced through the sialidase structure, the RIP motif and a conserved catalytic arginine triad that coordinates the carboxylate group of sialic acid (Buschiazzo & Alzari, 2008; Roggentin, Schauer, Hoyer, & Vimr, 1993). Bacterial enzymes are generally classified as the GH33 family of glycosyl hydrolases and those that have been structurally characterised display a characteristic six-blade propeller structure (Crennell, Garman, Laver, Vimr, & Taylor, 1994; Roggentin et al., 1993). In bacteria, the catalytic domains of sialidases are often preceded, followed or flanked on both sides in the protein structure by domains that often help align the target glycoconjugates via the use of lectin-like domains or carbohydrate-binding modules (CBM). One well-known example being that of the sialidase of *Vibrio cholerae* which contains two lectin-like domains (classified as CBM40) that appear to coordinate its cellular ligands (Crennell et al., 1994; Moustafa et al., 2004). Structural data have also revealed other CBM40-like domains, notably in the clostridial sialidase NanJ (Boraston, Ficko-Blean, & Healey, 2007).

The role of sialidases in the context of human–bacterial interactions is largely the specific cleavage of terminal α-2,3 and α-2,6 glycosidic linkages between sialic acid and underlying galactose of glucosamine residues at the terminal branches of glycans, i.e. exoglycosidases (Lewis & Lewis, 2012). In addition, some bacterial endo- or exosialidase enzymes are able to target 2,8-linked polysialic acid residues that are present in nerve tissue (Jungnickel et al., 2009). In terms of the oral cavity, the resident bacteria are exposed to salivary-derived mucins, such as the predominant MG1 and MG2, that are continuously produced and have been shown to be constituents of the salivary pellicle that coats the enamel of teeth, and are a major source of glycans orally, being particularly rich in sialic acid (Thomsson et al., 2002). As mentioned above, those bacteria that are in contact with mucosal surfaces will not only encounter sialylated epithelial mucins but also sialylated integral membrane proteins such as integrins (Yu et al., 2013), Toll-like receptors (TLRs) (Feng et al., 2012) and hormone receptors (e.g. EGFR) (Park et al., 2012) alongside sialylated glycosphingolipids, such as GM1 and GM3, that are linked via ceramide linkages to membrane lipids (Stanley & Cummings, 2009). It is therefore no surprise that as alluded to above several oral colonising bacteria possess sialidase enzymes. A search of the Human Oral Microbiome Database (HOMD) (Dewhirst et al.,

2010) and the CAZy database for GH33-containing proteins reveals that a range of oral species have confirmed and putative sialidases (Fig. 3; Lombard, Golaconda Ramulu, Drula, Coutinho, & Henrissat, 2014).

The association of sialidase activity with oral streptococcal species, such as the commensal and opportunistic pathogen *Streptococcus oralis* in particular, has been known for a number of years. Its role in nutrition in terms of growth on sialylated glycoproteins such as mucins and α-1-glycoprotein is well established through the work of Professor Beighton's group (Byers, Tarelli, Homer, & Beighton, 1999, 2000; Paddick, Brailsford, Kidd, & Beighton, 2005). These organisms are known to be able to grow on sialic acid as a sole carbon and energy source, with the detection of sialic acid catabolic activity and sialic acid uptake established in the years before its genome was sequenced but which contain homologues of the genetically well-characterised *S. pneumoniae* system (unpublished) (Byers, Homer, Tarelli, & Beighton, 1999; Byers, Homer, & Beighton, 1996). It is also of note that the genes encoding *S. oralis* sialidase enzymes are closely related to and indeed display evidence of recombination with those from the highly pathogenic *S. pneumoniae*, where sialidase activity is considered a significant virulence factor both in terms of nutritional capability and in terms of adhesion and immune modulation (Chang, Uchiyama, Varki, & Nizet, 2012; King, Whatmore, & Dowson, 2005; Uchiyama et al., 2009).

Up to now, we have considered sialidase activity in the context of interactions with salivary proteins, by what we might consider commensal early-colonising Gram-positive bacteria. However, it has been known for sometime that sialidase activity is associated with a range of Gram-negative late-colonising species that we might consider of more systemic significance. The studies of Moncla et al. (Moncla, Braham, & Hillier, 1990; Moncla, Braham, Rabe, & Hillier, 1991) performed a thorough assessment of whole-cell sialidase levels using the widely accepted and specific methyl–umbelliferyl substrates and revealed sialidase activity among a range of *Porphyromonas*, *Prevotella*, *Capnocytophaga* and *Tannerella* strains which has also been borne out in terms of observation of sialidase genes in the newly sequenced genomes of these organisms. In fact, the association of sialidase with *Tannerella* seems to have been present in the human population for several hundred years with recent analysis of medieval skeleton derived calculus DNA revealing the presence of a sialidase gene cluster in a genome reconstruction from these samples and the derived metagenomic datasets (Warinner et al., 2014). In addition, although not considered a traditional oral pathogen, the potential fourth member of the 'red-complex' of

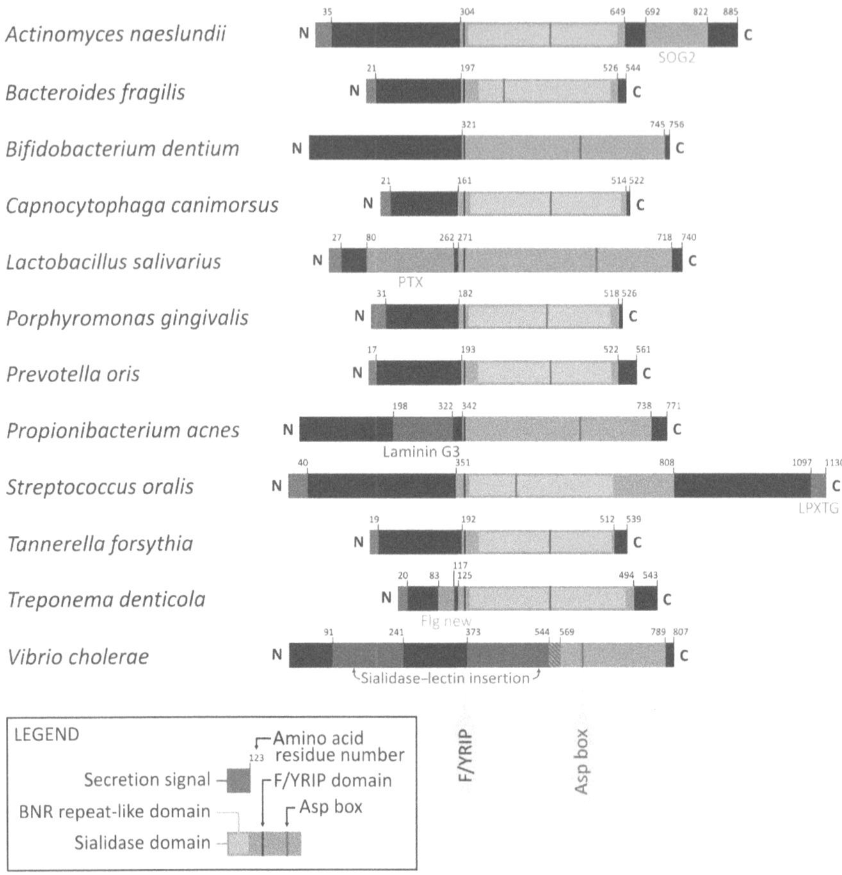

Figure 3 A schematic representation of the various domains and conserved regions of bacterial neuraminidases presented to scale and aligned by their Y/FRIP regions. The source of sialidases, along with their accession numbers in parentheses, is as follows: *Actinomyces naeslundii* Howell 279 (GenBank: EJN85732.1), *Bacteroides fragilis* NCTC 9343 (NanH; NCBI Ref: YP_211442.1), *Bifidobacterium dentium* ATCC 27679 (GenBank: EFM40399.1), *Capnocytophaga canimorsus* Cc5 (GenBank: AEK22601.1), *Lactobacillus salivarius* NIAS840 (GenBank: EGL97992.1), *Porphyromonas gingivalis* ATCC 33277 (NCBI Ref: YP_001929724.1), *Prevotella oris* F0302 (GenBank: EFB32834.1), *Propionibacterium acnes* 266 (NanA; GenBank: AEE71914.1), *Streptococcus oralis* Uo5 (NanA; GenBank: CBZ00210.1), *Tannerella forsythia* ATCC 43037 (GenBank: AEW22573.1), *Treponema denticola* ATCC 35405 (TDE0471; NCBI Ref: NP_971085.1) and *Vibrio cholerae* N16961 (NanH; NCBI Ref: NP_231419.1). Previously characterised accessory domains of these sialidases are indicated on the diagram.

periodontal pathogens, namely the anaerobe *F. alocis* has been reported to display sialidase activity using an indirect sialidase assay (Aruni, Roy, & Fletcher, 2011); however, the genome sequence does not seem to contain any putative GH33 family members meaning this observation is confusing and merits further investigation.

The increase in the amount of genome sequence and molecular techniques available for the mutagenesis and manipulation of periodontal pathogens have lead to a step-change in our molecular understanding of virulence factors in these organisms. In the case of sialidases, this has lead to the establishment that sialidase activity seems to be a major trait of periodontal pathogens and the ability to harvest and utilise sialic acid at the host interface is key to their colonisation and potential virulence in the subgingival environment (Stafford, Roy, Honma, & Sharma, 2012).

If we consider first the most well-studied sialic acid acquisition system in this group, namely, that of the red-complex pathogen *T. forsythia*, it is clear that this organism is very well evolved to harvest and utilise this rich source of nutrition. *T. forsythia* possesses a large sialic acid transport, catabolic and harvesting locus in its genome that contains not only the *nanH* sialidase gene but also an accompanying putative β-hexosaminidase that may well cleave the underlying glucosamine or galactosamine moieties exposed by sialidase action (as illustrated in Fig. 4) (Roy, Douglas, & Stafford, 2010; Roy et al., 2012; Thompson, Homer, Rao, Booth, & Hosie, 2009). In addition, *T. forsythia* contains a novel sialic acid transport system that utilises the TonB–ExbB–ExbD complex to energise transport of sialic acid across the *T. forsythia* outer membrane. This is achieved via the use of a SusCD family transporter pair homologue called NanOU, where NanO is a transmembrane β-barrel located in the outer membrane and a NanU membrane associated and secreted high-affinity sialic acid-binding protein (Fig. 4; Phansopa et al., 2013). The presence of a Tetraticopeptide Repeat structure and its homology with SusD proteins, which are often part of larger multiprotein carbohydrate harvesting sytems, leads us to postulate that it might form a stable or transient complex with the NanH sialidase, and potential β-hexosaminidase and adjacent putative sialate-9-*O*-acetylesterase. The role of sialic acid in the lifestyle of *T. forsythia* seems well founded since mutants in the *nanH* gene are less-efficient biofilm formers on glycoprotein surfaces and indeed are unable to grow on sialylated growth substrates such as mucins or whole human saliva while also displaying reduced adhesion to and invasion of epithelial cells (Honma, Mishima, & Sharma, 2011; Roy, Honma, Douglas, Sharma, & Stafford, 2011). In fact even the biochemistry of the

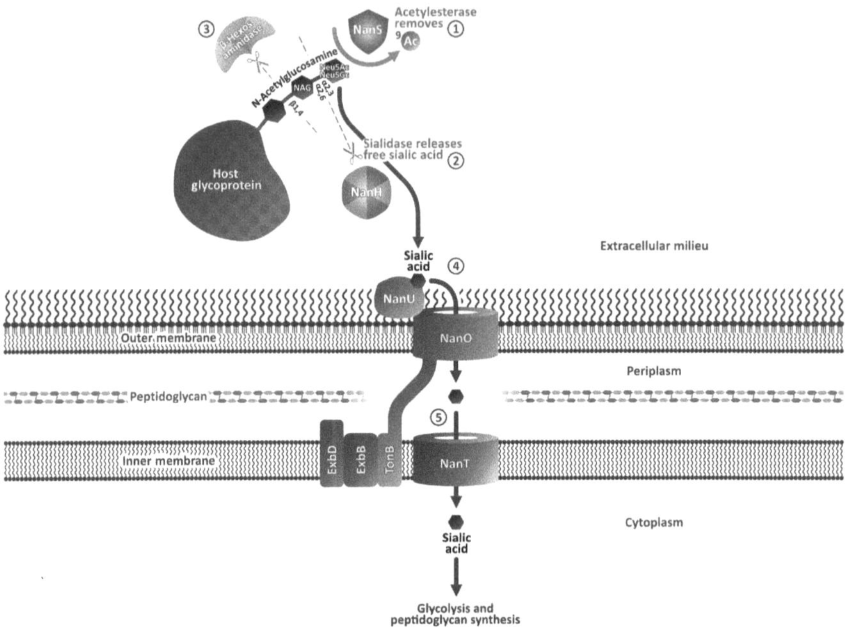

Figure 4 Schematic representing the sialic acid uptake and harvesting system present in *Tannerella forsythia*. We envisage that a putative sialate-9-*O*-acetylesterase (NanS) removes the 9-*O*-acetyl group from diacetylated sialloglycoproteins (1) before sialic acid, Neu5Ac (or Gc) is then released by the action of the sialidase NanH (2). A range of β-hexosaminidase enzymes may then attack underlying sugars (3) before Neu5Ac is bound by surface localised NanU which passes Neu5Ac through the NanO TonB-dependent transporter (4) to traverse the outer membrane before the NanT permease (5) takes Neu5Ac across the inner membrane before it is acted upon by the catabolic pathway internally.

sialidase of *T. forsythia* seems well adapted to its environment with evidence that not only is recombinant sialidase highly stable but also it has a pH optimum that is more neutral than the majority of isolated sialidases to date at pH 7.5 (A. Frey & G. Stafford, personal communication). In addition, the NanH sialidase of *T. forsythia* seems to possess an extended N-terminal domain that may represent a novel sialic acid-binding CBM (Fig. 3). There is, however, one surprising aspect to the observation that *T. forsythia* is able to use sialic acid for growth, which is that until 2010 it was considered that one could only maintain *T. forsythia* in the presence of the cell wall monomer *N*-acetyl muramic acid (Wyss, 1989). We now of course know that, at least *in vitro*, sialic acid in its free monomeric form or as a sialloconjugate sugar such as siallyllactose or as the terminal linkage on glycoproteins can be used as an alternative. However, given its apparent importance in the adhesion step of biofilm formation (Roy et al., 2010) and cell interactions (Honma et al., 2011) and the possibility that sialidase could contribute to the immune modulation *T. forsythia* is capable of, it seems that the intimate relationship with sialic acid might be an indication of a more pathogenic lifestyle. This seems to be borne out further by the recent sequencing of the unculturable *Tannerella* BU063 isolate, that is associate with less severe gingival disease, does not contain the sialic acid uptake and sialidase operon and may also have an intact NAM production system (Beall et al., 2014). This evidence gives us the tantalising idea that sialic acid as an adhesive molecule and nutritional source is an evolutionary pathogenic adaptation by *T. forsythia* to cope with the environment in which it resides.

While the most well-studied system in relation to life using sialic acid as a growth substrate is that of *T. forsythia*, recent years has seen the characterisation of sialidase activity and its molecular basis in both of the other red-complex periodontal pathogens, *P. gingivalis* and *T. denticola*. If we take the keystone pathogen *P. gingivalis*, the first report of the molecular basis of sialidase activity was by Aruni, Vanterpool, et al. (2011) in which they used an indirect sialidase activity assay to characterise mutants in three genes putatively possessing sialidase activity: *PG0352*, *PG0778* and *PG1724*. However, of these only the *PG0352* gene possesses the typical sialidase signature motifs and indeed like *T. forsythia* NanH seems to have a novel putative CBM at its N-terminal end, but which does not have significant homology at the primary amino acid level to the putative CBM of *T. forsythia nanH* (Fig. 3). In support of this recombinant *PG0352* has activity against human glycoproteins (Li et al., 2012) while a chromosomal inactivation mutant of this gene displays reduced adherence and invasion of oral epithelial cells (Aruni, Vanterpool, et al., 2011), biofilm formation and reduced virulence in a mouse infection model (Li et al., 2012). In addition, *PG0352*-deficient mutants display altered morphology and reduced capsule formation compared to the parent strain (Li et al., 2012). In contrast to *T. forsythia*, *P. gingivalis* does not seem to use sialic acid as a nutrient source directly as it does not possess any genes homologous to sialic acid catabolic pathway genes. Perhaps *P. gingivalis* incorporates Neu5Ac into its capsule, or requires Neu5Ac as a precursor for synthesis of other capsular sugars. This mutant also displayed decreased resistance to human serum, which is perhaps unsurprising due to the alteration in capsule morphology, though the sialidase could also contribute to alterations in the immune response (Stamatos et al., 2010). Since *P. gingivalis* probably lacks the ability to synthesise sialic acid, sialidase activity is likely to be key in the acquisition of exogenous sialic acid, which it might then place on its surface. In keeping with this idea several glycosylated proteins have now been identified in *P. gingivalis* (Kishi et al., 2012) while Curtis et al. (2002) showed that the secreted gingipains contain sialic acid as part of a large oligosaccharide attachment. One final explanation for possessing sialidase activity could be that like other species *P. gingivalis* cleaves sialic acid from human proteins to improve their proteolytic susceptibility, an idea suggested by the role of sialidase in some commensal species, namely those of the Gram-positive genera *Actinomyces*, *Corynebacterium* and *Propionibacterium*, where sialidase activity has been shown to desialylate human sIgA, making it more susceptible to proteolysis by organisms containing IgA proteases, or in the case of *P. gingivalis*, gingipains (Frandsen, 1994; Reinholdt et al., 1990). These data may also allude to the function of other sialidases in species that do not seem to contain sialic acid uptake or catabolic genes, i.e., the release of sialic acid and the subsequent exposure of both underlying sugars and changes to protein conformations to benefit the wider oral microbiota.

The shared trait between *P. gingivalis* and *T. forsythia* of the importance of sialidases is also the case for the third of the red-complex pathogens *T. denticola* (Fig. 3). In *T. denticola* sialidase activity has also been shown to modulate virulence in a mouse skin disease model where a sialidase-deficient mutant caused smaller abscesses (Kurniyati, Zhang, Zhang, & Li, 2013). In addition, sialidase was shown to protect against complement activity and surface complement deposition and it displayed reduced ability to grow using sialic acid containing substrates. This observation is, however, at odds with its lack of any clear sialic acid catabolic genes in its genome sequence (Stafford et al., 2012). However, the metabolism of spirochaetes is poorly understood and it is possible that another metabolic route is present.

When considered together, it seems clear that oral bacteria have an intimate relationship with host-derived sialic acid, both not only from the point of view of nutrition but also in terms of potentially modulating the immune response. For example, sialic acid is known to be a marker of self (i.e., an SAMP—self-associated molecular pattern) that dampens immune responses, and in fact modulation of its levels on TLRs is key to responses to LPS (Feng et al., 2012; Ray et al., 2010; Stamatos et al., 2010). There is also a growing body of evidence that the action of human and possibly more pertinently bacterial sialidases affects the response of macrophages to TLR ligands such as LPS and this might exacerbate the immune response and potentially contribute to the pathology observed (Chang et al., 2012).

It is thus clear that interaction with this sugar plays a key role in biofilm formation and virulence *in vivo*. However, it is also of note that in contrast to some other sialidases, particularly those of other Gram-negative organisms such as the periplasmic N-glycan targeting systems from *Capnocytophaga* (Renzi et al., 2011) and *Bacteriodes thetaiotaomicron* (Nihira et al., 2013), the sialidases of periodontal pathogens appear to be secreted either to be surface attached or released from the cell surface into the surrounding environment. At present, little is known of how sialidases might be secreted by either the oral Gram-negative Bacteroidetes (*T. forsythia* or *P. gingivalis*) or *T. denticola*. One option in the case of *T. forsythia* and *P. gingivalis* is that sialidase and other transport proteins are secreted via outer membrane vesicles, which have been shown to be enriched in certain types of proteins and are known to be produced by other Bacteroidetes. These contain both surface attached and luminal proteins and one could imagine proteins being released from the cell in this manner, in fact *B. thetaiotaomicron* has recently been shown to release OMVs enriched in acidic glyosidases, including sialidase, while *V. cholerae* also contains sialidase as part of its OMV proteome (Altindis, Fu, & Mekalanos, 2014; Elhenawy, Debelyy, & Feldman, 2014). However, at least in the case of *P. gingivalis*, a very recent study failed to detect sialidase in the OMV proteome of *P. gingivalis* (Veith et al., 2014) suggesting that another route may be responsible. Perhaps, sialidases cross the inner membrane via their Sec-dependent signals and are then secreted across the outer membrane by an as yet unknown route, although it is unlikely to be the Type IX secretion system as they lack CTD signal regions (see Section 4.2).

While sialidases are clearly key to the exploitation of human-derived sialic acid, normally targeted at the dominant conjugated form, Neu5Ac, several important alternative forms of sialic acid are important in the human body. For example, while Neu5Ac makes 50–60% of the sialic acids within the continuum of the digestive tract, the second-most common form has a second acetyl group at the C9 position of sialic acid (i.e., Neu5,9Ac) making up approximately 20% of conjugated sialic acid (Fig. 2; Robbe et al., 2003). This form of sialic acid has also been shown to be present in saliva and on salivary-derived glycoproteins (Haverkamp, Schauer, & Wember, 1976; Martin, Verhagen, & Varki, 2003). We also have evidence that oral cancer cell lines express Neu5,9Ac antigens (T. Farmilo & G. P. Stafford, personal communication), a situation that is not surprising given the importance of Neu5,9Ac levels in cancers such as ovarian tumours (Toba, Tenno, & Kurosaka, 2000). In terms of bacterial nutrition, it is known that diacetylated sialic acids are more resistant to sialidase cleavage while catabolic lyases catabolise Neu5,9Ac less efficiently (Krüger, Schauer, & Traving, 2001). Given this potential source of sialic acid present in the oral cavity and the lower efficiency of its utilisation, an organism that possessed the ability to harvest Neu5,9Ac from oral sources would have a competitive advantage. To achieve this, possession of a sialate-9-O-acetylesterase would allow removal of the 9-acetyl group and allow sialidases to then harvest the remaining Neu5,9Ac (Fig. 4). The sialic acid utilising locus of *T. forsythia* contains a candidate sialate-9-O-acetylesterase (9-O-Acase), which we have preliminary data indicating that it has activity against Neu5,9Ac-containing substrates (T. Farmilo, unpublished). Homologues of this enzyme include the NanS from *E. coli* which is a periplasmic enzyme containing an SGNH hydrolase domain used to process Neu5,9Ac once it has traversed the outer membrane (Steenbergen, Jirik, & Vimr, 2009). The putative NanS from *T. forsythia* is much larger and possesses only limited homology with NanS from *E. coli*. Bioinformatic searches reveal that 9-O-Acase encoding genes are present in a range of Bacteroidetes, including several oral spp. such as *Prevotella salivae*, *Prevotella denticola* and *Prevotella oris* while also being present in several oral *Capnocytophaga* spp. Unsurprisingly, this ability to harvest myriad forms of sialic acid extends to targeting nonhuman Neu5Gc, which enters the body through dietary sources but is known to be an important xeno-auto-antigen that might contribute to cancer progression (Pham et al., 2012). It is likely that these adaptations indicate the sialic acid specialism of a number of these organisms and indicate their propensity for carbohydrate foraging, much like their gut dwelling relatives such as *Bacteroides fragilis* and *B. thetaiotaomicron*.

4.1.2.2 The role of other glycosidases in oral bacteria

While most focus in this chapter and indeed in the scientific literature has fallen on sialic acids there are a myriad of glycosidase activities targeted at other glycan groups present in the human oral cavity (Beighton, Radford, & Naylor, 1992; Braham & Moncla, 1992; Maiden, Tanner, & Macuch, 1996). These include the broad and widespread family of enzymes targeting glucosamine, galactosamine, galactose that are usually present as β-linked glycoconjugates that are often termed hexosaminidases (Slámová, Bojarová, Petrásková, & Kren, 2010). These enzymes generally fall into the GH20 family but also contain many members from GH3 and 84. A bioinformatics survey of CAZy (Lombard et al., 2014) for putative GH3, 20 and 84 enzymes reveals candidate genes exist in oral *Streptococci*,

Aggregatibacter and members of the genera *Capnocytophaga*, *Bifidobacterium*, *Treponema*, *Prevotella*, *Porphyromonas* and *Tannerella*. It seems no coincidence that this list of organisms aligns well with those known to interact with sialic acid or that they are all effective oral colonisers.

Of these organisms the first to be studied in any depth were the oral streptococci, including *S. oralis* where their ability to sequentially degrade and utilise the sugars contained in the glycans of human α-1-glycoprotein revealed β-*N*-galactosidase and β-*N*-acetylglucosaminidase alongside apparent α-fucosidase and α-mannosidase activities, all of which contributed to growth on this glycoprotein substrate (Byers, Tarelli, et al., 1999; Byers, Tarelli, Homer, Hambley, & Beighton, 1999). Among the other oral streptocooci, the GcnA *N*-acetyl-β-D-glucosaminidase from *S. gordonii* has been shown to act on β-linked galactosamine as well as on glucosamine, however, a mutant in this gene did not show reduced ability to grow on glycoprotein substrates, presumably because its genome has now been shown to encode two other putatively GH20 enzymes (Langley et al., 2008). This nicely illustrates the case for several oral bacteria, with *T. forsythia* and *P. gingivalis* as prime examples that produce three to four predicted β-hexosaminidases (Roy et al., 2012), several of which have been enzymatically characterised (Hughes, Malki, Loo, Tanner, & Ganeshkumar, 2003; Lovatt & Roberts, 1994). Their role in pathogenesis is less well understood, however, with the only evidence of this activity influencing virulence traits being the inhibition of *T. forsythia* biofilm formation by the hexosaminidase inhibitor PUGNAc (Haltiwanger, Grove, & Philipsberg, 1998; Roy et al., 2012). We would also postulate that this might influence interactions with human cells but this has yet to be examined. One example where there is a clear example of the function of a bacterially secreted β-hexosaminidase is the Dispersin B, GH20 β-*N*-glucosaminidase, from *Aggregatibacter actinomycetemcomitans*, which plays a key role in turnover and dispersal of biofilms via degradation of polysaccharide polymers in the biofilm matrix of not only *Aggregatibacter* but also several other spp. including *E. coli* and *Stapylococcus aureus* (Manuel et al., 2007).

As mentioned above, several oral bacteria possess the ability to target a wide range of sugars with glycosidases and while many of these have not been studied in any detail, there is one class of these that warrants further consideration, namely those that target fucose moieties. Fucose makes up a significant proportion of the human glycome being a major component of blood group A, B, O antigens and the Lewis X and Lewis A antigens and is thus present in significant levels in epithelial layers and in mucins

(Becker & Lowe, 2003). It is also known to be important to innate and adaptive immunity and displays altered surface exposure in certain cancers with most of the effects thought to come about due to changes in cell–cell interaction phenotypes. As with sialic acid and hexose sugars, human-dwelling bacteria often produce secreted enzymes targeting the α-1,2/3/6 linked fucose present in humans that are categorised as GH29 glycosyl hydrolases (Lombard et al., 2014). One group of bacteria well known for their ability to release and utilise bound fucose is the Bacteroidetes. It is thus no surprise that fucosidase activity has thus been detected in periodontal communities, including in a range of *Tannerella*, *Porphyromonas* and *Prevotella* species (Beighton et al., 1992; Braham & Moncla, 1992; Moncla et al., 1991). It is also of note that fucose may be an important adhesive molecule for *P. gingivalis* fimbriae (Sojar & Smith, 2012) and is a key component of the glycan groups present on the surface layer (S-layer) of *Tannerella* and related species of the Bacteroidetes (Coyne et al., 2013; Posch et al., 2011, 2013). In the case of *Tannerella*, it is particularly intriguing given the parallels between this glycosylation system and that present in gut bacteria such as *B. fragilis* where the ability to fucosylate its surface is key to its survival *in vivo* (Coyne, Reinap, Lee, & Comstock, 2005). While it seems that *T. forsythia* produces a fucosidase enzyme, its genome does not contain obvious homologues of fucose catabolism genes (G. P. Stafford, personal communication), raising the possibility that this organism harvests fucose in order to salvage it for presentation on its surface proteins. This is in contrast to putative fucose catabolic pathways present in oral *Bifidobacteria* and *Fusobacterium* species identified using HOMD (Dewhirst et al., 2010). In addition, the release of free fucose by the oral microbiota may influence host cell behaviour in a similar way to how *B. thetaiotaomicron* stimulates gut cells to present more fucose on their surface in response to bacterial manipulation of host (Hooper, Xu, Falk, Midtvedt, & Gordon, 1999).

When considered overall the ability of oral bacteria, much like those in the gut, to utilise host-conjugated sugars as either carbon sources or structural capsular components seems unsurprising. However, the mechanisms and more importantly the role that these abilities play in colonisation and infection are only now beginning to be elucidated and one hopes to present a range of opportunities for the design of novel inhibitors and antimicrobials.

4.2. Protein secretion in the oral context

In common with most environmental and human-dwelling bacteria, the modulated secretion of proteins is important and widespread among oral bacteria. Currently, the literature has classified nine bacterial protein secretion systems to date with studies ongoing on this important area of research. This applies not only to the secretion of proteinaceous virulence factors (Type 1,2,3,5,7,9) but also to protein secretion systems employed in the construction of surface structures such as pili and fimbriae (Type 2,4), curli (Type 8) and flagella (Flagella Type 3). Of these nine secretion systems, six are present in Gram-negative bacteria where the challenge for protein secretion is the traversing of two membranes, while in Gram-positive bacteria the challenge is presented by only one membrane but in some cases there is difficulty to cross outer cell walls. Of the Gram-negative systems, Types 1, 3, 4 and 6 are known to secrete proteins from the cytoplasm directly to the outside of the cell, whereas Type 2 transport requires either the Sec-dependent or Twin-arginine (Tat) systems for traversing the cytoplasmic membrane of Gram-positive bacteria or the inner membrane into the periplasm of Gram-negative bacteria before other systems facilitate export to the exterior surface or milieu. Type II secretin or Type 5 autotransporters aiding crossing of the outer membrane are such examples of this outer membrane export (Tseng, Tyler, & Setubal, 2009).

While protein secretion for the Gram-positive oral streptococci is not well studied, there are some notable and surprising examples of specialised protein secretion systems in these organisms that warrant further investigation. The first of these is the presence of a specific accessory secretion system that seems to be dedicated to the secretion of a highly glycosylated surface protein and suggested virulence adhesin from the oral commensal and putative endocarditis pathogen *S. gordonii* called GspB (gordonii surface protein B) (Jakubovics, Keerigan, et al., 2005; Seepersaud, Bensing, Yen, & Sullam, 2010). This system contains homologues of the SecA and SecY proteins but requires a range of accessory proteins (Asp) for the efficient secretion of GspB. It also seems to be present in a range of streptococcal spp. alongside *S. aureus* suggesting that like many other secretion systems their employment for a range of substrates may be revealed over time (Bensing & Sullam, 2002; Seepersaud et al., 2010).

Protein secretion of certain virulence factors in *Mycobacterium tuberculosis* across the inner and mycobacterial membranes involves the Type VII secretion system, and this or a variant thereof is present in a range of Gram-positive bacteria (Schneewind & Missiakas, 2012). In agreement with this finding, a recent study assessing potential virulence determinants of *S. mutans* revealed potential Type 7 secretion systems with homologues of the key Type 7 secretion protein gene *essB* being found in the genome sequences of *S. oralis*, *S. mitis* and *S. intermedius* strains raising the possibility that this is an important virulence factor secreting system in the oral streptococci (Dewhirst et al., 2010; Palmer et al., 2013).

In terms of secreted virulence factors and toxins probably the two best-studied examples are the leukotoxin (LtxA) of the oral pathogen *A. actinomycetemcomitans*, which is one contributing species to aggressive periodontitis and the recently discovered T9SS that relates to secretion of gingipains in *P. gingivalis*. The LtxA is a member of the repeat-in-toxin (RTX) family of bacterial endotoxins and is known to lyse white blood cells during the infection process (Taichman, Dean, & Sanderson, 1980; Tsai, McArthur, Baehni, Hammond, & Taichman, 1979), and more recently, it has been shown to lyse erythrocytes. Recent evidence has suggested the need for sialic acid residues as recognition molecules for the LtxA protein to bind to the erythrocytes and white blood cells and cause cell destruction (Munksgaard et al., 2014). The biogenesis and secretion of the protein itself is determined by the *ltx* operon, designated *ltxCABD* which also produces LtxB and LtxD proteins with high homology to the HlyA and HlyD proteins involved in α-hemolysin toxin secretion in *E. coli*, respectively. The LtxA toxin protein is acylated with a short-chain fatty acid by LtxC (Fong et al., 2011) before being translocated across the inner membrane via LtxB, an inner membrane ATPase, and out of the cell through a channel likely to be formed by the TolC-like outer membrane protein (OMP), TdeA and LtxD, that crosses the periplasmic space formed by these two proteins. Deletion of these genes generates a mutant which fails to form this type I secretion complex and cannot secrete LtxA, thereby losing a major part of its pathogenicity (Crosby & Kachlany, 2007; Isaza, Duncan, Kaplan, & Kachlany, 2008). Other virulence factors that allow pathogenesis of *A. actinomycetemcomitans* include various adhesins, for example, EmaA (Yu, Ruiz, Lenox, & Mintz, 2008), an extracellular matrix protein adhesin and the epithelial cell adhesins, ApiA and Aae (Rose et al., 2003; Yue et al., 2007). These adhesins utilise the Type 5a (Aae) and Type 5c autotransporter type secretion system (EmaA, ApiA), which comprises a C-terminal β-barrel domain in the outer membrane (5a) or which forms a trimer that forms a β-barrel (5c) and is inserted into the outer membrane with each subunit donated by one monomer of the adhesin to allow surface presentation of the N-terminal passenger domains that in the case of EmaA binds collagen and for Aae mediates its virulence functions (Henderson, Navarro-Garcia, Desvaux, Fernandez, & Ala'Aldeen, 2004; Rose et al., 2003; Ruiz, Lenox, Radermacher, & Mintz, 2006).

As discussed throughout this chapter, oral anaerobic Gram-negative bacteria of the Bacteroidetes are key organisms in several oral infections. To bring about these infections and indeed colonise the oral cavity, it is no surprise that they employ several secreted and surface-presented proteins. Unusually, many of these are large glycosylated proteins, including the major gingipain virulence factors of the periodontal pathogen *P. gingivalis*. It appears from recent studies that within the Bacteroidetes phylum, a novel secretion system known as the Por Secretion system (PorSS), and now named Type IX (or T9SS as we will refer to it here) is employed not only to direct proteins to the cell surface but also to attach them via a novel modification. The presence of T9SS was first suggested over a decade ago, through the work in Eric Reynolds' laboratory who observed extensive carbohydrate modification to the membrane-associated gingipains, RgpB with the elevated molecular weight form of the protein showing reactivity to the mAb-1B5 antibody which recognises A-LPS (phosphorylated, branched mannan repeated units attached to a lipid A core) (Curtis et al., 1999; Rangarajan et al., 2008; Veith et al., 2002). In addition, this modification was coupled to the identification of the presence of a conserved C-terminal domain (CTD) common on a range of secreted proteins (Chen et al., 2011; Glew et al., 2012; O'Brien-Simpson et al., 2001), suggesting that the existence of a protein secretion system is involved not only in transporting proteins across the membrane but also in anchoring the proteins to the cell surface (Veith et al., 2013).

Since then similar CTD domains have been found on all proteins secreted via the T9SS system, it is ~80 amino acids in length, with five distinct motifs, A–E (Seers et al., 2006). Of these, there are highly conserved features in three of these regions (B, D and E) that are present across different species in the Bacteroidetes phylum, such as in *P. gingivalis*, *T. forsythia* and *Cytophaga hutchinsonii* (Veith et al., 2013). Using protein sequence databases, the current number of putative CTD-containing proteins stands at around 682 proteins found within a wide species distribution of 87 *Bacteroidetes* species (Veith et al., 2013), with varying ecological niches from soil bacteria (e.g. *C. hutchinsonii*) to host microbes found in the intestines (*B. fragilis* and *P. distasonis*) and oral cavity (*P. gingivalis*, *T. forsythia* and *P. intermedia*). However, as the number of the bacterial genomes become increasingly elucidated, this number is likely to grow. Within this, around 34 proteins of the *P. gingivalis* genome (Seers et al., 2006) and 37 proteins from the *T. forsythia* genome (Nguyen, Travis, & Potempa, 2007) are likely to be targeted for secretion via T9SS. One of the first T9SS proteins to be identified was PorT, which was found to be responsible for the translocation of proteins across the outer membrane with Δ*porT* mutants leading to an accumulation of gingipains and Hemagglutinin HagA in the periplasm (Sato et al., 2005). Comparative genome analysis by Sato et al. (2010) identified 11 additional proteins, including Sov and PG0027, involved in the secretion of gingipains, leading to their identification of participation in the T9SS system (Sato et al., 2010). Six of these genes, PorK, PorL, PorM, PorN, PorW and Sov, are related to the gliding motility proteins found in the soil bacterium, *Flavobacterium johnsoniae* (Sato et al., 2010), suggesting a link between the protein secretion system and gliding motility proteins (Chen et al., 2011).

Based on the accumulated evidence, we can summarise the sequence of events in the T9SS system, using the secretion of the arginine gingipain RgpB as an exemplar, as follows (illustrated in Fig. 5). First, in common with several Gram-negative secretion systems, secretion of proteins across the inner membrane seems to occur via the sec system, as determined by the presence of the N-terminal Sec-dependent signals in the gingipains and all other CTD proteins predicted to date (Takii, Kadowaki, Baba, Tsukuba, & Yamamoto, 2005). In agreement with this deletion of the CTD results in the accumulation of the protein in the periplasm, indicating that they are able to traverse the inner membrane independently of the T9SS (Seers et al., 2006) and that the T9SS is only necessary for secretion across the outer membrane. Once in the periplasm, the protein substrates may be held in an intermediate form, although whether they are fully folded or protected by a chaperone specific for CTD domains or a general periplasmic mechanism is not known at this point. The next stages are unclear, for example, it might be that the T9SS substrate might engage PorT, a protein thought to reside either on the periplasmic face of the inner membrane or embedded in the outer membrane, and whose function is not fully understood but is assumed to be involved in CTD protein maturation (Chen et al., 2011). In the periplasm, RgpB is also processed into its mature forms by the removal of the Prodomains (Glew et al., 2012; Zhou, Gao, Hunter, Potempa, & Nguyen, 2013), and one assumes is also when gingipains are glycosylated, before translocation across the outer membrane via a mechanism that involves the Sov protein (Saiki & Konishi, 2010). At the cell surface, PorV (also known as lptO) deacylates the LPS lipid A of A-LPS (which is synthesised by PorR, Paramonov et al., 2005; Chen et al., 2011), while in parallel the PorU CTD-specific peptidase (PG0027) cleaves the CTD from the secreted protein (Glew et al., 2012), which is then conjugated to A-LPS possibly in a process which is not understood but may be coupled to the interaction of these two proteins (Curtis et al., 1999; Saiki & Konishi,

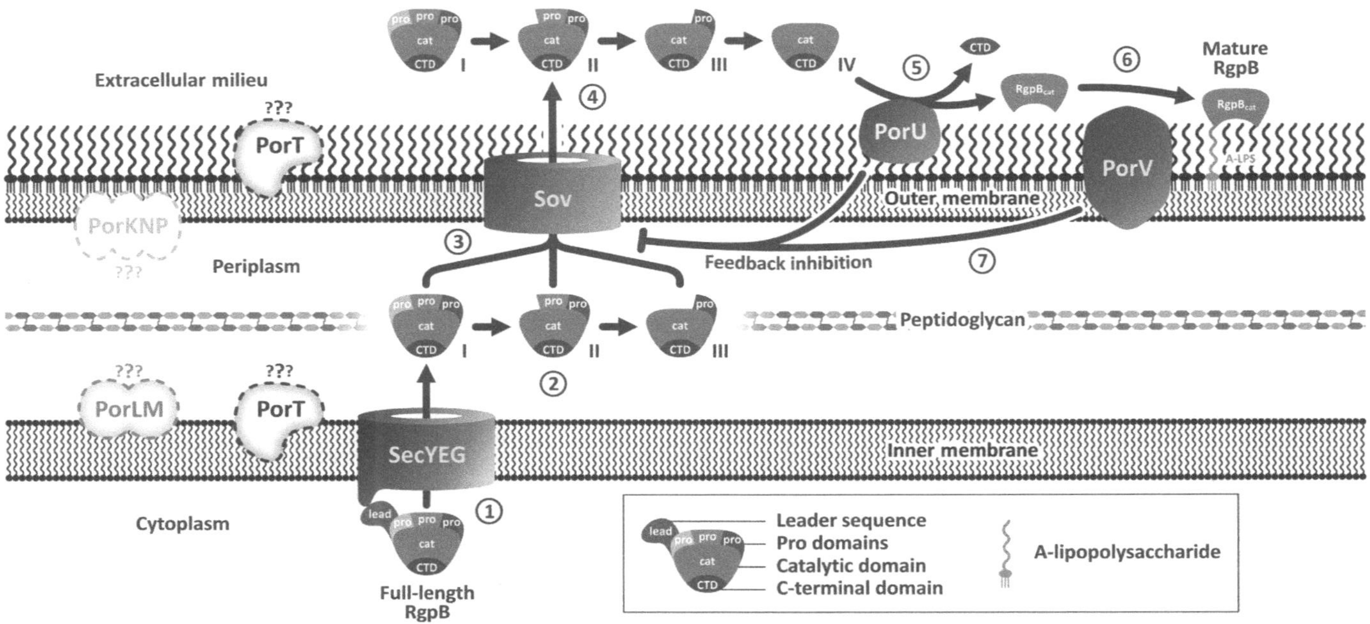

Figure 5 Summary of T9SS using RgpB gingipain as an exemplar. After secretion of RgpB across the inner membrane in a sec-dependent manner (1), its leader sequence is removed and it is processed via removal of prodomains both in the periplasm and on the cell surface (2). While PorT is key to the process of secretion with a postulated role in CTD protein maturation and its role is not clear at present since it has been found in the inner and outer membranes (3). These events are followed by translocation across the outer membrane in a process that involves the predicted Sov porin protein (4). These isoforms are then further processed to remove their prodomains and their CTD (C-terminal domain) by PorU (5) before attachment to A-LPS (A-lipopolysaccharide) (6) via PorV (LptO). There is also evidence that disruption of either CTD secretion and cleavage or A-LPS conjugation causes feedback inhibition on the whole system (7). Other T9SS proteins which are thought to be involved in the process in an undefined manner are also shown with dotted lines. Further details can be found in the text.

2014; Slakeski et al., 2011). The conjugation of the CTD proteins connected to the A-LPS creates an electron dense surface layer (EDSL) which is not present in mutants lacking T9SS components (Glew et al., 2012), or if CTD proteins proposed to make up the EDSL such as the gingipains are deleted (Chen et al., 2011). Knowledge of the mechanism of how this complex system functions to both secrete and attach substrates to LPS is still in its infancy with several perplexing aspects to the system, not least the fact that modification seems to depend on sequence-independent factors and that some substrates seem not to be modified or attached to A-LPS (e.g. PorV), suggesting that different substrate classes may exist. In addition, the observation that blockage of either A-LPS synthesis or CTD secretion genetically inhibits the other system suggests an intimate feedback between the systems that has not been elucidated thus far.

What is clear, however, is that the T9SS system is crucial in the virulence of oral bacteria such as *P. gingivalis* and *T. forsythia* as various virulence factors are secreted through the T9SS system. As previously mentioned, the gingipains (RgpA, RgpB and Kgp) are all secreted through T9SS, alongside the agglutinin proteins (HAs). Together, these proteins act as an iron acquisition complex, with the HAs binding to red blood cells, allowing the gingipains to digest surface proteins, releasing the haemoglobin (Shoji et al., 2002). Further digestion would release the iron and other nutrients required for *P. gingivalis* growth. Surplus iron may be stored on the surface in μ-oxo dimers (Smalley, Thomas, Birss, Withnall, & Silver, 2004), resulting in the black pigmentation that is observed in wild-type *P. gingivalis*. Gingipains are also essential for other aspects of the virulence of *P. gingivalis* as they are involved in the inactivation of particular host response systems, such as cytokines as well as the acquisition of other nutritional compounds (Guo et al., 2010; Murakami et al., 2004).

In addition to the gingipains, the other major virulence proteins known or surmised to be reliant on the T9SS are the BspA and surface (S-) layer proteins TfsA and TfsB of *T. forsythia* (Sharma, 2010). The TfsAB proteins make up a crystalline array on the surface of *T. forsythia* that is thought to be key for the virulence by functioning as a protective coating, facilitating cell adhesion and manipulating the immune system (Posch et al., 2011; Sekot et al., 2012; Settem, Honma, Nakajima, et al., 2013). The S-layer of *T. forsythia* is unique, being made up of two high molecular weight CTD-containing glycoproteins, whose theoretical molecular mass is much smaller than their mass estimated when subject to 2D SDS–PAGE, mainly due to the presence of a complex glycan made up of mannosamine, fucose and pseudaminic acid that was elucidated by Posch et al. (2011). The secretion of these two proteins which predominantly make up the S-layer is likely to be dependent on the CTD secretion system (Posch et al., 2012; Veith et al., 2009). However, the extensive glycosylation of the S-layer proteins is known to occur through an analogous system to the general glycosylation pathway of gut *Bacteroidetes* such as *B. fragilis* (Posch et al., 2013), in which its glycans seem to be added intracellularly in the periplasm before secretion to the outer surface of the cell. In keeping with this idea, a mutant that alters the glycan of the S-layer proteins of *T. forsythia* (lacking the *wecC* gene) produces an altered glycan but still seems to secrete the S-layer protein to the surface of the cell (Settem, Honma, Nakajima, et al., 2013).

Taken together, these data seem to indicate that the T9SS system might have evolved to secrete highly glycosylated proteins but that the T9SS itself is not responsible for addition of all of the extensive glycans present on many of its substrates. For example, in addition to the S-layer proteins of *T. forsythia*, the gingipains of *P. gingivalis* are modified with a range of sugars, not all of which have been associated with either A-LPS or the T9SS system, i.e., fucose, rhamnose, sialic acid and *N*-acetylglucosamine (Curtis et al., 1999), and indeed it would be surprising if the T9SS managed all of these modifications. However, this is one of many open questions surrounding the T9SS system, including whether we can find a way to inhibit its function as a means to aid treatment and prevention of periodontal disease. It is also perplexing why bacteria seem to utilise the T9SS system for two seemingly distant purposes, namely, the construction of a functional gliding motility system in *Flavobacterium* spp. and the secretion of surface attached virulence factors in human pathogens. One answer may be that evolution has co-opted these systems for both purposes in much the same way that the flagella system of eubacteria uses a specialist Type 3 secretion system to assemble its flagella while a very similar Type 3 secretion system is also used to inject toxic virulence proteins into human and plant hosts of bacteria (Erhardt, Namba, Hughes, & Injectisome, 2010). It is also of note that there seems to be little role for Type 3 secretion in virulence factor secretion of oral bacteria. What will be of interest is to see how widespread the T9SS proves to be and whether it has been co-opted for any other functions.

5. SURFACE ADHESINS AS COLONISATION FACTORS OF ORAL BACTERIA

In common with those from other environments in the body, oral bacteria rely on their ability to adhere to host surfaces either directly or via components of the extracellular milieu that coat surfaces while also interacting with other coinhabiting bacteria, in order to facilitate their colonisation of the oral cavity. In addition, colonisation of mucosal surfaces is closely linked to promoting interactions with human cell receptors on epithelial and other cell types.

The nature of these adhesive molecules varies with nonspecific factors such as lipids, smaller carbohydrates and extracellular DNA acting to facilitate microbial adhesion in some circumstances due to the physicochemical properties that they confer on microbial outer surfaces, such as hydrophobicity and net charge (Flemming et al., 2007; Sumei et al., 2014; Whitchurch, Tolker-Nielsen, Ragas, & Mattick, 2002). In contrast, proteinaceous adhesins tend to facilitate adhesion via specific interactions with coated host surfaces, matrix proteins, or cellular receptors. These modes of attachment are thus key not only to both plaque biofilm formation but also to the initial stages of interactions with epithelial cells that are the first stage in events such as cellular invasion and innate immune responses that are also key to how oral bacterial survive and prosper in the oral environment.

In this section, we will review the specific types of interaction by proteinaceous adhesins in the context of the differing surfaces that they encounter.

5.1. Attachment to hard tissues

At the tooth surface, enamel or hydroxyapatite is rapidly coated with salivary glycoprotein (Section 2.1) enabling attachment of early-coloniser species, among which *S. gordonii*, *Streptococcus sanguinis* predominate. Streptococci also colonise soft tissue surfaces, including the periodontium, lingual and buccal epithelium, partly because these surfaces are also subject to pellicle formation. *S. gordonii* utilises Hsa, SspA and SspB during adhesion to the pellicle (Table 1), with *S. sanguinis* utilising pili composed of three components—PilA, PilB and PilC (Table 1)—which has been shown to bind salivary amylase (Okahashi et al., 2011).

S. mutans, considered a primary etiological agent of caries, utilises a protein homologue of the *S. gordonii* proteins SspA and SspB called SpaP (also termed Pac, Table 1) to bind salivary glycoproteins enabling its attachment to the tooth surface. These proteins all belong to the AgI/II family of adhesins present in the majority of oral streptococci and have been shown to be important in adherence to SAG, a large oligomeric protein complex composed of the glycoprotein gp340, secretory IgA and other proteins (Jakubovics, Strömberg, et al., 2005; Loimaranta et al., 2009). The Ag

I/II primary structure possesses distinct regions based on sequence analysis: first, a sec-dependent signal peptide, followed by an alanine-rich repeat (A-) domain that precedes a variable domain that is presented at the cell-distal end of the protein and a proline-rich (P-) domain with which the A-domain interacts intimately to form a twisted fibrillar structure (Larson et al., 2011, 2010). Structure-based approaches have highlighted similarities between the V-domains of *S. mutans* SpaP and *S. gordonii* SspB adhesins, as well as lectin-like folds suggesting putative carbohydrate-binding functions (Forsgren, Lamont, & Persson, 2009; Troffer-Charlier, Ogier, Moras, & Cavarelli, 2002). Indeed, the Hsa adhesin of *S. gordonii* binds cellular sialic acid (Jakubovics et al., 2009). At the C-terminal end of the protein lies the C-domain and a cell wall anchoring (CWA) region, containing an LP-X-TG amino acid motif, required for conjugation to the cell wall via sortase activity. The C-domain is also responsible for ligand binding, for example, the SspB C-domain binds the minor fimbriae (Mfa) of *P. gingivalis* (Brooks et al., 1997). Structural conservation of this domain of the AgI/II proteins exists between SspB and SpaP of *S. mutans* and the AspA adhesin of *S. pyogenes* but also with collagen-binding proteins such as Cna from *S. aureus*, indicating the formation of a domain evolved for protein–protein interactions (Forsgren, Lamont, & Persson, 2010; Hall, Nylander, Jenkinson, & Persson, 2014). In addition, it seems that these fibrillar structures are stabilised by covalent bond formation in the form of isopeptide bond triads (Forsgren et al., 2010).

Another important coloniser of host surfaces are *Actinomyces* species which are known to utilise Type I and Type II pili for a variety of adhesive functions. These are encoded by the (*fimQ-fimP-srtC1*) and (*fimB-fimA-srtC2*) gene clusters, respectively, and are key to attachment to salivary-coated surfaces (Mishra et al., 2007). The Type 1 fimbriae have been shown to mediate adherence of *Actinomyces oris* (formerly *naeslundii*) to proline-rich proteins (PRPs) and statherin found in the salivary pellicle (Gibbons, Hay, Cisar, & Clark, 1988; Yeung, Chassy, & Cisar, 1987).

5.2. Interbacterial attachments

In addition to the importance of adhesion and interaction with host surfaces, it is well known that interactions with other members of the oral microbial community are key to biofilm formation with various mechanisms including well-reviewed quorum sensing based phenomena being well documented (Jakubovics, 2010; Kolenbrander et al., 2006; Wright et al., 2013). As a result here, we will largely focus on the role of surface properties in the physical interactions of bacteria with each other.

In addition to the SspB adhesin described above, streptococci also possess other adhesins capable of binding microbes. For example, *S. gordonii* utilises two distinct fibril-forming proteins, consisting of polymerised CshA or CshB working alongside SspB (Table 1) to mediate attachment to a variety of microorganisms including *A. oris* and *Candida albicans* (Holmes et al., 1996; Jakubovics, Kerrigan, et al., 2005; McNab et al., 1996; Silverman, Nobbs, Vickerman, Barbour, & Jenkinson, 2010). It is however of note that the bacterial and human protein interacting domains appear to be distinct, indicating that the organisms may be able to bind both simultaneously if necessary. *Actinomyces* itself is also able to mediate binding to other species via its Type 2 fimbriae, which is composed of the fimbrial shaft protein FimA and a tip protein, either FimB or coaggregation factor A (CafA) (Mishra et al., 2011; Reardon-Robinson et al., 2014). These Type 2 fimbriae bind to the disaccharide motifs GalNacβ1-3Gal and GalNacβ1-3GalNac, most likely via the tip protein CafA (Mishra et al., 2010; Reardon-Robinson et al., 2014). These disaccharide motifs are found in the receptor polysaccharides of streptococci such as *S. oralis*, as well as being part of host factors (see below), allowing *Actinomyces* to coaggregate with *S. oralis* (Cisar, Sandberg, Abeygunawardana, Reddy, & Bush, 1995).

While early colonisers often target their adhesins at human matrix and salivary proteins, many later colonisers utilise the surface antigens and biofilm matrix components of other bacteria as potential surfaces for attachment, a process that is involved in the formation of pathogenic complexes (as discussed in Section 3.4). Of the intermediate colonisers, the periodontal pathogens *Fusobacterium nucleatum* and *A. actinomycetemcomitans* have undergone extensive research with *Fusobacteria* spp. considered efficient bridging organisms in plaque biofilms due to their adhesive nature. *F. nucleatum* is composed of five distinct subspecies, with *F. nucleatum* subsp. *nucleatum* (*F. nucleatum nucleatum*) and subsp. *polymorphum* (*F. nucleatum polymorphum*) receiving the bulk of research attention. *F. nucleatum nucleatum* adhesion has been shown to rely on lactose-dependent adhesins during coaggregation with a variety of Gram-negative oral species (and also host cells) (Kolenbrander & Andersen, 1989). Membrane capture approaches and analysis of 16S ribosomal DNA have also been used to highlight the ability of *F. nucleatum nucleatum* to adhere to a number of different species enriched from saliva, including *Gemella*, *Granulicatella*, *Neisseria* and *Peptostreptococcus* species, as well as nonculturable unclassified organisms

(Wang, He, et al., 2011). The OMP FomA is thought to mediate binding to many of these organisms, a theory first highlighted by abrogation of *P. gingivalis–F. nucleatum* aggregation using anti-FomA antibodies (Kinder & Holt, 1993). The importance of FomA during integration into the bacterial community and in disease has been further confirmed in FomA vaccine studies, where antibodies produced against FomA resulted in decreased bacterial coaggregation and abscess formation in a murine model (Liu et al., 2010). FomA has also been shown to bind statherin-derived peptide, a component of the pellicle (and as a result, oral biofilm matrices), providing another mechanism by which FomA enables Fusobacterial integration into the oral microbiota.

In contrast to *F. nucleatum nucleatum*, *F. nucleatum polymorphum* utilises an arginine-sensitive mechanism for coaggregation: Arginine based pull-down assays and mass spectrometry identified candidate proteins potentially responsible for *F. nucleatum* binding to Gram-positive early colonisers (Edwards, Grossman, & Rudney, 2007). Construction of mutants deficient in one of these candidates followed by coaggregation studies revealed that the OMP RadD was largely responsible for aggregation with several early-coloniser streptococci (Kaplan et al., 2009). Thus, RadD can be considered key to *F. nucleatum polymorphum* initial attachment and integration into the oral microbiota. Alongside these surface adhesins, *F. nucleatum* has also been shown to respond to sensing pathways via Autoinducer (AI)-2 which was shown to induce biofilm formation in mono-and dual-species biofilms of *F. nucleatum*, *P. gingivalis* and *T. forsythia in vitro*, an effect abrogated by quorum sensing inhibitors (Jang, Choi, Lee, Jun, & Choi, 2013). Given the nature of *F. nucleatum* as a pathogenic bridging organism, enabling other pathogens to colonise the periodontium, the ability to reduce its biofilm and complex formation with other pathogens represents an exciting prospect for periodontal disease therapy.

Another well-known member of the oral biofilm is the aptly named *A. actinomycetemcomitans* which integrates into the oral microbiota by attachment using its Flp pili, a polymer of Flp-1 and Flp-2 proteins, part of the *tad* (tight adherence) operon (Clock et al., 2008; Inoue et al., 1998). These pili are instrumental in adhesion to early colonisers, but are also considered non-specific adhesins; when cultured, strains expressing a complete tad operon adhere to liquid culture vessels or form large aggregates, hence this organism's name and which reform rapidly after disruption by agitation, while strains deficient in the *tad* operon do not display such strong, rapid adhesion (Kachlany et al., 2000). It is therefore likely that these pili enable attachment to a wide variety of surfaces, including host and bacterial cell membranes, and components of the ECM.

The late colonisers are often associated with severe disease states, as seen in the periodontal pathogens *T. forsythia*, *P. gingivalis* and *T. denticola*, together forming the red-complex (see Section 3.4). All three of these organisms are able to attach to intermediate and early colonisers, as well as each other (reviewed in Rickard, Gilbert, High, Kolenbrander, & Handley, 2003; Suzuki, Yoneda, & Hirofuji, 2013), facilitating their integration into the oral microbiota. These organisms are also able to interact with the host via a number of processes mediated by attachment to components of host cell membranes (see Section 5.3). *P. gingivalis* possesses two fimbriae; major and minor, also termed long and short, which are composed primarily of multimeric subunits FimA and MfaI, respectively (ref). Major fimbriae are considered important for attachment as they have been shown to bind salivary proteins and epithelial cells (see below), as well as surface GAPDH of streptococci (Maeda et al., 2004). Minor fimbriae are considered more important for adhesion to other bacteria as they are known to play a role in attachment to *S. gordonii* through interaction with the SspB AgI/II adhesin (Lamont et al., 2002; Park et al., 2005). Polymeric Mfa1 forms the bulk of the fibrillar structure of minor fimbriae (Kuboniwa et al., 2009); however, various accessory proteins are associated with Mfa1:Mfa2 has been shown to associate with Mfa1 at the cell membrane, suggesting its role as an 'anchor' for minor fimbriae, and mutant strains deficient in Mfa2 produced much longer filaments, highlighting its role in regulation of fimbrial length (Hasegawa et al., 2009). A third protein, Mfa3, has recently been shown to localise at the tip of the minor fimbriae, and is important during biofilm formation, with *Δmfa3*, *ΔfimA* mutants displaying similar biofilm phenotypes to *Δmfa1*, *ΔfimA* mutants (Y. Hasegawa et al., 2013).

T. forsythia does not possess any visible pili-like appendages in TEM or contain putative pilus or fimbrial genes in its genome sequence. However, it does possess the major surface antigen and adhesin BspA (Sharma et al., 1998) that is involved in coaggregation with *F. nucleatum*, *T. denticola* and *P. gingivalis*, attachment to oral epithelial cells (Mishima & Sharma, 2011), as well as immune system modulation (Myneni et al., 2012). BspA consists of four domains: The N-terminal, leucine-rich repeats (LRRs), bacterial Ig-like domains, and the CTD. In the context of interspecies attachment, *T. forsythia* mutants deficient in BspA show a twofold decrease in aggregation with *F. nucleatum* compared to their parent strain, but do not show a significant decrease in biofilm formation *per se* (Sharma, Inagaki, Sigurdson, & Kuramitsu, 2005). The LRRs of BspA have been shown to bind to gp340 (see Section 5.1) (Loimaranta et al., 2009), which not only is part of SAG but also forms part of the biofilm matrix. BspA is also involved during mixed species infection of epithelial cells with *T. forsythia* as shown by BspA mutants displaying decreased attachment and invasion of oral epithelial cells in both the presence and absence of *P. gingivalis* (or *P. gingivalis* outer membrane vesicles) compared to wild-type *T. forsythia* (Inagaki, Onishi, Kuramitsu, & Sharma, 2006). However, the study did not show decreased coaggregation of BspA mutants with *P. gingivalis*, in contrast to the coaggregation studies with *F. nucleatum*. Taken together, these studies point to a mechanism by which *T. forsythia* coaggregates with *F. nucleatum*, allowing initial colonisation, which is followed by association with epithelial cells through a BspA-dependent pathway that acts synergistically with *P. gingivalis*, or *P. gingivalis* outer membrane vesicles (Inagaki et al., 2006). Further research into how bacteria cooperate to promote host cell association and biofilm formation will hopefully yield promising targets for therapeutic purposes.

Of the red-complex bacteria arguably the least known about is *T. denticola*. Despite this, the recently identified *T. denticola* major membrane protein TDE2508 has been shown to play a role in biofilm formation and adherence to host epithelial cells (mentioned later). A *T. denticola tde2508* deletion mutant strain was found to have dramatically increased biofilm formation, suggesting that TDE2508 may play a role in the regulation of this process although nothing is known of any potential mechanism involved (Abiko, Nagano, Yoshida, & Yoshimura, 2014). Other surface proteins that may play a role in interbacterial interactions of *T. denticola* include LrrA, a surface-associated LRR protein that seems to influence interactions with human cells, influences motility, and mediates interaction with its fellow red-complex periodontal pathogen *T. forsythia* with which the LrrA N-terminal region interacts via BspA of *T. forsythia* (Ikegami et al., 2004). *T. denticola* also seems to possess the means to interact with *P. gingivalis* fimbriae via interaction of its chymotrypsin-like protease (CTLP), also known as dentilisin, suggesting a role for CTLP in bacterial coaggregation (Hashimoto et al., 2003). Overall the accumulated data further support the intimate relationship between oral bacteria and indicate that the complexes associated with disease represent physical groupings, an idea supported by *in vivo* staining of plaque biofilms (Zijnge et al., 2010).

5.3. Attachment to host cell surfaces

Early colonisers may attach to soft tissue surfaces due to the salivary pellicle, but they also possess adhesins that mediate direct binding to host cell surfaces. *S. sanguis* pili (composed of PilA, PilB and PilC subunits) are one such adhesin, which bind not only salivary amylase but also fibronectin, which is present in soluble and a cellular integrin (α5β1) associated form that forms an adhesion target for host–pathogen interactions (Okahashi et al., 2010, 2011). Indeed, pili-deficient mutant strains display a nearly twofold reduction in attachment to oral epithelial cells (Okahashi et al., 2010). In addition, CshA and CshB adhesins of *S. gordonii* also bind fibronectin (McNab et al., 1996), potentially enabling their attachment to host cells, and given that *S. sanguinis* possesses homologues of *cshA* and *cshB* genes, this would explain why attachment to epithelial cells was not completely abrogated in pili-deficient strains. This targeting of fibronectin is also employed by the Hsa adhesin of *S. gordonii* which binds fibronectin via its sialic acid glycan groups (Jakubovics et al., 2009). This targeting of glycans is also employed by the fellow early-colonising *Actinomyces* which bind to the disaccharide motifs GalNacβ1-3Gal and GalNacβ1-3GalNac as mentioned previously, and that are exposed by the removal of terminal of sialic acid residues on many cell types including epithelial and immune cells (Mishra et al., 2011; Ruhl, Cisar, & Sandberg, 2000).

In keeping with the role of pili/fimbrial adhesins, *P. gingivalis* targets the fibronectin-binding integrin α5β1 via its major adhesin, the FimA major fimbriae, for binding and invasion of oral epithelial cells and osteoclasts (Yilmaz, Watanabe, & Lamont, 2002; Zhang, Ju, Rigney, & Tribble, 2013). This interaction and the subsequent invasion of cells is considered a key process in the pathogenesis of periodontal pathogens, both in terms of the disease and potentially as a means of immune evasion and persistence in the oral cavity (Rudney, Chen, & Sedgewick, 2001). In addition, it has been shown that the fimbrial accessory proteins FimC and FimD allow binding of *P. gingivalis* to fibronectin directly (alongside collagen) (Pierce et al., 2009). The importance of the major fimbriae is highlighted by the fact that *P. gingivoalis* strains possess multiple variants of the fimA gene in the population, with variations coding for slightly different proteins. Based on variation of *fimA* genes, the major fimbriae have been classified into six types, types I–V and Ib (Amano, Nakagawa, Okahashi, & Hamada, 2004). Strains expressing Type II major fimbriae are associated with more severe periodontitis compared to strains expressing other types and were found to have greater attachment and invasiveness during infection of host cells (Amano et al., 2004). Meanwhile, in periodontally healthy patients positive for *P. gingivalis* colonisation, type I fimbriae strains were the most prevalent (Amano et al., 2004), inferring that variation in major fimbriae influences the outcome of *P. gingivalis* colonisation but might also define pathotypes. In addition to the major fimbriae, *P. gingivalis* possess a minor fimbria (*mfa*) that may also play a role in host cell attachment, as evidenced by mutant strains of *P. gingivalis* displaying reduced cellular invasion abilities of oral epithelial cells (Umemoto & Hamada, 2003).

The use of surface adhesins of filamentous nature is a common theme in oral bacteria and while the presence of surface appendages on the surface of the bridging periodontal bacterium *F. nucleatum* has not been confirmed it produces a unique surface adhesin called FadA that forms a long fibrous protein *in vitro* (Nithianantham et al., 2009; Témoin, Wu, Wu, Shoham, & Han, 2012). This adhesin has been shown to mediate interactions with not only human oral epithelial cells (Han et al., 2005) but also endothelial and placental cells (Fardini et al., 2011; Ikegami, Chung, & Han, 2009; Xu et al., 2007). This interaction is known to be defined by adhesion to cellular cadherins, which increases permeability of monolayers of endothelial cells via loosening of intercellular junctions (Fardini et al., 2011) but is also thought to promote induction of colorectal cancer in humans (Rubinstein et al., 2013).

Like *F. nucleatum*, *A. actinomycetemcomitans* can also attach directly to epithelial cells via a surface adhesin, in this case the autotransporter family Aae protein (Rose et al., 2003). The host receptor for Aae is currently unknown, though the adhesin can attach to buccal cells, as well as gingival fibroblasts, but not pharyngeal, palatal, tongue, bronchial, or cervical epithelial cells (Fine et al., 2005). This might imply a receptor expressed only in the periodontium (and buccal cells). It is tempting to speculate that the tissue and host specificity of Aae highlight this protein as a niche adaptation, and possibly a therapeutic target, and indeed a recent study revealed that segments of the protein that interact with human cells can be used to design blocking peptides with therapeutic potential (Fine et al., 2010). *A. actinomycetemcomitans* also possesses another surface adhesin, EmaA (Mintz, 2004), which binds collagen (type V) via presentation as part of a fibrillar protein on its surface. EmaA structure and function are also related to their glycosylation via pathways shared with LPS glycan biosynthesis and attachment (Tang & Mintz, 2010; Tang, Ruiz, & Mintz, 2012).

The attachment of glycans to bacterial surface proteins has only recently come to prominence but is being shown to be both widespread and crucial to function of surface proteins from flagella to S-layers (Settem, Honma, Nakajima, et al., 2013). It is therefore no surprise that the heavily glycosylated S-layer proteins (TfsAB) of *T. forsythia* play a role in host cell attachment as well as the immune modulation reviewed elsewhere (Section 4.2) (Settem, Honma, Nakajima, et al., 2013). Mutant strains lacking one or both Tfs proteins have shown reduced adherence or invasion of oral epithelial cells (Sakakibara et al., 2007). Furthermore, S-layer-deficient mutants also show decreased hemagglutination compared to wild type. Given that bleeding commonly occurs in periodontal pockets, attachment to red blood cells could be a useful mechanism of adherence allowing *T. forsythia* to colonise, or do so more extensively. Hemagglutination by *T. forsythia* appears to be sensitive to sialyllactose (Murakami et al., 2002), again highlighting the importance of host sialoglyconjugates in *T. forsythia* attachment as highlighted elsewhere in this chapter. In addition to host cell binding, S-layer-deficient *T. forsythia* mutants show decreased coaggregation with the early-coloniser *S. sanguinis* compared to wild-type *T. forsythia*, indicating a role in interbacterial interactions also (Shimotahira et al., 2013).

As mentioned above, many oral bacterial target collagen, including *S. mutans* via *cnm* and *cbm* (Nakano et al., 2010; Nomura et al., 2012) while CbdA of *S. gordonii* (Moses et al., 2013) performs a similar function. In keeping with this idea the Major Surface protein (Msp) porin of the periodontal pathogen *T. denticola* has been shown to bind a range of human extracellular matrix proteins including collagen I, laminin and fibrinogen (Edwards et al., 2005; Fenno et al., 1997). However, Msp is not the only surface factor of *T. denticola* with potential roles in adhesion to host surfaces. Its lipooligosaccharide is also known to mediate binding to epithelial cells, but also laminin, fibronectin and collagen (Grenier, 2013). The LrrA protein, mentioned above in the context of interspecies interactions also seems to play a role in epithelial cell interactions, while the reduced motility of an *lrrA* strain also showed abrogation of tissue penetration due to the importance of motility in this process, although this might also be due to a reduced ability to interact with cellular components (Ikegami et al., 2004). This dual role of adhesins in host and interbacterial interactions is also displayed by dentilisin of *T. denticola*, as mentioned previously, which mediates not only *P. gingivalis* interactions but also fibrinogen interactions alongside FhbB that may be involved in colonisation (Bamford et al., 2007; Hellwage, Ku, & Zipfel, 1997; Mcdowell et al., 2005). It is also noteworthy that FhbB binds Factor H and Factor H-like protein I that may also mediate colonisation or immune evasion through as yet unknown mechanisms.

6. STRESS RESPONSES OF IMPORT IN COLONISATION AND INFECTION BY ORAL BACTERIA

As discussed above (Section 2.1), bacteria residing in the oral environment are continually exposed to a range of different stressors. These include changes that might be described as physical challenges, such as pH fluctuations (discussed in Section 3.1), temperature and shear forces. In addition, other stresses include oxidative stresses from exogenous sources such as inflammatory cells but also endogenous radical production within biofilms (Boles & Singh, 2008). The free radicals produced are known to cause damage to oxygen-sensitive enzymes and DNA (Miller & Britigan, 1997). In addition, as with most bacteria, oral organisms are subject to fluctuations in the availability of nutrients and micronutrients such as trace elements including transition metals that are key to the proper functioning of respiratory chains and many catalytic processes (Cabiscol, Tamarit, & Ros, 2000).

For both the aerobic and anaerobic residents of the oral cavity, the production of and resistance to free radicals such as superoxide, hydrogen peroxide and hydroxyl radicals is a key challenge. It is thus no surprise that several of these organisms, much like other bacteria have developed specialised oxidative stress resistance mechanisms as adaptations to life in the oral environment. In addition, several studies have revealed the importance of intact oxidative stress systems to plaque biofilm formation (Henry, McKenzie, Robles, & Fletcher, 2012; Honma, Mishima, Inagaki, & Sharma, 2009).

Indeed, it is within plaque biofilm that some of the sources of oxidative stress originate. For example, the oral commensal *S. sanguinis* is well known to produce hydrogen peroxide in biofilms, in a mechanism which it has been proposed functions to suppress growth of more pathogenic species such as *S. mutans* and represents an example of peer-stress aligned to competitiveness in the oral biofilm (Ryan & Kleinberg, 1995; Zheng et al., 2013). As one might expect, and since organisms like *S. mutans* are still present in health as well as cariogenic environments, organisms that live alongside peroxide producers have evolved resistance strategies to allow them to persist. Along these lines, recent work has identified a glutathione synthetase, GshAB, that plays a role in *S. mutans* resistance to peroxide but also in competing with *S. sanguinis* (Zheng et al., 2013). Further evidence of the importance of these mechanisms in competition in the face of oxidative stress is evidenced by recent microarray studies revealing increases in stress responsive gene expression in microbial competition assays (Liu, Wu, Huang, Merritt, & Qi, 2011). In addition, *S. mutans* contains active oxygen and peroxide detoxifying pathways in the form of superoxide dismutase (*sod*) and Alkyl hydroperoxidase (*ahpCF*) while also containing a ferrous iron detoxifying system (*dpr*) that are known to be key to oxidative stress resistance and are under the control of a number of stress responsive transcriptional pathways while also being influenced by levels of NADH-oxidase and LuxS-type quorum sensing mechanisms (Bitoun, Nguyen, Fan, Burne, & Wen, 2011; Derr et al., 2012; Fujishima et al., 2013; Wen & Burne, 2004; Zhang et al., 2012).

The importance of these kind of detoxification pathways in the oral context is widespread with several oxidative stress resistance pathways upregulated in biofilm across a range of pathogenic species including *T. forsythia* (Pham et al., 2010), *P. gingivalis* (reviewed in Henry et al., 2012) and *F. nucleatum* (Steeves, Potrykus, Barnett, & Bearne, 2011). Among these anaerobic pathogens, the OxyR sensing and regulatory system seems to be key with several groups, characterising its importance in oxidative stress resistance and linking this to their ability to form biofilm (Diaz et al., 2006; Honma et al., 2009). As with *S. mutans*, these systems also seem to be important in mixed species cultures with a proteomic study of model mixed species biofilms of *S. gordonii*, *P. gingivalis* and *F. nucleatum* displaying marked increases in oxidative stress response proteins (Hendrickson et al., 2012).

As described above, one of the main outcomes of persistent exposure to oxidative stress and free radicals is damage to DNA (Boles & Singh, 2008). It is therefore no surprise to see that several oral organisms possess specific systems to combat these effects. The most common repair mechanism involves Base Excision Repair (BER) systems which are responsible for the removal of the mutagenic base 8-oxo-7,8-dihydroxyguanine (8-oxo-G) that are commonly generated by intracellular free radicals (David, O'Shea, & Kundu, 2007). For example, a typical BER system exists in the well studied and genetically diverse *S. mutans* which acts to mediate DNA repair in this organism (Gonzalez, Faustoferri, & Quivey, 2012). Of note is that deletion of this system increases *S. mutans* mutability and DNA damage but conversely seems to improve its short-term ability to resist changes in environmental stressors meaning that this organism employs a trade off between genetic stability and mutability to allow responses to changes in its habitat (Gonzalez et al., 2012). However, BER systems are not the only mode of repair of 8-oxo-G lesions, with the recent discovery of an atypical DNA repair system in *P. gingivalis* a novel example that forms part of a range of mechanisms existing in this organism (Robles, Reid, Roy, & Fletcher, 2011). In this atypical system, a novel zinc-finger containing protein (PG1037) that binds to the mutagenic base 8-oxo-G identified as part of a potential DNA repair operon (Henry, Aruni, Sandberg, & Fletcher, 2013; Henry, Sandberg, Zhang, & Fletcher, 2008). Of further note is that *PG1037* is an essential gene in *P. gingivalis*, and although deletion was not possible, deletion of the two genes coexpressed with *PG1037* resulted in strains more susceptible to DNA damage (Henry et al., 2013; Klein et al., 2012).

In addition to pH (above) and oxidative stressors, bacteria are also exposed to fluctuations in the levels of nutrients with what we might call trace elements key to their lifestyle since many are employed as enzyme cofactors or as key components of their respiratory chains. This is further complicated by the fact that excess intracellular free-ferrous (Fe^{2+}) iron is also a well-known generator of free radicals. In fact, several iron acquisition and storage systems have been characterised in the oral context with most focus being on the well studied periodontal pathogen *P. gingivalis* (reviewed in Lewis, 2010). In *P. gingivalis*, iron is acquired by several means involving gingipain proteases (see above), hemin-binding systems and a FeoB-like ferrous iron uptake system that also plays a role in acquiring Manganese to aid in oxidative stress resistance (Moelling et al., 2007). It is also of note that the FeoB system was altered in expression levels in naturally occurring hyper-invasive subtypes of *P gingivalis*, indicating it might also play a role in iron acquisition or stress resistance inside human cells (Suwannakul, Stafford, Whawell, & Douglas, 2010).

While not usually considered a stressor, we feel that in light of recent studies on the effect of stress hormones on the gut microbiota and in fact feedback from the gut microbiome to our own neural systems, with suggestions that mood can be affected (Cryan & Dinan, 2012), we will briefly review the potential influence of stress hormones on oral bacteria. The gold-standard biological marker of psychological stress is an increase is the hormone cortisol, levels of which are raised when measured in saliva (Bakri, Douglas, & Rawlinson, 2013). It has also been considered for some time that increased stress is a mitigating factor reducing the success of periodontal treatment (Vettore, Quintanilha, Monteiro da Silva, Lamarca, & Leão, 2005). It is therefore of interest that recent studies have identified that cortisol seems to affect the growth of a range of periodontal pathogens *in vitro*

(Akcalı et al., 2014; Jentsch, März, & Krüger, 2013), while other hormones such as adrenaline seem to increase expression of several potential virulence factors (Graziano et al., 2013). While these data are somewhat limited and do not all agree with each other, it would be of great interest to discern if the oral microbiota does alter in response to long-term psychological stress on the subject.

7. SUMMARY AND FUTURE PERSPECTIVES

In summary, it is clear from the information gathered in this chapter that oral bacteria are well adapted to their environmental niches and that we are continually gaining more insight into the mechanisms at play in their persistent colonisation of the mouth. As this level of mechanistic and molecular level knowledge grows, we are entering an era where we can generate novel treatments to improve oral health. For example, increased structural knowledge of important potential virulence factors, such as the novel protease karilysin of *T. forsythia* or streptococcal adhesins should allow rational design of inhibitory compounds or peptides that might produce novel drugs to interfere with colonisation or inhibit cellular damage (Guevara et al., 2013). These ideas extend to the exciting work going on the T9SS, increased knowledge of which might allow the design of small peptide or screening of natural product library inhibitors that might allow inhibition of secretion of gingipains and other virulence factors, much in the way that small peptides have been designed to inhibit Type 3 secretion of Enterohemorrhagic *E. coli* (EHEC) (Larzábal et al., 2010; Zetterström et al., 2013). In keeping with this idea, the targeting of the interactions of oral bacteria with the host glycome also seems to be a promising area for development of new antimicrobials. In addition, our increased knowledge of the molecular basis of the host microbiome and the interactions therein in health as well as disease presents other opportunities for pre- and pro-biotic approaches that hopefully will yield new opportunities to manipulate the composition of our oral microbiota to maintain healthy rather than dysbiotic communities and thus provide knowledge to help prevent as well as address disease. It is therefore undoubtedly an exciting time to be a molecular oral microbiologist and we can expect the next few years to provide further molecular insight in this area and open new avenues for treatment of oral diseases.

ACKNOWLEDGEMENTS

Work in the group is by The Dunhill Medical Trust [Grant number: R185/0211] to G. S. and I. D., a BBSRC Case partnership award to G. S., and University of Sheffield PhD scholarships (T. F. and K. N.).

REFERENCES

Abiko, Y., Nagano, K., Yoshida, Y., & Yoshimura, F. (2014). Major membrane protein TDE2508 regulates adhesive potency in *Treponema denticola*. *PLoS One*, *9*, e89051.

Abusleme, L., Dupuy, A. K., Dutzan, N., Silva, N., Burleson, J. A., Strausbaugh, L. D., et al. (2013). The subgingival microbiome in health and periodontitis and its relationship with community biomass and inflammation. *The ISME Journal*, 7, 1016–1025.

Ajdić, D., McShan, W. M., McLaughlin, R. E., Savić, G., Chang, J., Carson, M. B., et al. (2002). Genome sequence of *Streptococcus mutans* UA159, a cariogenic dental pathogen. *Proceedings of the National Academy of Sciences of the United States of America*, *99*, 14434–14439.

Akcalı, A., Huck, O., Buduneli, N., Davideau, J.-L., Köse, T., & Tenenbaum, H. (2014). Exposure of *Porphyromonas gingivalis* to cortisol increases bacterial growth. *Archives of Oral Biology*, *59*, 30–34.

Altindis, E., Fu, Y., & Mekalanos, J. J. (2014). Proteomic analysis of *Vibrio cholerae* outer membrane vesicles. *Proceedings of the National Academy of Sciences of the United States of America*, *111*, E1548–E1556. http://dx.doi.org/10.1073/pnas.1403683111.

Amano, A., Nakagawa, I., Okahashi, N., & Hamada, N. (2004). Variations of Porphyromonas gingivalis fimbriae in relation to microbial pathogenesis. *Journal of Periodontal Research*, *39*(2), 136–142.

Aruni, A. W., Roy, F., & Fletcher, H. M. (2011). Filifactor alocis has virulence attributes that can enhance its persistence under oxidative stress conditions and mediate invasion of epithelial cells by *Porphyromonas gingivalis*. *Infection and Immunity*, *79*, 3872–3886.

Aruni, W., Vanterpool, E., Osbourne, D., Roy, F., Muthiah, A., Dou, Y., et al. (2011). Sialidase and sialoglycoproteases can modulate virulence in *Porphyromonas gingivalis*. *Infection and Immunity*, *79*, 2779–2791.

Bakri, I., Douglas, C. W. I., & Rawlinson, A. (2013). The effects of stress on periodontal treatment: A longitudinal investigation using clinical and biological markers. *Journal of Clinical Periodontology*, *40*, 955–961.

Baliban, R. C., Sakellari, D., Li, Z., DiMaggio, P. A., Garcia, B. A., & Floudas, C. A. (2012). Novel protein identification methods for biomarker discovery via a proteomic analysis of periodontally healthy and diseased gingival crevicular fluid samples. *Journal of Clinical Periodontology*, *39*, 203–212.

Bamford, C. V., Fenno, J. C., Jenkinson, H. F., & Dymock, D. (2007). The chymotrypsin-like protease complex of *Treponema denticola* ATCC 35405 mediates fibrinogen adherence and degradation. *Infection and Immunity*, *75*, 4364–4372.

Beall, C. J., Campbell, A. G., Dayeh, D. M., Griffen, A. L., Podar, M., & Leys, E. J. (2014). Single cell genomics of uncultured, health-associated *Tannerella* BU063 (Oral Taxon 286) and comparison to the closely related pathogen *Tannerella forsythia*. *PLoS One*, *9*, e89398.

Becker, D. J., & Lowe, J. B. (2003). Fucose: Biosynthesis and biological function in mammals. *Glycobiology*, *13*, 41R–53R.

Becker, M. R., Paster, B. J., Leys, E. J., Moeschberger, M. L., Kenyon, S. G., Galvin, J. L., et al. (2002). Molecular analysis of bacterial species associated with childhood caries. *Journal of Clinical Microbiology*, *40*, 1001–1009.

Beighton, D. (2005). The complex oral microflora of high-risk individuals and groups and its role in the caries process. *Community Dentistry and Oral Epidemiology, 33*, 248–255.

Beighton, D., Radford, J. R., & Naylor, M. N. (1992). Glycosidase activities in gingival crevicular fluid in subjects with adult periodontitis or gingivitis. *Archives of Oral Biology, 37*, 343–348.

Bensing, B. A., Rubens, C. E., & Sullam, P. M. (2001). Genetic loci of *Streptococcus mitis* that mediate binding to human platelets. *Infection and Immunity, 69*, 1373–1380.

Bensing, B. A., & Sullam, P. M. (2002). An accessory sec locus of *Streptococcus gordonii* is required for export of the surface protein GspB and for normal levels of binding to human platelets. *Molecular Microbiology, 44*, 1081–1094.

Bickel, M., & Cimasoni, G. (1985). The pH of human crevicular fluid measured by a new microanalytical technique. *Journal of Periodontal Research, 20*, 35–40.

Bitoun, J. P., Nguyen, A. H., Fan, Y., Burne, R. A., & Wen, Z. T. (2011). Transcriptional repressor Rex is involved in regulation of oxidative stress response and biofilm formation by *Streptococcus mutans*. *FEMS Microbiology Letters, 320*, 110–117.

Boles, B. R., & Singh, P. K. (2008). Endogenous oxidative stress produces diversity and adaptability in biofilm communities. *Proceedings of the National Academy of Sciences of the United States of America, 105*, 12503–12508.

Boraston, A. B., Ficko-Blean, E., & Healey, M. (2007). Carbohydrate recognition by a large sialidase toxin from Clostridium perfringens. *Biochemistry, 46*, 11352–11360.

Borgnakke, W. S., Ylöstalo, P. V., Taylor, G. W., & Genco, R. J. (2013). Effect of periodontal disease on diabetes: systematic review of epidemiologic observational evidence. *Journal of Clinical Periodontology, 40*(Suppl. 1), S135–S152.

Bowen, W. H., & Koo, H. (2011). Biology of *Streptococcus mutans*-derived glucosyltransferases: Role in extracellular matrix formation of cariogenic biofilms. *Caries Research, 45*, 69–86.

Bradshaw, D. J., & Marsh, P. D. (1998). Analysis of pH-driven disruption of oral microbial communities in vitro. *Caries Research, 32*, 456–462.

Braham, P. H., & Moncla, B. J. (1992). Rapid presumptive identification and further characterization of *Bacteroides forsythus*. *Journal of Clinical Microbiology, 30*, 649–654.

Brockhausen, I., Schachter, H., & Stanley, P. (2009). O-GalNAc glycans. In A. Varki, R. Cummings, & J. Esko (Eds.), *Essentials in glycobiology* (2nd ed.). Cold Spring Harbor (NY): Cold Spring Harbor Laboratory Press.

Brooks, W., Demuth, D. R., Gil, S., & Lamont, R. J. (1997). Identification of a *Streptococcus gordonii* SspB domain that mediates adhesion to *Porphyromonas gingivalis*. *Infection and Immunity, 65*, 3753–3758.

Brown, L. J., Johns, A. B., & Wall, T. P. (2002). The economics of periodontal. *Periodontology, 2000*(29), 223–234.

Burne, R. A., Zeng, L., Ahn, S. J., Palmer, S. R., Liu, Y., Lefebure, T., et al. (2012). Progress dissecting the oral microbiome in caries and health. *Advances in Dental Research, 24*, 77–80.

Buschiazzo, A., & Alzari, P. M. (2008). Structural insights into sialic acid enzymology. *Current Opinion in Chemical Biology, 12*, 565–572.

Byers, H. L., Homer, K. A., & Beighton, D. (1996). Utilization of sialic acid by viridans streptococci. *Journal of Dental Research, 75*, 1564–1571.

Byers, H. L., Homer, K. A., Tarelli, E., & Beighton, D. (1999). N-acetylneuraminic acid transport by *Streptococcus oralis* strain AR3. *Journal of Medical Microbiology, 48*, 375–381.

Byers, H. L., Tarelli, E., Homer, K. A., & Beighton, D. (1999). Sequential deglycosylation and utilization of the N-linked, complex-type glycans of human alpha1-acid glycoprotein mediates growth of *Streptococcus oralis*. *Glycobiology, 9*, 469–479.

Byers, H. L., Tarelli, E., Homer, K. A., & Beighton, D. (2000). Isolation and characterisation of sialidase from a strain of *Streptococcus oralis*. *Journal of Medical Microbiology, 49*, 235–244.

Byers, H. L., Tarelli, E., Homer, K. A., Hambley, H., & Beighton, D. (1999). Growth of Viridans streptococci on human serum 1-acid glycoprotein. *Journal of Dental Research, 78*, 1370–1380.

Byrne, D. P., Potempa, J., Olczak, T., & Smalley, J. W. (2013). Evidence of mutualism between two periodontal pathogens: Co-operative haem acquisition by the HmuY haemophore of *Porphyromonas gingivalis* and the cysteine protease interpain A (InpA) of *Prevotella intermedia*. *Molecular Oral Microbiology, 28*, 219–229.

Byrne, D. P., Wawrzonek, K., Jaworska, A., Birss, A. J., Potempa, J., & Smalley, J. W. (2010). Role of the cysteine protease interpain A of *Prevotella intermedia* in breakdown and release of haem from haemoglobin. *The Biochemical Journal, 425*, 257–264.

Cabiscol, E., Tamarit, J., & Ros, J. (2000). Oxidative stress in bacteria and protein damage by reactive oxygen species. *International Microbiology: The Official Journal of the Spanish Society for Microbiology, 3*, 3–8.

Carneiro, L. G., Venuleo, C., Oppenheim, F. G., & Salih, E. (2012). Proteome data set of human gingival crevicular fluid from healthy periodontium sites by multidimensional protein separation and mass spectrometry. *Journal of Periodontal Research, 47*, 248–262.

Chang, Y., Uchiyama, S., Varki, A., & Nizet, V. (2012). Leukocyte inflammatory responses provoked by pneumococcal sialidase. *mBio, 3*, e00220. http://dx.doi.org/10.1128/mBio.00220-11.

Chattoraj, P., Banerjee, A., Biswas, S., & Biswas, I. (2010). ClpP of *Streptococcus mutans* differentially regulates expression of genomic islands, mutacin production, and antibiotic tolerance. *Journal of Bacteriology, 192*, 1312–1323.

Chen, T., Nakayama, K., Belliveau, L., & Duncan, M. J. (2001). *Porphyromonas gingivalis* gingipains and adhesion to epithelial cells. *Infection and Immunity, 69*, 3048–3056.

Chen, Y.-Y., Peng, B., Yang, Q., Glew, M. D., Veith, P. D., Cross, K. J., et al. (2011). The outer membrane protein LptO is essential for the O-deacylation of LPS and the co-ordinated secretion and attachment of A-LPS and CTD proteins in *Porphyromonas gingivalis*. *Molecular Microbiology, 79*, 1380–1401.

Cisar, J. O., Sandberg, A. L., Abeygunawardana, C., Reddy, G. P., & Bush, C. A. (1995). Lectin recognition of host-like saccharide motifs in streptococcal cell wall polysaccharides. *Glycobiology, 5*, 655–662.

Clais, S., Boulet, G., Kerstens, M., Horemans, T., Teughels, W., Quirynen, M., et al. (2014). Importance of biofilm formation and dipeptidyl peptidase IV for the pathogenicity of clinical *Porphyromonas gingivalis* isolates. *Pathogens and Disease, 70*, 408–413.

Clark, W. B., Beem, J. E., Nesbitt, W. E., Cisar, J. O., Tseng, C. C., & Levine, M. J. (1989). Pellicle receptors for *Actinomyces viscosus* type 1 fimbriae in vitro. *Infection and Immunity, 57*, 3003–3008.

Clock, S. A., Planet, P. J., Perez, B. A., & Figurski, D. H. (2008). Outer membrane components of the Tad (tight adherence) secreton of *Aggregatibacter actinomycetemcomitans*. *Journal of Bacteriology, 190*, 980–990.

Cornejo, O. E., Lefébure, T., Bitar, P. D. P., Lang, P., Richards, V. P., Eilertson, K., et al. (2013). Evolutionary and population genomics of the cavity causing bacteria *Streptococcus mutans*. *Molecular Biology and Evolution, 30*, 881–893.

Coyne, M. J., Fletcher, C. M., Chatzidaki-Livanis, M., Posch, G., Schaffer, C., & Comstock, L. E. (2013). Phylum-wide general protein O-glycosylation system of the Bacteroidetes. *Molecular Microbiology, 88*, 772–783.

Coyne, M. J., Reinap, B., Lee, M. M., & Comstock, L. E. (2005). Human symbionts use a host-like pathway for surface fucosylation. *Science (New York, N.Y.), 307*, 1778–1781.

Crennell, S., Garman, E., Laver, G., Vimr, E., & Taylor, G. (1994). Crystal structure of Vibrio cholerae neuraminidase reveals dual lectin-like domains in addition to the catalytic domain. *Structure (London, England: 1993), 2*, 535–544.

Crosby, J. A., & Kachlany, S. C. (2007). TdeA, a TolC-like protein required for toxin and drug export in *Aggregatibacter (Actinobacillus) actinomycetemcomitans*. *Gene*, *388*, 83–92.
Cryan, J. F., & Dinan, T. G. (2012). Mind-altering microorganisms: The impact of the gut microbiota on brain and behaviour. *Nature Reviews. Neuroscience*, *13*, 701–712.
Curtis, M. A., Aduse Opoku, J., Rangarajan, M., Gallagher, A., Sterne, J. A. C., Reid, C. R., et al. (2002). Attenuation of the virulence of *Porphyromonas gingivalis* by using a specific synthetic Kgp protease inhibitor. *Infection and Immunity*, *70*, 6968–6975.
Curtis, M. A., Aduse-Opoku, J., & Rangarajan, M. (2001). Cysteine proteases of *Porphyromonas gingivalis*. *Critical Reviews in Oral Biology and Medicine*, *12*, 192–216.
Curtis, M. A., Thickett, A., Slaney, J. M., Rangarajan, M., Aduse-Opoku, J., Shepherd, P., et al. (1999). Variable carbohydrate modifications to the catalytic chains of the RgpA and RgpB proteases of *Porphyromonas gingivalis* W50. *Infection and Immunity*, *67*, 3816–3823.
Curtis, M. A., Zenobia, C., & Darveau, R. P. (2011). The relationship of the oral microbiotia to periodontal health and disease. *Cell Host & Microbe*, *10*, 302–306.
Cvitkovitch, D. G., Boyd, D. A., Thevenot, T., & Hamilton, I. R. (1995). Glucose transport by a mutant of *Streptococcus mutans* unable to accumulate sugars via the phosphoenolpyruvate phosphotransferase system. *Journal of Bacteriology*, *177*, 2251–2258.
Dale, B. D., & Page Fredericks, L. (2004). Antimicrobial peptides in the oral environment: Expression and function in health and disease. In R. L. Gallo (Ed.), *Antimicrobial peptides in human health and disease* (pp. 223–251). Wymondham, United Kingdom: Horizon Bioscience.
David, S. S., O'Shea, V. L., & Kundu, S. (2007). Base-excision repair of oxidative DNA damage. *Nature*, *447*, 941–950.
Derr, A. M., Faustoferri, R. C., Betzenhauser, M. J., Gonzalez, K., Marquis, R. E., & Quivey, R. G. (2012). Mutation of the NADH oxidase gene (nox) reveals an overlap of the oxygen- and acid-mediated stress responses in *Streptococcus mutans*. *Applied and Environmental Microbiology*, *78*, 1215–1227.
Dewhirst, F. E., Chen, T., Izard, J., Paster, B. J., Tanner, A. C. R., Yu, W.-H., et al. (2010). The human oral microbiome. *Journal of Bacteriology*, *192*, 5002–5017.
Di Benedetto, A., Gigante, I., Colucci, S., & Grano, M. (2013). Periodontal disease: Linking the primary inflammation to bone loss. *Clinical & Developmental Immunology*, *2013*, 503754.
Diaz, P. I., Slakeski, N., Reynolds, E. C., Morona, R., Rogers, A. H., & Kolenbrander, P. E. (2006). Role of oxyR in the oral anaerobe *Porphyromonas gingivalis*. *Journal of Bacteriology*, *188*, 2454–2462.
Du Toit, D. F. (2003). The tongue: Structure and function relevant to disease and oral health. *SADJ: Journal of the South African Dental Association = tydskrif van die Suid-Afrikaanse Tandheelkundige Vereniging*, *58*, 375–376. 380–383.
Edwards, A. M., Grossman, T. J., & Rudney, J. D. (2007). Association of a high-molecular weight arginine-binding protein of *Fusobacterium nucleatum* ATCC 10953 with adhesion to secretory immunoglobulin A and coaggregation with *Streptococcus cristatus*. *Oral Microbiology and Immunology*, *22*, 217–224.
Edwards, A. M., Jenkinson, H. F., Woodward, M. J., & Dymock, D. (2005). Binding properties and adhesion-mediating regions of the major sheath protein of Treponema denticola ATCC 35405. *Infection and Immunity*, *73*, 2891–2898.
Eggert, F. M., Drewell, L., Bigelow, J. A., Speck, J. E., & Goldner, M. (1991). The pH of gingival crevices and periodontal pockets in children, teenagers and adults. *Archives of Oral Biology*, *36*, 233–238.
Eke, P. I., Dye, B. A., Wei, L., Thornton-Evans, G. O., & Genco, R. J. (2012). Prevalence of periodontitis in adults in the United States: 2009 and 2010. *Journal of Dental Research*, *91*, 914–920.

Elhenawy, W., Debelyy, M. O., & Feldman, M. F. (2014). Preferential packing of acidic glycosidases and proteases into Bacteroides outer membrane vesicles. *mBio*, *5*, e00909–e00914. http://dx.doi.org/10.1128/mBio.00909-14.

El-Kheshen, M. (2008). Plaque plague. *The Dentist*, *2008*, 48–50.

Engels-Deutsch, M., Rizk, S., & Haïkel, Y. (2011). *Streptococcus mutans* antigen I/II binds to α5β1 integrins via its A-domain and increases β1 integrins expression on periodontal ligament fibroblast cells. *Archives of Oral Biology*, *56*, 22–28.

Erhardt, M., Namba, K., Hughes, K. T., & Injectisome, T. I. I. I. (2010). Bacterial nanomachines: The flagellum and type III injectisome. *Cold Spring Harbor Perspectives in Biology*, *2*, a000299.

Fardini, Y., Wang, X., Témoin, S., Nithianantham, S., Lee, D., Shoham, M., et al. (2011). Fusobacterium nucleatum adhesin FadA binds vascular endothelial cadherin and alters endothelial integrity. *Molecular Microbiology*, *82*, 1468–1480.

Fedi, P. F., & Killoy, W. J. (1992). Temperature differences at periodontal sites in health and disease. *Journal of Periodontology*, *63*, 24–27.

Feng, C., Stamatos, N. M., Dragan, A. I., Medvedev, A., Whitford, M., Zhang, L., et al. (2012). Sialyl residues modulate LPS-mediated signaling through the Toll-like receptor 4 complex. *PLoS One*, *7*, e32359.

Fenno, J. C., Wong, G. W., Hannam, P. M., Müller, K. H., Leung, W. K., & McBride, B. C. (1997). Conservation of msp, the gene encoding the major outer membrane protein of oral *Treponema* spp. *Journal of Bacteriology*, *179*, 1082–1089.

Fine, D. H., Kaplan, J. B., Furgang, D., Karched, M., Velliyagounder, K., & Yue, G. (2010). Mapping the epithelial-cell-binding domain of the *Aggregatibacter actinomycetemcomitans* autotransporter adhesin Aae. *Microbiology (Reading, England)*, *156*, 3412–3420.

Fine, D. H., Velliyagounder, K., Furgang, D., & Kaplan, J. B. (2005). The *Actinobacillus actinomycetemcomitans* autotransporter adhesin Aae exhibits specificity for buccal epithelial cells from humans and old world primates. *Infection and Immunity*, *73*, 1947–1953.

Fitzpatrick, R. E., Wijeyewickrema, L. C., & Pike, R. N. (2009). The gingipains: Scissors and glue of the periodontal pathogen, *Porphyromonas gingivalis*. *Future Microbiology*, *4*, 471–487.

Flemming, H.-C., Neu, T. R., & Wozniak, D. J. (2007). The EPS matrix: The "house of biofilm cells". *Journal of Bacteriology*, *189*, 7945–7947.

Fong, K. P., Tang, H.-Y., Brown, A. C., Kieba, I. R., Speicher, D. W., Boesze-Battaglia, K., et al. (2011). *Aggregatibacter actinomycetemcomitans* leukotoxin is post-translationally modified by addition of either saturated or hydroxylated fatty acyl chains. *Molecular Oral Microbiology*, *26*, 262–276.

Forsgren, N., Lamont, R. J., & Persson, K. (2009). Crystal structure of the variable domain of the *Streptococcus gordonii* surface protein SspB. *Protein Science: A Publication of the Protein Society*, *18*, 1896–1905.

Forsgren, N., Lamont, R. J., & Persson, K. (2010). Two intramolecular isopeptide bonds are identified in the crystal structure of the *Streptococcus gordonii* SspB C-terminal domain. *Journal of Molecular Biology*, *397*, 740–751.

Frandsen, E. V. (1994). Carbohydrate depletion of immunoglobulin A1 by oral species of gram-positive rods. *Oral Microbiology and Immunology*, *9*, 352–358.

Fujimura, S., Ueda, O., Shibata, Y., & Hirai, K. (2003). Isolation and properties of a tripeptidyl peptidase from a periodontal pathogen *Prevotella nigrescens*. *FEMS Microbiology Letters*, *219*, 305–309.

Fujishima, K., Kawada-Matsuo, M., Oogai, Y., Tokuda, M., Torii, M., & Komatsuzawa, H. (2013). dpr and sod in *Streptococcus mutans* are involved in coexistence with S. sanguinis, and PerR is associated with resistance to H2O2. *Applied and Environmental Microbiology*, *79*, 1436–1443.

Fujiwara, T., Hoshino, T., Ooshima, T., & Hamada, S. (2002). Differential and quantitative analyses of mRNA expression of glucosyltransferases from *Streptococcus mutans* MT8148. *Journal of Dental Research*, *81*, 109–113.

Fukuda, M. (2002). Roles of mucin-type O-glycans in cell adhesion. *Biochimica et Biophysica Acta*, *1573*, 394–405.

Galgut, P. N. (2001). The relevance of pH to gingivitis and periodontitis. *Journal of the International Academy of Periodontology*, *3*, 61–67.

Garberoglio, R., & Brännström, M. (1976). Scanning electron microscopic investigation of human dentinal tubules. *Archives of Oral Biology*, *21*, 355–362.

Gibbons, R. J., Hay, D. I., Cisar, J. O., & Clark, W. B. (1988). Adsorbed salivary proline-rich protein 1 and statherin: Receptors for type 1 fimbriae of *Actinomyces viscosus* T14V-J1 on apatitic surfaces. *Infection and Immunity*, *56*, 2990–2993.

Glew, M. D., Veith, P. D., Peng, B., Chen, Y.-Y., Gorasia, D. G., Yang, Q., et al. (2012). PG0026 is the C-terminal signal peptidase of a novel secretion system of *Porphyromonas gingivalis*. *The Journal of Biological Chemistry*, *287*, 24605–24617.

Goetting-Minesky, M. P., Godovikova, V., Li, J. J., Seshadrinathan, S., Timm, J. C., Kamodia, S. S., et al. (2013). Conservation and revised annotation of the *Treponema denticola* prcB-prcA-prtP locus encoding the dentilisin (CTLP) protease complex. *Molecular Oral Microbiology*, *28*, 181–191.

Gonzalez, K., Faustoferri, R. C., & Quivey, R. G. (2012). Role of DNA base excision repair in the mutability and virulence of *Streptococcus mutans*. *Molecular Microbiology*, *85*, 361–377.

Graziano, T. S., Closs, P., Poppi, T., Franco, G. C., Cortelli, J. R., Groppo, F. C., et al. (2013). Catecholamines promote the expression of virulence and oxidative stress genes in *Porphyromonas gingivalis*. *Journal of Periodontal Research*, *49*, 660–669.

Grenier, D. (2013). Binding properties of Treponema denticola lipooligosaccharide. *Journal of Oral Microbiology*, *5*, 3–7.

Griffen, A. L., Beall, C. J., Campbell, J. H., Firestone, N. D., Kumar, P. S., Yang, Z. K., et al. (2012). Distinct and complex bacterial profiles in human periodontitis and health revealed by 16S pyrosequencing. *The ISME Journal*, *6*, 1176–1185.

Griswold, A. R., Chen, Y.-Y. M., & Burne, R. A. (2004). Analysis of an agmatine deiminase gene cluster in *Streptococcus mutans* UA159. *Journal of Bacteriology*, *186*, 1902–1904.

Guevara, T., Ksiazek, M., Skottrup, P. D., Cerdà-Costa, N., Trillo-Muyo, S., de Diego, I., et al. (2013). Structure of the catalytic domain of the *Tannerella forsythia* matrix metallopeptidase karilysin in complex with a tetrapeptidic inhibitor. *Acta Crystallographica. Section F, Structural Biology and Crystallization Communications*, *69*, 472–476.

Guo, Y., Nguyen, K.-A., & Potempa, J. (2010). Dichotomy of gingipains action as virulence factors: From cleaving substrates with the precision of a surgeon's knife to a meat chopper-like brutal degradation of proteins. *Periodontology*, *2000*(54), 15–44.

Hajishengallis, G., Darveau, R. P., & Curtis, M. A. (2012). The keystone-pathogen hypothesis. *Nature Reviews. Microbiology*, *10*, 717–725.

Hajishengallis, G., & Lamont, R. J. (2012). Beyond the red complex and into more complexity: The polymicrobial synergy and dysbiosis (PSD) model of periodontal disease etiology. *Molecular Oral Microbiology*, *27*, 409–419.

Hajishengallis, G., Liang, S., Payne, M. A., Hashim, A., Jotwani, R., Eskan, M. A., et al. (2011). Low-abundance biofilm species orchestrates inflammatory periodontal disease through the commensal microbiota and complement. *Cell Host & Microbe*, *10*, 497–506.

Hall, M., Nylander, Å., Jenkinson, H. F., & Persson, K. (2014). Structure of the C-terminal domain of AspA (antigen I/II-family) protein from *Streptococcus pyogenes*. *FEBS Open Bio*, *4*, 283–289.

Haltiwanger, R. S., Grove, K., & Philipsberg, G. A. (1998). Modulation of O-linked N-acetylglucosamine levels on nuclear and cytoplasmic proteins in vivo using the peptide O-GlcNAc-beta-N-acetylglucosaminidase inhibitor O-(2-acetamido-2-deoxy-D-glucopyranosylidene)amino-N-phenylcarbamate. *The Journal of Biological Chemistry, 273*, 3611–3617.

Han, Y. W., Ikegami, A., Rajanna, C., Kawsar, H. I., Zhou, Y., Li, M., et al. (2005). Identification and characterization of a novel adhesin unique to oral fusobacteria. *Journal of Bacteriology, 187*, 5330–5340.

Han, T. K., Zhang, C., & Dao, M. L. (2006). Identification and characterization of collagen-binding activity in *Streptococcus mutans* wall-associated protein: a possible implication in dental root caries and endocarditis. *Biochemical and Biophysical Research Communications, 343*(3), 787–792.

Hasegawa, Y., Iwami, J., Sato, K., Park, Y., Nishikawa, K., Atsumi, T., et al. (2009). Anchoring and length regulation of *Porphyromonas gingivalis* Mfa1 fimbriae by the downstream gene product Mfa2. *Microbiology (Reading, England), 155*, 3333–3347.

Hasegawa, Y., Nagano, K., Ikai, R., Izumigawa, M., Yoshida, Y., Kitai, N., et al. (2013). Localization and function of the accessory protein Mfa3 in *Porphyromonas gingivalis* Mfa1 fimbriae. *Molecular Oral Microbiology, 28*, 467–480.

Hashimoto, M., Ogawa, S., Asai, Y., Takai, Y., & Ogawa, T. (2003). Binding of *Porphyromonas gingivalis* fimbriae to *Treponema denticola* dentilisin. *FEMS Microbiology Letters, 226*, 267–271.

Hashioka, K., Yamasaki, M., Nakane, A., Horiba, N., & Nakamura, H. (1992). The relationship between clinical symptoms and anaerobic bacteria from infected root canals. *Journal of Endodontics, 18*, 558–561.

Haverkamp, J., Schauer, R., & Wember, M. (1976). Neuraminic acid derivatives newly discovered in humans: N-acetyl-9-O-L-lactoylneuraminic acid, N,9-O-Diacetylneuraminic acid and N-acetyl-2,3-dehydro-2-deoxyneuraminic acid. *Hoppe-Seyler's Zeitschrift für Physiologische Chemie, 357*, 1699–1705.

Hayes, M. L., Carter, E. C., & Griffiths, S. J. (1983). The acidogenic microbial composition of dental plaque from caries-free and caries-prone people. *Archives of Oral Biology, 28*, 381–386.

Hellwage, J., Ku, S., & Zipfel, P. F. (1997). The human complement regulatory factor-H-like protein 1, which represents a truncated form of factor H, displays cell-attachment activity. *The Biochemical Journal, 326*, 321–327.

Helmerhorst, E. J., & Oppenheim, F. G. (2007). Saliva: A dynamic proteome. *Journal of Dental Research, 86*, 680–693.

Henderson, I. R., Navarro-Garcia, F., Desvaux, M., Fernandez, R. C., & Ala'Aldeen, D. (2004). Type V protein secretion pathway: The autotransporter story. *Microbiology and Molecular Biology Reviews, 68*, 692–744.

Hendrickson, E. L., Wang, T., Dickinson, B. C., Whitmore, S. E., Wright, C. J., Lamont, R. J., et al. (2012). Proteomics of *Streptococcus gordonii* within a model developing oral microbial community. *BMC Microbiology, 12*, 211.

Henry, L. G., Aruni, W., Sandberg, L., & Fletcher, H. M. (2013). Protective role of the PG1036-PG1037-PG1038 operon in oxidative stress in *Porphyromonas gingivalis* W83. *PLoS One, 8*, e69645.

Henry, L. G., McKenzie, R. M. E., Robles, A., & Fletcher, H. M. (2012). Oxidative stress resistance in *Porphyromonas gingivalis*. *Future Microbiology*, 7, 497–512.

Henry, L. G., Sandberg, L., Zhang, K., & Fletcher, H. M. (2008). DNA repair of 8-oxo-7,8-dihydroguanine lesions in *Porphyromonas gingivalis*. *Journal of Bacteriology, 190*, 7985–7993.

Hojo, K., Nagaoka, S., Ohshima, T., & Maeda, N. (2009). Bacterial interactions in dental biofilm development. *Journal of Dental Research, 88*, 982–990.

Holmes, A. R., McNab, R., & Jenkinson, H. F. (1996). Candida albicans binding to the oral bacterium *Streptococcus gordonii* involves multiple adhesin-receptor interactions. *Infection and Immunity*, *64*, 4680–4685.

Honma, K., Mishima, E., Inagaki, S., & Sharma, A. (2009). The OxyR homologue in *Tannerella forsythia* regulates expression of oxidative stress responses and biofilm formation. *Microbiology*, *155*, 1912–1922.

Honma, K., Mishima, E., & Sharma, A. (2011). Role of *Tannerella forsythia* NanH sialidase in epithelial cell attachment. *Infection and Immunity*, *79*, 393–401.

Hooper, L. V., Xu, J., Falk, P. G., Midtvedt, T., & Gordon, J. I. (1999). A molecular sensor that allows a gut commensal to control its nutrient foundation in a competitive ecosystem. *Proceedings of the National Academy of Sciences of the United States of America*, *96*, 9833–9838.

Hsiao, W. W., Li, K. L., Liu, Z., Jones, C., Fraser-Liggett, C. M., & Fouad, A. F. (2012). Microbial transformation from normal oral microbiota to acute endodontic infections. *BMC Genomics*, *13*, 345.

Hughes, C. V., Malki, G., Loo, C. Y., Tanner, A. C. R., & Ganeshkumar, N. (2003). Cloning and expression of alpha-D-glucosidase and N-acetyl-beta-glucosaminidase from the periodontal pathogen, *Tannerella forsythensis (Bacteroides forsythus)*. *Oral Microbiology and Immunology*, *18*, 309–312.

Humphrey, S. P., & Williamson, R. T. (2001). A review of saliva: Normal composition, flow, and function. *The Journal of Prosthetic Dentistry*, *85*, 162–169.

Ikegami, A., Chung, P., & Han, Y. W. (2009). Complementation of the fadA mutation in *Fusobacterium nucleatum* demonstrates that the surface-exposed adhesin promotes cellular invasion and placental colonization. *Infection and Immunity*, 77, 3075–3079.

Ikegami, A., Honma, K., Sharma, A., & Kuramitsu, H. K. (2004). Multiple functions of the leucine-rich repeat protein LrrA of *Treponema denticola*. *Infection and Immunity*, *72*, 4619–4627.

Inagaki, S., Onishi, S., Kuramitsu, H. K., & Sharma, A. (2006). *Porphyromonas gingivalis* vesicles enhance attachment, and the leucine-rich repeat BspA protein is required for invasion of epithelial cells by "Tannerella forsythia". *Infection and Immunity*, *74*(9), 5023–5028.

Inoue, T., Tanimoto, I., Ohta, H., Kato, K., Murayama, Y., & Fukui, K. (1998). Molecular characterization of low-molecular-weight component protein, Flp, in *Actinobacillus actinomycetemcomitans* fimbriae. *Microbiology and Immunology*, *42*, 253–258.

Isaza, M. P., Duncan, M. S., Kaplan, J. B., & Kachlany, S. C. (2008). Screen for leukotoxin mutants in *Aggregatibacter actinomycetemcomitans*: Genes of the phosphotransferase system are required for leukotoxin biosynthesis. *Infection and Immunity*, *76*, 3561–3568.

Jakubovics, N. S. (2010). Talk of the town: Interspecies communication in oral biofilms. *Molecular Oral Microbiology*, *25*, 4–14.

Jakubovics, N. S., Brittan, J. L., Dutton, L. C., & Jenkinson, H. F. (2009). Multiple adhesin proteins on the cell surface of *Streptococcus gordonii* are involved in adhesion to human fibronectin. *Microbiology (Reading, England)*, *155*, 3572–3580.

Jakubovics, N. S., Kerrigan, S. W., Nobbs, A. H., Stro, N., Van Dolleweerd, C. J., Cox, D. M., et al. (2005). Functions of cell surface-anchored antigen I/II family and hsa polypeptides in interactions of *Streptococcus gordonii* with host receptors. *Infection and Immunity*, *73*, 6629–6638.

Jakubovics, N. S., Shields, R. C., Rajarajan, N., & Burgess, J. G. (2013). Life after death: The critical role of extracellular DNA in microbial biofilms. *Letters in Applied Microbiology*, *57*, 467–475.

Jakubovics, N. S., Strömberg, N., van Dolleweerd, C. J., Kelly, C. G., & Jenkinson, H. F. (2005). Differential binding specificities of oral streptococcal antigen I/II family adhesins for human or bacterial ligands. *Molecular Microbiology*, *55*, 1591–1605.

Jang, Y.-J., Choi, Y.-J., Lee, S.-H., Jun, H.-K., & Choi, B.-K. (2013). Autoinducer 2 of *Fusobacterium nucleatum* as a target molecule to inhibit biofilm formation of periodontopathogens. *Archives of Oral Biology, 58*, 17–27.

Jentsch, H. F. R., März, D., & Krüger, M. (2013). The effects of stress hormones on growth of selected periodontitis related bacteria. *Anaerobe, 24*, 49–54.

Jones, J. E., Dreyton, C. J., Flick, H., Causey, C. P., & Thompson, P. R. (2010). Mechanistic studies of agmatine deiminase from multiple bacterial species. *Biochemistry, 49*, 9413–9423.

Jungnickel, J., Brämer, C., Bronzlik, P., Lipokatic-Takacs, E., Weinhold, B., Gerardy-Schahn, R., et al. (2009). Level and localization of polysialic acid is critical for early peripheral nerve regeneration. *Molecular and Cellular Neurosciences, 40*, 374–381.

Jusko, M., Potempa, J., Karim, A. Y., Ksiazek, M., Riesbeck, K., Garred, P., et al. (2012). A metalloproteinase karilysin present in the majority of *Tannerella forsythia* isolates inhibits all pathways of the complement system. *Journal of Immunology (Baltimore, Md.: 1950), 188*, 2338–2349.

Kachlany, S. C., Planet, P. J., Bhattacharjee, M. K., Kollia, E., Desalle, R., Fine, D. H., et al. (2000). Nonspecific adherence by *Actinobacillus actinomycetemcomitans* requires genes widespread in bacteria and archaea. *Journal of Bacteriology, 182*, 6169–6176.

Kachlany, S. C., Planet, P. J., Desalle, R., Fine, D. H., & Figurski, D. H. (2001). *Actinobacillus actinomycetemcomitans*: From plaque to plague to pond scum. *Trends in Microbiology, 9*, 429–437.

Kamma, J. J., Nakou, M., Gmur, R., & Baehni, P. C. (2004). Microbiological profile of early onset/aggressive periodontitis patients. *Oral Microbiology and Immunology, 19*, 314–321.

Kaplan, C. W., Lux, R., Haake, S. K., & Shi, W. (2009). The *Fusobacterium nucleatum* outer membrane protein RadD is an arginine-inhibitable adhesin required for inter-species adherence and the structured architecture of multispecies biofilm. *Molecular Microbiology, 71*, 35–47.

Kaur, R., Gilbert, S. C., Sheehy, E. C., & Beighton, D. (2013). Salivary levels of Bifidobacteria in caries-free and caries-active children. *International Journal of Paediatric Dentistry/The British Paedodontic Society [and] the International Association of Dentistry for Children, 23*, 32–38.

Kaur, S., White, S., & Bartold, P. M. (2013). Periodontal disease and rheumatoid arthritis: A systematic review. *Journal of Dental Research, 92*, 399–408.

Kerrigan, S. W., Jakubovics, N. S., Keane, C., Maguire, P., Wynne, K., Jenkinson, H. F., et al. (2007). Role of *Streptococcus gordonii* surface proteins SspA/SspB and Hsa in platelet function. *Infection and Immunity, 75*, 5740–5747.

Kim, T.-S., Obst, C., Zehaczek, S., & Geenen, C. (2008). Detection of bone loss with different X-ray techniques in periodontal patients. *Journal of Periodontology, 79*, 1141–1149.

Kinane, J. A., Benakanakere, M. R., Zhao, J., Hosur, K. B., & Kinane, D. F. (2012). *Porphyromonas gingivalis* influences actin degradation within epithelial cells during invasion and apoptosis. *Cellular Microbiology, 14*, 1085–1096.

Kinder, S. A., & Holt, S. C. (1993). Localization of the *Fusobacterium nucleatum* T18 adhesin activity mediating coaggregation with *Porphyromonas gingivalis* T22. *Journal of Bacteriology, 175*, 840–850.

King, S. J., Whatmore, A. M., & Dowson, C. G. (2005). NanA, a neuraminidase from Streptococcus pneumoniae, shows high levels of sequence diversity, at least in part through recombination with *Streptococcus oralis*. *Journal of Bacteriology, 187*, 5376–5386.

Kishi, M., Hasegawa, Y., Nagano, K., Nakamura, H., Murakami, Y., & Yoshimura, F. (2012). Identification and characterization of novel glycoproteins involved in growth and biofilm formation by *Porphyromonas gingivalis*. *Molecular Oral Microbiology, 27*, 458–470.

Kistler, J. O., Booth, V., Bradshaw, D. J., & Wade, W. G. (2013). Bacterial community development in experimental gingivitis. *PLoS One*, *8*, e71227.

Klein, B. A., Tenorio, E. L., Lazinski, D. W., Camilli, A., Duncan, M. J., & Hu, L. T. (2012). Identification of essential genes of the periodontal pathogen *Porphyromonas gingivalis*. *BMC Genomics*, *13*, 578.

Knutsson, G., Jontell, M., & Bergenholtz, G. (1994). Determination of plasma proteins in dentinal fluid from cavities prepared in healthy young human teeth. *Archives of Oral Biology*, *39*, 185–190.

Kolenbrander, P. E., & Andersen, R. N. (1989). Inhibition of coaggregation between *Fusobacterium nucleatum* and *Porphyromonas (Bacteroides) gingivalis* by lactose and related sugars. *Infection and Immunity*, *57*, 3204–3209.

Kolenbrander, P. E., Palmer, R. J., Rickard, A. H., Jakubovics, N. S., Chalmers, N. I., & Diaz, P. I. (2006). Bacterial interactions and successions during plaque development. *Periodontology*, *2000*(42), 47–79.

Koziel, J., Karim, A. Y., Przybyszewska, K., Ksiazek, M., Rapala-Kozik, M., Nguyen, K.-A., et al. (2010). Proteolytic inactivation of LL-37 by karilysin, a novel virulence mechanism of *Tannerella forsythia*. *Journal of Innate Immunity*, *2*, 288–293.

Krüger, D., Schauer, R., & Traving, C. (2001). Characterization and mutagenesis of the recombinant N-acetylneuraminate lyase from *Clostridium perfringens*: Insights into the reaction mechanism. *European Journal of Biochemistry/FEBS*, *268*, 3831–3839.

Kuboniwa, M., Amano, A., Hashino, E., Yamamoto, Y., Inaba, H., Hamada, N., et al. (2009). Distinct roles of long/short fimbriae and gingipains in homotypic biofilm development by *Porphyromonas gingivalis*. *BMC Microbiology*, *9*, 105.

Kung, R. T., Ochs, B., & Goodson, J. M. (1990). Temperature as a periodontal diagnostic. *Journal of Clinical Periodontology*, *17*, 557–563.

Kurniyati, K., Zhang, W., Zhang, K., & Li, C. (2013). A surface-exposed neuraminidase affects complement resistance and virulence of the oral spirochaete *Treponema denticola*. *Molecular Microbiology*, *89*, 842–856.

Lamont, R. J., El-Sabaeny, A., Park, Y., Cook, G. S., Costerton, J. W., & Demuth, D. R. (2002). Role of the *Streptococcus gordonii* SspB protein in the development of *Porphyromonas gingivalis* biofilms on streptococcal substrates. *Microbiology (Reading, England)*, *148*, 1627–1636.

Langley, D. B., Harty, D. W. S., Jacques, N. A., Hunter, N., Guss, J. M., & Collyer, C. A. (2008). Structure of N-acetyl-beta-D-glucosaminidase (GcnA) from the endocarditis pathogen *Streptococcus gordonii* and its complex with the mechanism-based inhibitor NAG-thiazoline. *Journal of Molecular Biology*, *377*, 104–116.

Larson, M. R., Rajashankar, K. R., Crowley, P. J., Kelly, C., Mitchell, T. J., Brady, L. J., et al. (2011). Crystal structure of the C-terminal region of *Streptococcus mutans* antigen I/II and characterization of salivary agglutinin adherence domains. *The Journal of Biological Chemistry*, *286*, 21657–21666.

Larson, M. R., Rajashankar, K. R., Patel, M. H., Robinette, R. A., Crowley, P. J., Michalek, S., et al. (2010). Elongated fibrillar structure of a streptococcal adhesin assembled by the high-affinity association of alpha- and PPII-helices. *Proceedings of the National Academy of Sciences of the United States of America*, *107*, 5983–5988.

Larzábal, M., Mercado, E. C., Vilte, D. A., Salazar-González, H., Cataldi, A., & Navarro-Garcia, F. (2010). Designed coiled-coil peptides inhibit the type three secretion system of enteropathogenic *Escherichia coli*. *PLoS One*, *5*, e9046.

Le Goff, A., Bunetel, L., Mouton, C., & Bonnaure-Mallet, M. (1997). Evaluation of root canal bacteria and their antimicrobial susceptibility in teeth with necrotic pulp. *Oral Microbiology and Immunology*, *12*, 318–322.

Lee, S.-W., Sabet, M., Um, H.-S., Yang, J., Kim, H. C., & Zhu, W. (2006). Identification and characterization of the genes encoding a unique surface (S-) layer of Tannerella forsythia. *Gene*, *371*, 102–111.

Lee, Y. H., Zimmerman, J. N., Custodio, W., Xiao, Y., Basiri, T., Hatibovic-Kofman, S., et al. (2013). Proteomic evaluation of acquired enamel pellicle during in vivo formation. *PLoS One*, *8*, e67919.

Lemos, J. A., & Burne, R. A. (2008). A model of efficiency: Stress tolerance by *Streptococcus mutans*. *Microbiology (Reading, England)*, *154*, 3247–3255.

Len, A. C. L., Harty, D. W. S., & Jacques, N. A. (2004a). Proteome analysis of *Streptococcus mutans* metabolic phenotype during acid tolerance. *Microbiology (Reading, England)*, *150*, 1353–1366.

Len, A. C. L., Harty, D. W. S., & Jacques, N. A. (2004b). Stress-responsive proteins are upregulated in *Streptococcus mutans* during acid tolerance. *Microbiology (Reading, England)*, *150*, 1339–1351.

Levine, M. J., Reddy, M. S., Tabak, L. A., Loomis, R. E., Bergey, E. J., Jones, P. C., et al. (1987). Structural aspects of salivary glycoproteins. *Journal of Dental Research*, *66*, 436–441.

Lewis, J. P. (2010). Metal uptake in host-pathogen interactions: Role of iron in *Porphyromonas gingivalis* interactions with host organisms. *Periodontology*, *2000*(52), 94–116.

Lewis, J. P., Dawson, J. A., Hannis, J. C., Macrina, F. L., & Muddiman, D. (1999). Hemoglobinase activity of the lysine gingipain protease (Kgp) of *Porphyromonas gingivalis* W83. *Journal of Bacteriology*, *181*, 4905–4913.

Lewis, A. L., & Lewis, W. G. (2012). Host sialoglycans and bacterial sialidases: A mucosal perspective. *Cellular Microbiology*, *14*, 1174–1182.

Li, Y., He, J., He, Z., Zhou, Y., Yuan, M., Xu, X., et al. (2014). Phylogenetic and functional gene structure shifts of the oral microbiomes in periodontitis patients. *The ISME Journal*, *8*, 1879–1891.

Li, L., Hsiao, W. W. L., Nandakumar, R., Barbuto, S. M., Mongodin, E. F., Paster, B. J., et al. (2010). Analyzing endodontic infections by deep coverage pyrosequencing. *Journal of Dental Research*, *89*, 980–984.

Li, C., Kurniyati, Hu, B., Bian, J., Sun, J., Zhang, W., et al. (2012). Abrogation of neuraminidase reduces biofilm formation, capsule biosynthesis, and virulence of *Porphyromonas gingivalis*. *Infection and Immunity*, *80*, 3–13.

Li, Y. H., Tang, N., Aspiras, M. B., Lau, P. C., Lee, J. H., Ellen, R. P., et al. (2002). A quorum-sensing signaling system essential for genetic competence in *Streptococcus mutans* is involved in biofilm formation. *Journal of Bacteriology*, *184*, 2699–2708.

Liu, P.-F., Shi, W., Zhu, W., Smith, J. W., Hsieh, S.-L., Gallo, R. L., et al. (2010). Vaccination targeting surface FomA of *Fusobacterium nucleatum* against bacterial co-aggregation: Implication for treatment of periodontal infection and halitosis. *Vaccine*, *28*, 3496–3505.

Liu, J., Wu, C., Huang, I.-H., Merritt, J., & Qi, F. (2011). Differential response of *Streptococcus mutans* towards friend and foe in mixed-species cultures. *Microbiology (Reading, England)*, *157*, 2433–2444.

Loe, H., Theilade, E., & Jensen, S. B. (1960). Experimental gingivitis in man. *The Journal of Periodontology*, *36*, 177–187.

Loimaranta, V., Hytönen, J., Pulliainen, A. T., Sharma, A., Tenovuo, J., Strömberg, N., et al. (2009). Leucine-rich repeats of bacterial surface proteins serve as common pattern recognition motifs of human scavenger receptor gp340. *The Journal of Biological Chemistry*, *284*, 18614–18623.

Loimaranta, V., Jakubovics, N. S., Hytönen, J., Finne, J., Jenkinson, H. F., & Strömberg, N. (2005). Fluid- or surface-phase human salivary scavenger protein gp340 exposes different bacterial recognition properties. *Infection and Immunity*, *73*, 2245–2252.

Lombard, V., Golaconda Ramulu, H., Drula, E., Coutinho, P. M., & Henrissat, B. (2014). The carbohydrate-active enzymes database (CAZy) in 2013. *Nucleic Acids Research*, *42*, D490–D495.

Lovatt, A., & Roberts, I. S. (1994). Cloning and expression in *Escherichia coli* of the nahA gene from *Porphyromonas gingivalis* indicates that P-N-acetylhexosaminidase is an outer-membrane-associated lipoprotein. *Microbiology*, *140*, 3399–3406.

Love, R. M., & Jenkinson, H. F. (2002). Invasion of dentinal tubules by oral bacteria. *Critical Reviews in Oral Biology and Medicine: An Official Publication of the American Association of Oral Biologists, 13*, 171–183.

Love, R. M., McMillan, M. D., Park, Y., & Jenkinson, H. F. (2000). Coinvasion of dentinal tubules by *Porphyromonas gingivalis* and *Streptococcus gordonii* depends upon binding specificity of streptococcal antigen I/II adhesin. *Infection and Immunity, 68*, 1359–1365.

Maeda, K., Nagata, H., Kuboniwa, M., Kataoka, K., Nishida, N., Tanaka, M., et al. (2004). Characterization of binding of *Streptococcus oralis* glyceraldehyde-3-phosphate dehydrogenase to Porphyromonas gingivalis major fimbriae. *Infection and Immunity, 72*, 5475–5477.

Maiden, M. F., Tanner, A., & Macuch, P. J. (1996). Rapid characterization of periodontal bacterial isolates by using fluorogenic substrate tests. *Journal of Clinical Microbiology, 34*, 376–384.

Manuel, S. G. A., Ragunath, C., Sait, H. B. R., Izano, E. A., Kaplan, J. B., & Ramasubbu, N. (2007). Role of active-site residues of dispersin B, a biofilm-releasing beta-hexosaminidase from a periodontal pathogen, in substrate hydrolysis. *The FEBS Journal, 274*, 5987–5999.

Marcotte, H., & Lavoie, M. C. (1998). Oral microbial ecology and the role of salivary immunoglobulin A. *Microbiology and Molecular Biology Reviews, 62*, 71–109.

Marsh, P. D. (2003). Are dental diseases examples of ecological catastrophes? *Microbiology (Reading, England), 149*, 279–294.

Martin, L. T., Verhagen, A., & Varki, A. (2003). Recombinant influenza C hemagglutinin-esterase as a probe for sialic acid 9-O-acetylation. *Methods in Enzymology, 363*, 489–498.

Mcdowell, J. V., Lankford, J., Stamm, L., Sadlon, T., Gordon, D. L., & Marconi, R. T. (2005). Demonstration of factor H-like protein 1 binding to *Treponema denticola*, a pathogen associated with periodontal disease in humans. *Infection and Immunity, 73*, 7126–7132.

McLaughlin, R. E., & Ferretti, J. J. (1996). The multiple-sugar metabolism (msm) gene cluster of *Streptococcus mutans* is transcribed as a single operon. *FEMS Microbiology Letters, 140*, 261–264.

McNab, R., Holmes, A. R., Clarke, J. M., Tannock, G. W., & Jenkinson, H. F. (1996). Cell surface polypeptide CshA mediates binding of *Streptococcus gordonii* to other oral bacteria and to immobilized fibronectin. *Infection and Immunity, 64*, 4204–4210.

McNab, R., Jenkinson, H. F., Loach, D. M., & Tannock, G. W. (1994). Cell-surface-associated polypeptides CshA and CshB of high molecular mass are colonization determinants in the oral bacterium *Streptococcus gordonii*. *Molecular Microbiology, 14*, 743–754.

McNeill, K., & Hamilton, I. R. (2003). Acid tolerance response of biofilm cells of *Streptococcus mutans*. *FEMS Microbiology Letters, 221*, 25–30.

Miller, R. A., & Britigan, B. E. (1997). Role of oxidants in microbial pathophysiology. *Clinical Microbiology Reviews, 10*, 1–18.

Mintz, K. P. (2004). Identification of an extracellular matrix protein adhesin, EmaA, which mediates the adhesion of *Actinobacillus actinomycetemcomitans* to collagen. *Microbiology (Reading, England), 150*, 2677–2688.

Mishima, E., & Sharma, A. (2011). *Tannerella forsythia* invasion in oral epithelial cells requires phosphoinositide 3-kinase activation and clathrin-mediated endocytosis. *Microbiology, 157*, 2382–2391.

Mishra, A., Das, A., Cisar, J. O., & Ton-That, H. (2007). Sortase-catalyzed assembly of distinct heteromeric fimbriae in *Actinomyces naeslundii*. *Journal of Bacteriology, 189*, 3156–3165.

Mishra, A., Devarajan, B., Reardon, M. E., Dwivedi, P., Krishnan, V., Cisar, J. O., et al. (2011). Two autonomous structural modules in the fimbrial shaft adhesin FimA mediate

Actinomyces interactions with streptococci and host cells during oral biofilm development. *Molecular Microbiology, 81*, 1205–1220.

Mishra, A., Wu, C., Yang, J., Cisar, J. O., Das, A., & Ton-That, H. (2010). The *Actinomyces oris* type 2 fimbrial shaft FimA mediates co-aggregation with oral streptococci, adherence to red blood cells and biofilm development. *Molecular Microbiology, 77*, 841–854.

Mitchell, J., & Sullam, P. M. (2009). *Streptococcus mitis* phage-encoded adhesins mediate attachment to {alpha}2-8-linked sialic acid residues on platelet membrane gangliosides. *Infection and Immunity, 77*, 3485–3490.

Moelling, C., Oberschlacke, R., Ward, P., Karijolich, J., Borisova, K., Bjelos, N., et al. (2007). Metal-dependent repression of siderophore and biofilm formation in *Actinomyces naeslundii. FEMS Microbiology Letters, 275*, 214–220.

Moen, K., Brun, J. G., Valen, M., Skartveit, L., Eribe, E. K. R., Olsen, I., et al. (2006). Synovial inflammation in active rheumatoid arthritis and psoriatic arthritis facilitates trapping of a variety of oral bacterial DNAs. *Clinical and Experimental Rheumatology, 24*, 656–663.

Moncla, B. J., Braham, P., & Hillier, S. L. (1990). Sialidase (neuraminidase) activity among gram-negative anaerobic and capnophilic bacteria. *Journal of Clinical Microbiology, 28*, 422–425.

Moncla, B. J., Braham, P., Rabe, L. K., & Hillier, S. L. (1991). Rapid presumptive identification of black-pigmented gram-negative anaerobic bacteria by using 4-methylumbelliferone derivatives. *Journal of Clinical Microbiology, 29*, 1955–1958.

Moses, P. J., Power, D. A., Jesionowski, A. M., Jenkinson, H. F., Pantera, E. A., & Vickerman, M. M. (2013). *Streptococcus gordonii* collagen-binding domain protein CbdA may enhance bacterial survival in instrumented root canals ex vivo. *Journal of Endodontics, 39*, 39–43.

Moustafa, I., Connaris, H., Taylor, M., Zaitsev, V., Wilson, J. C., Kiefel, M. J., et al. (2004). Sialic acid recognition by *Vibrio cholerae* neuraminidase. *The Journal of Biological Chemistry, 279*, 40819–40826.

Moynihan, P. J., Ferrier, S., Blomley, S., Wright, W. G., & Russell, R. R. (1998). Acid production from lactulose by dental plaque bacteria. *Letters in Applied Microbiology, 27*, 173–177.

Munksgaard, P. S., Skals, M., Reinholdt, J., Poulsen, K., Jensen, M. R., Yang, C., et al. (2014). Sialic acid residues are essential for cell lysis mediated by leukotoxin from *Aggregatibacter actinomycetemcomitans. Infection and Immunity, 82*, 2219–2228.

Munson, M. A., Banerjee, A., Watson, T. F., & Wade, W. G. (2004). Molecular analysis of the microflora associated with dental caries. *Journal of Clinical Microbiology, 42*, 3023–3029.

Murakami, Y., Masuda, T., Imai, M., Iwami, J., Nakamura, H., Noguchi, T., et al. (2004). Analysis of major virulence factors in *Porphyromonas gingivalis* under various culture temperatures using specific antibodies. *Microbiology and Immunology, 48*, 561–569.

Murakami, M., Ohtake, T., Dorschner, R. A., & Gallo, R. L. (2002). Cathelicidin antimicrobial peptides are expressed in salivary glands and saliva. *Journal of Dental Research, 81*, 845–850.

Myneni, S. R., Settem, R. P., Connell, T. D., Keegan, A. D., Gaffen, S. L., & Sharma, A. (2011). TLR2 signaling and Th2 responses drive *Tannerella forsythia*-induced periodontal bone loss. *Journal of Immunology (Baltimore, Md.: 1950), 187*, 501–509.

Myneni, S. R., Settem, R. P., Sojar, H. T., Malone, J. P., Vuokko, L., Nakajima, T., et al. (2012). Identification of a unique TLR2-interacting peptide motif in a microbial leucine-rich repeat protein. *Biochemical and Biophysical Research Communications, 423*(3), 577–582.

Nagaoka, S., Liu, H. J., Minemoto, K., & Kawagoe, M. (1995). Microbial induction of dentinal caries in human teeth in vitro. *Journal of Endodontics, 21*, 546–551.

Nakagaki, H., Sekine, S., Terao, Y., Toe, M., Tanaka, M., Ito, H.-O., et al. (2010). *Fusobacterium nucleatum* envelope protein FomA is immunogenic and binds to the salivary statherin-derived peptide. *Infection and Immunity, 78*, 1185–1192.
Nakajo, K., Takahashi, N., & Beighton, D. (2010). Resistance to acidic environments of caries-associated bacteria: *Bifidobacterium dentium* and *Bifidobacterium longum*. *Caries Research, 44*, 431–437.
Nakano, K., Nomura, R., Taniguchi, N., Lapirattanakul, J., Kojima, A., Naka, S., et al. (2010). Molecular characterization of *Streptococcus mutans* strains containing the cnm gene encoding a collagen-binding adhesin. *Archives of Oral Biology, 55*, 34–39.
Nguyen, K.-A., Travis, J., & Potempa, J. (2007). Does the importance of the C-terminal residues in the maturation of RgpB from *Porphyromonas gingivalis* reveal a novel mechanism for protein export in a subgroup of Gram-Negative bacteria? *Journal of Bacteriology, 189*, 833–843.
Nihira, T., Suzuki, E., Kitaoka, M., Nishimoto, M., Ohtsubo, K., & Nakai, H. (2013). Discovery of β-1,4-D-mannosyl-N-acetyl-D-glucosamine phosphorylase involved in the metabolism of N-glycans. *The Journal of Biological Chemistry, 288*, 27366–27374.
Nithianantham, S., Xu, M., Yamada, M., Ikegami, A., Shoham, M., & Han, Y. W. (2009). Crystal structure of FadA adhesin from *Fusobacterium nucleatum* reveals a novel oligomerization motif, the leucine chain. *The Journal of Biological Chemistry, 284*, 3865–3872.
Nobbs, A. H., Jenkinson, H. F., & Jakubovics, N. S. (2011). Stick to your gums: Mechanisms of oral microbial adherence. *Journal of Dental Research, 90*, 1271–1278.
Nobbs, A. H., Shearer, B. H., Drobni, M., Jepson, M. A., & Jenkinson, H. F. (2007). Adherence and internalization of *Streptococcus gordonii* by epithelial cells involves beta1 integrin recognition by SspA and SspB (antigen I/II family) polypeptides. *Cellular Microbiology, 9*, 65–83.
Nomura, R., Nakano, K., Naka, S., Nemoto, H., Masuda, K., Lapirattanakul, J., et al. (2012). Identification and characterization of a collagen-binding protein, Cbm, in *Streptococcus mutans*. *Molecular Oral Microbiology, 27*, 308–323.
O'Brien-Simpson, N. M., Paolini, R. A., Hoffmann, B., Slakeski, N., Dashper, S. G., & Reynolds, E. C. (2001). Role of RgpA, RgpB, and Kgp proteinases in virulence of *Porphyromonas gingivalis* W50 in a murine lesion model. *Infection and Immunity, 69*, 7527–7534.
Oda, H., Saiki, K., Tonosaki, M., Yajima, A., & Konishi, K. (2009). Participation of the secreted dipeptidyl and tripeptidyl aminopeptidases in asaccharolytic growth of *Porphyromonas gingivalis*. *Journal of Periodontal Research, 44*, 362–367.
Offenbacher, S. (1996). Periodontal diseases: Pathogenesis. *Annals of Periodontology/The American Academy of Periodontology, 1*, 821–878.
Ohara-Nemoto, Y., Rouf, S. M. A., Naito, M., Yanase, A., Tetsuo, F., Ono, T., et al. (2014). Identification and characterization of prokaryotic dipeptidyl-peptidase 5 from *Porphyromonas gingivalis*. *The Journal of Biological Chemistry, 289*, 5436–5448.
Ohara-Nemoto, Y., Shimoyama, Y., Kimura, S., Kon, A., Haraga, H., Ono, T., et al. (2011). Asp- and Glu-specific novel dipeptidyl peptidase 11 of *Porphyromonas gingivalis* ensures utilization of proteinaceous energy sources. *The Journal of Biological Chemistry, 286*, 38115–38127.
Okahashi, N., Nakata, M., Sakurai, A., Terao, Y., Hoshino, T., Yamaguchi, M., et al. (2010). Pili of oral *Streptococcus sanguinis* bind to fibronectin and contribute to cell adhesion. *Biochemical and Biophysical Research Communications, 391*, 1192–1196.
Okahashi, N., Nakata, M., Terao, Y., Isoda, R., Sakurai, A., Sumitomo, T., et al. (2011). Pili of oral *Streptococcus sanguinis* bind to salivary amylase and promote the biofilm formation. *Microbial Pathogenesis, 50*, 148–154.

Okahashi, N., Sasakawa, C., & Yoshikawa, M. (1989). Molecular characterization of a surface protein antigen gene from serotype c *Streptococcus mutans*, implicated in dental caries. *Molecular Microbiology, 3*, 673–678.

Okuda, K., Hanada, N., Usui, Y., Takeuchi, H., Koba, H., Nakao, R., et al. (2010). Inhibition of *Streptococcus mutans* adherence and biofilm formation using analogues of the SspB peptide. *Archives of Oral Biology, 55*, 754–762, Elsevier Ltd.

Oppenheim, F. G., Salih, E., Siqueira, W. L., Zhang, W., & Helmerhorst, E. J. (2007). Salivary proteome and its genetic polymorphisms. *Annals of the New York Academy of Sciences, 1098*, 22–50.

Paddick, J. S., Brailsford, S. R., Kidd, E. A. M., & Beighton, D. (2005). Phenotypic and genotypic selection of microbiota surviving under dental restorations. *Applied and Environmental Microbiology, 71*, 2467–2472. http://dx.doi.org/10.1128/AEM.71.5.2467.

Palmer, S. R., Miller, J. H., Abranches, J., Zeng, L., Lefebure, T., Richards, V. P., et al. (2013). Phenotypic heterogeneity of genomically-diverse isolates of *Streptococcus mutans*. *PLoS One, 8*, e61358.

Paramonov, N., Rangarajan, M., Hashim, A., Gallagher, A., Aduse-Opoku, J., Slaney, J. M., et al. (2005). Structural analysis of a novel anionic polysaccharide from *Porphyromonas gingivalis* strain W50 related to Arg-gingipain glycans. *Molecular Microbiology, 58*, 847–863.

Park, Y., Simionato, M. R., Sekiya, K., Murakami, Y., James, D., Chen, W., et al. (2005). Short Fimbriae of *Porphyromonas gingivalis* and Their Role in Coadhesion with Streptococcus gordonii. *Infection and Immunity, 73*, 3983–3989.

Park, J.-J., Yi, J. Y., Jin, Y. B., Lee, Y.-J., Lee, J.-S., Lee, Y.-S., et al. (2012). Sialylation of epidermal growth factor receptor regulates receptor activity and chemosensitivity to gefitinib in colon cancer cells. *Biochemical Pharmacology, 83*, 849–857.

Perez, B. A., Planet, P. J., Kachlany, S. C., Tomich, M., Fine, D. H., & Figurski, D. H. (2006). Genetic analysis of the requirement for flp-2, tadV, and rcpB in *Actinobacillus actinomycetemcomitans* biofilm formation. *Journal of Bacteriology, 188*, 6361–6375.

Pham, T., Gregg, C. J., Karp, F., Chow, R., Padler-karavani, V., Cao, H., et al. (2012). Evidence for a novel human-specific xeno-auto-antibody response against vascular endothelium. *Blood, 114*, 5225–5235.

Pham, T. K., Roy, S., Noirel, J., Douglas, I., Wright, P. C., Stafford, G. P., et al. (2010). A quantitative proteomic analysis of biofilm adaptation by the periodontal pathogen *Tannerella forsythia*. *Proteomics, 10*, 3130–3141.

Phansopa, C., Roy, S., Rafferty, J. B., Douglas, C. W. I., Pandhal, J., Wright, P. C., et al. (2013). Structural and functional characterisation of NanU, a novel high-affinity sialic acid-inducible binding protein of oral and gut-dwelling Bacteroidetes spp. *The Biochemical Journal, 458*, 499–511.

Pierce, D. L., Nishiyama, S., Liang, S., Wang, M., Triantafilou, M., Triantafilou, K., et al. (2009). Host adhesive activities and virulence of novel fimbrial proteins of *Porphyromonas gingivalis*. *Infection and Immunity, 77*, 3294–3301.

Pihlstrom, B. L., Michalowicz, B. S., & Johnson, N. W. (2005). Periodontal diseases. *Lancet, 366*, 1809–1820.

Pike, R. N., Potempa, J., McGraw, W., Coetzer, T. H., & Travis, J. (1996). Characterization of the binding activities of proteinase-adhesin complexes from *Porphyromonas gingivalis*. *Journal of Bacteriology, 178*, 2876–2882.

Poole, S., Singhrao, S. K., Kesavalu, L., Curtis, M. A., & Crean, S. (2013). Determining the presence of periodontopathic virulence factors in short-term postmortem Alzheimer's disease brain tissue. *Journal of Alzheimer's Disease, 36*, 665–677.

Posch, G., Pabst, M., Brecker, L., Altmann, F., Messner, P., & Schäffer, C. (2011). Characterization and scope of S-layer protein O-glycosylation in *Tannerella forsythia*. *The Journal of Biological Chemistry, 286*, 38714–38724.

Posch, G., Pabst, M., Neumann, L., Coyne, M. J., Altmann, F., Messner, P., et al. (2013). "Cross-glycosylation" of proteins in Bacteroidales species. *Glycobiology, 23*, 568–577.

Posch, G., Sekot, G., Friedrich, V., Megson, Z. A., Koerdt, A., Messner, P., et al. (2012). Glycobiology aspects of the periodontal pathogen *Tannerella forsythia*. *Biomolecules, 2*, 467–482.

Potempa, J., Mikolajczyk-Pawlinska, J., Brassell, D., Nelson, D., Thøgersen, I. B., Enghild, J. J., et al. (1998). Comparative properties of two cysteine proteinases (gingipains R), the products of two related but individual genes of *Porphyromonas gingivalis*. *The Journal of Biological Chemistry, 273*, 21648–21657.

Puntervoll, P., Ruud, M., Bruseth, L. J., Kleivdal, H., Høgh, B. T., Benz, R., et al. (2002). Structural characterization of the fusobacterial non-specific porin FomA suggests a 14-stranded topology, unlike the classical porins. *Microbiology, 148*, 3395–3403.

Qian, H., & Dao, M. L. (1993). Inactivation of the *Streptococcus mutans* wall-associated protein A gene (wapA) results in a decrease in sucrose-dependent adherence and aggregation. *Infection and Immunity, 61*(12), 5021–5028.

Quivey, R. G., Kuhnert, W. L., & Hahn, K. (2001). Genetics of acid adaptation in oral streptococci. *Critical Reviews in Oral Biology and Medicine: An Official Publication of the American Association of Oral Biologists, 12*, 301–314.

Rangarajan, M., Aduse-Opoku, J., Paramonov, N., Hashim, A., Bostanci, N., Fraser, O. P., et al. (2008). Identification of a second lipopolysaccharide in *Porphyromonas gingivalis* W50. *Journal of Bacteriology, 190*, 2920–2932.

Ray, S., Jayanth, P., Franchuk, S., Finlay, T., Seyrantepe, V., Beyaert, R., et al. (2010). Neu1 desialylation of sialyl alpha-2,3-linked beta-galactosyl residues of TOLL-like receptor 4 is essential for receptor activation and cellular signaling. *Cellular Signalling, 22*, 314–324.

Reardon-Robinson, M. E., Wu, C., Mishra, A., Chang, C., Bier, N., Das, A., et al. (2014). Pilus hijacking by a bacterial coaggregation factor critical for oral biofilm development. *Proceedings of the National Academy of Sciences of the United States of America, 111*, 3835–3840.

Reinholdt, J., Tomana, M., Mortensen, S. B., & Kilian, M. (1990). Molecular aspects of immunoglobulin A1 degradation by oral streptococci. *Infection and Immunity, 58*, 1186–1194.

Renzi, F., Manfredi, P., Mally, M., Moes, S., Jenö, P., Cornelis, G. R., et al. (2011). The N-glycan glycoprotein deglycosylation complex (Gpd) from *Capnocytophaga canimorsus* deglycosylates human IgG. *PLoS Pathogens*, 7, e1002118.

Rickard, A. H., Gilbert, P., High, N. J., Kolenbrander, P. E., & Handley, P. S. (2003). Bacterial coaggregation: An integral process in the development of multi-species biofilms. *Trends in Microbiology, 11*, 94–100.

Robbe, C., Capon, C., Maes, E., Rousset, M., Zweibaum, A., Zanetta, J.-P., et al. (2003). Evidence of regio-specific glycosylation in human intestinal mucins: Presence of an acidic gradient along the intestinal tract. *The Journal of Biological Chemistry, 278*, 46337–46348.

Robles, A. G., Reid, K., Roy, F., & Fletcher, H. M. (2011). *Porphyromonas gingivalis* mutY is involved in the repair of oxidative stress-induced DNA mispairing. *Molecular Oral Microbiology, 26*, 175–186.

Roggentin, P., Schauer, R., Hoyer, L. L., & Vimr, E. R. (1993). The sialidase superfamily and its spread by horizontal gene transfer. *Molecular Microbiology, 9*, 915–921.

Rose, J. E., Meyer, D. H., & Fives-Taylor, P. M. (2003). Aae, an autotransporter involved in adhesion of *Actinobacillus actinomycetemcomitans* to epithelial cells. *Infection and Immunity, 71*, 2384–2393.

Roy, S., Douglas, C. W. I., & Stafford, G. P. (2010). A novel sialic acid utilization and uptake system in the periodontal pathogen *Tannerella forsythia*. *Journal of Bacteriology, 192*, 2285–2293.

Roy, S., Honma, K., Douglas, C. W. I., Sharma, A., & Stafford, G. P. (2011). Role of sialidase in glycoprotein utilization by *Tannerella forsythia*. *Microbiology (Reading, England)*, *157*, 3195–3202.

Roy, S., Phansopa, C., Stafford, P., Honma, K., Douglas, C. W. I., Sharma, A., et al. (2012). Beta-hexosaminidase activity of the oral pathogen *Tannerella forsythia* influences biofilm formation on glycoprotein substrates. *FEMS Immunology and Medical Microbiology*, *65*, 116–120.

Rubinstein, M. R., Wang, X., Liu, W., Hao, Y., Cai, G., & Han, Y. W. (2013). *Fusobacterium nucleatum* promotes colorectal carcinogenesis by modulating E-cadherin/β-catenin signaling via its FadA adhesin. *Cell Host & Microbe*, *14*, 195–206.

Rudney, J. D., Chen, R., & Sedgewick, G. J. (2001). Intracellular *Actinobacillus actinomycetemcomitans* and *Porphyromonas gingivalis* in buccal epithelial cells collected from human subjects. *Infection and Immunity*, *69*, 2700–2707.

Ruggiero, S., Cosgarea, R., Potempa, J., Potempa, B., Eick, S., & Chiquet, M. (2013). Cleavage of extracellular matrix in periodontitis: Gingipains differentially affect cell adhesion activities of fibronectin and tenascin-C. *Biochimica et Biophysica Acta*, *1832*, 517–526.

Ruhl, S., Cisar, J. O., & Sandberg, A. L. (2000). Identification of polymorphonuclear leukocyte and HL-60 cell receptors for adhesins of *Streptococcus gordonii* and *Actinomyces naeslundii*. *Infection and Immunity*, *68*, 6346–6354.

Ruiz, T., Lenox, C., Radermacher, M., & Mintz, K. P. (2006). Novel surface structures are associated with the adhesion of *Actinobacillus actinomycetemcomitans* to collagen. *Infection and Immunity*, *74*, 6163–6170.

Russell, R. R., Aduse-Opoku, J., Sutcliffe, I. C., Tao, L., & Ferretti, J. J. (1992). A binding protein-dependent transport system in *Streptococcus mutans* responsible for multiple sugar metabolism. *The Journal of Biological Chemistry*, *267*, 4631–4637.

Ryan, C. S., & Kleinberg, I. (1995). Bacteria in human mouths involved in the production and utilization of hydrogen peroxide. *Archives of Oral Biology*, *40*, 753–763.

Saiki, K., & Konishi, K. (2010). The role of Sov protein in the secretion of gingipain protease virulence factors of *Porphyromonas gingivalis*. *FEMS Microbiology Letters*, *302*, 166–174.

Saiki, K., & Konishi, K. (2014). *Porphyromonas gingivalis* C-terminal signal peptidase PG0026 and HagA interact with outer membrane protein PG27/LptO. *Molecular Oral Microbiology*, *29*, 32–44.

Sakakibara, J., Nagano, K., Murakami, Y., Higuchi, N., Nakamura, H., Shimozato, K., et al. (2007). Loss of adherence ability to human gingival epithelial cells in S-layer protein-deficient mutants of *Tannerella forsythensis*. *Microbiology*, *153*, 866–876.

Santiago, B., MacGilvray, M., Faustoferri, R. C., & Quivey, R. G. (2012). The branched-chain amino acid aminotransferase encoded by ilvE is involved in acid tolerance in *Streptococcus mutans*. *Journal of Bacteriology*, *194*, 2010–2019.

Sasaki, Y., Nogami, E., Maeda, M., Nakanishi-Matsui, M., & Iwamoto-Kihara, A. (2014). A unique F-type H^+-ATPase from *Streptococcus mutans*: An active H^+ pump at acidic pH. *Biochemical and Biophysical Research Communications*, *443*, 677–682.

Sato, K., Naito, M., Yukitake, H., Hirakawa, H., Shoji, M., McBride, M. J., et al. (2010). A protein secretion system linked to bacteroidete gliding motility and pathogenesis. *Proceedings of the National Academy of Sciences of the United States of America*, *107*, 276–281.

Sato, K., Sakai, E., Veith, P. D., Shoji, M., Kikuchi, Y., Yukitake, H., et al. (2005). Identification of a new membrane-associated protein that influences transport/maturation of gingipains and adhesins of *Porphyromonas gingivalis*. *The Journal of Biological Chemistry*, *280*, 8668–8677.

Schaefer, A. S., Richter, G. M., Groessner-Schreiber, B., Noack, B., Nothnagel, M., El Mokhtari, N.-E., et al. (2009). Identification of a shared genetic susceptibility locus for coronary heart disease and periodontitis. *PLoS Genetics*, *5*, e1000378.

Schätzle, M., Löe, H., Bürgin, W., Anerud, A., Boysen, H., & Lang, N. P. (2003). Clinical course of chronic periodontitis. I. Role of gingivitis. *Journal of Clinical Periodontology*, *30*, 887–901.

Schenkein, H. A., Fletcher, H. M., Bodnar, M., & Macrina, F. L. (1995). Increased opsonization of a prtH-defective mutant of *Porphyromonas gingivalis* W83 is caused by reduced degradation of complement-derived opsonins. *Journal of Immunology*, *154*, 5331–5337.

Schneewind, O., & Missiakas, D. M. (2012). Protein secretion and surface display in Gram-positive bacteria. *Philosophical Transactions of the Royal Society of London. Series B, Biological Sciences*, *367*, 1123–1139.

Sciotti, M. A., Yamodo, I., Klein, J. P., & Ogier, J. A. (1997). The N-terminal half part of the oral streptococcal antigen I/IIf contains two distinct binding domains. *FEMS Microbiology Letters*, *153*, 439–445.

Seepersaud, R., Bensing, B. A., Yen, Y. T., & Sullam, P. M. (2010). Asp3 mediates multiple protein-protein interactions within the accessory Sec system of *Streptococcus gordonii*. *Molecular Microbiology*, *78*, 490–505.

Seers, C. A., Slakeski, N., Veith, P. D., Nikolof, T., Chen, Y.-Y., Dashper, S. G., et al. (2006). The RgpB C-terminal domain has a role in attachment of RgpB to the outer membrane and belongs to a novel C-terminal-domain family found in *Porphyromonas gingivalis*. *Journal of Bacteriology*, *188*, 6376–6386.

Sekot, G., Posch, G., Oh, Y. J., Zayni, S., Mayer, H. F., Pum, D., et al. (2012). Analysis of the cell surface layer ultrastructure of the oral pathogen *Tannerella forsythia*. *Archives of Microbiology*, *194*, 525–539.

Settem, R. P., Honma, K., Nakajima, T., Phansopa, C., Roy, S., Stafford, G. P., et al. (2013). A bacterial glycan core linked to surface (S)-layer proteins modulates host immunity through Th17 suppression. *Mucosal Immunology*, *6*, 415–426.

Settem, R. P., Honma, K., Stafford, G. P., & Sharma, A. (2013). Protein-linked glycans in periodontal bacteria: Prevalence and role at the immune interface. *Frontiers in Microbiology*, *4*, 310.

Sharma, A. (2010). Virulence mechanisms of *Tannerella forsythia*. *Periodontology*, *2000*(54), 106–116.

Sharma, A., Inagaki, S., Sigurdson, W., & Kuramitsu, H. K. (2005). Synergy between *Tannerella forsythia* and *Fusobacterium nucleatum* in biofilm formation. *Oral Microbiology and Immunology*, *20*(1), 39–42.

Sharma, A., Sojar, H. T., Glurich, I., Honma, K., Kuramitsu, H. K., & Genco, R. J. (1998). Cloning, expression, and sequencing of a cell surface antigen containing a leucine-rich repeat motif from Bacteroides forsythus ATCC 43037. *Infection and Immunity*, *66*, 5703–5710.

Sheng, J., & Marquis, R. E. (2006). Enhanced acid resistance of oral streptococci at lethal pH values associated with acid-tolerant catabolism and with ATP synthase activity. *FEMS Microbiology Letters*, *262*, 93–98.

Shibata, Y., Kawada-Matsuo, M., Shirai, Y., Saito, N., Li, D., & Yamashita, Y. (2011). *Streptococcus mutans* diacylglycerol kinase homologue: A potential target for anti-caries chemotherapy. *Journal of Medical Microbiology*, *60*, 625–630.

Shibata, Y., Miwa, Y., Hirai, K., & Fujimura, S. (2003). Purification and partial characterization of a dipeptidyl peptidase from *Prevotella intermedia*. *Oral Microbiology and Immunology*, *18*, 196–198.

Shibata, Y., van der Ploeg, J. R., Kozuki, T., Shirai, Y., Saito, N., Kawada-Matsuo, M., et al. (2009). Kinase activity of the dgk gene product is involved in the virulence of *Streptococcus mutans*. *Microbiology (Reading, England)*, *155*, 557–565.

Shimotahira, N., Oogai, Y., Kawada-Matsuo, M., Yamada, S., Fukutsuji, K., Nagano, K., et al. (2013). The surface layer of *Tannerella forsythia* contributes to serum resistance and oral bacterial coaggregation. *Infection and Immunity*, *81*, 1198–1206.

Shoji, M., Ratnayake, D. B., Shi, Y., Kadowaki, T., Yamamoto, K., Yoshimura, F., et al. (2002). Construction and characterization of a nonpigmented mutant of *Porphyromonas gingivalis*: Cell surface polysaccharide as an anchorage for gingipains. *Microbiology (Reading, England)*, *148*, 1183–1191.

Silverman, R. J., Nobbs, A. H., Vickerman, M. M., Barbour, M. E., & Jenkinson, H. F. (2010). Interaction of *Candida albicans* cell wall Als3 protein with *Streptococcus gordonii* SspB adhesin promotes development of mixed-species communities. *Infection and Immunity*, *78*, 4644–4652.

Siqueira, J. F., & Rôças, I. N. (2009). Diversity of endodontic microbiota revisited. *Journal of Dental Research*, *88*, 969–981.

Slakeski, N., Seers, C. A., Ng, K., Moore, C., Cleal, S. M., Veith, P. D., et al. (2011). C-terminal domain residues important for secretion and attachment of RgpB in *Porphyromonas gingivalis*. *Journal of Bacteriology*, *193*, 132–142.

Slámová, K., Bojarová, P., Petrásková, L., & Kren, V. (2010). β-N-acetylhexosaminidase: What's in a name...? *Biotechnology Advances*, *28*, 682–693.

Smalley, J. W., Silver, J., Birss, A. J., Withnall, R., & Titler, P. J. (2003). The haem pigment of the oral anaerobes *Prevotella nigrescens* and *Prevotella intermedia* is composed of iron(III) protoporphyrin IX in the monomeric form. *Microbiology (Reading, England)*, *149*, 1711–1718.

Smalley, J. W., Thomas, M. F., Birss, A. J., Withnall, R., & Silver, J. (2004). A combination of both arginine- and lysine-specific gingipain activity of *Porphyromonas gingivalis* is necessary for the generation of the micro-oxo bishaem-containing pigment from haemoglobin. *The Biochemical Journal*, *379*, 833–840.

Socransky, S. S., Haffajee, A. D., Cugini, M. A., Smith, C., & Kent, R. L. (1998). Microbial complexes in subgingival plaque. *Journal of Clinical Periodontology*, *25*, 134–144.

Sojar, H. T., & Smith, D. F. (2012). *Porphyromonas gingivalis* fimbriae carbohydrate specificity assessment by glycomics. *FEMS Immunology and Medical Microbiology*, *66*, 83–87.

Stafford, P., Higham, J., Pinnock, A., Murdoch, C., Douglas, C. W. I., Stafford, G. P., et al. (2013). Gingipain-dependent degradation of mammalian target of rapamycin pathway proteins by the periodontal pathogen *Porphyromonas gingivalis* during invasion. *Molecular Oral Microbiology*, *28*, 366–378.

Stafford, G., Roy, S., Honma, K., & Sharma, A. (2012). Sialic acid, periodontal pathogens and *Tannerella forsythia*: Stick around and enjoy the feast! *Molecular Oral Microbiology*, *27*, 11–22.

Stamatos, N. M., Carubelli, I., van de Vlekkert, D., Bonten, E. J., Papini, N., Feng, C., et al. (2010). LPS-induced cytokine production in human dendritic cells is regulated by sialidase activity. *Journal of Leukocyte Biology*, *88*, 1227–1239.

Stanley, P., & Cummings, R. (2009). Structures common to different glycans. In A. Varki, R. Cummings, & J. Esko (Eds.), *Essentials in glycobiology* (2nd ed.). Cold Spring Harbor (NY): Cold Spring Harbor Laboratory Press.

Steenbergen, S. M., Jirik, J. L., & Vimr, E. R. (2009). YjhS (NanS) is required for *Escherichia coli* to grow on 9-O-acetylated N-acetylneuraminic acid. *Journal of Bacteriology*, *191*, 7134–7139.

Steeves, C. H., Potrykus, J., Barnett, D. A., & Bearne, S. L. (2011). Oxidative stress response in the opportunistic oral pathogen *Fusobacterium nucleatum*. *Proteomics*, *11*, 2027–2037.

Sumei, L., Klein, M. I., Heim, K. P., Fan, Y., Bitoun, J. P., Ahn, S.-J., et al. (2014). *Streptococcus mutans* eDNA is up-regulated during growth in biofilms, actively released via membrane vesicles, and influenced by components of the protein secretion machinery. *Journal of Bacteriology*, *196*, 2355–2366.

Suwannakul, S., Stafford, G. P., Whawell, S. A., & Douglas, C. W. I. (2010). Identification of bistable populations of *Porphyromonas gingivalis* that differ in epithelial cell invasion. *Microbiology (Reading, England)*, *156*, 3052–3064.

Suzuki, N., Yoneda, M., & Hirofuji, T. (2013). Mixed red-complex bacterial infection in periodontitis. *International Journal of Dentistry*, *2013*, 587279.

Tabak, L. A. (1995). In defense of the oral cavity: Structure, biosynthesis, and function of salivary mucins. *Annual Review of Physiology*, *57*, 547–564.

Taichman, N. S., Dean, R. T., & Sanderson, C. J. (1980). Biochemical and morphological characterization of the killing of human monocytes by a leukotoxin derived from *Actinobacillus actinomycetemcomitans*. *Infection and Immunity*, *28*, 258–268.

Takahashi, N. (2003). Acid-neutralizing activity during amino acid fermentation by *Porphyromonas gingivalis, Prevotella intermedia* and *Fusobacterium nucleatum*. *Oral Microbiology and Immunology*, *18*, 109–113.

Takahashi, Y., Konishi, K., Cisar, J. O., & Yoshikawa, M. (2002). Identification and characterization of hsa, the gene encoding the sialic acid-binding adhesin of *Streptococcus gordonii* DL1. *Infection and Immunity*, *70*, 1209–1218.

Takahashi, N., & Nyvad, B. (2011). The role of bacteria in the caries process: Ecological perspectives. *Journal of Dental Research*, *90*, 294–303.

Takahashi, N., Sato, T., & Yamada, T. (2000). Metabolic pathways for cytotoxic end product formation from glutamate- and aspartate-containing peptides by *Porphyromonas gingivalis*. *Journal of Bacteriology*, *182*, 4704–4710.

Takahashi, Y., Yajima, A., Cisar, J. O., & Konishi, K. (2004). Functional analysis of the *Streptococcus gordonii* DL1 sialic acid-binding adhesin and its essential role in bacterial binding to platelets. *Infection and Immunity*, *72*, 3876–3882.

Takii, R., Kadowaki, T., Baba, A., Tsukuba, T., & Yamamoto, K. (2005). A functional virulence complex composed of gingipains, adhesins, and lipopolysaccharide shows high affinity to host cells and matrix proteins and escapes recognition by host immune systems. *Infection and Immunity*, *73*, 883–893.

Tang, G., & Mintz, K. P. (2010). Glycosylation of the collagen adhesin EmaA of *Aggregatibacter actinomycetemcomitans* is dependent upon the lipopolysaccharide biosynthetic pathway. *Journal of Bacteriology*, *192*, 1395–1404.

Tang, G., Ruiz, T., & Mintz, K. P. (2012). O-polysaccharide glycosylation is required for stability and function of the collagen adhesin EmaA of *Aggregatibacter actinomycetemcomitans*. *Infection and Immunity*, *80*, 2868–2877.

Tao, L., Sutcliffe, I. C., Russell, R. R., & Ferretti, J. J. (1993). Transport of sugars, including sucrose, by the msm transport system of *Streptococcus mutans*. *Journal of Dental Research*, *72*, 1386–1390.

Témoin, S., Wu, K. L., Wu, V., Shoham, M., & Han, Y. W. (2012). Signal peptide of FadA adhesin from *Fusobacterium nucleatum* plays a novel structural role by modulating the filament's length and width. *FEBS Letters*, *586*, 1–6.

Thompson, H., Homer, K. A., Rao, S., Booth, V., & Hosie, A. H. (2009). An orthologue of Bacteroides fragilis NanH is the principal sialidase in *Tannerella forsythia*. *Journal of Bacteriology*, *191*, 3623–3628.

Thomsson, K. A., Prakobphol, A., Leffler, H., Reddy, M. S., Levine, M. J., Fisher, S. J., et al. (2002). The salivary mucin MG1 (MUC5B) carries a repertoire of unique oligosaccharides that is large and diverse. *Glycobiology*, *12*, 1–14.

Toba, S., Tenno, M., & Kurosaka, A. (2000). An O-acetylated sialyl-Tn is involved in ovarian cancer-associated antigenicity. *Biochemical and Biophysical Research Communications*, *271*, 281–286.

Troffer-Charlier, N., Ogier, J., Moras, D., & Cavarelli, J. (2002). Crystal structure of the V-region of *Streptococcus mutans* antigen I/II at 2.4 A resolution suggests a sugar preformed binding site. *Journal of Molecular Biology*, *318*, 179–188.

Tsai, C. C., McArthur, W. P., Baehni, P. C., Hammond, B. F., & Taichman, N. S. (1979). Extraction and partial characterization of a leukotoxin from a plaque-derived Gram-negative microorganism. *Infection and Immunity*, *25*, 427–439.

Tseng, T.-T., Tyler, B. M., & Setubal, J. C. (2009). Protein secretion systems in bacterial-host associations, and their description in the Gene Ontology. *BMC Microbiology, 9*(Suppl. 1), S2.

Uchiyama, S., Carlin, A. F., Khosravi, A., Weiman, S., Banerjee, A., Quach, D., et al. (2009). The surface-anchored NanA protein promotes pneumococcal brain endothelial cell invasion. *The Journal of Experimental Medicine, 206*, 1845–1852.

Umemoto, T., & Hamada, N. (2003). Characterization of biologically active cell surface components of a periodontal pathogen. The roles of major and minor fimbriae of *Porphyromonas gingivalis*. *Journal of Periodontology, 74*, 119–122.

Van Hijum, S. A. F. T., Kralj, S., Ozimek, L. K., Dijkhuizen, L., & van Geel-Schutten, I. G. H. (2006). Structure-function relationships of glucansucrase and fructansucrase enzymes from lactic acid bacteria. *Microbiology and Molecular Biology Reviews, 70*, 157–176.

Vavasseur, F., Dole, K., Yang, J., Matta, K. L., Myerscough, N., Corfield, A., et al. (1994). O-glycan biosynthesis in human colorectal adenoma cells during progression to cancer. *European Journal of Biochemistry/FEBS, 222*, 415–424.

Veith, P. D., Chen, Y.-Y., Gorasia, D. G., Chen, D., Glew, M. D., O'Brien-Simpson, N. M., et al. (2014). *Porphyromonas gingivalis* outer membrane vesicles exclusively contain outer membrane and periplasmic proteins and carry a cargo enriched with virulence factors. *Journal of Proteome Research, 13*, 2420–2432.

Veith, P. D., Nor Muhammad, N. A., Dashper, S. G., Likić, V. A., Gorasia, D. G., Chen, D., et al. (2013). Protein substrates of a novel secretion system are numerous in the bacteroidetes phylum and have in common a cleavable C-terminal secretion signal, extensive post-translational modification, and cell-surface attachment. *Journal of Proteome Research, 12*, 4449–4461.

Veith, P. D., O'Brien-Simpson, N. M., Tan, Y., Djatmiko, D. C., Dashper, S. G., & Reynolds, E. C. (2009). Outer membrane proteome and antigens of Tannerella forsythia. *Journal of Proteome Research, 8*, 4279–4292.

Veith, P. D., Talbo, G. H., Slakeski, N., Dashper, S. G., Moore, C., Paolini, R. A., et al. (2002). Major outer membrane proteins and proteolytic processing of RgpA and Kgp of *Porphyromonas gingivalis* W50. *The Biochemical Journal, 363*, 105–115.

Vettore, M., Quintanilha, R. S., Monteiro da Silva, A. M., Lamarca, G. A., & Leão, A. T. T. (2005). The influence of stress and anxiety on the response of non-surgical periodontal treatment. *Journal of Clinical Periodontology, 32*, 1226–1235.

Vroom, J. M., De Grauw, K. J., Gerritsen, H. C., Bradshaw, D. J., Marsh, P. D., Watson, G. K., et al. (1999). Depth penetration and detection of pH gradients in biofilms by two-photon excitation microscopy. *Applied and Environmental Microbiology, 65*, 3502–3511.

Wang, B.-Y., Deutch, A., Hong, J., & Kuramitsu, H. K. (2011). Proteases of an early colonizer can hinder *Streptococcus mutans* colonization in vitro. *Journal of Dental Research, 90*, 501–505.

Wang, R., He, X., Hu, W., Lux, R., Li, J., Zhou, X., et al. (2011). Analysis of interspecies adherence of oral bacteria using a membrane binding assay coupled with polymerase chain reaction-denaturing gradient gel electrophoresis profiling. *International Journal of Oral Science, 3*, 90–97.

Wang, M., Krauss, J. L., Domon, H., Hosur, K. B., Liang, S., Magotti, P., et al. (2010). Microbial hijacking of complement-toll-like receptor crosstalk. *Science Signaling, 3*, ra11.

Warinner, C., Rodrigues, J. F. M., Vyas, R., Trachsel, C., Shved, N., Grossmann, J., et al. (2014). Pathogens and host immunity in the ancient human oral cavity. *Nature Genetics, 46*, 336–344.

Webb, A. J., Homer, K. A., & Hosie, A. H. F. (2008). Two closely related ABC transporters in *Streptococcus mutans* are involved in disaccharide and/or oligosaccharide uptake. *Journal of Bacteriology, 190*, 168–178.

Wen, Z. T., & Burne, R. A. (2004). LuxS-mediated signaling in *Streptococcus mutans* is involved in regulation of acid and oxidative stress tolerance and biofilm formation. *Journal of Bacteriology*, *186*, 2682–2691.
Whitchurch, C. B., Tolker-Nielsen, T., Ragas, P. C., & Mattick, J. S. (2002). Extracellular DNA required for bacterial biofilm formation. *Science (New York, N.Y.)*, *295*, 1487.
WHO: Fact sheet N°318. (2012). *Oral health*. Retrieved from, http://www.who.int/mediacentre/factsheets/fs318/en/.
Wilkins, J. C., Homer, K. A., & Beighton, D. (2002). Analysis of *Streptococcus mutans* proteins modulated by culture under acidic conditions. *Applied and Environmental Microbiology*, *68*, 2382–2390.
Wright, C. J., Burns, L. H., Jack, A. A., Back, C. R., Dutton, L. C., Nobbs, A. H., et al. (2013). Microbial interactions in building of communities. *Molecular Oral Microbiology*, *28*, 83–101.
Wyss, C. (1989). Dependence of proliferation of *Bacteroides forsythus* on exogenous N-acetylmuramic acid. *Infection and Immunity*, *57*, 1757–1759.
Xu, M., Yamada, M., Li, M., Liu, H., Chen, S. G., & Han, Y. W. (2007). FadA from *Fusobacterium nucleatum* utilizes both secreted and nonsecreted forms for functional oligomerization for attachment and invasion of host cells. *The Journal of Biological Chemistry*, *282*, 25000–25009.
Yajima, A., Urano-Tashiro, Y., Shimazu, K., Takashima, E., Takahashi, Y., & Konishi, K. (2008). Hsa, an adhesin of *Streptococcus gordonii* DL1, binds to alpha2-3-linked sialic acid on glycophorin A of the erythrocyte membrane. *Microbiology and Immunology*, *52*, 69–77.
Yamada, T., Igarashi, K., & Mitsutomi, M. (1980). Evaluation of cariogenicity of glycosylsucrose by a new method of measuring pH under human dental plaque in situ. *Journal of Dental Research*, *59*, 2157–2162.
Yamashita, Y., & Shibata, Y. (2011). Acid stress survival mechanisms of the cariogenic bacterium *Streptococcus mutans*. In J. Agboola (Ed.), *Relevant perspectives in global environmental change* (pp. 39–54). Rijeka, Croatia: Intech Open.
Yeung, M. K., Chassy, B. M., & Cisar, J. O. (1987). Cloning and expression of a type 1 fimbrial subunit of *Actinomyces viscosus* T14V. *Journal of Bacteriology*, *169*, 1678–1683.
Yilmaz, O., Watanabe, K., & Lamont, R. J. (2002). Involvement of integrins in fimbriae-mediated binding and invasion by Porphyromonas gingivalis. *Cellular Microbiology*, *4*, 305–314.
Yoshida, A., & Kuramitsu, H. K. (2002). Multiple *Streptococcus mutans* genes are involved in biofilm formation. *Applied and Environmental Microbiology*, *68*, 6283–6291.
Yu, S., Fan, J., Liu, L., Zhang, L., Wang, S., & Zhang, J. (2013). Caveolin-1 up-regulates integrin α2,6-sialylation to promote integrin α5β1-dependent hepatocarcinoma cell adhesion. *FEBS Letters*, *587*, 782–787.
Yu, C., Ruiz, T., Lenox, C., & Mintz, K. P. (2008). Functional mapping of an oligomeric autotransporter adhesin of *Aggregatibacter actinomycetemcomitans*. *Journal of Bacteriology*, *190*, 3098–3109.
Yue, G., Kaplan, J. B., Furgang, D., Mansfield, K. G., & Fine, D. H. (2007). A second *Aggregatibacter actinomycetemcomitans* autotransporter adhesin exhibits specificity for buccal epithelial cells in humans and Old World primates. *Infection and Immunity*, *75*, 4440–4448.
Zalewska, A., Zwierz, K., Zółkowski, K., & Gindzieński, A. (2000). Structure and biosynthesis of human salivary mucins. *Acta Biochimica Polonica*, *47*, 1067–1079.
Zeituni, A. E., McCaig, W., Scisci, E., Thanassi, D. G., & Cutler, C. W. (2010). The native 67-kilodalton minor fimbria of *Porphyromonas gingivalis* is a novel glycoprotein with DC-SIGN-targeting motifs. *Journal of Bacteriology*, *192*, 4103–4110.
Zetterström, C. E., Hasselgren, J., Salin, O., Davis, R. A., Quinn, R. J., Sundin, C., et al. (2013). The resveratrol tetramer (-)-hopeaphenol inhibits type III secretion in the

gram-negative pathogens Yersinia pseudotuberculosis and *Pseudomonas aeruginosa*. *PLoS One*, *8*, e81969.

Zhang, T., Ding, Y., Li, T., Wan, Y., Li, W., Chen, H., et al. (2012). A Fur-like protein PerR regulates two oxidative stress response related operons dpr and metQIN in Streptococcus suis. *BMC Microbiology*, *12*, 85.

Zhang, W., Ju, J., Rigney, T., & Tribble, G. (2013). Integrin α5β1-fimbriae binding and actin rearrangement are essential for *Porphyromonas gingivalis* invasion of osteoblasts and subsequent activation of the JNK pathway. *BMC Microbiology*, *13*, 5.

Zheng, X., Zhang, K., Zhou, X., Liu, C., Li, M., Li, Y., et al. (2013). Involvement of gshAB in the interspecies competition within oral biofilm. *Journal of Dental Research*, *92*, 819–824.

Zhou, X.-Y., Gao, J.-L., Hunter, N., Potempa, J., & Nguyen, K.-A. (2013). Sequence-independent processing site of the C-terminal domain (CTD) influences maturation of the RgpB protease from *Porphyromonas gingivalis*. *Molecular Microbiology*, *89*, 903–917.

Zhu, L., Kreth, J., Cross, S. E., Gimzewski, J. K., Shi, W., & Qi, F. (2006). Functional characterization of cell-wall-associated protein WapA in *Streptococcus mutans*. *Microbiology (Reading, England)*, *152*(Pt 8), 2395–2404.

Zijnge, V., van Leeuwen, M. B. M., Degener, J. E., Abbas, F., Thurnheer, T., Gmür, R., et al. (2010). Oral biofilm architecture on natural teeth. *PLoS One*, *5*, e9321.

Zilm, P. S., Mira, A., Bagley, C. J., & Rogers, A. H. (2010). Effect of alkaline growth pH on the expression of cell envelope proteins in *Fusobacterium nucleatum*. *Microbiology (Reading, England)*, *156*, 1783–1794.

CHAPTER SEVEN

Virulence Factors of Uropathogenic *E. coli* and Their Interaction with the Host

Petra Lüthje, Annelie Brauner[1]

Department of Microbiology, Tumor and Cell Biology, Division of Clinical Microbiology, Karolinska Institutet and Karolinska University Hospital, Stockholm, Sweden

[1]Corresponding author: e-mail address: annelie.brauner@ki.se

Contents

Abstract

Urinary tract infections (UTIs) belong to the most common infectious diseases worldwide. The most frequently isolated pathogen from uncomplicated UTIs is *Escherichia coli*.

Advances in Microbial Physiology
ISSN 0065-2911
http://dx.doi.org/10.1016/bs.ampbs.2014.08.006

To establish infection in the urinary tract, *E. coli* has to overcome several defence strategies of the host, including the urine flow, exfoliation of urothelial cells, endogenous antimicrobial factors and invading neutrophils. Thus, uropathogenic *E. coli* (UPEC) harbour a number of virulence and fitness factors enabling the bacterium to resist and overcome these different defence mechanisms. There is no particular factor which allows the identification of UPEC among the commensal faecal flora apart from the ability to enter the urinary tract and cause an infection. Many of potential virulence or fitness factors occur moreover with high redundancy. Fimbriae are inevitable for adherence to and invasion into the host cells; the type 1 pilus is an established virulence factor in UPEC and indispensable for successful infection of the urinary tract. Flagella and toxins promote bacterial dissemination, while different iron-acquisition systems allow bacterial survival in the iron-limited environment of the urinary tract. The immune response to UPEC is primarily mediated by toll-like receptors recognising lipopolysaccharide, flagella and other structures on the bacterial surface. UPEC have the capacity to subvert this immune response of the host by means of actively impacting on pro-inflammatory signalling pathways, or by physical masking of immunogenic structures. The large repertoire of bacterial virulence and fitness factors in combination with host-related differences results in a complex interaction between host and pathogen in the urinary tract.

1. INTRODUCTION

Urinary tract infections (UTIs) belong to the most common infectious diseases worldwide. Uncomplicated, community-acquired UTIs are caused by uropathogenic *Escherichia coli* (UPEC) strains in about 80% of the cases, and this species is also frequently isolated from such hospital-acquired infections. UTIs can be limited to the lower urinary tract (cystitis) or involve the kidney (acute pyelonephritis). Bacteria might eventually spread to the bloodstream and cause life-threatening septicaemia. Most patients are women, and half of all women at the age of 32 years have experienced a UTI at least once (Foxman & Brown, 2003). In around 25% of these young, otherwise healthy women, a cystitis will recur within 6 months and a substantial number of women will suffer from three or more episodes a year (recurrent UTI) (Foxman et al., 2000). Acute UTIs of the lower and upper urinary tract in otherwise healthy, premenopausal and non-pregnant women without any anatomical abnormalities in the urinary tract, are considered uncomplicated, and can easily be treated with antibiotics, even though emerging antimicrobial resistances make treatments increasingly difficult (Hooton, 2012; Wang, Nizran, Malone, & Riley, 2013). However, infection might also spread and cause infection of the kidneys (acute pyelonephritis), but upper UTIs are less common than infections of the lower urinary tract (Czaja, Scholes, Hooton, & Stamm, 2007). Beside symptomatic infections, *E. coli* might be found in the urine in high numbers without causing any symptoms, referred to as asymptomatic bacteriuria (ABU). In healthy, non-pregnant women, these infections are generally not treated (Nicolle, Mayhew, & Bryan, 1987; Ouslander et al., 1995; Schneeberger, Kazemier, & Geerlings, 2014) and might even prevent infections by strains causing symptomatic disease (Roos, Ulett, Schembri, & Klemm, 2006).

2. PATHOGENESIS OF URINARY TRACT INFECTION

Infections of the urinary tract occur when *E. coli* ascend via the urethra and succeed in colonising the bladder. The bowl but also the vaginal and periurethral mucosa serve as reservoir for such infections (Moreno et al., 2008; Navas-Nacher et al., 2001; Yamamoto et al., 1997). Within the urinary tract, *E. coli* needs to overcome several lines of defence (Chromek & Brauner, 2008); the mechanical force of the urine flow, antimicrobial substances in the urine produced and secreted by the epithelial cells lining the urinary bladder; eradication by neutrophils; and the excretion with exfoliating superficial cells. Thus, tight and irreversible adherence to the urothelial cells is the first step in establishing an infection, followed by bacterial invasion into the host cells and multiplication (Justice et al., 2004). Upon contact with bacteria, the host reacts with an inflammatory response (Khalil et al., 1997; Samuelsson, Hang, Wullt, Irjala, & Svanborg, 2004). While the intracellular compartment offers a niche for *E. coli* to replicate protected from the initiated host defence, active suppression of the immune reaction is another strategy for *E. coli* to evade defence mechanisms. Consequently, virulence and fitness factors found in UPEC strains support different steps in uropathogenesis and might therefore link particular factors to certain manifestations of the disease.

Adherence to the host cell is the pre-requisite for a successful infection. The major receptors for *E. coli* binding to the urothelial surfaces of the lower urinary tract are the uroplakins (UPs), expressed by the differentiated umbrella cells exposed at the luminal surface of the bladder and ureters (Wu, Kong, Pellicer, Kreibich, & Sun, 2009). These mannose-containing receptors are readily recognised by type 1 fimbriae (Zhou et al., 2001), which allow tight adhesion in face of the mechanical stresses present in the bladder (Thomas, Nilsson, Forero, Sokurenko, & Vogel, 2004). After adherence, a small proportion of bacteria will enter the cell. Initiated by binding to β1 integrins (Eto, Jones, Sundsbak, & Mulvey, 2007), *E. coli* exploits several structural components of the host cell including the actin cytoskeleton, microtubules and lipid rafts to enter the cell (Dhakal & Mulvey, 2009; Duncan, Li, Shin, Carson, & Abraham, 2004; Martinez, Mulvey, Schilling, Pinkner, & Hultgren, 2000). Besides type 1 fimbriae, with their functional linkage to the lower urinary tract, P fimbriae are among the best investigated adhesins in UPEC. Defined by their receptor specificity, P fimbriae relate to the upper urinary tract. In addition, UPEC expresses a multitude of other adhesins, which might be of specific relevance in different regions of the urinary tract.

Once adhered and taken up by the cells, *E. coli* enters the cytoplasm and multiplies rapidly within the host cell. Eventually, the bacterial colonies grow so large that they appear as protruding 'pods' towards the lumen of the urinary bladder (Anderson et al., 2003). These so-called intracellular bacterial communities (IBCs) are the hallmark of the acute stage of infection when bacteria multiply excessively during several generations of IBCs (Justice et al., 2004). Eventually, bacteria emerge the IBCs to again colonise the urothelium and to invade new cells. Due to the loss of superficial cells in the process of exfoliation, less differentiated cells in lower layers of the urothelium are now exposed at the luminal surface and to bacteria. Within these cells, *E. coli* do not multiply, probably inhibited by the denser network of actin (Eto, Sundsbak, & Mulvey, 2006). Referring to the dormant state of the bacteria, these aggregates have therefore been named quiescent intracellular reservoirs (QIRs) (Mysorekar & Hultgren, 2006). Upon differentiation of the host cell and retraction of the actin network to basolateral regions of the cell, bacteria re-enter an active stage, start multiplying and cause a new UTI episode.

An instant reaction of the host cell to the bacterial attack is the apoptosis-dependent shedding of infected cells with the urine (Mulvey et al., 1998; Thumbikat, Berry, Schaeffer, & Klumpp, 2009; Thumbikat, Berry, Zhou, et al., 2009). This process is induced by the interaction between the type 1 fimbrial-adhesin FimH with UPs. While adhesion is mediated primarily through UPIa, UPIIIa plays a pivotal role for exfoliation, since this UP is the only major UP which possesses a cytoplasmic signalling domain (Thumbikat, Berry, Schaeffer, et al., 2009; Thumbikat, Berry, Zhou, et al., 2009).

UPEC infection induces an upregulation of cytokines and chemokines by binding to pathogen recognition receptors. Toll-like receptor (TLR) 4 and TLR5 appear the most important receptors for this response, with TLR4 recognising lipopolysaccharide (LPS) as the major component of the cell wall in Gram negative bacteria (Park & Lee, 2013); and TLR5 recognising flagella (Smith et al., 2003). Apart from the well-known interaction between LPS and TLR4, P fimbriae and type 1 fimbriae induce and modulate the TLR4-mediated response. While induction of a pro-inflammatory response by type 1 fimbriae requires LPS (Schilling, Mulvey, Vincent, Lorenz, & Hultgren, 2001; Schilling et al., 2003), P fimbriae induces a TLR4-response independently of this factor (Frendeus et al., 2001; Hedlund et al., 1999). Binding of these fimbriae to their receptors induces a release of ceramide, which was found to act as agonist on TLR4 and thus allowed an inflammatory reaction independently of LPS and the TLR4 co-receptor CD14 (Fischer et al., 2007). Various other adhesins as well as other virulence factors such as toxins promote this immune induction or act immunogenic themselves; on the other hand, bacterial components might prevent *E. coli* from immune recognition or even actively reduce an immune response.

From the bladder, bacteria may ascend via the ureters to the kidneys to cause an acute pyelonephritis; in the worst scenario, bacteria might even enter the bloodstream (urosepticaemia). *E. coli* uses flagella to swim against the urine flow via the ureter to the kidney (Lane, Alteri, Smith, & Mobley, 2007). Flagella-mediated motility is also involved in further dissemination of bacteria with the bloodstream (Lane et al., 2007). Overall however, flagella provide only a modest fitness advantage in the urinary tract in an animal model (Lane et al., 2005; Wright, Seed, & Hultgren, 2005). Referring to the opposed function of adhesive fimbriae and motility-mediating flagella, these two groups of appendages are, in general, reversely regulated (Simms & Mobley, 2008a, 2008b).

The production of toxins might likewise help bacteria to spread within the host tissue by disrupting cellular integrity. Bacteria gain moreover access to nutrient from lysed host cells. In contrast, tissue damage provokes a strong inflammatory response and thus might eventually help the host terminating the infection.

E. coli can be separated in four major phylogenetic lineages, A, B1, B2 and D (Herzer, Inouye, Inouye, & Whittam, 1990). The majority of extra-intestinal pathogenic *E. coli* including UPEC strains belong to group B2 and, to a lesser extent, to group D; while commensal isolates primarily belong to phylogenetic group A (Picard et al., 1999). Isolates of the phylogenetic group B2 tend to carry more virulence factors and in general less resistance genes. In contrast, group A isolates are less virulent but more resistant (Johnson, Kuskowski, Owens, Gajewski, & Winokur, 2003; Moreno et al., 2006).

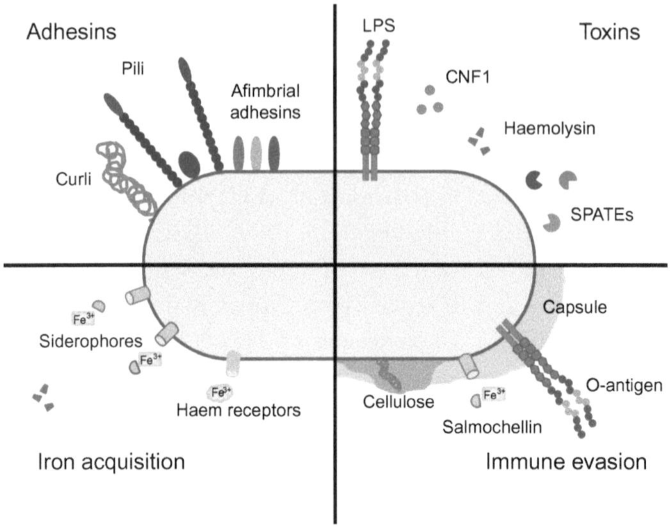

Figure 1 Virulence and fitness factors of uropathogenic *E. coli*. *E. coli* employs different strategies to infect the urinary tract, to resist immune defences of the host and to persist.

In this review, we aim to present UPEC factors in the context of their function within the urinary tract which contribute to urovirulence (virulence factors) or provide a competitive advantage in the urinary tract (fitness factors). The vast number of virulence or fitness factors in UPEC and the uncertain delineation between the two groups do however not allow covering all aspects here (Fig. 1).

3. ADHESINS

Adherence to host cells can be mediated by fimbriae but also afimbrial adhesins. Fimbriae, or pili, are complex structures and thus encoded by gene clusters coding for fimbrial subunits, assembly and secretion machinery. Pathogenic but also commensal *E. coli* harbour numerous different operons coding for fimbriae in the genome, most of them belonging to the usher-chaperon family (Welch et al., 2002; Wurpel, Beatson, Totsika, Petty, & Schembri, 2013). These fimbriae contain a rod, composed of several hundreds to thousands of major subunits; and the adhesive tip, formed by single or few minor subunits. Within the periplasmic space, the chaperon facilitates folding of the subunits, which are then assembled and secreted by the usher in the outer membrane (Waksman & Hultgren, 2009).

A large proportion of non-fimbrial adhesins belong to the group of autotransporter proteins. An autotransporter protein is composed of different domains, allowing its own transport across the bacterial membranes (Henderson, Navarro-Garcia, & Nataro, 1998). A signal sequence directs secretion of the protein across the inner membrane and is then removed from the polyprotein; a translocation (β) domain inserts into the outer membrane to form a pore and mediates transport of the passenger (α) domain. In contrast to the transporter domain, the passenger domain is highly variable and determines the function of the protein. After secretion, the passenger domain can remain at the cell surface to act as an adhesin, or is released as toxin to the surrounding (Henderson & Nataro, 2001). Sequences coding for autotransporter proteins are widely distributed among commensal and *E. coli* belonging to different pathotypes, but some are concentrated in UPEC strains, indicating a role in urovirulence (Restieri, Garriss, Locas, & Dozois, 2007).

In addition, UPEC expresses other types of adhesins, such as the amyloid fibres curli and the afimbrial adhesins of the Dr adhesin family. Moreover, structures which are primarily associated to other functions may also promote adhesion and invasion. For example, flagella might, partly independent from their role for mobility, mediate entry into cells of the renal collecting duct (Pichon et al., 2009). Also some toxins contribute to bacterial adherence to and uptake by the host cells *in vitro* and have a possible role as adhesin or invasin *in vivo*.

3.1. Type 1 fimbriae

Type 1 fimbriae are a definite virulence factor in UPEC, although the wide distribution of these pili is not restricted to pathogenic strains (Norinder, Koves, Yadav, Brauner, & Svanborg, 2012; Vejborg, Hancock, Schembri, & Klemm, 2011). Type 1 fimbriae are essential to colonise the urinary tract in mice and humans (Bahrani-Mougeot et al., 2002; Connell et al., 1996), but alone not enough to establish a long-term infection (Mulvey, Schilling, & Hultgren, 2001). The adhesive tip of these fimbriae, FimH, binds to α-D-mannosylated proteins such as UPs, which are expressed by the differentiated urothelium and thus relates this adhesin to the lower urinary tract. Due to its binding affinity, FimH-mediated adhesion or aggregation of eukaryotic cells (yeast or red blood cells) is inhibited by mannose.

Binding of FimH to its receptors mediates adhesion, invasion and also promotes the formation of IBCs (Eto et al., 2007; Wright, Seed, & Hultgren, 2007; Zhou et al., 2001), reflecting the crucial function of these fimbriae for bacterial colonisation. The importance of this host–pathogen interaction is further demonstrated by the high susceptibility of premenopausal women to UPEC infection (Foxman & Brown, 2003; Sonnex, 1998). Oestrogen increases the expression of the major fimbrial receptors UPI (Luthje et al., 2013), and this linkage explains the direct correlation between menstrual cycle and the fluctuations in the receptivity to UTI among young women (Hooton, Winter, Tiu, & Stamm, 1996).

While gene distribution and fimbriae expression *in vitro* is similar among pathogens and commensal faecal strains, differences in the regulation of fimbrial expression may influence bacterial virulence in the urinary tract (Bahrani-Mougeot et al., 2002; Bryan et al., 2006; Schwan, 2011). Interestingly, strains isolated from ABU do not produce functional type 1 fimbriae when grown in urine despite the presence of the encoding gene clusters (Roos, Nielsen, & Klemm, 2006). In addition, the adhesin FimH or its signal peptide show minor structural differences between commensals and pathogens (Chen et al., 2009; Ronald et al., 2008; Schwartz et al., 2013), and moreover variations between strains from lower or upper UTI (Schwartz et al., 2013). In particular, the superior ability of cystitis strain UTI89 to form IBCs, to cause high bladder titres and to persist in the mouse urinary tract could be transferred to pyelonephritis strain CFT073 by replacing the cognate FimH by the UTI89 variant (Schwartz et al., 2013). However, other factors than FimH played a role in the efficient colonisation of the bladder by *E. coli* UTI89.

3.2. P fimbriae

P fimbriae are classically associated with pyelonephritis. While type 1 fimbriae recognise mannosylated receptors, the PapG adhesin of P fimbriae binds to Galα1-4Gal moieties in glycolipids of the host cell membrane and P fimbrial binding is therefore not inhibited by mannose. Depending on neighbouring carbohydrates, Galα1-4Gal is recognised by different classes of PapG. Their binding affinities influence the preference of PapG variants to different tissues (Stromberg, Nyholm, Pascher, & Normark, 1991; Stromberg et al., 1990). The PapGIII variant (or PrsG) is preferably associated to cystitis in human and kidney infections in dogs (Johnson, O'Bryan, et al., 2000; Johnson, Russo, Brown, & Stapleton, 1998), based on the binding specificity to globopentaosylceramide (Forssman antigen), the predominant receptor in the canine but not the human kidney. The PapGII variant binds to globotetraosylceramide, which is predominant in the human kidney, and thus PapGII is associated to pyelonephritis and bacteremia in human

(Johnson, 1998; Johnson, Kuskowski, et al., 2005; Otto, Sandberg, Marklund, Ulleryd, & Svanborg, 1993). The PapGI variant preferably binds to globotriaosylceramide (GbO_3), but the relevance of this affinity for disease is not well documented. In addition to these three major PapG classes, minor variants of PapGI and the novel PapGIV variant has been described (Johnson, Stell, Kaster, Fasching, & O'Bryan, 2001; Manning et al., 2001). The binding specificity of these adhesins and the relevance during UTI are however not known.

While these experimentally described and structurally characterised affinities of the different PapG classes to their specific isoreceptors are generally supported by the distribution of the corresponding genes in UPEC isolated from different manifestations of the disease (Daigle, Harel, Fairbrother, & Lebel, 1994; Johanson, Plos, Marklund, & Svanborg, 1993), animal studies show that P fimbriae are dispensable for successful colonisation of the kidneys (Hagberg et al., 1983; Mobley et al., 1993). Some studies indicate a role for persistence in the lower urinary tract (Ejrnaes et al., 2011; Luo et al., 2012; Norinder et al., 2011), the mechanism of this association however remains unclear.

3.3. Curli fimbriae

Curli fimbriae belong to the class of amyloids and are composed of a large number of the major subunit CsgA, which polymerises after transport across the outer cell membrane with help of the nucleator subunit CsgB to form the functional fibre exposed at the bacterial surface (Barnhart & Chapman, 2006). Together with cellulose, curli represent one of the major biofilm components in *E. coli* and other Enterobacteriaceae (Romling, 2005) and are frequently expressed by pathogenic as well as commensal isolates (Kai-Larsen et al., 2010; Norinder et al., 2012). Since biofilm formation is a typical reaction of bacteria to protect themselves from adverse environmental conditions, curli expression is promoted by limitation of nutrients and salts, at reduced oxygen tension and at temperature below 30 °C (Gerstel & Romling, 2001; Olsen, Arnqvist, Hammar, Sukupolvi, & Normark, 1993; Romling, Sierralta, Eriksson, & Normark, 1998). However, many pathogenic but also commensal strains can express curli at 37 °C under defined conditions *in vitro* (Kudinha et al., 2013; Lim, Pinkner, & Cegelski, 2014; Ramos et al., 2011), and during infection in humans (Bian, Brauner, Li, & Normark, 2000; Kai-Larsen et al., 2010). This observation indicate that, in addition to supporting bacterial survival outside the host, curli are relevant virulence factors.

Curli act as adhesins, mediate invasion into host cells and induce strong immune responses. Curli interact with many host proteins including serum and contact-phase proteins, which might promote bacterial dissemination and entrance into the bloodstream (summarised in Barnhart & Chapman, 2006). Curliated strains produced a more progressive acute infection in a mouse model (Kai-Larsen et al., 2010), and in concordance, curliated UPEC strains are more likely to cause urosepticaemia than non-curliated strains (Hung, Marschall, Burnham, Byun, & Henderson, 2014).

Curli fimbriae are recognised by TLR2/TLR1 and initiate a NF-κB dependent pro-inflammatory response *in vitro* and during experimental sepsis and UTI in mice (Bian, Yan, Hansson, Thoren, & Normark, 2001; Bian et al., 2000; Kai-Larsen et al., 2010; Tukel et al., 2010, 2005). The curli-dependent, strong immune induction however provokes rapid recruitment of neutrophils and elimination from the mouse kidneys (Kai-Larsen et al., 2010). This, for the bacterium adverse effect of curli is attenuated in the presence of cellulose, allowing UPEC to persist in the urinary tract without losing the advantage during the acute phase of infection depending on the adhesive properties of curli (Kai-Larsen et al., 2010). Thus, the partly opposing functions of curli and cellulose are compensated when both structures are expressed together. The advantage of the combined expression is illustrated by the association of this phenotype to UPEC strains (Kai-Larsen et al., 2010), and more severe infections (Kudinha et al., 2013) in children, and to persisting UPEC strains in adult women (Norinder et al., 2011).

Due to the infection-promoting properties of curli, these fibres have been target for therapeutic approaches. Major attempts have been made to inhibit polymerisation of CsgA monomers to functional fibres (Cegelski et al., 2009). Interestingly, such inhibitory action on fibre formation is provided by the endogenous antimicrobial peptide LL-37 (Kai-Larsen et al., 2010), which plays a pivotal role in the defence of the urinary tract (Chromek et al., 2006). At the same time, binding of LL-37 by curli prevents its bactericidal activity and thus curli confer increased resistance against this peptide, possibly contributing to the advantage of curliated bacteria during the initial phase of infection.

3.4. Afa/Dr adhesins

E. coli strains harbouring operons coding for Afa/Dr adhesins are referred to as diffusely adherent *E. coli* and might be considered a separate group of diarrheagenic *E. coli*. They are typically associated with intestinal infections especially in young children, but *E. coli* expressing these adhesins are frequently found among strains from extra-intestinal infections including UTI (Le Bouguenec & Servin, 2006). Interestingly, despite similar genetic organisation of the gene operons, the genes within this family code for afimbrial as well as fimbrial adhesins with similar receptor specificities. Afa/Dr$^+$ adhesins recognise the decay-accelerating factor (DAF; CD55), while a lower number of Afa/Dr$^-$ adhesins do not (Nowicki, Selvarangan, & Nowicki, 2001). Together with collagen type IV, DAF is thought to drive the renal tropism of Afa/Dr$^+$ strains (Goluszko, Moseley, et al., 1997). The complete receptor repertoire of all Afa/Dr adhesin subtypes has not been fully identified, but different binding specificities may influence the association of certain strains to different tissues (Le Bouguenec et al., 2001). For example, *afaE5*, *afaE1* and *draE/afaE3* dominate in cystitis strains (Zhang et al., 1997), while the Afa/Dr$^-$ subtype *afaE8* has been found highly concentrated among UPEC from pyelonephritis (Le Bouguenec et al., 2001).

Apart from their association to diarrheagic diseases in children, pregnant women appear predisposed to infections by Afa/Dr adhesin-expressing *E. coli* (Goluszko et al., 2001; Hart et al., 2001; Pham et al., 1995). The association of Afa/Dr$^+$ adhesins to pregnancy has been suggested to rely on progesterone-mediated upregulation of DAF (Kaul et al., 1996; Nowicki et al., 2001). Interestingly, also β1 integrins are involved in adherence and invasion of these strains (Guignot et al., 2001; Kansau et al., 2004; Plancon et al., 2003). This receptor in turn is upregulated by oestrogen in the urinary tract (Luthje et al., 2013) and might thus additionally promote infection in women before menopause.

DAF and CD66e, another receptor binding to Afa/Dr adhesins, have GPI anchors and binding to these receptors induces signalling events eventually leading to bacterial uptake into the host cell via a zipper-like mechanism (Goluszko, Popov, et al., 1997; Goluszko et al., 1999). In addition, β1 integrins confer bacterial internalisation involving the microtubules system and caveolae (Guignot et al., 2001; Korotkova et al., 2008). Within the host cell, Afa/Dr appears to promote intracellular multiplication in a similar manner as has been described for type 1 fimbriae (Goluszko, Popov, et al., 1997; Jouve et al., 1997). This function might relate to the association of these adhesins to persistence in the host (Goluszko, Moseley, et al., 1997) and thus recurrent infections (Foxman et al., 1995; Qin et al., 2013; Selvarangan et al., 2004).

3.5. F1C/S fimbriae

F1C and S fimbriae are common among uropathogens but less frequently found than type 1 or P fimbriae. The fimbriae are genetically related as the nucleotide sequences share a high degree of homology (Schmoll et al., 1990), but display different receptor affinities. F1C fimbriae bind to galactosylceramide (GalCer2)-containing receptors expressed by epithelial cells throughout the urinary tract and to globotriaosylceramides (asialo-GM_2, GgO_3Cer) in the kidney (Backhed et al., 2002; Khan et al., 2000). Even though this receptor specificity suggest an association of these fimbriae to the kidneys (Backhed et al., 2002), the prevalence of F1C fimbriae is too low to confirm this relation in clinical strains (Ikaheimo et al., 1994; Pere, Leinonen, Vaisanen-Rhen, Rhen, & Korhonen, 1985; Pere, Nowicki, Saxen, Siitonen, & Korhonen, 1987; Qin et al., 2013; Siitonen, Martikainen, Ikaheimo, Palmgren, & Makela, 1993; Tarchouna, Ferjani, Ben-Selma, & Boukadida, 2013). Interestingly however, the ABU *E. coli* strain 83972, despite the presence of the encoding genes lacks functional F1C fimbriae (Roos, Schembri, Ulett, & Klemm, 2006).

S fimbriae bind to α-sialyl-2-3-β-galactose (NeuAc-α 2,3-Gal)-containing receptors on erythrocytes (Parkkinen, Rogers, Korhonen, Dahr, & Finne, 1986) and renal tubular epithelial cells (Korhonen et al., 1986; Kreft et al., 1995). Such residues are also found on UPIII, which is highly expressed in the bladder epithelium (Malagolini, Cavallone, Wu, & Serafini-Cessi, 2000). In contrast to type 1 and P fimbriae, S fimbrial binding is inhibited by the Tamm–Horsfall protein (Parkkinen, Virkola, & Korhonen, 1988), a glycoprotein produced by the epithelial cells of the distal tubuli, which might reduce their strength as virulence factor in the kidney (Parkkinen et al., 1988). The relation of S fimbriae to sepsis and meningitis in contrast might be based on the affinity of S fimbriae to the extracellular matrix protein laminin (Johnson, Oswald, O'Bryan, Kuskowski, & Spanjaard, 2002; Parkkinen et al., 1986; Virkola, Parkkinen, Hacker, & Korhonen, 1993).

3.6. F9 and type 3 fimbriae

The role of these recently described fimbriae in UTI remains to be elucidated (Ulett, Mabbett, Fung, Webb, & Schembri, 2007). Type 3 fimbriae are present in virtually all clinical *Klebsiella pneumoniae* isolates, but only few *E. coli* carry these genes (Stahlhut et al., 2013). Both F9 and type 3 fimbriae have been found to promote biofilm formation *in vitro* and might thus be relevant in the context of catheter-associated UTI (Ong et al., 2008; Ulett, Mabbett, et al., 2007). In addition, Galβ1-3GlcNAc residues, which are highly abundant in the mucosal surfaces of bladder and kidney, have been identified as receptors for F9 fimbriae (Wurpel et al., 2014). However, an interaction with host cells could not be established *in vitro* (Ulett, Mabbett, et al., 2007). Moreover, functional fimbriae appear to be expressed only at low temperature (Wurpel et al., 2014), further questioning the role of these fimbriae during UTI.

3.7. Antigen 43

The surface-exposed Antigen 43 (Ag43) adhesin is the best characterised adhesive autotransporter protein in *E. coli*. The ability of self-recognition leads to autoaggregation of Ag43-expressing cells, promotes the formation of biofilm and adhesion to host cells (Charbonneau & Mourez, 2007; Sherlock, Dobrindt, Jensen, Munk Vejborg, & Klemm, 2006). Ag43 proteins are encoded by *agn43* (or *flu*) genes. Different gene variants display pronounced differences within the passenger domain, which determines the functional properties of the Ag43 protein. The uropathogenic type strain CFT073 harbours two *agn43* variants, designated *fluA* and *fluB* (Klemm, Hjerrild, Gjermansen, & Schembri, 2004), which are widely distributed also among other UPEC strains (Restieri et al., 2007). Compared to Ag43b (encoded by *fluB*), Ag43a (encoded by *fluA*) mediates stronger autoaggregation and biofilm formation *in vitro*, and promotes long-term colonisation and persistence in a mouse model of UTI (Ulett, Valle, et al., 2007). Ag43-mediated cell aggregation protects *E. coli* from neutrophil killing *in vitro* (Fexby et al., 2007), which might have an impact on bacterial clearance *in vivo*. Furthermore, the expression of Ag43 by *E. coli* within IBCs suggests an impact on bacterial intracellular survival and persistence (Anderson et al., 2003). While adhesion via specific receptors is often associated with increased inflammatory response, eventually leading to the elimination of the pathogen, Ag43 appears not to be highly immunogenic and might even confer immune protection. In particular, urine IL-8 levels during UTI in children were lower when infected with *fluA*-positive strains, especially in comparison with levels induced by *fluB*-carrying isolates (Luthje & Brauner, 2010). Strikingly, different studies show an association of *agn43* genes to UPEC strains persisting in the urinary tract (Ejrnaes et al., 2011; Luthje & Brauner, 2010; Norinder et al., 2011), further supporting the additional role of Ag43 as immune evasion mechanism.

3.8. Uropathogenic *E. coli* autotransporter

Several autotransporter proteins with adhesive functions, designated uropathogenic *E. coli* autotransporter (Upa), have recently been described in *E. coli*, but only little is known about their relevance for uropathogenesis. These proteins promote the formation of biofilm [UpaC (Allsopp, Beloin, Ulett, et al., 2012), UpaG (Valle et al., 2008), UpaH (Allsopp et al., 2010)] and adhesion to different extracellular matrix proteins [UpaB (Allsopp, Beloin, Ulett, et al., 2012), UpaG (Totsika et al., 2012; Valle et al., 2008), UpaH (Allsopp, Beloin, Moriel, et al., 2012)] *in vitro*. For UpaB, a role during early colonisation of the urinary tract has been suggested (Allsopp, Beloin, Ulett, et al., 2012), while no advantage was conferred by the remaining Upa proteins.

4. TOXINS

Toxins help the pathogen spreading into deeper tissues after disrupting cell integrity; to gain access to nutrients inside the host cell; or to destroy immune effector cells and thus evade their potential antibacterial activity. Toxic activity however is likely to result in a strong inflammatory reaction, in response to necrosis or to the toxin itself.

4.1. Endotoxin

LPS is the major component of the cell wall in Gram negative bacteria and highly immunogenic. Since it is bound to the bacterial cell, in contrast to toxins which are secreted into the surrounding, it is also referred to as endotoxin. LPS consists of the hydrophobic lipid A, located in the outer leaflet of the outer membrane; the core polysaccharides and repeats of O-antigen subunits, which are exposed at the surface of the bacteria and constitute the major immunogen (Wang & Quinn, 2010). Lipid A is highly conserved and mediates the toxic effects of LPS. The core polysaccharides and especially the O-antigens are more variable in structure and function. Certain O-antigens are common among UPEC such as O1, O2, O4, O6, O7, O8, O15, O16, O18, O21, O22, O25, O75 and O83, and are related to specific virulence gene profiles (Blanco, Blanco, Alonso, & Blanco, 1994, 1996; Li et al., 2010). In UPEC, O-antigens can even confer immunomodulatory functions. These aspects of LPS will be discussed in sections 6.1 and 6.2.

4.2. α-Haemolysin

α-Haemolysin (HlyA) is a secreted, pore-forming toxin, named after its lytic activity against erythrocytes. However, HlyA is cytotoxic also towards various nucleated cell types, including immune, endothelial and epithelial cells in the urinary tract (Island et al., 1998; Mobley et al., 1990). Serum confers partial protection against the toxic activity, but especially neutrophils are attacked by HlyA under physiological conditions, suggesting a protective role of HlyA against neutrophil-mediated killing (Bhakdi et al., 1989). The detrimental effect of HlyA on neutrophils during infection with *E. coli* was impressively demonstrated in a zebra fish model (Wiles, Bower, Redd, & Mulvey, 2009); however, the relevance of this finding during UTI in mammals has not been demonstrated.

More recently, an interaction of HlyA with natural killer (NK) cells in the urinary bladder has been identified (Gur et al., 2013). After bacterial binding to NK cells via type 1 fimbriae, NK cells are killed by HlyA. Since NK cells promote secretion of TNF-α in response to infection, this action of HlyA suppresses the pro-inflammatory response to UPEC. HlyA might also directly reduce cytokine production in various immune (Bhushan et al., 2011; Konig & Konig, 1993) and epithelial cells (Hilbert et al., 2012). This function is closely related to the cytotoxic effect (Hilbert et al., 2012) and the underlying mechanisms have not fully been elucidated (Wiles & Mulvey, 2013). Thus, it remains unclear whether the effect on cytokine secretion is specific or rather an observation related to cell death as has been reported for NK cells (Gur et al., 2013).

In addition to haemorrhage, HlyA-expressing *E. coli* induce a pronounced exfoliation early during infection (Smith, Rasmussen, Grande, Conran, & O'Brien, 2008). Interestingly, this reaction is not directly related to the cytotoxic effect of HlyA. In contrast, HlyA stimulates activity of serine proteases and caspases, which then mediate the degradation of paxillin (important to stabilise cell–cell contacts) and induce apoptosis, respectively (Dhakal & Mulvey, 2012). This indirect proteolytic activity might also contribute to the proposed anti-inflammatory action of HlyA (Dhakal & Mulvey, 2012). While exfoliation removes pathogens from the urinary tract, it also promotes the dissemination of bacteria and facilitates bacterial entry into newly exposed, less differentiated cells of the urothelium. Within these cells, UPEC forms dormant reservoirs, QIRs, for recurrent infections. Moreover, *hlyA* is upregulated in UPEC within IBC (Berry, Klumpp, & Schaeffer, 2009; Reigstad, Hultgren, & Gordon, 2007), the pre-requisite for efficient multiplication in the bladder and establishment of persisting reservoirs.

Overall, HlyA does not contribute to bacterial colonisation in experimental UTIs (Hilbert et al., 2012; Smith et al., 2008). The assessment of its contribution to virulence among clinical strains is hampered by the frequently combined occurrence of HlyA and cytotoxic necrotising factor (CNF) 1 (Birosova et al., 2004; Brauner, Katouli, Tullus, & Jacobson, 1990; Real et al., 2007). Clinically, haemolytic UPEC strains are related to haemorrhage infections and pyelonephritis (Brauner et al., 1990) and are associated with a strong pro-inflammatory response (Garcia, Ventura, Smith, Merrell, & O'Brien, 2013; Real et al., 2007). In contrast, *hlyA*-positive strains are also more frequently found among strains persisting in the urinary tract (Ejrnaes et al., 2011; Savoia, Millesimo, & Fontana, 1994).

4.3. Cytotoxic necrotising factor 1

The presence of CNF1 is closely linked to haemolytic UPEC (Blanco et al., 1992; Landraud, Gibert, Popoff, Boquet, & Gauthier, 2003), due to the genetic organisation of the *cnf1* and *hlyA* genes. While *hlyA* can be located within a pathogenicity island without *cnf1*, *cnf1* is always linked to *hlyA* due to its combined location on pathogenicity island II_{J96} (Landraud et al., 2003).

The CNF1 toxin activates Rho GTPases (Boquet, 2001). In consequence, CNF1 induces a number of cellular changes based on rearrangements of the actin cytoskeleton. A transient activation followed by degradation of Rho GTPases, in particular Rac, promotes bacterial internalisation into urothelial cells (Doye et al., 2002; Falzano et al., 1993; Hertting et al., 2008), corresponding to an advantage of CNF1-expressing *E. coli* in early colonisation of the bladder (Rippere-Lampe, O'Brien, Conran, & Lockman, 2001).

CNF1 exhibits a profound pro-inflammatory activity in bladder epithelial cells and monocytes *in vitro* (Falzano et al., 2003; Hertting et al., 2008), but the pro-inflammatory effect is less evident *in vivo* (Garcia et al., 2013; Rippere-Lampe et al., 2001). In contrast, CNF1 exerts inhibitory effects on neutrophils, reducing phagocytosis and antimicrobial activity (Davis, Carvalho, Rasmussen, & O'Brien, 2006; Davis, Rasmussen, & O'Brien, 2005).

CNF1 has also been suggested to induce apoptosis specifically in uroepithelial cells and thus might promote exfoliation during UTI (Mills, Meysick, & O'Brien, 2000). However, apoptosis or increased cell shedding could not be confirmed in other studies (Falzano et al., 2006; Smith, Grande, Rasmussen, & O'Brien, 2006) or in animal experiments

(Johnson, Drachenberg, et al., 2000; Johnson, O'Bryan, et al., 2000; Smith et al., 2008).

4.4. Serine protease autotransporters of the Enterobacteriaceae

Autotransporter proteins can, in addition to acting as adhesins, display toxic functions. Several such toxins have been identified in bacteria of the family Enterobacteriaceae, which are consequently referred to as serine protease autotransporters of the Enterobacteriaceae (SPATEs) (Henderson & Nataro, 2001).

Three toxins of the SPATE family have been detected in the pyelonephritis type strain *E. coli* CFT073, Sat (secreted autotransporter protein) (Guyer, Henderson, Nataro, & Mobley, 2000), Pic (protease involved in colonisation) (Heimer, Rasko, Lockatell, Johnson, & Mobley, 2004; Parham et al., 2004) and Vat [vacuolating autotransporter toxin; or Tsh (temperature-sensitive hemagglutinating factor)] (Heimer et al., 2004). Pic has previously been found in *Shigella* and enteroaggregative *E. coli*, and degrades mucine; Vat has first been described in avian pathogenic *E. coli* and agglutinates haem and acts as haemoglobinase (Henderson & Nataro, 2001). It has not yet been proven that Pic and Vat from UPEC exhibit the same activities (Heimer et al., 2004).

All three genes are widely distributed and preferably found among UPEC strains compared to commensal strains (Guyer et al., 2000; Parham et al., 2005), and the presence of *vat* has been identified as a strong predictor for high virulence in UPEC (Spurbeck et al., 2012). All three SPATEs are expressed during UTI in the mouse, but an advantage for colonisation of the human urinary tract has not been demonstrated (Guyer et al., 2000).

5. IRON-ACQUISITION SYSTEMS

The availability of iron is extremely restricted in the urinary tract and thus bacteria have to be equipped with systems to survive in this limited environment. Iron ions are highly toxic and nearly insoluble and thus bacteria must deal with protein-bound iron sources from the host or haem, the most abundant source for iron in the host. Downregulation of iron-binding proteins such as lactoferrin or transferrin is thus a typical reaction of the host to bacterial infection. Bacteria also produce their own iron-complexing proteins, referred to as siderophores, to acquire iron.

The vital necessity of iron acquisition for *E. coli* in the urinary tract is illustrated by a strong upregulation of genes coding for iron-acquisition systems during UTI (Hagan, Lloyd, Rasko, Faerber, & Mobley, 2010; Snyder et al., 2004); as well as the presence of these genes in strains causing ABU, otherwise lacking a large proportion of virulence factors commonly found in UPEC (Roos, Ulett, et al., 2006; Watts et al., 2012). However, the high redundancy of iron-acquisition systems in UPEC but also commensal *E. coli* makes it difficult to establish the contribution of single systems to urovirulence or to classify them as virulence factors (Garcia, Brumbaugh, & Mobley, 2011).

5.1. Haem receptors ChuA and Hma

A large proportion of the host's iron is found within erythrocytes, as part of haem within haemoglobin. This iron can be reached after lysis of erythrocytes with haemolysin. Haem receptors in the outer membrane bind haem, which is then taken up in the periplasm and further transported to the cytosol. The two haem receptors ChuA and Hma found in *E. coli* provide significant advantages in co-infection models, especially for colonisation of the kidneys (Garcia et al., 2011; Hagan & Mobley, 2009). Iron acquisition systems, in particular ChuA, are highly upregulated in *E. coli* during intracellular replication within IBCs in the bladder epithelium (Reigstad et al., 2007). Deletion of *chuA* resulted in the development of smaller IBCs compared to those formed by the wild-type strain. Multiplication within IBCs is important for UPEC to successfully colonise the bladder and to eventually establish persisting reservoirs. Thus, *chuA* and other iron-related genes are highly concentrated in strains causing recurrent infections (Ejrnaes et al., 2011).

5.2. Siderophores

Siderophores are secreted iron-chelating molecules which are then, loaded with iron, taken up by the bacterial cell via specific receptors at the outer membrane (Garenaux, Caza, & Dozois, 2011). Four siderophore systems have been investigated in UPEC in the context of infection; enterobactin and its receptor FebA, salmochelin and IroN, aerobactin and IutA, and yersiniabactin and FyuA. The redundancy of siderophore systems in UPEC makes it however complicated to identify certain systems as virulence factors while others might not confer that property (Garcia et al., 2011). While none of them is indispensable for a successful infection, *iroN* (Russo et al., 2002) and *iutA* mutants (Torres, Redford, Welch, & Payne, 2001) are outcompeted by wild-type strains during co-infection. In addition, results from independent epidemiological studies illustrate the importance of these iron-acquisition systems for UPEC persistence in the urinary tract, as iron-related genes are highly concentrated among persisting strains compared to those from sporadic infections (Ejrnaes et al., 2011; Luo et al., 2012; Soto et al., 2006).

The enterobactin system is expressed by virtually all *E. coli* strains and consists of the siderophore enterobactin and its receptor FepA. Binding of enterobactin to its receptors is outcompeted by the mammalian protein lipocalin-2, which is upregulated in response to infection and thus counteracts the bacterial attempt of iron acquisition (Fischbach et al., 2006). Salmochelin represents a glucosylated form of enterobactin and binding to its receptor IroN is not inhibited by lipocalin-2 (Fischbach et al., 2006). Thus, in addition to iron acquisition, salmochelin can be considered as immune evasion strategy. In contrast to enterobactin, salmochelin is associated with pathogenic strains.

The receptor IroN has been suggested to mediate bacterial adherence to urothelial cells (Feldmann, Sorsa, Hildinger, & Schubert, 2007), although these results have been challenged (Leveille et al., 2006; Russo et al., 2002). This is interesting however, since Iha, which shows high homology to siderophore receptors, was initially considered to function as adhesin in UPECs (Johnson, Jelacic, et al., 2005; Tarr et al., 2000). The homology to FepA and IroN however prompted to investigate whether Iha might also function as siderophore receptor in *E. coli*. In fact, Iha serves as an alternative receptor for enterobactin (Leveille et al., 2006). Despite the dual function of Iha, the contribution to urovirulence remains uncertain (Johnson, Jelacic, et al., 2005; Leveille et al., 2006).

6. IMMUNE EVASION MECHANISMS

Upon infection, the host elicits a strong pro-inflammatory response, followed by influx of neutrophils and bacterial elimination. In contrast to commensal strains, UPEC may have the capacity to suppress this induction of pro-inflammatory mediators. An alternative strategy is escaping from immune recognition, either spatially by hiding intracellular, or by covering immunogenic surface structures, for example, by the formation of biofilm. Some virulence and fitness factors might exhibit such properties which protect UPEC from the immune response of the host. For example, salmochelin is not bound by the host protein lipocalin-2 and thus not inhibited in its iron-sequestering function. However, since other functions dominate in those factors (or group of factors), the additional aspect of immune evasion is discussed in the respective section 5.

6.1. Immune suppression

Specific O-antigens of LPS are responsible for the ability of UPEC to dampen the induction of cytokines and chemokines in epithelial cells (Billips, Schaeffer, & Klumpp, 2008; Hunstad, Justice, Hung, Lauer, & Hultgren, 2005). Disruption of the *rfa* operon, in particular the deletion of *waaL*, coding for the O-antigen ligase, neutralises the ability of UPEC strains to inhibit the induction of interleukin (IL) 6 and IL-8 (Billips et al., 2008; Hunstad et al., 2005). *In vivo*, this is accompanied with increased neutrophil recruitment and elimination of the bacteria from the urinary tract (Billips et al., 2008). Defects in the peptidoglycan synthesis are similarly related to increased immune induction by UPEC *in vitro*, but have compared to O-antigens a less-pronounced effect *in vivo* (Billips et al., 2008).

The direct modification of surface antigens has therefore a major impact on the host response evoked by the pathogen. The activity of two periplasmic proteins, SurA and YbcL, can influence UPEC immunogenicity in an indirect manner (Hunstad et al., 2005; Lau, Loughman, & Hunstad, 2012). SurA acts as chaperon and participates in folding and assembly of outer membrane porins (Bitto & McKay, 2003). In particular, OmpA, OmpC, OmpF and LamB are dependent on SurA activity in *E. coli*, and interestingly, the assembly of P and type 1 fimbriae (Justice et al., 2005). This appears surprising considering that these structures act as immunogens during infection. Also YbcL has been related to immune evasion (Lau et al., 2012). The *ybcL* gene exists in different variants, which are ubiquitously distributed among *E. coli*. However, only one YbcL variant identified so far is able to reduce neutrophil transmigration *in vitro* and in the mouse bladder (Lau et al., 2012). This variant is found in high concentration among UPEC strains but to a far lesser extent among commensal strains (Lau et al., 2012). Interestingly, in this study all pyelonephritis strains and the majority of ABU strains but less strains from acute or recurrent cystitis carried the immune-suppressing YbcL variant (Lau et al., 2012).

6.2. Serum resistance and protection against phagocytes

E. coli is rapidly killed in serum by the complement system, or opsonised bacteria are recognised and killed by phagocytic cells. However, pathogenic strains have evolved mechanisms to resist these attacks, which allows them to cause more invasive infections than sensitive strains.

The majority of genetic determinants for serum resistance is related to LPS biosynthesis (Phan et al., 2013), and serum resistance is associated to certain O-antigens frequently found among UPEC (Hughes, Phillips, & Roberts, 1982; Jacobson, Ostenson, Tullus, & Brauner, 1992). For example, the pyelonephritis strains CFT073 and 536 belong to the O6 serotype, and the cystitis strains UTI89 and NU14 to the O18 serotype. Both O-antigen variants, in particular O6, have been associated to serum resistance (Hughes et al., 1982; Jacobson et al., 1992).

Capsular polysaccharides might contribute to serum persistence, depending on the specific K-antigen (Buckles et al., 2009; Burns & Hull, 1998; Sarkar, Ulett, Totsika, Phan, & Schembri, 2014; Schneider et al., 2004). For example, the K2 capsule of CFT073 provides an advantage during colonisation of the urinary tract, possibly due to the contribution of the K2 capsule to serum resistance (Buckles et al., 2009; Sarkar et al., 2014). In contrast, expression of the K15 capsule of strain 536 was irrelevant for serum resistance but still contributes to pathogenesis in the urinary tract (Schneider et al., 2004). In contrast to the modifying role of capsular polysaccharides with respect to serum resistance, K- and O-antigen contribute similarly to the protection against phagocytosis and killing by neutrophils and monocytes (Burns & Hull, 1999; Sarkar et al., 2014).

6.3. Biofilm formation and extracellular matrix components

Biofilms are multicellular communities, composed of bacterial cells embedded in an extracellular matrix (Costerton, Stewart, & Greenberg, 1999). Within these biofilms, bacteria are protected from adverse environmental conditions, including antimicrobial treatment and endogenous host defence mechanisms (Fux, Costerton, Stewart, & Stoodley, 2005). The ability to form biofilm could therefore be considered the sum of adhesion, production of extracellular matrix and growth characteristics.

The ability of bacteria to form biofilm is associated to the pathogenesis of several diseases (Costerton et al., 1999). Also uropathogenic *E. coli* form more biofilm *in vitro* compared to commensal faecal isolates, demonstrating the contribution of biofilm to *E. coli* urovirulence (Kai-Larsen et al., 2010). Moreover, increased biofilm formation as well as the production of exopolysaccharides is associated to more severe and persisting infections (Ejrnaes et al., 2011; Norinder et al., 2011; Salo et al., 2009; Soto et al., 2006; Tapiainen, Hanni, Salo, Ikaheimo, & Uhari, 2014).

Biofilm formation constitutes a major problem in association with urinary catheters, but biofilm-like bacterial formations occur also in association to cells and tissues. The formation of biofilm-like intracellular aggregates in epithelial cells of the urinary bladder plays a central role during the acute stage of cystitis and for the recurrence of infections (Justice et al., 2004; Mysorekar & Hultgren, 2006). Surprisingly, persisting UPEC isolates do not consistently form more biofilms *in vitro* than isolates from sporadic infections (Ejrnaes et al., 2011; Soto et al., 2006; Tapiainen et al., 2014). It should however be noted that *in vitro* biofilm formation does not necessarily reflect the *in vivo* situation.

Several factors contributing to the formation of IBCs and persistence have been discussed above. In addition, also exopolysaccharides are involved in bacterial persistence (Anderson et al., 2003). For example in the cystitis strain UTI89, the ability to generate multiple generations of IBCs is critically dependent on its polysaccharide capsule (Anderson, Goller, Justice, Hultgren, & Seed, 2010). *E. coli* UTI89 expresses the K1 serotype. This K-antigen protects from killing by neutrophils (Sarkar et al., 2014), which is an important defence mechanism during bacterial colonisation of the bladder. In addition, capsular polysaccharides promotes the formation of solid IBCs, which better resist the neutrophilic attack. Whether other K-antigens confer similar advantages has not been investigated; of note however, an *E. coli* strain isolated from IBCs in the urine of an infected child was K1 positive (Robino et al., 2013).

Exopolysaccharides, in general, promote long-term colonisation and possibly persistence of UPEC in the urinary tract, in particular, the capsular polysaccharides (Bahrani-Mougeot et al., 2002), poly-*N*-acetyl glucosamine (Subashchandrabose, Smith, Spurbeck, Kole, & Mobley, 2013) and cellulose (Kai-Larsen et al., 2010). Cellulose reduces the immune response to curliated bacteria probably by covering these highly immunogenic structures. Similar functions might be anticipated for other exopolysaccharides.

7. CONCLUSION

The infection process involves several steps in which *E. coli* interacts with the host cell, each promoted by different virulence factors. While type 1 fimbriae are a pre-requisite for an infection of the urinary tract, several other factors might be dispensable but nevertheless confer an advantage during a particular stage of infection. Therefore together with factors of the host, the combination of bacterial virulence and fitness factors expressed by one particular strain might predict the fate of infection.

REFERENCES

Allsopp, L. P., Beloin, C., Moriel, D. G., Totsika, M., Ghigo, J. M., & Schembri, M. A. (2012). Functional heterogeneity of the UpaH autotransporter protein from uropathogenic *Escherichia coli*. *Journal of Bacteriology*, *194*(21), 5769–5782.

Allsopp, L. P., Beloin, C., Ulett, G. C., Valle, J., Totsika, M., Sherlock, O., et al. (2012). Molecular characterization of UpaB and UpaC, two new autotransporter proteins of uropathogenic *Escherichia coli* CFT073. *Infection and Immunity*, *80*(1), 321–332.

Allsopp, L. P., Totsika, M., Tree, J. J., Ulett, G. C., Mabbett, A. N., Wells, T. J., et al. (2010). UpaH is a newly identified autotransporter protein that contributes to biofilm formation and bladder colonization by uropathogenic *Escherichia coli* CFT073. *Infection and Immunity*, *78*(4), 1659–1669.

Anderson, G. G., Goller, C. C., Justice, S., Hultgren, S. J., & Seed, P. C. (2010). Polysaccharide capsule and sialic acid-mediated regulation promote biofilm-like intracellular bacterial communities during cystitis. *Infection and Immunity*, *78*(3), 963–975.

Anderson, G. G., Palermo, J. J., Schilling, J. D., Roth, R., Heuser, J., & Hultgren, S. J. (2003). Intracellular bacterial biofilm-like pods in urinary tract infections. *Science*, *301*(5629), 105–107.

Backhed, F., Alsen, B., Roche, N., Angstrom, J., von Euler, A., Breimer, M. E., et al. (2002). Identification of target tissue glycosphingolipid receptors for uropathogenic, F1C-fimbriated *Escherichia coli* and its role in mucosal inflammation. *The Journal of Biological Chemistry*, *277*(20), 18198–18205.

Bahrani-Mougeot, F. K., Buckles, E. L., Lockatell, C. V., Hebel, J. R., Johnson, D. E., Tang, C. M., et al. (2002). Type 1 fimbriae and extracellular polysaccharides are preeminent uropathogenic *Escherichia coli* virulence determinants in the murine urinary tract. *Molecular Microbiology*, *45*(4), 1079–1093.

Barnhart, M. M., & Chapman, M. R. (2006). Curli biogenesis and function. *Annual Reviews in Microbiology*, *60*, 131–147.

Berry, R. E., Klumpp, D. J., & Schaeffer, A. J. (2009). Urothelial cultures support intracellular bacterial community formation by uropathogenic *Escherichia coli*. *Infection and Immunity*, *77*(7), 2762–2772.

Bhakdi, S., Greulich, S., Muhly, M., Eberspacher, B., Becker, H., Thiele, A., et al. (1989). Potent leukocidal action of *Escherichia coli* hemolysin mediated by permeabilization of target cell membranes. *Journal of Experimental Medicine*, *169*(3), 737–754.

Bhushan, S., Hossain, H., Lu, Y., Geisler, A., Tchatalbachev, S., Mikulski, Z., et al. (2011). Uropathogenic E. coli induce different immune response in testicular and peritoneal macrophages: Implications for testicular immune privilege. *PLoS One*, *6*(12), e28452.

Bian, Z., Brauner, A., Li, Y., & Normark, S. (2000). Expression of and cytokine activation by *Escherichia coli* curli fibers in human sepsis. *The Journal of Infectious Diseases*, *181*(2), 602–612.

Bian, Z., Yan, Z. Q., Hansson, G. K., Thoren, P., & Normark, S. (2001). Activation of inducible nitric oxide synthase/nitric oxide by curli fibers leads to a fall in blood pressure during systemic *Escherichia coli* infection in mice. *The Journal of Infectious Diseases*, *183*(4), 612–619.

Billips, B. K., Schaeffer, A. J., & Klumpp, D. J. (2008). Molecular basis of uropathogenic *Escherichia coli* evasion of the innate immune response in the bladder. *Infection and Immunity*, *76*(9), 3891–3900.

Birosova, E., Siegfried, L., Kmet'ova, M., Makara, A., Ostro, A., Gresova, A., et al. (2004). Detection of virulence factors in alpha-haemolytic *Escherichia coli* strains isolated from various clinical materials. *Clinical Microbiology and Infection*, *10*(6), 569–573.

Bitto, E., & McKay, D. B. (2003). The periplasmic molecular chaperone protein SurA binds a peptide motif that is characteristic of integral outer membrane proteins. *The Journal of Biological Chemistry*, *278*(49), 49316–49322.

Blanco, M., Blanco, J. E., Alonso, M. P., & Blanco, J. (1994). Virulence factors and O groups of *Escherichia coli* strains isolated from cultures of blood specimens from urosepsis and non-urosepsis patients. *Microbiología, 10*(3), 249–256.

Blanco, M., Blanco, J. E., Alonso, M. P., & Blanco, J. (1996). Virulence factors and O groups of *Escherichia coli* isolates from patients with acute pyelonephritis, cystitis and asymptomatic bacteriuria. *European Journal of Epidemiology, 12*(2), 191–198.

Blanco, J., Blanco, M., Alonso, M. P., Blanco, J. E., Gonzalez, E. A., & Garabal, J. I. (1992). Characteristics of haemolytic *Escherichia coli* with particular reference to production of cytotoxic necrotizing factor type 1 (CNF1). *Research in Microbiology, 143*(9), 869–878.

Boquet, P. (2001). The cytotoxic necrotizing factor 1 (CNF1) from *Escherichia coli*. *Toxicon, 39*(11), 1673–1680.

Brauner, A., Katouli, M., Tullus, K., & Jacobson, S. H. (1990). Production of cytotoxic necrotizing factor, verocytotoxin and haemolysin by pyelonephritogenic *Escherichia coli*. *European Journal of Clinical Microbiology and Infectious Diseases, 9*(10), 762–767.

Bryan, A., Roesch, P., Davis, L., Moritz, R., Pellett, S., & Welch, R. A. (2006). Regulation of type 1 fimbriae by unlinked FimB- and FimE-like recombinases in uropathogenic *Escherichia coli* strain CFT073. *Infection and Immunity, 74*(2), 1072–1083.

Buckles, E. L., Wang, X., Lane, M. C., Lockatell, C. V., Johnson, D. E., Rasko, D. A., et al. (2009). Role of the K2 capsule in *Escherichia coli* urinary tract infection and serum resistance. *The Journal of Infectious Diseases, 199*(11), 1689–1697.

Burns, S. M., & Hull, S. I. (1998). Comparison of loss of serum resistance by defined lipopolysaccharide mutants and an acapsular mutant of uropathogenic *Escherichia coli* O75: K5. *Infection and Immunity, 66*(9), 4244–4253.

Burns, S. M., & Hull, S. I. (1999). Loss of resistance to ingestion and phagocytic killing by O (−) and K(−) mutants of a uropathogenic *Escherichia coli* O75:K5 strain. *Infection and Immunity, 67*(8), 3757–3762.

Cegelski, L., Pinkner, J. S., Hammer, N. D., Cusumano, C. K., Hung, C. S., Chorell, E., et al. (2009). Small-molecule inhibitors target *Escherichia coli* amyloid biogenesis and biofilm formation. *Nature Chemical Biology, 5*(12), 913–919.

Charbonneau, M. E., & Mourez, M. (2007). Functional organization of the autotransporter adhesin involved in diffuse adherence. *Journal of Bacteriology, 189*(24), 9020–9029.

Chen, S. L., Hung, C. S., Pinkner, J. S., Walker, J. N., Cusumano, C. K., Li, Z., et al. (2009). Positive selection identifies an in vivo role for FimH during urinary tract infection in addition to mannose binding. *Proceedings of the National Academy of Sciences of the United States of America, 106*(52), 22439–22444.

Chromek, M., & Brauner, A. (2008). Antimicrobial mechanisms of the urinary tract. *Journal of Molecular Medicine (Berlin, Germany), 86*(1), 37–47.

Chromek, M., Slamova, Z., Bergman, P., Kovacs, L., Podracka, L., Ehren, I., et al. (2006). The antimicrobial peptide cathelicidin protects the urinary tract against invasive bacterial infection. *Nature Medicine, 12*(6), 636–641.

Connell, I., Agace, W., Klemm, P., Schembri, M., Marild, S., & Svanborg, C. (1996). Type 1 fimbrial expression enhances *Escherichia coli* virulence for the urinary tract. *Proceedings of the National Academy of Sciences of the United States of America, 93*(18), 9827–9832.

Costerton, J. W., Stewart, P. S., & Greenberg, E. P. (1999). Bacterial biofilms: A common cause of persistent infections. *Science, 284*(5418), 1318–1322.

Czaja, C. A., Scholes, D., Hooton, T. M., & Stamm, W. E. (2007). Population-based epidemiologic analysis of acute pyelonephritis. *Clinical Infectious Diseases, 45*(3), 273–280.

Daigle, F., Harel, J., Fairbrother, J. M., & Lebel, P. (1994). Expression and detection of pap-, sfa-, and afa-encoded fimbrial adhesin systems among uropathogenic *Escherichia coli*. *Canadian Journal of Microbiology, 40*(4), 286–291.

Davis, J. M., Carvalho, H. M., Rasmussen, S. B., & O'Brien, A. D. (2006). Cytotoxic necrotizing factor type 1 delivered by outer membrane vesicles of uropathogenic *Escherichia coli* attenuates polymorphonuclear leukocyte antimicrobial activity and chemotaxis. *Infection and Immunity, 74*(8), 4401–4408.

Davis, J. M., Rasmussen, S. B., & O'Brien, A. D. (2005). Cytotoxic necrotizing factor type 1 production by uropathogenic *Escherichia coli* modulates polymorphonuclear leukocyte function. *Infection and Immunity, 73*(9), 5301–5310.

Dhakal, B. K., & Mulvey, M. A. (2009). Uropathogenic *Escherichia coli* invades host cells via an HDAC6-modulated microtubule-dependent pathway. *Journal of Biolo gy Chemistry, 284*(1), 446–454.

Dhakal, B. K., & Mulvey, M. A. (2012). The UPEC pore-forming toxin alpha-hemolysin triggers proteolysis of host proteins to disrupt cell adhesion, inflammatory, and survival pathways. *Cell Host & Microbe, 11*(1), 58–69.

Doye, A., Mettouchi, A., Bossis, G., Clement, R., Buisson-Touati, C., Flatau, G., et al. (2002). CNF1 exploits the ubiquitin-proteasome machinery to restrict Rho GTPase activation for bacterial host cell invasion. *Cell, 111*(4), 553–564.

Duncan, M. J., Li, G., Shin, J. S., Carson, J. L., & Abraham, S. N. (2004). Bacterial penetration of bladder epithelium through lipid rafts. *The Journal of Biological Chemistry, 279*(18), 18944–18951.

Ejrnaes, K., Stegger, M., Reisner, A., Ferry, S., Monsen, T., Holm, S. E., et al. (2011). Characteristics of *Escherichia coli* causing persistence or relapse of urinary tract infections: Phylogenetic groups, virulence factors and biofilm formation. *Virulence, 2*(6), 528–537.

Eto, D. S., Jones, T. A., Sundsbak, J. L., & Mulvey, M. A. (2007). Integrin-mediated host cell invasion by type 1-piliated uropathogenic *Escherichia coli*. *PLoS Pathogens, 3*(7), e100.

Eto, D. S., Sundsbak, J. L., & Mulvey, M. A. (2006). Actin-gated intracellular growth and resurgence of uropathogenic *Escherichia coli*. *Cellular Microbiology, 8*(4), 704–717.

Falzano, L., Filippini, P., Travaglione, S., Miraglia, A. G., Fabbri, A., & Fiorentini, C. (2006). *Escherichia coli* cytotoxic necrotizing factor 1 blocks cell cycle G2/M transition in uroepithelial cells. *Infection and Immunity, 74*(7), 3765–3772.

Falzano, L., Fiorentini, C., Donelli, G., Michel, E., Kocks, C., Cossart, P., et al. (1993). Induction of phagocytic behaviour in human epithelial cells by *Escherichia coli* cytotoxic necrotizing factor type 1. *Molecular Microbiology, 9*(6), 1247–1254.

Falzano, L., Quaranta, M. G., Travaglione, S., Filippini, P., Fabbri, A., Viora, M., et al. (2003). Cytotoxic necrotizing factor 1 enhances reactive oxygen species-dependent transcription and secretion of proinflammatory cytokines in human uroepithelial cells. *Infection and Immunity, 71*(7), 4178–4181.

Feldmann, F., Sorsa, L. J., Hildinger, K., & Schubert, S. (2007). The salmochelin siderophore receptor IroN contributes to invasion of urothelial cells by extraintestinal pathogenic *Escherichia coli* in vitro. *Infection and Immunity, 75*(6), 3183–3187.

Fexby, S., Bjarnsholt, T., Jensen, P. O., Roos, V., Hoiby, N., Givskov, M., et al. (2007). Biological Trojan horse: Antigen 43 provides specific bacterial uptake and survival in human neutrophils. *Infection and Immunity, 75*(1), 30–34.

Fischbach, M. A., Lin, H., Zhou, L., Yu, Y., Abergel, R. J., Liu, D. R., et al. (2006). The pathogen-associated iroA gene cluster mediates bacterial evasion of lipocalin 2. *Proceedings of the National Academy of Sciences of the United States of America, 103*(44), 16502–16507.

Fischer, H., Ellstrom, P., Ekstrom, K., Gustafsson, L., Gustafsson, M., & Svanborg, C. (2007). Ceramide as a TLR4 agonist; a putative signalling intermediate between sphingolipid receptors for microbial ligands and TLR4. *Cellular Microbiology, 9*(5), 1239–1251.

Foxman, B., & Brown, P. (2003). Epidemiology of urinary tract infections: Transmission and risk factors, incidence, and costs. *Infectious Disease Clinics of North America, 17*(2), 227–241.

Foxman, B., Gillespie, B., Koopman, J., Zhang, L., Palin, K., Tallman, P., et al. (2000). Risk factors for second urinary tract infection among college women. *American Journal of Epidemiology, 151*(12), 1194–1205.

Foxman, B., Zhang, L., Tallman, P., Palin, K., Rode, C., Bloch, C., et al. (1995). Virulence characteristics of *Escherichia coli* causing first urinary tract infection predict risk of second infection. *The Journal of Infectious Diseases*, *172*(6), 1536–1541.

Frendeus, B., Wachtler, C., Hedlund, M., Fischer, H., Samuelsson, P., Svensson, M., et al. (2001). *Escherichia coli* P fimbriae utilize the Toll-like receptor 4 pathway for cell activation. *Molecular Microbiology*, *40*(1), 37–51.

Fux, C. A., Costerton, J. W., Stewart, P. S., & Stoodley, P. (2005). Survival strategies of infectious biofilms. *Trends in Microbiology*, *13*(1), 34–40.

Garcia, E. C., Brumbaugh, A. R., & Mobley, H. L. (2011). Redundancy and specificity of *Escherichia coli* iron acquisition systems during urinary tract infection. *Infection and Immunity*, *79*(3), 1225–1235.

Garcia, T. A., Ventura, C. L., Smith, M. A., Merrell, D. S., & O'Brien, A. D. (2013). Cytotoxic necrotizing factor 1 and hemolysin from uropathogenic *Escherichia coli* elicit different host responses in the murine bladder. *Infection and Immunity*, *81*(1), 99–109.

Garenaux, A., Caza, M., & Dozois, C. M. (2011). The Ins and Outs of siderophore mediated iron uptake by extra-intestinal pathogenic *Escherichia coli*. *Veterinary Microbiology*, *153*(1–2), 89–98.

Gerstel, U., & Romling, U. (2001). Oxygen tension and nutrient starvation are major signals that regulate agfD promoter activity and expression of the multicellular morphotype in Salmonella typhimurium. *Environmental Microbiology*, *3*(10), 638–648.

Goluszko, P., Moseley, S. L., Truong, L. D., Kaul, A., Williford, J. R., Selvarangan, R., et al. (1997). Development of experimental model of chronic pyelonephritis with *Escherichia coli* O75:K5:H-bearing Dr fimbriae: Mutation in the dra region prevented tubulointerstitial nephritis. *The Journal of Clinical Investigation*, *99*(7), 1662–1672.

Goluszko, P., Niesel, D., Nowicki, B., Selvarangan, R., Nowicki, S., Hart, A., et al. (2001). Dr operon-associated invasiveness of *Escherichia coli* from pregnant patients with pyelonephritis. *Infection and Immunity*, *69*(7), 4678–4680.

Goluszko, P., Popov, V., Selvarangan, R., Nowicki, S., Pham, T., & Nowicki, B. J. (1997). Dr fimbriae operon of uropathogenic *Escherichia coli* mediate microtubule-dependent invasion to the HeLa epithelial cell line. *The Journal of Infectious Diseases*, *176*(1), 158–167.

Goluszko, P., Selvarangan, R., Popov, V., Pham, T., Wen, J. W., & Singhal, J. (1999). Decay-accelerating factor and cytoskeleton redistribution pattern in HeLa cells infected with recombinant *Escherichia coli* strains expressing Dr family of adhesins. *Infection and Immunity*, *67*(8), 3989–3997.

Guignot, J., Bernet-Camard, M. F., Pous, C., Plancon, L., Le Bouguenec, C., & Servin, A. L. (2001). Polarized entry of uropathogenic Afa/Dr diffusely adhering *Escherichia coli* strain IH11128 into human epithelial cells: Evidence for alpha5beta1 integrin recognition and subsequent internalization through a pathway involving caveolae and dynamic unstable microtubules. *Infection and Immunity*, *69*(3), 1856–1868.

Gur, C., Coppenhagen-Glazer, S., Rosenberg, S., Yamin, R., Enk, J., Glasner, A., et al. (2013). Natural killer cell-mediated host defense against uropathogenic E. coli is counteracted by bacterial hemolysinA-dependent killing of NK cells. *Cell Host & Microbe*, *14*(6), 664–674.

Guyer, D. M., Henderson, I. R., Nataro, J. P., & Mobley, H. L. (2000). Identification of sat, an autotransporter toxin produced by uropathogenic *Escherichia coli*. *Molecular Microbiology*, *38*(1), 53–66.

Hagan, E. C., Lloyd, A. L., Rasko, D. A., Faerber, G. J., & Mobley, H. L. (2010). *Escherichia coli* global gene expression in urine from women with urinary tract infection. *PLoS Pathogens*, *6*(11), e1001187.

Hagan, E. C., & Mobley, H. L. (2009). Haem acquisition is facilitated by a novel receptor Hma and required by uropathogenic *Escherichia coli* for kidney infection. *Molecular Microbiology*, *71*(1), 79–91.

Hagberg, L., Hull, R., Hull, S., Falkow, S., Freter, R., & Svanborg Eden, C. (1983). Contribution of adhesion to bacterial persistence in the mouse urinary tract. *Infection and Immunity*, *40*(1), 265–272.
Hart, A., Nowicki, B. J., Reisner, B., Pawelczyk, E., Goluszko, P., Urvil, P., et al. (2001). Ampicillin-resistant *Escherichia coli* in gestational pyelonephritis: Increased occurrence and association with the colonization factor Dr adhesin. *The Journal of Infectious Diseases*, *183*(10), 1526–1529.
Hedlund, M., Wachtler, C., Johansson, E., Hang, L., Somerville, J. E., Darveau, R. P., et al. (1999). P fimbriae-dependent, lipopolysaccharide-independent activation of epithelial cytokine responses. *Molecular Microbiology*, *33*(4), 693–703.
Heimer, S. R., Rasko, D. A., Lockatell, C. V., Johnson, D. E., & Mobley, H. L. (2004). Autotransporter genes pic and tsh are associated with *Escherichia coli* strains that cause acute pyelonephritis and are expressed during urinary tract infection. *Infection and Immunity*, *72*(1), 593–597.
Henderson, I. R., & Nataro, J. P. (2001). Virulence functions of autotransporter proteins. *Infection and Immunity*, *69*(3), 1231–1243.
Henderson, I. R., Navarro-Garcia, F., & Nataro, J. P. (1998). The great escape: Structure and function of the autotransporter proteins. *Trends in Microbiology*, *6*(9), 370–378.
Hertting, O., Chromek, M., Slamova, Z., Kadas, L., Soderkvist, M., Vainumae, I., et al. (2008). Cytotoxic necrotizing factor 1 (CNF1) induces an inflammatory response in the urinary tract in vitro but not in vivo. *Toxicon*, *51*(8), 1544–1547.
Herzer, P. J., Inouye, S., Inouye, M., & Whittam, T. S. (1990). Phylogenetic distribution of branched RNA-linked multicopy single-stranded DNA among natural isolates of *Escherichia coli*. *Journal of Bacteriology*, *172*(11), 6175–6181.
Hilbert, D. W., Paulish-Miller, T. E., Tan, C. K., Carey, A. J., Ulett, G. C., Mordechai, E., et al. (2012). Clinical *Escherichia coli* isolates utilize alpha-hemolysin to inhibit in vitro epithelial cytokine production. *Microbes and Infection*, *14*(7–8), 628–638.
Hooton, T. M. (2012). Clinical practice. Uncomplicated urinary tract infection. *The New England Journal of Medicine*, *366*(11), 1028–1037.
Hooton, T. M., Winter, C., Tiu, F., & Stamm, W. E. (1996). Association of acute cystitis with the stage of the menstrual cycle in young women. *Clinical Infectious Diseases*, *23*(3), 635–636.
Hughes, C., Phillips, R., & Roberts, A. P. (1982). Serum resistance among *Escherichia coli* strains causing urinary tract infection in relation to O type and the carriage of hemolysin, colicin, and antibiotic resistance determinants. *Infection and Immunity*, *35*(1), 270–275.
Hung, C., Marschall, J., Burnham, C. A., Byun, A. S., & Henderson, J. P. (2014). The bacterial amyloid curli is associated with urinary source bloodstream infection. *PLoS One*, *9*(1), e86009.
Hunstad, D. A., Justice, S. S., Hung, C. S., Lauer, S. R., & Hultgren, S. J. (2005). Suppression of bladder epithelial cytokine responses by uropathogenic *Escherichia coli*. *Infection and Immunity*, *73*(7), 3999–4006.
Ikaheimo, R., Siitonen, A., Karkkainen, U., Mustonen, J., Heiskanen, T., & Makela, P. H. (1994). Community-acquired pyelonephritis in adults: Characteristics of E. coli isolates in bacteremic and non-bacteremic patients. *Scandinavian The Journal of Infectious Diseases*, *26*(3), 289–296.
Island, M. D., Cui, X., Foxman, B., Marrs, C. F., Stamm, W. E., Stapleton, A. E., et al. (1998). Cytotoxicity of hemolytic, cytotoxic necrotizing factor 1-positive and -negative *Escherichia coli* to human T24 bladder cells. *Infection and Immunity*, *66*(7), 3384–3389.
Jacobson, S. H., Ostenson, C. G., Tullus, K., & Brauner, A. (1992). Serum resistance in *Escherichia coli* strains causing acute pyelonephritis and bacteraemia. *APMIS*, *100*(2), 147–153.

Johanson, I. M., Plos, K., Marklund, B. I., & Svanborg, C. (1993). Pap, papG and prsG DNA sequences in *Escherichia coli* from the fecal flora and the urinary tract. *Microbial Pathogenesis, 15*(2), 121–129.

Johnson, J. R. (1998). papG alleles among *Escherichia coli* strains causing urosepsis: Associations with other bacterial characteristics and host compromise. *Infection and Immunity, 66*(9), 4568–4571.

Johnson, D. E., Drachenberg, C., Lockatell, C. V., Island, M. D., Warren, J. W., & Donnenberg, M. S. (2000). The role of cytotoxic necrotizing factor-1 in colonization and tissue injury in a murine model of urinary tract infection. *FEMS Immunology and Medical Microbiology, 28*(1), 37–41.

Johnson, J. R., Jelacic, S., Schoening, L. M., Clabots, C., Shaikh, N., Mobley, H. L., et al. (2005). The IrgA homologue adhesin Iha is an *Escherichia coli* virulence factor in murine urinary tract infection. *Infection and Immunity, 73*(2), 965–971.

Johnson, J. R., Kuskowski, M. A., Gajewski, A., Soto, S., Horcajada, J. P., Jimenez de Anta, M. T., et al. (2005). Extended virulence genotypes and phylogenetic background of *Escherichia coli* isolates from patients with cystitis, pyelonephritis, or prostatitis. *The Journal of Infectious Diseases, 191*(1), 46–50.

Johnson, J. R., Kuskowski, M. A., Owens, K., Gajewski, A., & Winokur, P. L. (2003). Phylogenetic origin and virulence genotype in relation to resistance to fluoroquinolones and/or extended-spectrum cephalosporins and cephamycins among *Escherichia coli* isolates from animals and humans. *The Journal of Infectious Diseases, 188*(5), 759–768.

Johnson, J. R., O'Bryan, T. T., Low, D. A., Ling, G., Delavari, P., Fasching, C., et al. (2000). Evidence of commonality between canine and human extraintestinal pathogenic *Escherichia coli* strains that express papG allele III. *Infection and Immunity, 68*(6), 3327–3336.

Johnson, J. R., Oswald, E., O'Bryan, T. T., Kuskowski, M. A., & Spanjaard, L. (2002). Phylogenetic distribution of virulence-associated genes among *Escherichia coli* isolates associated with neonatal bacterial meningitis in the Netherlands. *The Journal of Infectious Diseases, 185*(6), 774–784.

Johnson, J. R., Russo, T. A., Brown, J. J., & Stapleton, A. (1998). papG alleles of *Escherichia coli* strains causing first-episode or recurrent acute cystitis in adult women. *The Journal of Infectious Diseases, 177*(1), 97–101.

Johnson, J. R., Stell, A. L., Kaster, N., Fasching, C., & O'Bryan, T. T. (2001). Novel molecular variants of allele I of the *Escherichia coli* P fimbrial adhesin gene papG. *Infection and Immunity, 69*(4), 2318–2327.

Jouve, M., Garcia, M. I., Courcoux, P., Labigne, A., Gounon, P., & Le Bouguenec, C. (1997). Adhesion to and invasion of HeLa cells by pathogenic *Escherichia coli* carrying the afa-3 gene cluster are mediated by the AfaE and AfaD proteins, respectively. *Infection and Immunity, 65*(10), 4082–4089.

Justice, S. S., Hung, C., Theriot, J. A., Fletcher, D. A., Anderson, G. G., Footer, M. J., et al. (2004). Differentiation and developmental pathways of uropathogenic *Escherichia coli* in urinary tract pathogenesis. *Proceedings of the National Academy of Sciences of the United States of America, 101*(5), 1333–1338.

Justice, S. S., Hunstad, D. A., Harper, J. R., Duguay, A. R., Pinkner, J. S., Bann, J., et al. (2005). Periplasmic peptidyl prolyl cis-trans isomerases are not essential for viability, but SurA is required for pilus biogenesis in *Escherichia coli*. *Journal of Bacteriology, 187*(22), 7680–7686.

Kai-Larsen, Y., Luthje, P., Chromek, M., Peters, V., Wang, X., Holm, A., et al. (2010). Uropathogenic *Escherichia coli* modulates immune responses and its curli fimbriae interact with the antimicrobial peptide LL-37. *PLoS Pathogens, 6*(7), e1001010.

Kansau, I., Berger, C., Hospital, M., Amsellem, R., Nicolas, V., Servin, A. L., et al. (2004). Zipper-like internalization of Dr-positive *Escherichia coli* by epithelial cells is preceded by an adhesin-induced mobilization of raft-associated molecules in the initial step of adhesion. *Infection and Immunity, 72*(7), 3733–3742.

Kaul, A. K., Kumar, D., Nagamani, M., Goluszko, P., Nowicki, S., & Nowicki, B. J. (1996). Rapid cyclic changes in density and accessibility of endometrial ligands for *Escherichia coli* Dr fimbriae. *Infection and Immunity, 64*(2), 611–615.

Khalil, A., Brauner, A., Bakhiet, M., Burman, L. G., Jaremko, G., Wretlind, B., et al. (1997). Cytokine gene expression during experimental *Escherichia coli* pyelonephritis in mice. *The Journal of Urology, 158*(4), 1576–1580.

Khan, A. S., Kniep, B., Oelschlaeger, T. A., Van Die, I., Korhonen, T., & Hacker, J. (2000). Receptor structure for F1C fimbriae of uropathogenic *Escherichia coli. Infection and Immunity, 68*(6), 3541–3547.

Klemm, P., Hjerrild, L., Gjermansen, M., & Schembri, M. A. (2004). Structure-function analysis of the self-recognizing Antigen 43 autotransporter protein from *Escherichia coli. Molecular Microbiology, 51*(1), 283–296.

Konig, B., & Konig, W. (1993). Induction and suppression of cytokine release (tumour necrosis factor-alpha; interleukin-6, interleukin-1 beta) by *Escherichia coli* pathogenicity factors (adhesions, alpha-haemolysin). *Immunology, 78*(4), 526–533.

Korhonen, T. K., Parkkinen, J., Hacker, J., Finne, J., Pere, A., Rhen, M., et al. (1986). Binding of *Escherichia coli* S fimbriae to human kidney epithelium. *Infection and Immunity, 54*(2), 322–327.

Korotkova, N., Yarova-Yarovaya, Y., Tchesnokova, V., Yazvenko, N., Carl, M. A., Stapleton, A. E., et al. (2008). *Escherichia coli* DraE adhesin-associated bacterial internalization by epithelial cells is promoted independently by decay-accelerating factor and carcinoembryonic antigen-related cell adhesion molecule binding and does not require the DraD invasin. *Infection and Immunity, 76*(9), 3869–3880.

Kreft, B., Placzek, M., Doehn, C., Hacker, J., Schmidt, G., Wasenauer, G., et al. (1995). S fimbriae of uropathogenic *Escherichia coli* bind to primary human renal proximal tubular epithelial cells but do not induce expression of intercellular adhesion molecule 1. *Infection and Immunity, 63*(8), 3235–3238.

Kudinha, T., Johnson, J. R., Andrew, S. D., Kong, F., Anderson, P., & Gilbert, G. L. (2013). Genotypic and phenotypic characterization of *Escherichia coli* isolates from children with urinary tract infection and from healthy carriers. *The Pediatric Infectious Disease Journal, 32*(5), 543–548.

Landraud, L., Gibert, M., Popoff, M. R., Boquet, P., & Gauthier, M. (2003). Expression of cnf1 by *Escherichia coli* J96 involves a large upstream DNA region including the hlyCABD operon, and is regulated by the RfaH protein. *Molecular Microbiology, 47*(6), 1653–1667.

Lane, M. C., Alteri, C. J., Smith, S. N., & Mobley, H. L. (2007). Expression of flagella is coincident with uropathogenic *Escherichia coli* ascension to the upper urinary tract. *Proceedings of the National Academy of Sciences of the United States of America, 104*(42), 16669–16674.

Lane, M. C., Lockatell, V., Monterosso, G., Lamphier, D., Weinert, J., Hebel, J. R., et al. (2005). Role of motility in the colonization of uropathogenic *Escherichia coli* in the urinary tract. *Infection and Immunity, 73*(11), 7644–7656.

Lau, M. E., Loughman, J. A., & Hunstad, D. A. (2012). YbcL of uropathogenic *Escherichia coli* suppresses transepithelial neutrophil migration. *Infection and Immunity, 80*(12), 4123–4132.

Le Bouguenec, C., Lalioui, L., du Merle, L., Jouve, M., Courcoux, P., Bouzari, S., et al. (2001). Characterization of AfaE adhesins produced by extraintestinal and intestinal human *Escherichia coli* isolates: PCR assays for detection of Afa adhesins that do or do not recognize Dr blood group antigens. *Journal of Clinical Microbiology, 39*(5), 1738–1745.

Le Bouguenec, C., & Servin, A. L. (2006). Diffusely adherent *Escherichia coli* strains expressing Afa/Dr adhesins (Afa/Dr DAEC): Hitherto unrecognized pathogens. *FEMS Microbiology Letters*, *256*(2), 185–194.
Leveille, S., Caza, M., Johnson, J. R., Clabots, C., Sabri, M., & Dozois, C. M. (2006). Iha from an *Escherichia coli* urinary tract infection outbreak clonal group A strain is expressed in vivo in the mouse urinary tract and functions as a catecholate siderophore receptor. *Infection and Immunity*, *74*(6), 3427–3436.
Li, D., Liu, B., Chen, M., Guo, D., Guo, X., Liu, F., et al. (2010). A multiplex PCR method to detect 14 *Escherichia coli* serogroups associated with urinary tract infections. *Journal of Microbiological Methods*, *82*(1), 71–77.
Lim, J. Y., Pinkner, J. S., & Cegelski, L. (2014). Community behavior and amyloid-associated phenotypes among a panel of uropathogenic E. coli. *Biochemical and Biophysical Research Communications*, *443*(2), 345–350.
Luo, Y., Ma, Y., Zhao, Q., Wang, L., Guo, L., Ye, L., et al. (2012). Similarity and divergence of phylogenies, antimicrobial susceptibilities, and virulence factor profiles of *Escherichia coli* isolates causing recurrent urinary tract infections that persist or result from reinfection. *Journal of Clinical Microbiology*, *50*(12), 4002–4007.
Luthje, P., & Brauner, A. (2010). Ag43 promotes persistence of uropathogenic *Escherichia coli* isolates in the urinary tract. *Journal of Clinical Microbiology*, *48*(6), 2316–2317.
Luthje, P., Brauner, H., Ramos, N. L., Ovregaard, A., Glaser, R., Hirschberg, A. L., et al. (2013). Estrogen supports urothelial defense mechanisms. *Science Translational Medicine*, *5*(190), 190ra180.
Malagolini, N., Cavallone, D., Wu, X. R., & Serafini-Cessi, F. (2000). Terminal glycosylation of bovine uroplakin III, one of the major integral-membrane glycoproteins of mammalian bladder. *Biochimica et Biophysica Acta*, *1475*(3), 231–237.
Manning, S. D., Zhang, L., Foxman, B., Spindler, A., Tallman, P., & Marrs, C. F. (2001). Prevalence of known P-fimbrial G alleles in *Escherichia coli* and identification of a new adhesin class. *Clinical and Diagnist Laboratory Immunology*, *8*(3), 637–640.
Martinez, J. J., Mulvey, M. A., Schilling, J. D., Pinkner, J. S., & Hultgren, S. J. (2000). Type 1 pilus-mediated bacterial invasion of bladder epithelial cells. *The EMBO Journal*, *19*(12), 2803–2812.
Mills, M., Meysick, K. C., & O'Brien, A. D. (2000). Cytotoxic necrotizing factor type 1 of uropathogenic *Escherichia coli* kills cultured human uroepithelial 5637 cells by an apoptotic mechanism. *Infection and Immunity*, *68*(10), 5869–5880.
Mobley, H. L., Green, D. M., Trifillis, A. L., Johnson, D. E., Chippendale, G. R., Lockatell, C. V., et al. (1990). Pyelonephritogenic *Escherichia coli* and killing of cultured human renal proximal tubular epithelial cells: Role of hemolysin in some strains. *Infection and Immunity*, *58*(5), 1281–1289.
Mobley, H. L., Jarvis, K. G., Elwood, J. P., Whittle, D. I., Lockatell, C. V., Russell, R. G., et al. (1993). Isogenic P-fimbrial deletion mutants of pyelonephritogenic *Escherichia coli*: The role of alpha Gal(1–4) beta Gal binding in virulence of a wild-type strain. *Molecular Microbiology*, *10*(1), 143–155.
Moreno, E., Andreu, A., Pigrau, C., Kuskowski, M. A., Johnson, J. R., & Prats, G. (2008). Relationship between *Escherichia coli* strains causing acute cystitis in women and the fecal E. coli population of the host. *Journal of Clinical Microbiology*, *46*(8), 2529–2534.
Moreno, E., Prats, G., Sabate, M., Perez, T., Johnson, J. R., & Andreu, A. (2006). Quinolone, fluoroquinolone and trimethoprim/sulfamethoxazole resistance in relation to virulence determinants and phylogenetic background among uropathogenic *Escherichia coli*. *The Journal of Antimicrobial Chemotherapy*, *57*(2), 204–211.
Mulvey, M. A., Lopez-Boado, Y. S., Wilson, C. L., Roth, R., Parks, W. C., Heuser, J., et al. (1998). Induction and evasion of host defenses by type 1-piliated uropathogenic *Escherichia coli*. *Science*, *282*(5393), 1494–1497.

Mulvey, M. A., Schilling, J. D., & Hultgren, S. J. (2001). Establishment of a persistent *Escherichia coli* reservoir during the acute phase of a bladder infection. *Infection and Immunity, 69*(7), 4572–4579.

Mysorekar, I. U., & Hultgren, S. J. (2006). Mechanisms of uropathogenic *Escherichia coli* persistence and eradication from the urinary tract. *Proceedings of the National Academy of Sciences of the United States of America, 103*(38), 14170–14175.

Navas-Nacher, E. L., Dardick, F., Venegas, M. F., Anderson, B. E., Schaeffer, A. J., & Duncan, J. L. (2001). Relatedness of *Escherichia coli* colonizing women longitudinally. *Molecular Urology, 5*(1), 31–36.

Nicolle, L. E., Mayhew, W. J., & Bryan, L. (1987). Prospective randomized comparison of therapy and no therapy for asymptomatic bacteriuria in institutionalized elderly women. *The American Journal of Medicine, 83*(1), 27–33.

Norinder, B. S., Koves, B., Yadav, M., Brauner, A., & Svanborg, C. (2012). Do *Escherichia coli* strains causing acute cystitis have a distinct virulence repertoire? *Microbial Pathogenesis, 52*(1), 10–16.

Norinder, B. S., Luthje, P., Yadav, M., Kadas, L., Fang, H., Nord, C. E., et al. (2011). Cellulose and PapG are important for *Escherichia coli* causing recurrent urinary tract infection in women. *Infection, 39*(6), 571–574.

Nowicki, B., Selvarangan, R., & Nowicki, S. (2001). Family of *Escherichia coli* Dr adhesins: Decay-accelerating factor receptor recognition and invasiveness. *The Journal of Infectious Diseases, 183*(Suppl. 1), S24–S27.

Olsen, A., Arnqvist, A., Hammar, M., Sukupolvi, S., & Normark, S. (1993). The RpoS sigma factor relieves H-NS-mediated transcriptional repression of csgA, the subunit gene of fibronectin-binding curli in *Escherichia coli*. *Molecular Microbiology*, 7(4), 523–536.

Ong, C. L., Ulett, G. C., Mabbett, A. N., Beatson, S. A., Webb, R. I., Monaghan, W., et al. (2008). Identification of type 3 fimbriae in uropathogenic *Escherichia coli* reveals a role in biofilm formation. *Journal of Bacteriology, 190*(3), 1054–1063.

Otto, G., Sandberg, T., Marklund, B. I., Ulleryd, P., & Svanborg, C. (1993). Virulence factors and pap genotype in *Escherichia coli* isolates from women with acute pyelonephritis, with or without bacteremia. *Clinical Infectious Diseases, 17*(3), 448–456.

Ouslander, J. G., Schapira, M., Schnelle, J. F., Uman, G., Fingold, S., Tuico, E., et al. (1995). Does eradicating bacteriuria affect the severity of chronic urinary incontinence in nursing home residents? *Annals of Internal Medicine, 122*(10), 749–754.

Parham, N. J., Pollard, S. J., Desvaux, M., Scott-Tucker, A., Liu, C., Fivian, A., et al. (2005). Distribution of the serine protease autotransporters of the Enterobacteriaceae among extraintestinal clinical isolates of *Escherichia coli*. *Journal of Clinical Microbiology, 43*(8), 4076–4082.

Parham, N. J., Srinivasan, U., Desvaux, M., Foxman, B., Marrs, C. F., & Henderson, I. R. (2004). PicU, a second serine protease autotransporter of uropathogenic *Escherichia coli*. *FEMS Microbiology Letters, 230*(1), 73–83.

Park, B. S., & Lee, J. O. (2013). Recognition of lipopolysaccharide pattern by TLR4 complexes. *Experimental & Molecular Medicine, 45*, e66.

Parkkinen, J., Rogers, G. N., Korhonen, T., Dahr, W., & Finne, J. (1986). Identification of the O-linked sialyloligosaccharides of glycophorin A as the erythrocyte receptors for S-fimbriated *Escherichia coli*. *Infection and Immunity, 54*(1), 37–42.

Parkkinen, J., Virkola, R., & Korhonen, T. K. (1988). Identification of factors in human urine that inhibit the binding of *Escherichia coli* adhesins. *Infection and Immunity, 56*(10), 2623–2630.

Pere, A., Leinonen, M., Vaisanen-Rhen, V., Rhen, M., & Korhonen, T. K. (1985). Occurrence of type-1C fimbriae on *Escherichia coli* strains isolated from human extraintestinal infections. *Journal of General Microbiology, 131*(7), 1705–1711.

Pere, A., Nowicki, B., Saxen, H., Siitonen, A., & Korhonen, T. K. (1987). Expression of P, type-1, and type-1C fimbriae of *Escherichia coli* in the urine of patients with acute urinary tract infection. *The Journal of Infectious Diseases*, *156*(4), 567–574.

Pham, T., Kaul, A., Hart, A., Goluszko, P., Moulds, J., Nowicki, S., et al. (1995). dra-related X adhesins of gestational pyelonephritis-associated *Escherichia coli* recognize SCR-3 and SCR-4 domains of recombinant decay-accelerating factor. *Infection and Immunity*, *63*(5), 1663–1668.

Phan, M. D., Peters, K. M., Sarkar, S., Lukowski, S. W., Allsopp, L. P., Gomes Moriel, D., et al. (2013). The serum resistome of a globally disseminated multidrug resistant uropathogenic *Escherichia coli* clone. *PLoS Genetics*, *9*(10), e1003834.

Picard, B., Garcia, J. S., Gouriou, S., Duriez, P., Brahimi, N., Bingen, E., et al. (1999). The link between phylogeny and virulence in *Escherichia coli* extraintestinal infection. *Infection and Immunity*, *67*(2), 546–553.

Pichon, C., Hechard, C., du Merle, L., Chaudray, C., Bonne, I., Guadagnini, S., et al. (2009). Uropathogenic *Escherichia coli* AL511 requires flagellum to enter renal collecting duct cells. *Cellular Microbiology*, *11*(4), 616–628.

Plancon, L., Du Merle, L., Le Friec, S., Gounon, P., Jouve, M., Guignot, J., et al. (2003). Recognition of the cellular beta1-chain integrin by the bacterial AfaD invasin is implicated in the internalization of afa-expressing pathogenic *Escherichia coli* strains. *Cellular Microbiology*, *5*(10), 681–693.

Qin, X., Hu, F., Wu, S., Ye, X., Zhu, D., Zhang, Y., et al. (2013). Comparison of adhesin genes and antimicrobial susceptibilities between uropathogenic and intestinal commensal *Escherichia coli* strains. *PLoS One*, *8*(4), e61169.

Ramos, N. L., Dzung, D. T., Stopsack, K., Janko, V., Pourshafie, M. R., Katouli, M., et al. (2011). Characterisation of uropathogenic *Escherichia coli* from children with urinary tract infection in different countries. *European Journal of Clinical Microbiology & Infectious Diseases*, *30*(12), 1587–1593.

Real, J. M., Munro, P., Buisson-Touati, C., Lemichez, E., Boquet, P., & Landraud, L. (2007). Specificity of immunomodulator secretion in urinary samples in response to infection by alpha-hemolysin and CNF1 bearing uropathogenic *Escherichia coli*. *Cytokine*, *37*(1), 22–25.

Reigstad, C. S., Hultgren, S. J., & Gordon, J. I. (2007). Functional genomic studies of uropathogenic *Escherichia coli* and host urothelial cells when intracellular bacterial communities are assembled. *The Journal of Biological Chemistry*, *282*(29), 21259–21267.

Restieri, C., Garriss, G., Locas, M. C., & Dozois, C. M. (2007). Autotransporter-encoding sequences are phylogenetically distributed among *Escherichia coli* clinical isolates and reference strains. *Applied and Environmental Microbiology*, *73*(5), 1553–1562.

Rippere-Lampe, K. E., O'Brien, A. D., Conran, R., & Lockman, H. A. (2001). Mutation of the gene encoding cytotoxic necrotizing factor type 1 (cnf(1)) attenuates the virulence of uropathogenic *Escherichia coli*. *Infection and Immunity*, *69*(6), 3954–3964.

Robino, L., Scavone, P., Araujo, L., Algorta, G., Zunino, P., & Vignoli, R. (2013). Detection of intracellular bacterial communities in a child with *Escherichia coli* recurrent urinary tract infections. *Pathogens and Disease*, *68*(3), 78–81.

Romling, U. (2005). Characterization of the rdar morphotype, a multicellular behaviour in Enterobacteriaceae. *Cellular and Molecular Life Sciences*, *62*(11), 1234–1246.

Romling, U., Sierralta, W. D., Eriksson, K., & Normark, S. (1998). Multicellular and aggregative behaviour of Salmonella typhimurium strains is controlled by mutations in the agfD promoter. *Molecular Microbiology*, *28*(2), 249–264.

Ronald, L. S., Yakovenko, O., Yazvenko, N., Chattopadhyay, S., Aprikian, P., Thomas, W. E., et al. (2008). Adaptive mutations in the signal peptide of the type 1 fimbrial adhesin of uropathogenic *Escherichia coli*. *Proceedings of the National Academy of Sciences of the United States of America*, *105*(31), 10937–10942.

Roos, V., Nielsen, E. M., & Klemm, P. (2006). Asymptomatic bacteriuria *Escherichia coli* strains: Adhesins, growth and competition. *FEMS Microbiology Letters, 262*(1), 22–30.

Roos, V., Schembri, M. A., Ulett, G. C., & Klemm, P. (2006). Asymptomatic bacteriuria *Escherichia coli* strain 83972 carries mutations in the foc locus and is unable to express F1C fimbriae. *Microbiology, 152*(Pt 6), 1799–1806.

Roos, V., Ulett, G. C., Schembri, M. A., & Klemm, P. (2006). The asymptomatic bacteriuria *Escherichia coli* strain 83972 outcompetes uropathogenic E. coli strains in human urine. *Infection and Immunity, 74*(1), 615–624.

Russo, T. A., McFadden, C. D., Carlino-MacDonald, U. B., Beanan, J. M., Barnard, T. J., & Johnson, J. R. (2002). IroN functions as a siderophore receptor and is a urovirulence factor in an extraintestinal pathogenic isolate of *Escherichia coli*. *Infection and Immunity, 70*(12), 7156–7160.

Salo, J., Sevander, J. J., Tapiainen, T., Ikaheimo, I., Pokka, T., Koskela, M., et al. (2009). Biofilm formation by *Escherichia coli* isolated from patients with urinary tract infections. *Clinical Nephrology, 71*(5), 501–507.

Samuelsson, P., Hang, L., Wullt, B., Irjala, H., & Svanborg, C. (2004). Toll-like receptor 4 expression and cytokine responses in the human urinary tract mucosa. *Infection and Immunity, 72*(6), 3179–3186.

Sarkar, S., Ulett, G. C., Totsika, M., Phan, M. D., & Schembri, M. A. (2014). Role of capsule and O antigen in the virulence of uropathogenic *Escherichia coli*. *PLoS One, 9*(4), e94786.

Savoia, D., Millesimo, M., & Fontana, G. (1994). Evaluation of virulence factors and the adhesive capability of *Escherichia coli* strains. *Microbios, 80*(323), 73–81.

Schilling, J. D., Martin, S. M., Hunstad, D. A., Patel, K. P., Mulvey, M. A., Justice, S. S., et al. (2003). CD14- and Toll-like receptor-dependent activation of bladder epithelial cells by lipopolysaccharide and type 1 piliated *Escherichia coli*. *Infection and Immunity, 71*(3), 1470–1480.

Schilling, J. D., Mulvey, M. A., Vincent, C. D., Lorenz, R. G., & Hultgren, S. J. (2001). Bacterial invasion augments epithelial cytokine responses to *Escherichia coli* through a lipopolysaccharide-dependent mechanism. *Journal of Immunology, 166*(2), 1148–1155.

Schmoll, T., Morschhauser, J., Ott, M., Ludwig, B., van Die, I., & Hacker, J. (1990). Complete genetic organization and functional aspects of the *Escherichia coli* S fimbrial adhesion determinant: Nucleotide sequence of the genes sfa B, C, D, E, F. *Microbial Pathogenesis, 9*(5), 331–343.

Schneeberger, C., Kazemier, B. M., & Geerlings, S. E. (2014). Asymptomatic bacteriuria and urinary tract infections in special patient groups: Women with diabetes mellitus and pregnant women. *Current Opinion in Infectious Diseases, 27*(1), 108–114.

Schneider, G., Dobrindt, U., Bruggemann, H., Nagy, G., Janke, B., Blum-Oehler, G., et al. (2004). The pathogenicity island-associated K15 capsule determinant exhibits a novel genetic structure and correlates with virulence in uropathogenic *Escherichia coli* strain 536. *Infection and Immunity, 72*(10), 5993–6001.

Schwan, W. R. (2011). Regulation of genes in uropathogenic E. coli. *World Journal of Clinical Infectious Diseases, 1*(1), 17–25.

Schwartz, D. J., Kalas, V., Pinkner, J. S., Chen, S. L., Spaulding, C. N., Dodson, K. W., et al. (2013). Positively selected FimH residues enhance virulence during urinary tract infection by altering FimH conformation. *Proceedings of the National Academy of Sciences of the United States of America, 110*(39), 15530–15537.

Selvarangan, R., Goluszko, P., Singhal, J., Carnoy, C., Moseley, S., Hudson, B., et al. (2004). Interaction of Dr adhesin with collagen type IV is a critical step in *Escherichia coli* renal persistence. *Infection and Immunity, 72*(8), 4827–4835.

Sherlock, O., Dobrindt, U., Jensen, J. B., Munk Vejborg, R., & Klemm, P. (2006). Glycosylation of the self-recognizing *Escherichia coli* Ag43 autotransporter protein. *Journal of Bacteriology, 188*(5), 1798–1807.

Siitonen, A., Martikainen, R., Ikaheimo, R., Palmgren, J., & Makela, P. H. (1993). Virulence-associated characteristics of *Escherichia coli* in urinary tract infection: A statistical analysis with special attention to type 1C fimbriation. *Microbial Pathogenesis, 15*(1), 65–75.

Simms, A. N., & Mobley, H. L. (2008a). Multiple genes repress motility in uropathogenic *Escherichia coli* constitutively expressing type 1 fimbriae. *Journal of Bacteriology, 190*(10), 3747–3756.

Simms, A. N., & Mobley, H. L. (2008b). PapX, a P fimbrial operon-encoded inhibitor of motility in uropathogenic *Escherichia coli*. *Infection and Immunity, 76*(11), 4833–4841.

Smith, K. D., Andersen-Nissen, E., Hayashi, F., Strobe, K., Bergman, M. A., Barrett, S. L., et al. (2003). Toll-like receptor 5 recognizes a conserved site on flagellin required for protofilament formation and bacterial motility. *Nature Immunology, 4*(12), 1247–1253.

Smith, Y. C., Grande, K. K., Rasmussen, S. B., & O'Brien, A. D. (2006). Novel three-dimensional organoid model for evaluation of the interaction of uropathogenic *Escherichia coli* with terminally differentiated human urothelial cells. *Infection and Immunity, 74*(1), 750–757.

Smith, Y. C., Rasmussen, S. B., Grande, K. K., Conran, R. M., & O'Brien, A. D. (2008). Hemolysin of uropathogenic *Escherichia coli* evokes extensive shedding of the uroepithelium and hemorrhage in bladder tissue within the first 24 hours after intraurethral inoculation of mice. *Infection and Immunity, 76*(7), 2978–2990.

Snyder, J. A., Haugen, B. J., Buckles, E. L., Lockatell, C. V., Johnson, D. E., Donnenberg, M. S., et al. (2004). Transcriptome of uropathogenic *Escherichia coli* during urinary tract infection. *Infection and Immunity, 72*(11), 6373–6381.

Sonnex, C. (1998). Influence of ovarian hormones on urogenital infection. *Sexually Transmitted Infections, 74*(1), 11–19.

Soto, S. M., Smithson, A., Horcajada, J. P., Martinez, J. A., Mensa, J. P., & Vila, J. (2006). Implication of biofilm formation in the persistence of urinary tract infection caused by uropathogenic *Escherichia coli*. *Clinical Microbiology and Infection, 12*(10), 1034–1036.

Spurbeck, R. R., Dinh, P. C., Jr., Walk, S. T., Stapleton, A. E., Hooton, T. M., Nolan, L. K., et al. (2012). *Escherichia coli* isolates that carry vat, fyuA, chuA, and yfcV efficiently colonize the urinary tract. *Infection and Immunity, 80*(12), 4115–4122.

Stahlhut, S. G., Chattopadhyay, S., Kisiela, D. I., Hvidtfeldt, K., Clegg, S., Struve, C., et al. (2013). Structural and population characterization of MrkD, the adhesive subunit of type 3 fimbriae. *Journal of Bacteriology, 195*(24), 5602–5613.

Stromberg, N., Marklund, B. I., Lund, B., Ilver, D., Hamers, A., Gaastra, W., et al. (1990). Host-specificity of uropathogenic *Escherichia coli* depends on differences in binding specificity to Gal alpha 1-4Gal-containing isoreceptors. *The EMBO Journal, 9*(6), 2001–2010.

Stromberg, N., Nyholm, P. G., Pascher, I., & Normark, S. (1991). Saccharide orientation at the cell surface affects glycolipid receptor function. *Proceedings of the National Academy of Sciences of the United States of America, 88*(20), 9340–9344.

Subashchandrabose, S., Smith, S. N., Spurbeck, R. R., Kole, M. M., & Mobley, H. L. (2013). Genome-wide detection of fitness genes in uropathogenic *Escherichia coli* during systemic infection. *PLoS Pathogens, 9*(12), e1003788.

Tapiainen, T., Hanni, A. M., Salo, J., Ikaheimo, I., & Uhari, M. (2014). *Escherichia coli* biofilm formation and recurrences of urinary tract infections in children. *European Journal of Clinical Microbiology & Infectious Diseases, 33*(1), 111–115.

Tarchouna, M., Ferjani, A., Ben-Selma, W., & Boukadida, J. (2013). Distribution of uropathogenic virulence genes in *Escherichia coli* isolated from patients with urinary tract infection. *International Journal of Infectious Diseases, 17*(6), e450–e453.

Tarr, P. I., Bilge, S. S., Vary, J. C., Jr., Jelacic, S., Habeeb, R. L., Ward, T. R., et al. (2000). Iha: A novel *Escherichia coli* O157:H7 adherence-conferring molecule encoded on a

recently acquired chromosomal island of conserved structure. *Infection and Immunity*, *68*(3), 1400–1407.
Thomas, W. E., Nilsson, L. M., Forero, M., Sokurenko, E. V., & Vogel, V. (2004). Shear-dependent 'stick-and-roll' adhesion of type 1 fimbriated *Escherichia coli*. *Molecular Microbiology*, *53*(5), 1545–1557.
Thumbikat, P., Berry, R. E., Schaeffer, A. J., & Klumpp, D. J. (2009). Differentiation-induced uroplakin III expression promotes urothelial cell death in response to uropathogenic E. coli. *Microbes and Infection*, *11*(1), 57–65.
Thumbikat, P., Berry, R. E., Zhou, G., Billips, B. K., Yaggie, R. E., Zaichuk, T., et al. (2009). Bacteria-induced uroplakin signaling mediates bladder response to infection. *PLoS Pathogens*, *5*(5), e1000415.
Torres, A. G., Redford, P., Welch, R. A., & Payne, S. M. (2001). TonB-dependent systems of uropathogenic *Escherichia coli*: Aerobactin and heme transport and TonB are required for virulence in the mouse. *Infection and Immunity*, *69*(10), 6179–6185.
Totsika, M., Wells, T. J., Beloin, C., Valle, J., Allsopp, L. P., King, N. P., et al. (2012). Molecular characterization of the EhaG and UpaG trimeric autotransporter proteins from pathogenic *Escherichia coli*. *Applied and Environmental Microbiology*, *78*(7), 2179–2189.
Tukel, C., Nishimori, J. H., Wilson, R. P., Winter, M. G., Keestra, A. M., van Putten, J. P., et al. (2010). Toll-like receptors 1 and 2 cooperatively mediate immune responses to curli, a common amyloid from enterobacterial biofilms. *Cellular Microbiology*, *12*(10), 1495–1505.
Tukel, C., Raffatellu, M., Humphries, A. D., Wilson, R. P., Andrews-Polymenis, H. L., Gull, T., et al. (2005). CsgA is a pathogen-associated molecular pattern of Salmonella enterica serotype Typhimurium that is recognized by Toll-like receptor 2. *Molecular Microbiology*, *58*(1), 289–304.
Ulett, G. C., Mabbett, A. N., Fung, K. C., Webb, R. I., & Schembri, M. A. (2007). The role of F9 fimbriae of uropathogenic *Escherichia coli* in biofilm formation. *Microbiology*, *153*(Pt. 7), 2321–2331.
Ulett, G. C., Valle, J., Beloin, C., Sherlock, O., Ghigo, J. M., & Schembri, M. A. (2007). Functional analysis of antigen 43 in uropathogenic *Escherichia coli* reveals a role in long-term persistence in the urinary tract. *Infection and Immunity*, *75*(7), 3233–3244.
Valle, J., Mabbett, A. N., Ulett, G. C., Toledo-Arana, A., Wecker, K., Totsika, M., et al. (2008). UpaG, a new member of the trimeric autotransporter family of adhesins in uropathogenic *Escherichia coli*. *Journal of Bacteriology*, *190*(12), 4147–4161.
Vejborg, R. M., Hancock, V., Schembri, M. A., & Klemm, P. (2011). Comparative genomics of *Escherichia coli* strains causing urinary tract infections. *Applied and Environmental Microbiology*, 77(10), 3268–3278.
Virkola, R., Parkkinen, J., Hacker, J., & Korhonen, T. K. (1993). Sialyloligosaccharide chains of laminin as an extracellular matrix target for S fimbriae of *Escherichia coli*. *Infection and Immunity*, *61*(10), 4480–4484.
Waksman, G., & Hultgren, S. J. (2009). Structural biology of the chaperone-usher pathway of pilus biogenesis. *Nature Reviews Microbiology*, 7(11), 765–774.
Wang, A., Nizran, P., Malone, M. A., & Riley, T. (2013). Urinary tract infections. *Primary Care*, *40*(3), 687–706.
Wang, X., & Quinn, P. J. (2010). Lipopolysaccharide: Biosynthetic pathway and structure modification. *Progress in Lipid Research*, *49*(2), 97–107.
Watts, R. E., Totsika, M., Challinor, V. L., Mabbett, A. N., Ulett, G. C., De Voss, J. J., et al. (2012). Contribution of siderophore systems to growth and urinary tract colonization of asymptomatic bacteriuria *Escherichia coli*. *Infection and Immunity*, *80*(1), 333–344.
Welch, R. A., Burland, V., Plunkett, G., 3rd, Redford, P., Roesch, P., Rasko, D., et al. (2002). Extensive mosaic structure revealed by the complete genome sequence of

uropathogenic *Escherichia coli*. *Proceedings of the National Academy of Sciences of the United States of America*, *99*(26), 17020–17024.
Wiles, T. J., Bower, J. M., Redd, M. J., & Mulvey, M. A. (2009). Use of zebrafish to probe the divergent virulence potentials and toxin requirements of extraintestinal pathogenic *Escherichia coli*. *PLoS Pathogens*, *5*(12), e1000697.
Wiles, T. J., & Mulvey, M. A. (2013). The RTX pore-forming toxin alpha-hemolysin of uropathogenic *Escherichia coli*: Progress and perspectives. *Future Microbiology*, *8*(1), 73–84.
Wright, K. J., Seed, P. C., & Hultgren, S. J. (2005). Uropathogenic *Escherichia coli* flagella aid in efficient urinary tract colonization. *Infection and Immunity*, *73*(11), 7657–7668.
Wright, K. J., Seed, P. C., & Hultgren, S. J. (2007). Development of intracellular bacterial communities of uropathogenic *Escherichia coli* depends on type 1 pili. *Cellular Microbiology*, *9*(9), 2230–2241.
Wu, X. R., Kong, X. P., Pellicer, A., Kreibich, G., & Sun, T. T. (2009). Uroplakins in urothelial biology, function, and disease. *Kidney International*, *75*(11), 1153–1165.
Wurpel, D. J., Beatson, S. A., Totsika, M., Petty, N. K., & Schembri, M. A. (2013). Chaperone-usher fimbriae of *Escherichia coli*. *PLoS One*, *8*(1), e52835.
Wurpel, D. J., Totsika, M., Allsopp, L. P., Hartley-Tassell, L. E., Day, C. J., Peters, K. M., et al. (2014). F9 Fimbriae of uropathogenic *Escherichia coli* are expressed at low temperature and recognise Galbeta1-3GlcNAc-containing glycans. *PLoS One*, *9*(3), e93177.
Yamamoto, S., Tsukamoto, T., Terai, A., Kurazono, H., Takeda, Y., & Yoshida, O. (1997). Genetic evidence supporting the fecal-perineal-urethral hypothesis in cystitis caused by *Escherichia coli*. *The Journal of Urology*, *157*(3), 1127–1129.
Zhang, L., Foxman, B., Tallman, P., Cladera, E., Le Bouguenec, C., & Marrs, C. F. (1997). Distribution of drb genes coding for Dr binding adhesins among uropathogenic and fecal *Escherichia coli* isolates and identification of new subtypes. *Infection and Immunity*, *65*(6), 2011–2018.
Zhou, G., Mo, W. J., Sebbel, P., Min, G., Neubert, T. A., Glockshuber, R., et al. (2001). Uroplakin Ia is the urothelial receptor for uropathogenic *Escherichia coli*: Evidence from in vitro FimH binding. *Journal of Cell Science*, *114*(Pt. 22), 4095–4103.

AUTHOR INDEX

Note: Page numbers followed by "*f*" indicate figures and "*t*" indicate tables.

A

B

D

F

G

H

I

J

K

M

N

O

P

S

U

V

X

Y

Z

SUBJECT INDEX

Note: Page numbers followed by "*f*" indicate figures and "*t*" indicate tables.

C

D

F

G

H

I

L

M

T

U

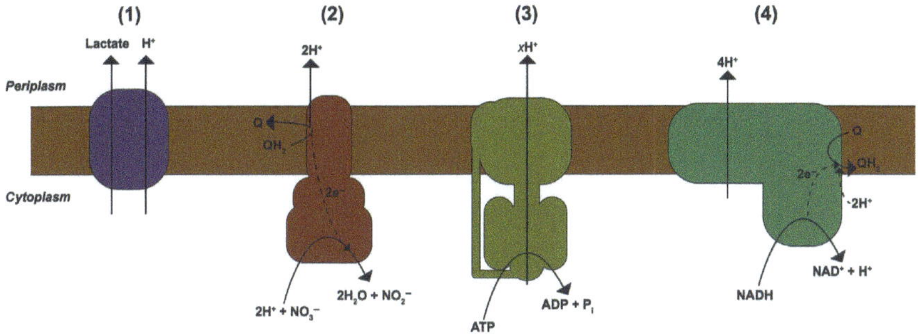

Gregory M. Cook *et al.*, Figure 1 *Mechanisms (1–4) by which a proton motive force can be generated in bacteria.* (1) Co-transport of protons driven by solute (lactate) symport into the periplasm. (2) Redox-loop separation of charge; quinol oxidation results in proton release into the periplasm by virtue of quinol site proximity to the periplasm, while electrons are transferred to reduce a terminal electron acceptor in the cytoplasm that results in neutralisation of charge. (3) Proton export driven by ATP hydrolysis, i.e., ATP synthase working in the reverse direction. (4) Proton translocation mediated by primary proton-pumping complexes.

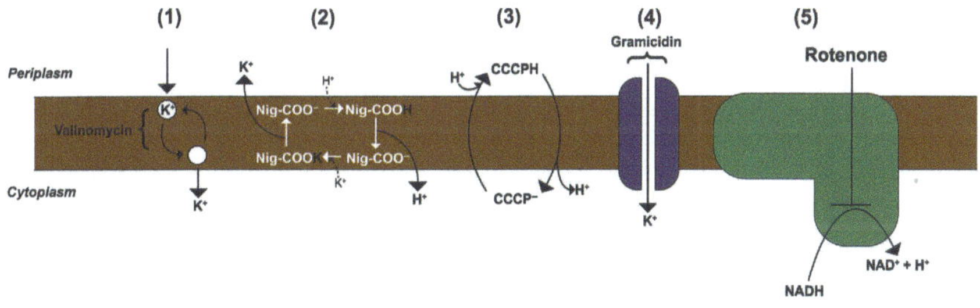

Gregory M. Cook *et al.*, Figure 2 *Traditional inhibitors of proton motive force generation.* (1) Valinomycin is an ionophore, selective for potassium ions, which equilibrates the potassium gradient—dissipating the $\Delta\psi$ (electrogenic). (2) Nigericin is a hydrophobic weak carboxylic acid, which can traverse the membrane as its either protonated acid or neutral salt. It dissipates chemical gradients (i.e. ΔpH) but maintains the charge (one positive charge exchanged for one positive charge—electroneutral). (3) Carbonyl cyanide *m*-chlorophenyl hydrazine (CCCP) is an electrogenic protonophore. $CCCP^-$ is driven to the periplasm by the $\Delta\psi$, while CCCPH is driven to the cytoplasm by the ΔpH. It can equilibrate both $\Delta\psi$ and ΔpH. (4) Gramicidin is a channel-forming ionophore, making the membrane more permeable to ions. (5) Rotenone inhibits primary proton pumping—preventing the initial generation of a proton motive force.

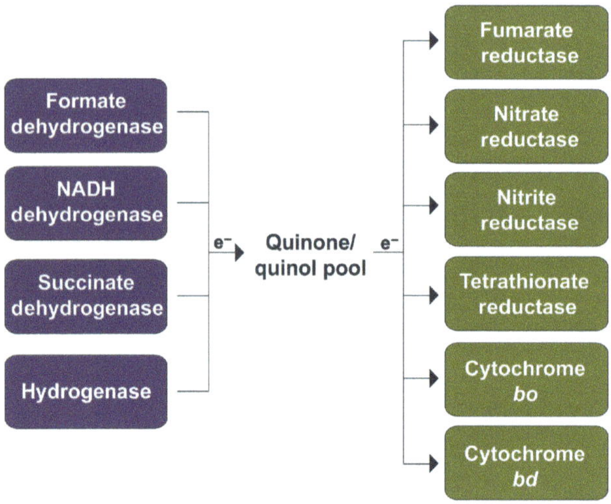

Gregory M. Cook *et al.*, Figure 3 *Generalised schematic overview of relevant electron transfer components.* Complexes indicated in blue oxidise various substrates to reduce quinones. The resulting quinol molecules can be oxidised to result in reduction of various terminal electron acceptors, mediated by the complexes indicated in green. For some electron transfer pathways intermediate complexes and molecules exist, for example, complex III will generally reduce cytochrome *c*, which will serve as the electron donor for complex IV. The complexes used, types of quinones, and intermediates thereof are highly variable between genera. Only complexes relevant to this review are indicated.

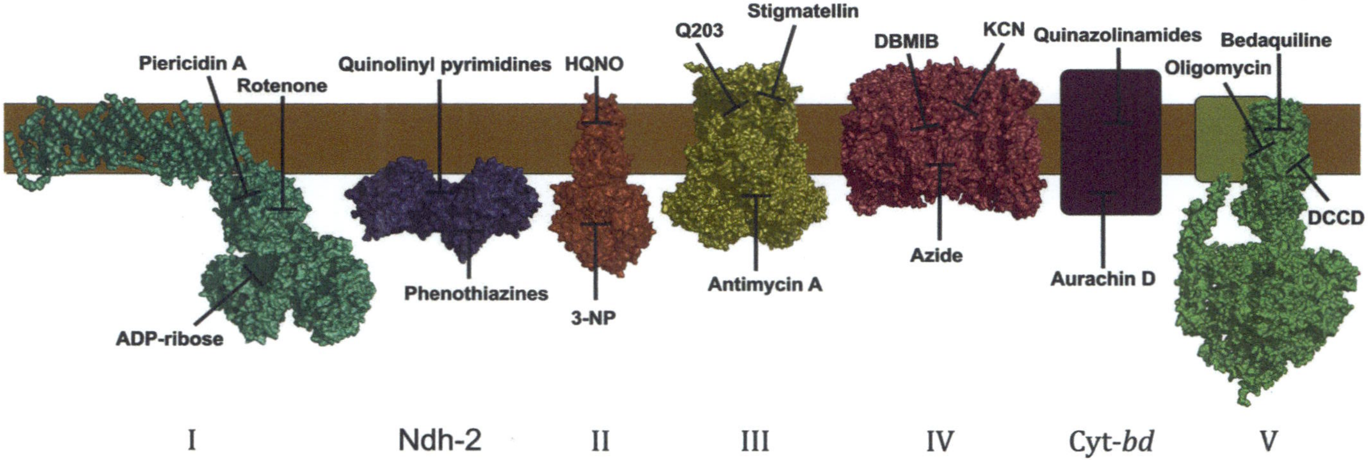

Gregory M. Cook *et al.*, Figure 4 *Diversity of electron transport chain inhibitors*. Structural surface representations of electron transport chain components are indicated where possible. Selected inhibitors of these complexes are indicated with flathead arrows and do not reflect the binding site of the inhibitors. Crystal structures were obtained from RCSB protein data bank from the following accession numbers: complex I, 3M9S; Ndh-2, 4NWZ; complex II, 2WDV; complex III, 3H1J; complex IV, 3ASN; and complex V, 4B2Q. Models were generated using the PyMOL molecular graphics system.

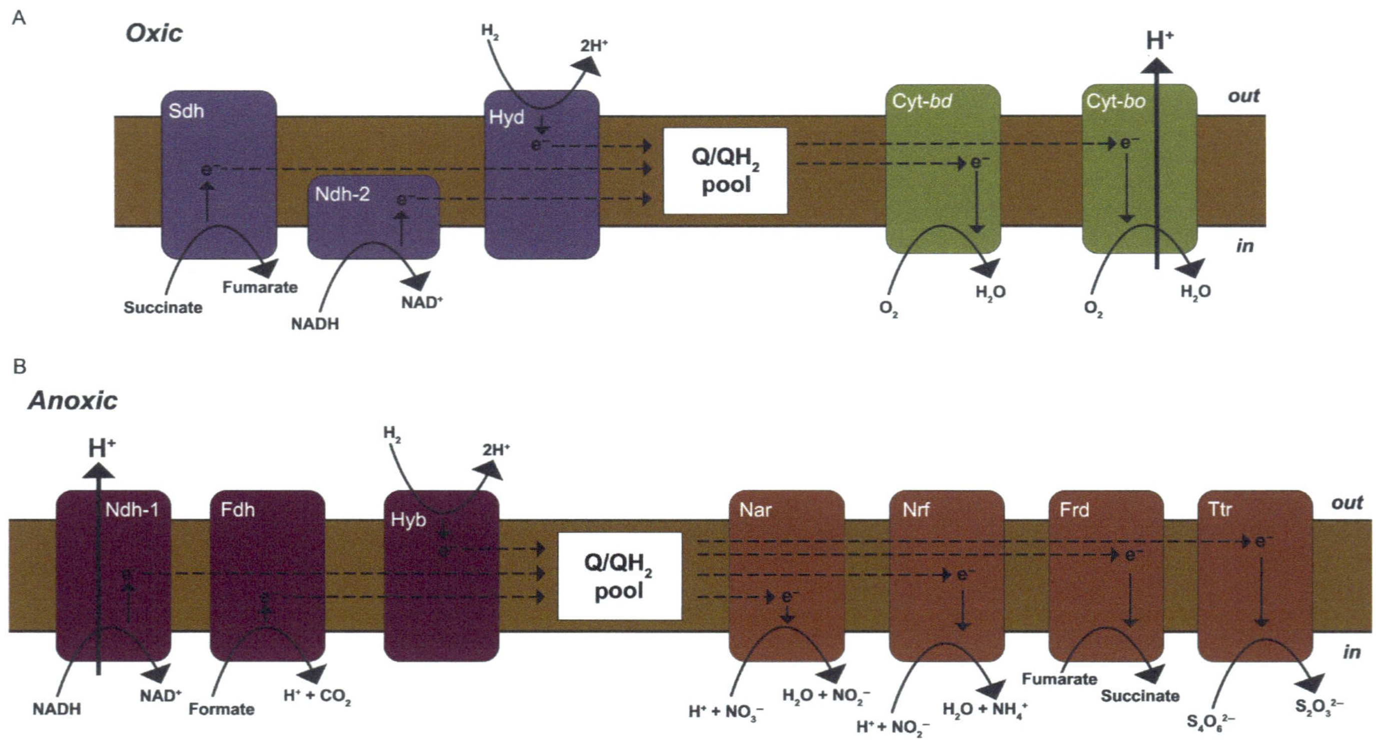

Gregory M. Cook *et al.*, Figure 5 *Diversity of electron transport chain composition under different oxygen tensions*. The major components of the *Salmonella* enterica electron transport chain under (A) oxic and (B) anoxic conditions. As with Fig. 3, not all intermediate steps in electron transport are indicated. While represented as oxic, cytochrome *bd* is generally expressed under microoxic conditions. Additionally, nitrate and nitrite reductases are used preferentially to the other anoxic electron acceptor complexes. This diversity serves as a general model for that which is observed in other bacteria, culturable under both oxic and anoxic conditions.

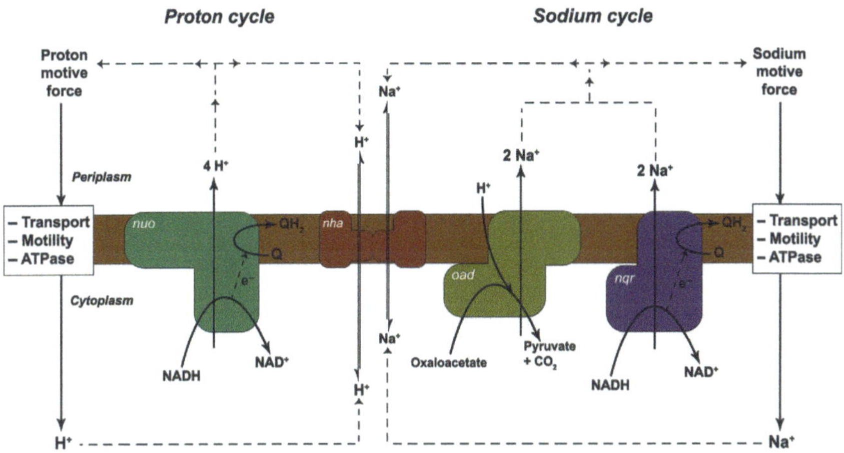

Gregory M. Cook *et al.*, Figure 6 *Generalised role of sodium and proton cycles at the level of the membrane.* Only representative proton- (*nuo*) and sodium- (*oad, nqr*) pumping complexes are shown for simplicity. The sodium and proton motive forces can be used to drive cellular processes such as flagellar motion and ATP synthesis. Antiporters provide an overlap between the two cycles as they can theoretically operate to generate one electrochemical gradient at the expense of the other, depending on cellular needs. Nha is used to refer to antiporters in general, but in reality, antiporters can operate on variable stoichiometry (not pictured) which results in either electrogenic or electroneutral exchange of H^+ and Na^+ depending on the protein.

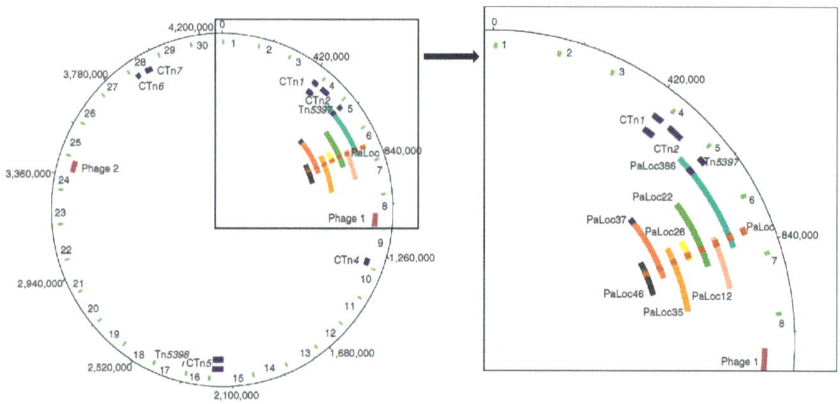

Adam P. Roberts *et al.*, Figure 2 The panel on the left shows a circular schematic representation of the *C. difficile* 630 chromosome. The first inner ring shows the position of the 30 regions of differing DNA, which were used to determine the source of DNA in transconjugants. The second inner ring shows the position of the PaLoc, conjugative transposons and prophage. The next eight inner rings show the extent of the transferred DNA in eight transconjugants, which are enlarged and labelled in the right-hand panel. *Reproduced from Brouwer et al. (2013).*

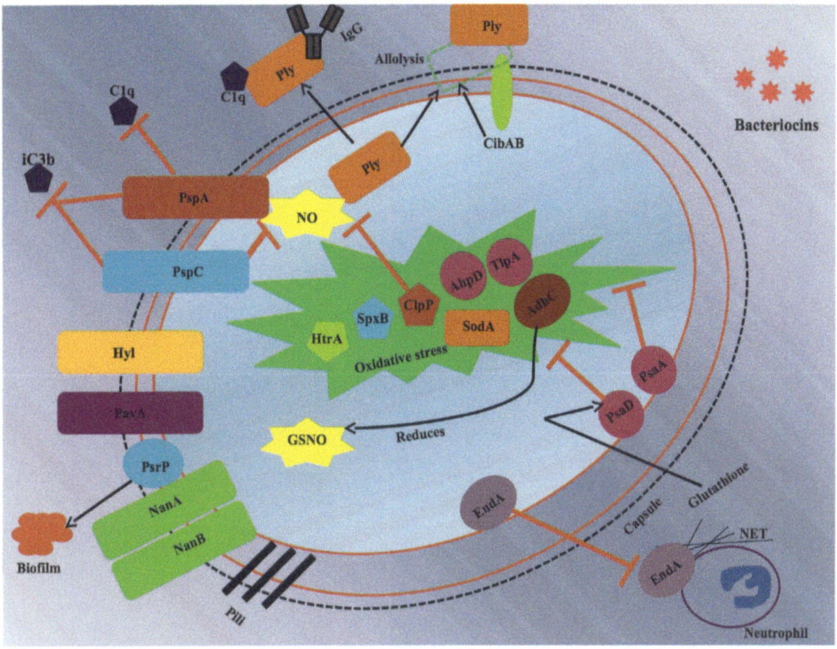

Joby Cole *et al.*, Figure 1 *Streptococcus pneumoniae* virulence factors that resist phagocytosis and antimicrobial killing. *Streptococcus pneumoniae* (Spn) has a polysaccharide capsule that helps it resist phagocytosis. Pneumococcal surface protein A (PspA) and Pneumococcal surface protein C (PspC) inhibit complement activation. Pneumococcal serine-rich repeat protein (PsrP) regulates biofilm formation. Spn synthesises virulence factors to resists oxidative and nitrosative stress. Superoxide dismutase A (SodA) allows Spn to resist superoxide ($O_2^{\cdot -}$). High temperature requirement A (HtrA), alkylhydroperoxidase A (AhpA), caseinolytic peptidase P (ClpP), thioredoxin-like protein A (TlpA), and SpxB (a pyruvate oxidase) help Spn resists hydrogen peroxide (H_2O_2). Pneumococcal surface adhesin A (PsaA) and Pneumococcal surface adhesin D (PsaD) also contribute to resistance to oxidative stress. Alcohol dehydrogenase C (AdhC) provides reduced glutathione required by PsaD. ClpP and PspC help Spn resist nitrosative stress. Endonuclease A (EndA) degrades neutrophil extracellular traps (NETs). Pneumolysin (Ply) prevents complement activation. Ply interacts with CibAB to induce allolysis. Neuraminidase A (NanA) and Neuraminidase B (NanB), Hyaluronate lyase (Hyl), Pneumococcal adherence and virulence factor A (PavA), and the pilus are involved in colonisation. Bacteriocins help survival in the nasopharynx.

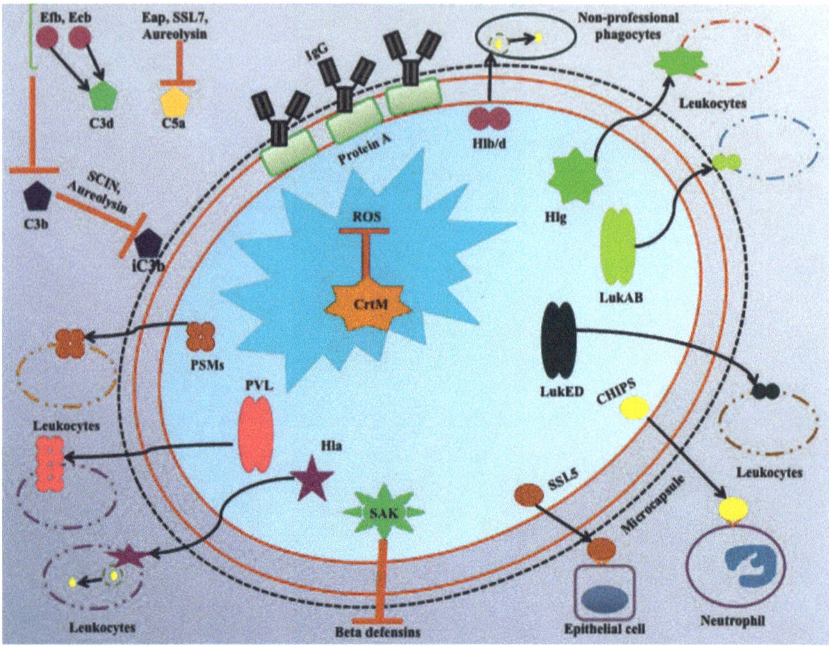

Joby Cole *et al.*, Figure 2 *Staphylococcus aureus* virulence factors that resist phagocytosis and antimicrobial killing. *S. aureus* expresses a range of virulence factors to inhibit phagocytosis and intracellular killing. Some strains express a microcapsule to impede phagocytosis. Surface expressed protein A (Spa) binds the Fc region of IgG inhibiting phagocytosis. The extracellular fibrinogen-binding protein (Efb) binds C3d blocking C3b containing convertases and binds fibrinogen preventing neutrophil activation. The extracellular complement-binding protein (Ecb) also binds C3d and blocks C3b containing convertases and affects C5 convertases. The staphylococcal complement inhibitor (SCIN) and aureolysin prevent complement activation. Superantigen-like protein 7 (SSL7) prevents complement activation by binding C5a. The chemotaxis inhibitory protein of *S. aureus* (CHIPs) prevents neutrophil chemotaxis. Superantigen-like protein 5 (SSL5) binds P-selectin glycoprotein on epithelial cell surfaces preventing neutrophil rolling. *S. aureus* synthesises a range of pore forming toxins to damage host cells. These include Panton–Valentine Leukocidin (PVL), alpha haemolysin (Hla), Phenol Soluble Modulins (PSMs), Leukocidin AB (LukAB), Leukocidin ED (LukED), and gamma haemolysin (Hlg). Expression of Hla of beta haemolysin (Hlb) and delta haemolysin (Hld) contribute to *S. aureus* escape from phagosomes. Carotenoid M (CrtM) helps *S. aureus* survive oxidative stress and staphylokinase (SAK) neutralises defensins.

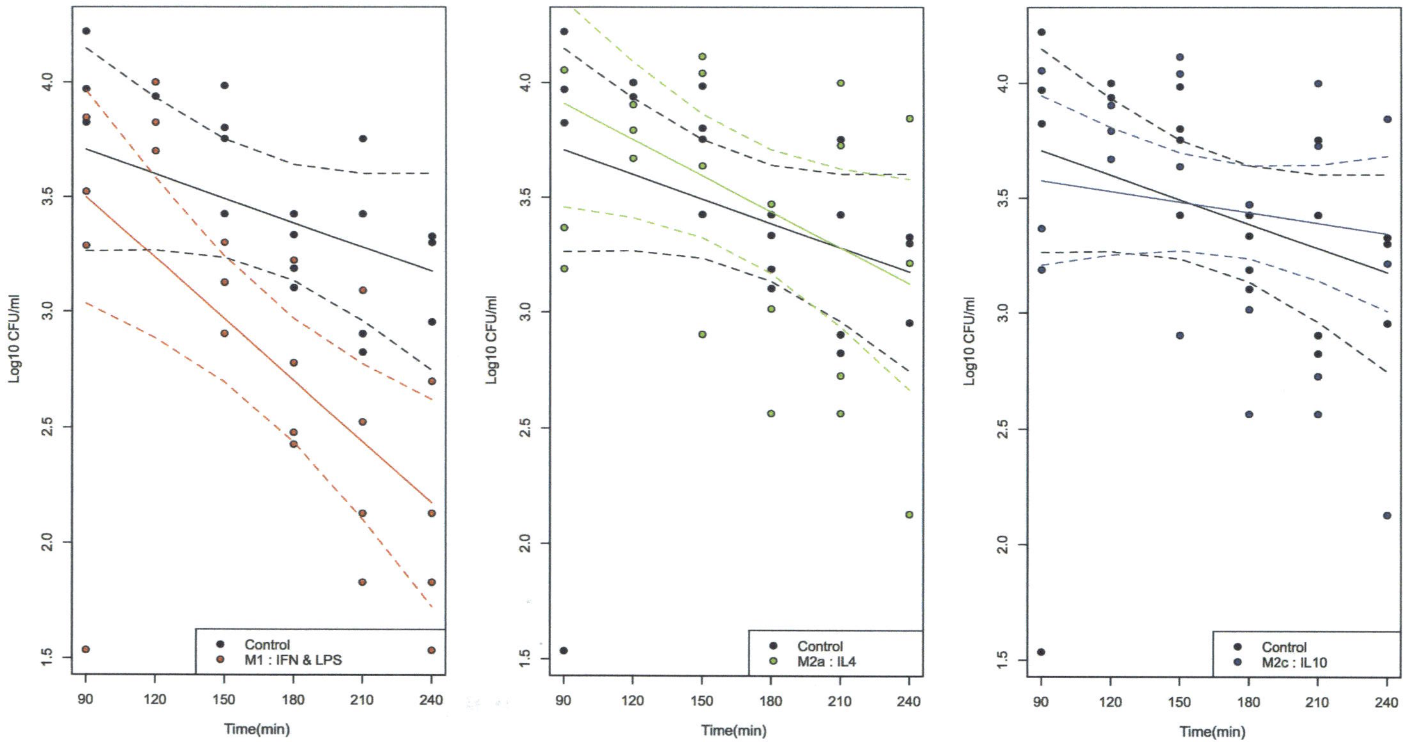

Joby Cole *et al.*, Figure 3 Early intracellular killing of *S. pneumoniae* by polarised human monocyte-derived macrophages (MDMs). Healthy donor monocytes were isolated from peripheral blood mononuclear cells and cultured in 24 well plates as previously described for 14 days to generate MDM and then challenged with opsonised serotype 14 *S. pneumoniae* at a multiplicity of infection of 10 (Dockrell, Lee, Lynch, & Read, 2001). For 72 h prior to infection, they were polarised towards an M1, M2a, or M2c phenotype with 50 ng/ml IFNγ and LPS, IL-4, or IL-10, respectively (Ambarus et al., 2012). Viable intracellular *S. pneumoniae* were quantified at various time points after bacterial challenge using standard gentamicin protection assays and the method of Miles and Misra, as previously described (Dockrell et al., 2001).

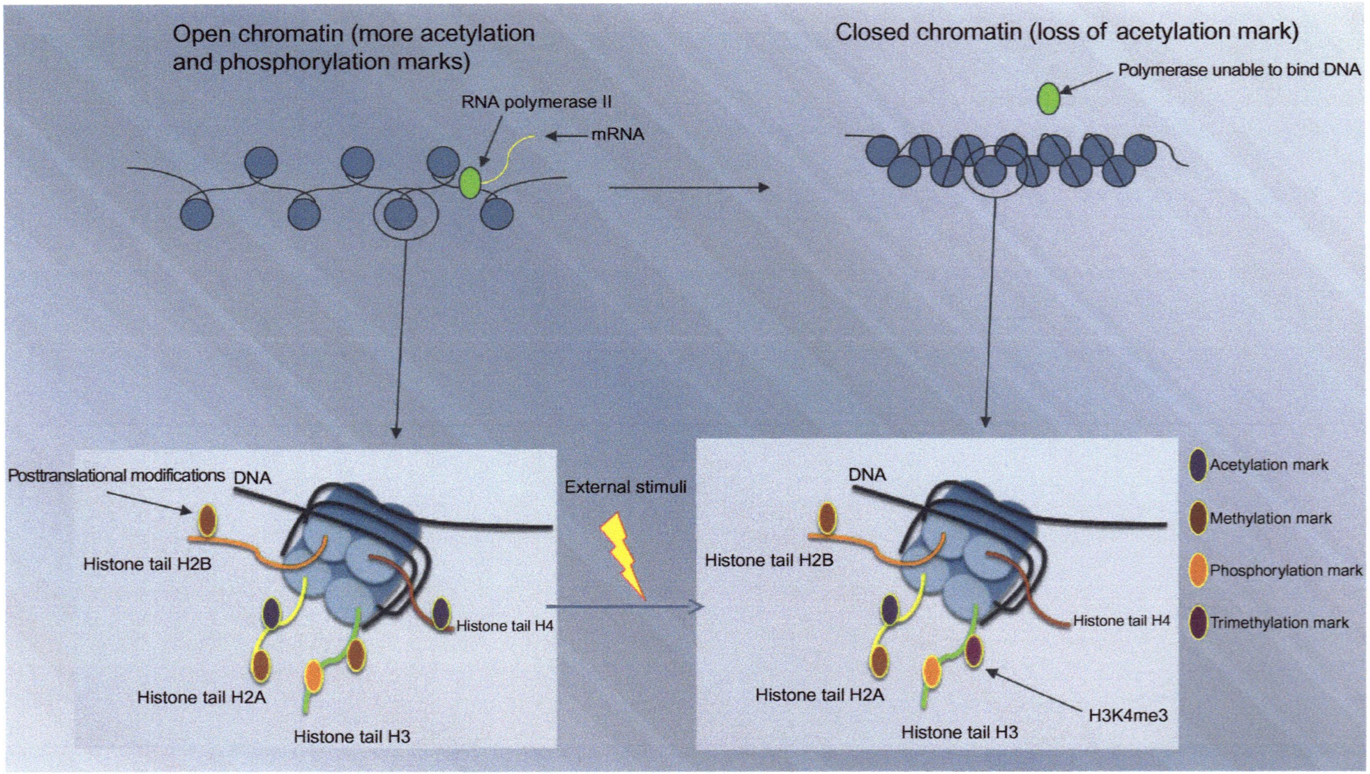

Joby Cole *et al.*, Figure 4 Posttranslational modifications of histone tails regulate macrophage gene transcription. A nucleosome is depicted composed of the four pairs of histones with a schematic representation of some representative posttranslational modifications of the histone tails. Following external stimulation, the levels of H3K4me3 are modified leading to a change in the conformation of the chromatin, in this example from euchromatin to heterochromatin thereby inhibiting access to the DNA sequence wound around it. This in turn leads to decreased transcription of the associated gene, in this example IL-6.

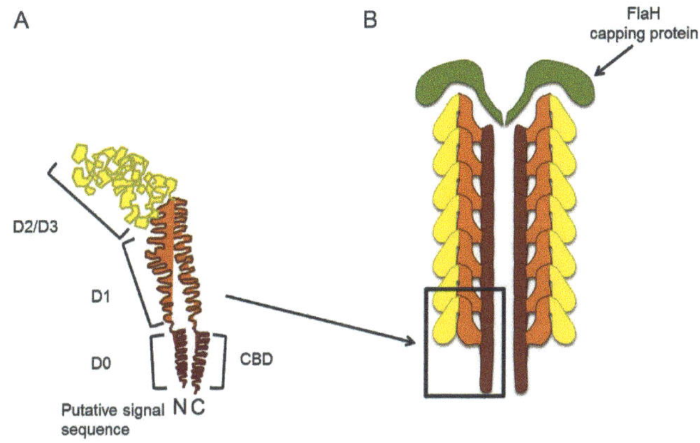

Rebecca Lowry *et al.*, Figure 1 Schematic diagram depicting the cross-section of a flagella filament and flagellin subunit. (A) A single flagellin subunit, showing the conserved D0 and D1 domains located at the C- and N-termini of the flagellin; this is the area recognised by the innate immune systems Toll-like receptor TLR-5. The variable D2/D3 domain is also indicated that in the *Aeromonas* is glycosylated with pseudaminic acid. The N-terminal D0 regions contain the putative signal sequence for flagellin secretion, whereas the C-terminal final 45 amino acids comprises the chaperone (FlaJ) binding domain (CBD). (B) Cross-section of part of a flagella filament showing a number of individual flagellin subunits polymerised into a filament showing the hollow tube through which flagellin subunits are secreted. The capping proteins (FlaH) help the flagellins polymerise into the filament at the distal end of the growing flagellum. The conserved parts of the flagellins the D0 and D1 domains are located inside the filament, whereas the variable glycosylated D2/D3 domain of the flagellin are exposed on the filament exterior.

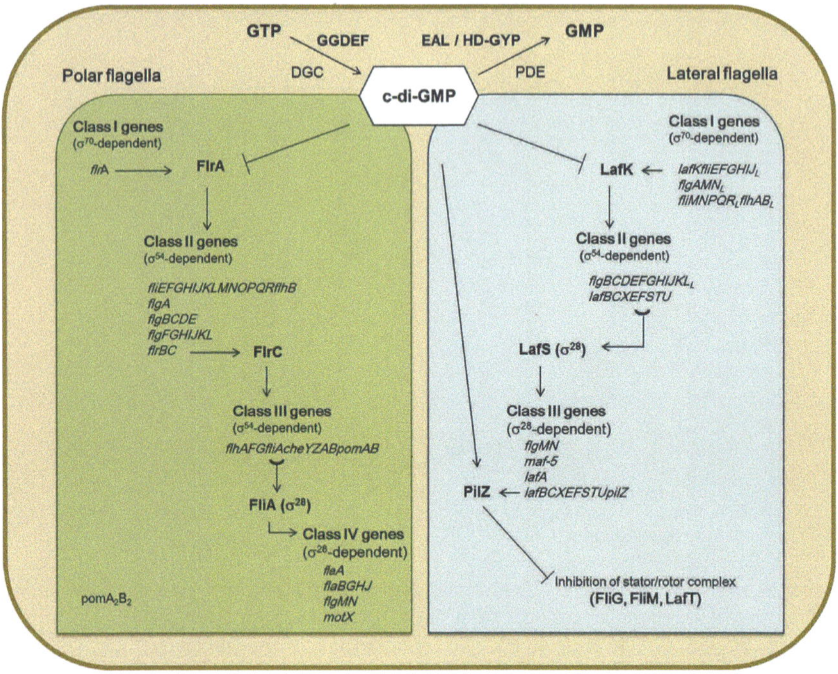

Rebecca Lowry *et al.*, Figure 3 The predicted role of cyclic-di-GMP on the previously determined polar and lateral flagella regulatory cascades in *Aeromonas* species. C-di-GMP is produced by GGDEF domain containing diguanylate cyclase (DGC) enzymes and is broken down by EAL or HD-GYP domain containing phosphodiesterases (PDE) proteins. The rise in the intracellular concentration of c-di-GMP in *Aeromonas* species has been demonstrated to inhibit both swimming (polar flagella dependent) and swarming (lateral flagella dependent) motility, while favouring sessile activity such as biofilm formation (Kozlova, Khajanchi, Sha, & Chopra, 2011). In related bacteria (*Vibrio, Pseudomonas*), c-di-GMP has been shown to bind to the enhancer-binding protein-response regulators FlrA and FleQ, respectively. These proteins are homologs of the aeromonad FlrA and LafK proteins that control the expression of the polar and lateral flagella gene expression cascades, respectively (Wilhelms et al., 2011, 2013). PilZ the c-di-GMP binding protein encoded within the lateral flagella gene cluster is postulated to modulate lateral flagella rotation through interaction with the rotor/stator complex analogous to YcgR of *E. coli*. The genes encoding the extra set of stator proteins (*$pomA_2B_2$*) are transcribed independently.

FSC
www.fsc.org
MIX
Papier | Fördert gute Waldnutzung
FSC® C083411

Kolibri 360 GmbH
Ferdinand-Jühlke-Straße 7
99095 Erfurt, Deutschland
produktsicherheit@kolibri360.de